Advances in Fuzzy Control

Studies in Fuzziness and Soft Computing

Editor-in-chief
Prof. Janusz Kacprzyk
Systems Research Institute
Polish Academy of Sciences
u. Newelska 6
01-447 Warsaw, Poland
E-mail: kacprzyk@ibspan.waw.pl

Vol. 3. A. Geyer-Schulz
Fuzzy Rule-Based Expert Systems and Genetic Machine Learning, 2nd ed. 1996
ISBN 3-7908-0964-0

Vol. 4. T. Onisawa and J. Kacprzyk (Eds.)
Reliability and Safety Analyses under Fuzziness, 1995
ISBN 3-7908-0837-7

Vol. 5. P. Bosc and J. Kacprzyk (Eds.)
Fuzziness in Database Management Systems, 1995
ISBN 3-7908-0858-X

Vol. 6. E.S. Lee and Q. Zhu
Fuzzy and Evidence Reasoning, 1995
ISBN 3-7908-0880-6

Vol. 7. B.A. Juliano and W. Bandler
Tracing Chains-of-Thought, 1996
ISBN 3-7908-0922-5

Vol. 8. F. Herrera and J.L. Verdegay (Eds.)
Genetic Algorithms and Soft Computing, 1996
ISBN 3-7908-0956-X

Vol. 9. M. Sato et al.
Fuzzy Clustering Models and Applications, 1997
ISBN 3-7908-1026-6

Vol. 10. L.C. Jain (Ed.)
Soft Computing Techniques in Knowledge-based Intelligent Engineering Systems, 1997
ISBN 3-7908-1035-5

Vol. 11. W. Mielczarski (Ed.)
Fuzzy Logic Techniques in Power Systems, 1998
ISBN 3-7908-1044-4

Vol. 12. B. Bouchon-Meunier (Ed.)
Aggregation and Fusion of Imperfect Information, 1998
ISBN 3-7908-1048-7

Vol. 13. E. Orłowska (Ed.)
Incomplete Information: Rough Set Analysis, 1998
ISBN 3-7908-1049-5

Vol. 14. E. Hisdal
Logical Structures for Representation of Knowledge and Uncertainty, 1998
ISBN 3-7908-1056-8

Vol. 15. G.J. Klir and M.J. Wierman
Uncertainty-Based Information, 1998
ISBN 3-7908-1073-8

Dimiter Driankov · Rainer Palm (Eds.)

Advances in Fuzzy Control

With 146 Figures
and 16 Tables

Springer-Verlag Berlin Heidelberg GmbH

Assoc. Prof. Dimiter Driankov
University of Linköping
Department of Computer Science
S-58183 Linköping, Sweden

Dr.-Ing. Rainer Palm
Siemens AG
Corporate Technology
Information and Communications
Dept. ZT IK4
D-81730 Munich, Germany

Library of Congress Cataloging-in-Publication Data
Die Deutsche Bibliothek – CIP-Einheitsaufnahme
Advances in fuzzy control; with 16 tables / Dimiter Driankov; Rainer Palm. – Heidelberg; New York: Physica-Verl., 1998
(Studies in fuzziness and soft computing; Vol. 16)

ISBN 978-3-662-11053-9 ISBN 978-3-7908-1886-4 (eBook)
DOI 10.1007/978-3-7908-1886-4

Originally published by Physica-Verlag Heidelberg in 1998.

Softcover reprint of the hardcover 1st edition 1998

Hardcover Design: Erich Kirchner, Heidelberg

SPIN 10663509 88/2202-5 4 3 2 1 0 – Printed on acid-free paper

Preface

During the past few years two principally different approaches to the design of fuzzy controllers have emerged: heuristics-based design and model-based design.

The main motivation for the heuristics-based design is given by the fact that many industrial processes are still controlled in one of the following two ways:

- The process is controlled manually by an experienced operator.
- The process is controlled by an automatic control system which needs manual, on-line "trimming" of its parameters by an experienced operator.

In both cases it is enough to translate in terms of a set of fuzzy if-then rules the operator's manual control algorithm or manual on-line trimming strategy in order to obtain an equally good, or even better, wholly automatic fuzzy control system. This implies that the design of a fuzzy controller can only be done *after* a manual control algorithm or trimming strategy exists.

It is admitted in the literature on fuzzy control that the heuristics-based approach to the design of fuzzy controllers is very difficult to apply to multiple-input/multiple-output control problems which represent the largest part of challenging industrial process control applications. Furthermore, the heuristics-based design lacks systematic and formally verifiable tuning techniques. Also, studies of the stability, performance, and robustness of a closed loop system incorporating a heuristics-based fuzzy controller can only be done via extensive simulations. Last but not least, there is a lack of systematic and easily verifiable knowledge acquisition techniques via which the heuristic knowledge constituting the basis of a manual control algorithm or a manual on-line trimming strategy can be extracted.

These difficulties of the heuristics-based approach explain the recent surge of interest in the model-based design of fuzzy controllers. Model-based fuzzy control uses a given conventional or a fuzzy open loop model of the plant under control in order to derive the set of fuzzy if-then rules constituting the corresponding fuzzy controller. Furthermore, of central interest are the consequent stability, performance, and robustness analysis of the resulting closed loop system involving a conventional model and a fuzzy controller, or a fuzzy model and a fuzzy controller. The major objective of the model-based fuzzy control is to use the full available range of existing linear and nonlinear design and analysis methods for the design of such fuzzy controllers which have better stability, performance, and robustness properties than the corresponding non-fuzzy controllers designed by the use of these same techniques.

This interest in the model-based design of fuzzy controllers is paralleled by a similar surge of interest in fuzzy identification.. That is, the derivation of *black box* fuzzy models of the plant under control, in terms of the identification of a set of fuzzy if-then rules (encoding structural knowledge) together with their corresponding parameters, by the use of clustering techniques, neural networks, genetic algorithms, or a mixture of these techniques.

We believe that research in the field of model-based design of fuzzy controllers has reached a sufficient level of maturity and should be suitably reported to a much larger group of practitioners and researchers. Out of this belief the present edited volume was born. The collection of papers represented in the volume does not cover all methods for model-based design of fuzzy controllers that can be found in the literature. Instead, only methods that constitute the major trends in this area, that have shown their relevance in solving practical problems, whose implementation can be easily automated, and last but not least, that can be especially interesting from the point of view of conventional modern control theory have been selected. Furthermore, the presentation of the methods presented in this volume is not attempted as a comprehensive discussion of all theoretical and implementation issues characteristic for a particular method. Instead, only important topics in connection with the principal control issues (e.g., stability, model inversion, robustness, learning, identification) and control paradigms (e.g., adaptive, predictive, and gain scheduled control) underlying a particular method are discussed.

It is our hope that by collecting such contributions in a single volume we will not only present the major themes in the design of model-based fuzzy controllers, but also encourage their use in practice and possibly call attention to still open research problems.

Munich, Germany, 1997

Dimiter Driankov
Rainer Palm

Table of Contents

Adaptive Fuzzy Control

Predictive Fuzzy Control

Gain Scheduled Fuzzy Control

Introduction

This edited volume assembles a collection of recent works in the field of model-based fuzzy control. Our goal is twofold:

1. To expose the field of model-based fuzzy control to conventional control theorists as a complement to the existing approaches to control of nonlinear systems and provide practicing control engineers with some algorithmic and practical aspects of the fuzzy controllers and their possible use in a variety of applications.

2. To emphasize the need for a more systematic and coherent theory of model-based fuzzy control by bringing together methods based on different control paradigms, but based on the same types of fuzzy models.

In what follows we will describe the general control engineering context in which model-based fuzzy control has a useful role to play.

What makes fuzzy controllers different ?

There are three main aspects of fuzzy controllers (FCs) that cannot be found in their non-fuzzy counterparts designed via conventional linear/nonlinear control methods:

1. The use of if-then rules

2. The universal approximation property

3. The ability to cope with setwise inputs.

The first aspect concerns the representation of expert's knowledge about the behavior of the plant under control and heuristic's based, manual control algorithms. The manual control algorithm, expressing available direct control or a supervisory control strategy is formulated in terms of fuzzy (if-then) rules. In the same way, the plant's behavior can also be expressed by a set of fuzzy rules. Then the major problem is to identify these fuzzy rules and their parameters so that the operator's control or supervisory actions and the plant's response are sufficiently well described and thus, certain control objectives and/or performance criteria can be achieved.

The identification of this type of fuzzy rules can be done in two ways:

1. Knowledge aquisition via the use of interviewing techniques. This type of identification has been applied successfully to the control of mainly single-input/single-output (SISO) plants, but is difficult to apply and verify for multiple-inputs/multiple-outputs (MIMO) control problems.

2. Black box type of identification via the use of clustering, neural nets, and genetic algorithms based techniques. The identified plant model can also accomodate additional if-then rules which do not stem from the input-output data, but are additional expert's knowledge about the plant.

The second aspect from above, the universal approximation property, means that a fuzzy system with product-based inference, centroid defuzzification, and Gaussian membership functions can approximate any real continuous function on a compact set to arbitrary accuracy. In the context of control, the approximation has to be done with a finite set of fuzzy rules employing triangular or trapezoidal membership functions and concerns a known/unknown transfer function within a finite state space. In this case certain approximation errors must be accepted and accounted for in the analysis and design stages. The approximation property is due to the overlap of the membership functions from the if-parts of the set of fuzzy rules. Because of this overlap, every single rule is influenced by its neighboring rules. The result is that every point in state space is approximated by a subset of fuzzy rules.

The third aspect, coping with setwise inputs, is relevant for control tasks where the controller inputs take fuzzy values instead of crisp (point wise) ones. In contrast to classical controllers, FCs can also deal with fuzzy values and even the mixture of crisp and fuzzy values becomes possible. Fuzzy values are qualitative "numbers" being obtained from different sources. One particular source is a qualitative statement of a human operator while controlling a plant like *Temperature is high.* Another source may originate from a sensor which provides information about the *intensity* of a physical signal with in a certain interval. Here, the *intensity or distribution* of the signal with respect to this interval is expressed by a membership function.

It has to be mentioned here that the third aspect is the one that is least explored in the design of FCs while the first two aspects consitute the core of any analysis and design method.

Model-based fuzzy control

During the past few years two principally different approaches to the design of fuzzy controllers have emerged: heuristics-based design and model-based design.

The main motivation for the heuristics-based design is given by the fact that many industrial processes are still controlled in one of the following two ways:

- The process is controlled manually by an experienced operator.
- The process is controlled by an automatic control system which needs manual, on-line "trimming" of its parameters by an experienced operator.

In both cases it is enough to translate in terms of a set of fuzzy rules the operator's manual control algorithm or manual on-line trimming strategy in order to obtain an equally good, or even better, wholly automatic fuzzy control system. This implies that the design of a fuzzy controller can only be done *after* a manual control algorithm or trimming strategy exists. It is admitted in the literature on fuzzy control that the heuristics-based approach to the design of fuzzy controllers is very difficult to apply to multiple-input/multiple-output control problems which represent the largest part of challenging industrial process control applications. Furthermore, the heuristics-based design lacks systematic and formally verifiable tuning techniques. Also, studies of the stability, performance, and robustness of a closed loop system incorporating a heuristics-based fuzzy controller can only be done via extensive simulations. Last but not least, there is a lack of systematic and easily verifiable knowledge acquisition techniques via which the heuristic knowledge constituting the basis of a manual control algorithm or a manual on-line trimming strategy can be extracted.

These difficulties of the heuristics-based approach explain the recent surge of interest in the model-based design of fuzzy controllers. Model-based fuzzy control uses a given conventional or a fuzzy open loop model of the plant under control in order to derive the set of fuzzy rules constituting the corresponding fuzzy controller. Furthermore, of central interest are the consequent stability, performance, and robustness analysis of the resulting closed loop system involving a conventional model and a fuzzy controller, or a fuzzy model and a fuzzy controller.

The articles in this edited volume present a collection of model-based techniques for the analysis and design of FCs that utilize the above mentioned aspects of fuzzy controllers in the context of principal control issues (e.g., stability, model inversion, robustness, learning, identification) and control paradigms (e.g., adaptive, predictive, and gain scheduled control)

Fuzzy models

The design objective in fuzzy control can be stated as follows:

> Given a model (heuristic or analytical) of the physical system to be controlled and specifications for its desired behavior, design a feedback control law in the form of a set of fuzzy rules such that the closed loop system exhibits the desired behavior.

The feedback control law is normally intended to solve the following two control problems: *nonlinear regulation (stabilization)* and *nonlinear tracking.* The specifications of desired behavior normally include stability, accuracy and response speed, and robustness, i.e., the accepted measures of performance for nonlinear control systems.

To achieve this design goal the articles collected in this volume use models of the process under control and/or controller that are of the following main types:

- *Takagi-Sugeno (TS) Fuzzy Models.*
- *Relational Fuzzy Models.*
- *Mamdani Fuzzy Models.*
- *Differential Equations Nonlinear Models.*
- *Modified TS Fuzzy Models approximating a given Differential Equations Nonlinear Model.*

Takagi-Sugeno fuzzy models of the process and/or controller are used in the contributions by Cao, Rees, and Feng (pp. 33–66); Marin (pp. 67–

102); Fischer and Isermann (pp. 103–128); Babuška, Sousa, and Verbruggen (pp. 129–154); Spooner and Passino (pp. 155–188); Škrjanc, Kavšek-Biasizzo, and Matko (pp. 337–356).

Relational fuzzy models of the process and/or controller are used in the contributions by Bourke and Fisher (pp. 283–316); de Oliveira, and Lemos (pp. 317–336).

Mamdani Fuzzy Models of the process and/or controller are used in the contributions by Jagannathan (pp. 225–262); Layne and Passino (pp. 263–282); Aracil, Gordillo, and Àlamo (pp. 11–32); Berstecher, Palm, and Unbehauen (pp. 189–224).

Modified TS Fuzzy Models approximating a given Differential Equations Nonlinear Model of the process are used in the contributions by Johansen, Hunt, and Gawthrop (pp. 357–376); Palm and Driankov (pp. 377–421).

Fuzzy controllers

Model-based fuzzy control has two major advantages over heuristics-based fuzzy control:

1. The study of stability and robustness is done in a strict, formal manner by using to the largest extent possible the available methods from linear/nonlinear control theory.

2. The explicit use of a model (conventional or fuzzy) allows for the design of fuzzy controllers utilizing a varity of control paradigms such as inverse-model based, adaptive, predictive, and gain scheduled control.

The articles collected in this edited volume highlight both of the above issues.

Stability analysis of fuzzy models

In the article by Aracil, Gordillo, and Àlamo (pp. 11–32) a nonlinear system consisting of a linear plant and a fuzzy controller of the Mamdani-type are considered. Only the two-dimensional case is studied, where at the most, point attractors and limit cycles can occur. For this case, the full range of state portrait morphologies is displayed. This gives a global picture of the behavior modes the system can show. Some emphasis is put on the case where the system is locally stable around the operating point, but is not globally stable, so large enough perturbations can lead it out of control. The analysis is implemented with the help of classical system analysis control techniques such as frequency domain graphical methods. In particular, these techniques can be used to predict the existence of undesirable limit cycles. The theoretical results reported are illustrated for four different plants exhibiting different state portrait morphologies and controlled by a Mamdani-type of a fuzzy controller.

In the article by Cao, Rees, and Feng (pp. 33–66) new quadratic stability results for plants represented by Takagi-Sugeno fuzzy models are presented. A constructive algorithm is developed to obtain the stabilizing feedback control law for such plants. The main contribution of this article is the so-called "equivalent principle": the design of a fuzzy control system, based on the Takagi-Sugeno fuzzy model of the plant, is equivalent to the design of a set of linear time-invariant "extreme" systems. Thus, the full power of linear control system theory can be utilized for the design of a fuzzy controller. Since certain boundary conditions are the key to the global stability of the fuzzy control system they are studied in detail in this contribution. The theoretical results reported are illustrated by an example concerning the stabilization of an inverted pendulum on a cart.

In the article by Marin (pp. 67–102) a new type of a fuzzy model is considered which bridges the gap between Mamdani and Takagi-Sugeno fuzzy models. This "hybrid" fuzzy model is obtained by the use of input-dependent fuzzy sets in the consequent part (then-part) of the fuzzy if-then rules. Using a quadratic parametrization of fuzzy sets, an exact analytical characterization of the model is derived. Furthermore, basic concepts of Lyapunov stability theory are used to define fuzzy stability criteria. The latter removes drastic distinction between stability and unstability. Then, based on the analytical characterization of a given hybrid fuzzy model, the author provides a sufficient condition, via the use of matrix inequalities, to check the fuzzy stability criteria. The theoretical results reported are illustrated by a study of the fuzzy equilibrium and the fuzzy attraction set of an electric circuit with a nonlinear resistence.

Inversion of fuzzy models

In the article by Fischer and Isermann (pp. 103–128) a robust fuzzy controller based on the inverse of a Takagi-Sugeno fuzzy model is introduced. The identification of both forward and inverse Takagi-Sugeno fuzzy models is reviewed. Several methods for the special task of inverse model estimation are presented. Besides the analytical inversion, training from measurement data is also considered. A control scheme, based on the idea of disturbance observation is introduced and thoroughly analyzed. The design procedure is described step by step: a Takagi-Sugeno fuzzy model is identified, the inverse fuzzy model is estimated and interpreted, and the guidelines for controller tuning obtained from the theoretical results are validated. Finally, the controller is applied to the control of the speed of a nonlinear cooling blast with variable dynamics.

In the article by Babuška, Sousa, and Verbruggen (pp. 129–154) an inverse fuzzy model is used whithin an internal model control scheme. Two inversion techniques are described for Takagi-Sugeno and Mamdani singleton fuzzy models, respectively. Both methods guarantee an exact inversion. The inverted fuzzy model is incorporated in the nonlinear internal model control scheme. To effectively deal with constraints, a predictive control strategy is applied. The optimal control sequence is sought by means of a branch-and-bound method. Chattering due to the discrete nature of the control input is avoided by combining the predictive control strategy with inverse-model control. An application of this technique to an air-conditioning system is described. The controller based on the fuzzy model is compared with standard linear predictive control and real-time control results are given.

Adaptive fuzzy control

In the article by Spooner and Passino (pp. 155–188) a stable direct adaptive controller is presented which uses Takagi-Sugeno fuzzy models, Mamdani fuzzy models, or neural networks to provide asymptotic tracking of a reference signal for a class of continuous-time nonlinear plants with poorly understood dynamics. This scheme allows for the inclusion prior knowledge about how to control the plant in terms of exact mathematical equations or linguistics. It is proved that with or without such knowledge the adaptive scheme can "learn" how to control the plant, provide for bounded internal signals, and achieve asymptotically stable tracking of a reference input. The performance of the adaptive scheme is demonstrated through the longitudinal control of an automobile within an automated lane.

In the article by Berstecher, Palm, and Unbehauen (pp. 189–224) the nonlinear transfer characteristic of a fuzzy sliding-mode controller is adapted to ensure the desired closed-loop behaviour of the plant in a changing environment. The nonlinear plant model is given in terms of a conventional, analytical model while the control element is a fuzzy sliding mode controller, i.e., of a Mamdani-type. The linguistically defined adaptation scheme is also given in terms of a Mamdani-type of a fuzzy controller. The detailed formal analysis of the adaptation scheme reveals that it is itself of the fuzzy sliding-mode type. The stability of the closed loop system, including the adaptation loop, and the convergence of the adaptation parameters are studied in detail by using Lyapunov stability. The theoretical results reported are illustrated by means of the control a two-link robot arm.

In the article by Jagannathan (pp. 225–262) the tracking control of a class of feedback linearizable nonlinear systems using a discrete-time Mamdani fuzzy controller is studied. A repeatable design algorithm and a stability proof for an adaptive fuzzy controller is presented. It uses basis functions defined via the Mamdani fuzzy model, unlike most standard adaptive control techniques that generate basis vectors by computing a regression matrix. With mild assumptions on the class of discrete-time nonlinear systems and using the adaptive fuzzy controller, the uniform ultimate boundedness of the closed-loop system is shown. The persistency of excitation condition is not required, certainty equivalence is not used, and a regression matrix is not computed. The result is a model-free universal fuzzy controller that works for any system in the given class of feedback linearizable nonlinear systems. Simulation results from the control of a planar two-link robot arm illustrate the theoretical results reported.

In the article by Layne and Passino (pp. 263–282) a "learning controller" is developed by synthesizing several basic ideas from fuzzy set and control theory, self-organizing control, and conventional adaptive control. A learning mechanism is utilized which observes the plant outputs and adjusts the membership functions of the fuzzy rules in a Mamdani-type of a fuzzy controller so that the overall system behaves like a "reference model". The effectiveness of this "fuzzy model reference learning controller" is illustrated by showing that it can achieve high performance learning control for a nonlinear time-varying rocket velocity control problem and a multi-input/multi-output two degree-of-freedom robot manipulator. Moreover, the authors summarize the wide variety of systems that this type of a learning fuzzy controller has been successfully applied to.

Predictive fuzzy control

In the article by Bourke and Fisher (pp. 283–316) the design of a fuzzy predictive controller that has a structure similar to conventional model-based predictive controllers is presented. The process, prediction, and controller models are given in terms of Fuzzy Relational Models. The latter type of fuzzy models are superior to rule based fuzzy models in that they are developed more quickly and easily and permit identification directly from input/output data. Another advantage is that predictive control, either one-step or multi-step ahead, can be implemented in a manner similar to classical predictive controllers. This article clearly demonstrates the parallels between conventional predictive control and fuzzy predictive control. The results reported are illustrated by the application of the proposed predictive fuzzy controller to the control of a simulated nonlinear, over-damped process with large process gain variations.

In the article by de Oliveira and Lemos (pp. 317–336) a simplified Fuzzy Relational Model of the plant for the purpose of adaptive predictive control is introduced. To address the issues of fast convergence rates for the identifier and the avoidance of drastic forgetting, a re-parametrization of the Fuzzy relational Model of the plant is proposed. It is shown that such re-parametrized fuzzy models are computationally efficient and have the property of being universal approximators with clear semantics. This reparametrization allows one to use the Recursive Least Squares algorithm for speeding up the identification of the model using on-line closed-loop observations. Experimental results obtained from the control of an electric furnace and illustrating the application of proposed adaptive predictive fuzzy controller framework are presented.

In the article by Škrjanc, Kavšek-Biasizzo, and Matko (pp. 337–356) a predictive model-reference fuzzy controller using a dynamic matrix is presented. The controller is based on the on-line computation of the dynamic matrix. This offers advantages in cases when the plant dynamics depends strongly on the operating point.The dynamic matrix is computed on the basis of a Takagi-Sugeno fuzzy model and its inverse whenever the operating point is changed. The design algorithm is presented in detail and simulated on a strongly nonlinear liquid level process. Also an implementation on real industrial-scale temperature plant is reported.

Gain scheduled fuzzy control

In the article by Johansen, Hunt, and Gawthrop (pp. 357–376) the design of a Takagi-Sugeno fuzzy gain schedulers with emphasis on transient performance is presented. The work is motivated by the fact that in conventional gain scheduling and in Takagi-Sugeno fuzzy models, based on a set of local linear models, linearization is performed at equilibrium points. This may result in poor transient performance of the control system during trajectory tracking and when the setpoint changes. The authors analyze the case when linearization and local control design are performed in transient operating regimes. The improvements in performance with such off-equilibrium linearization and control design are illustrated with detailed simulation-based examples.

In the article by Palm and Driankov (pp. 377–421) a modified version of the original Takagi-Sugeno fuzzy model is derived via Lyapunov linearization of a given, differential equations based, open loop nonlinear model. The linearization is performed at a finite number of appropriately selected set points in the fuzzy state space thus obtaining a set of local open loop linear models. The latter constitute the basis for an open loop linear Takagi-Sugeno fuzzy model that approximates the original open loop nonlinear model. Furthermore, the authors consider the control problem of stabilizing the so obtained open loop fuzzy model around an arbitrary set point. In this context, it is shown how to design a modified Takagi-Sugeno fuzzy controller which guarantees the local stability of the original nonlinear model. The stability margins in the case when the different local linear models are subject to unknown disturbances are determined. Finally, the use of the modified Takagi-Sugeno fuzzy controller for the purpose of gain scheduling is shown. The theoretical results reported are illustrated on the simulated control of a two-link robot arm.

Global Stability Analysis of Second-Order Fuzzy Systems

J. Aracil, F. Gordillo, and T. Álamo
Escuela Superior de Ingenieros
Universidad de Sevilla
Avenida Reina Mercedes s/n, 41012 Sevilla, Spain

1 Introduction

The design of fuzzy controllers (FCs) is an open problem in which the use of heuristics is one of the main design methodologies. In this way, available design methods do not give fully satisfactory results because they lack an analytical background which would allow a thorough analysis. Only analytical study will assure a good performance in all situations (or at least, in a large class of them). To introduce such analytical results, a concrete class of fuzzy control systems will be studied in this article. This class of systems has been chosen such that is at the same time simple in structure and general in the richness of the results obtained. This class is the one shown in Fig. 1 where there is a linear plant, represented by its transfer function $G(s)$, and a FC with the associated nonlinearity $u = \Phi(e, \dot{e}) = \Phi(-y, -\dot{y})$. The main objective of the analysis developed is to study an elementary system with a well known mathematical structure and of which all the behaviours it can exhibit can be fully analyzed. To that end a linear model for the plant has been adopted. All the nonlinearities in the system stand in the fuzzy controller. As will be seen, the qualitative analysis of this system allows to display a full variety of behaviours. For the sake of simplicity, in this article it will be assumed that (1) $G(s)$ is second-order; (2) $\Phi(-y, -\dot{y}) = -\Phi(y, \dot{y})$; and (3) out of the region of normal operation, the controller saturates. This unavoidable saturation will have important consequences in the following. Assumptions 2 and 3 are quite normal. However, the analysis developed and the kind of results reached can be extended to more general systems.

A system with a FC is a nonlinear dynamical system even if the plant is linear. Nonlinear systems display two main differences with regard to the

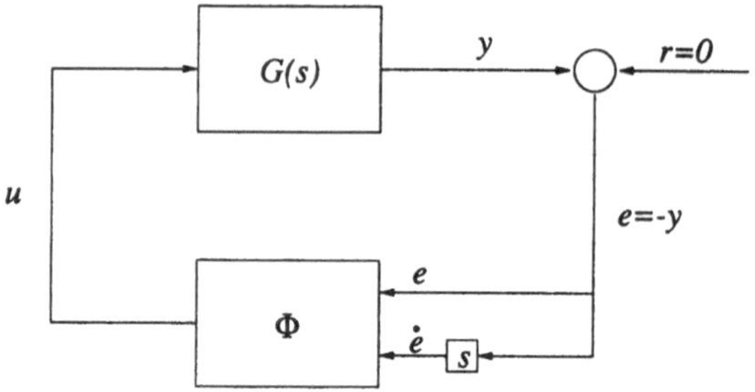

Figure 1: Structure of the system with a FC.

linear ones: (1) they can show many steady regimes, and not only the point attractor associated with the operating point; and (2) they can exhibit long-term behaviors that are more complex than point attractors, such as limit cycles and chaotic attractors. The search for these more complex behaviors is the goal of the qualitative analysis of nonlinear systems [10], [11], [14], [19], [21]. The application of these methods for analyzing nonlinear control systems has gained some spread [3], [20].

Since a system with a FC is a nonlinear system, it has to be checked if it shows any of these behavior characteristics. However, the first step is to analyze the stability of the operating point. There are several methods available for that [7], [22], [23]. Nevertheless, their results do not assure global stability. Once the stability of the operating point has been guaranteed, it should be analyzed if it has other long-term behaviors. Notice that the local stability of the operating point does not guarantee global stability. For example, the existence of other unstable equilibria will imply that even if some trajectories in the state space tend to the stable equilibrium (that is, these trajectories are in the attraction basin of that point), other trajectories will diverge to the other attractors or to infinity. Another example is the existence of an unstable limit cycle surrounding a stable equilibrium and limiting its attraction basin. In both cases, if the system undergoes a perturbation that leads it out of its basin, it will become unstable even if the operating point remains stable. That is why this kind of situation has to be treated with care. This kind of problem is found when an unstable plant is stabilized at the operating point by means of an appropriate FC, as it will be seen later. In this article, the existence of equilibrium points is analyzed with the tools introduced in [4], while the existence of limit cycles is studied with the well-known first harmonic balance analysis. The describing function method has been applied to fuzzy control systems in [1], [2] and [13], but with a different approach.

The approach proposed in this article was inspired by the bifurcation analysis of a second-order linear system with saturation in the feedback loop. This analysis has been reported in other papers [5], [15], [16], [20], and leads to the bifurcation diagram shown in Fig. 2. The diagram presents the four different state portraits the system has in the parameter space (a_1, a_2). Parameters a_1 and a_2 are the parameters in the open loop characteristic polynomial $s^2 + a_1 s + a_2$. Region A of Fig. 2 is defined by $a_1 > 0$ and $a_2 > 0$, so the system is open loop stable. In that region, the closed loop system has a single equilibrium point. Moreover, the system is globally stable. But in the other three regions, where the open loop system is no longer stable (at least one of the parameters (a_1, a_2) is negative), the closed loop system has other equilibria so that, even when the operating point is stable, the system loses its global stability. In region B, there is an unstable limit cycle that gives rise to a bounded attraction basin. In region C, there are two saddles so that some trajectories go to infinity and the system is not globally stable. In region D, both unstable limit cycles and saddles coexist.

It should be mentioned that the boundaries between these regions are given by bifurcations. When a_2 changes from positive to negative, taking the value $a_2 = 0$, a pitchfork bifurcation at infinity P_∞ is produced. In the same way, a Hopf bifurcation at infinity B_∞ occurs at the boundary of regions A and B. A more complex phenomenon, a double saddle connection (DSC) bifurcation, takes place at the boundary of regions C and D. These bifurcations are described in detail in [5], [15], [16], [20].

As it will be shown later, these same qualitative portraits are displayed by the type of fuzzy control systems analyzed in the present article. That is mostly due to the fact that these qualitative behaviors are related to the saturation characteristic present both in the system analyzed in [5], [20] and in the FC studied here. The main contribution of this article is to establish this fact. In Sect. 4, it will be shown how the pitchfork bifurcation at infinity P_∞ and the Hopf bifurcation at infinity B_∞ occur in the FC analyzed in this article.

The article is organized as follows. In Sect. 2, the study of equilibrium points is presented. In Sect. 3, the existence of limit cycles is analyzed. Then in Sect. 4, the richness of the morphology of state portraits of a second-order fuzzy control system is displayed by means of concrete examples. This last section describes how the bifurcations that give rise to the boundaries between the different regions are produced. In Sect. 5 we present some conclusions.

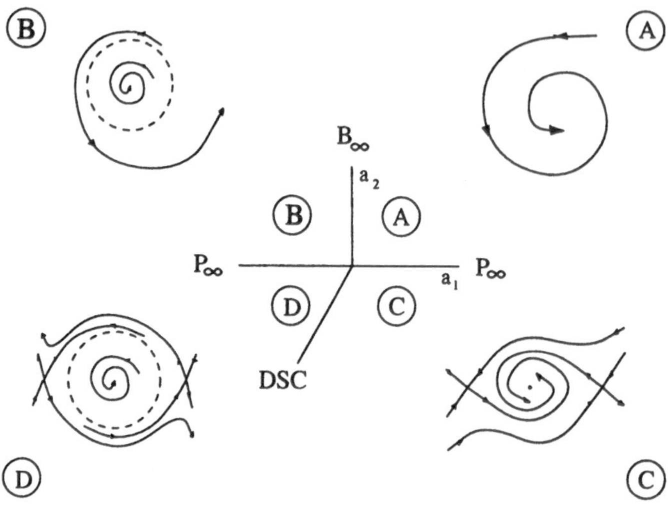

Figure 2: Bifurcation diagram of a second-order feedback system with a saturation.

2 Analysis of equilibrium points

In this article, a control system such as the one shown in Fig. 1 is assumed, where there is a linear plant, given by

$$\begin{aligned} \dot{x} &= Ax + Bu, \\ y &= Cx, \end{aligned}$$

with $x \in \mathbb{R}^2$. The following transfer function is associated with this system

$$G(s) = C(sI - A)^{-1}B = \frac{b_1 s + b_2}{s^2 + a_1 s + a_2}. \tag{1}$$

The FC controller is given by the nonlinearity

$$u = \Phi(e, \dot{e}) = \Phi(-y, -\dot{y}),$$

and it is further assumed that Φ is an odd function such that $\Phi(-y, -\dot{y}) = -\Phi(y, \dot{y})$.

The equilibrium points of the system are given by the condition $\dot{x} = 0$, which leads to $Ax = -Bu = B\Phi(y, \dot{y})$. As to the equilibrium $\dot{y} = 0$, we have

$$Ax = B\Phi(y, 0). \tag{2}$$

Consider the case where (A, B, C) is given in the canonical control form:

$$A = \begin{pmatrix} 0 & 1 \\ -a_2 & -a_1 \end{pmatrix}, \quad B = \begin{pmatrix} 0 \\ 1 \end{pmatrix}, \quad C = \begin{pmatrix} b_2 & b_1 \end{pmatrix}.$$

Then,

$$Ax = \begin{pmatrix} x_2 \\ -a_2x_1 - a_1x_2 \end{pmatrix}, \qquad B\Phi(y,0) = \begin{pmatrix} 0 \\ \Phi(y,0) \end{pmatrix}.$$

The equilibrium equations (2) lead to

$$x_2 = 0, \tag{3}$$
$$-a_2x_1 - a_1x_2 = \Phi(y,0). \tag{4}$$

Equation (3) leads to the conclusion that the equilibrium points are located in the subspace defined by $x_2 = 0$. Furthermore, (4) gives $-a_2x_1 = \Phi(y,0)$, but since $y = b_2x_1$ for $x_2 = 0$, it is concluded that

$$-a_2y = b_2\Phi(y,0). \tag{5}$$

This last equation can be solved graphically as shown in Fig. 3. From

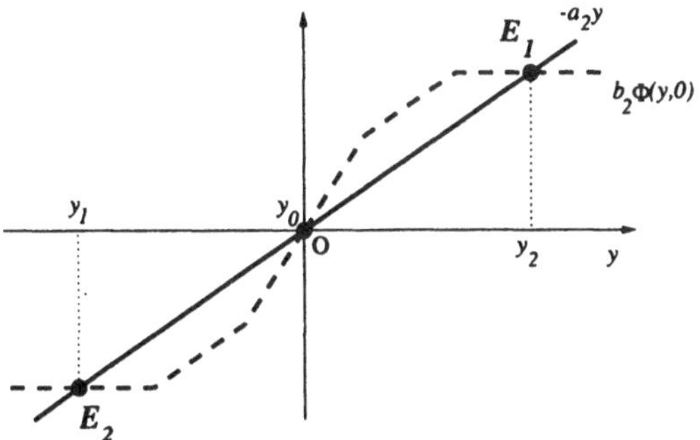

Figure 3: Graphical resolution of the equilibrium equation.

this figure, it is clear that the system has a single equilibrium for values of parameters a_2 and b_2 such that the straight line $-a_2y$ cuts the curve $b_2\Phi(y,0)$ only at the origin. The system displays three equilibria in the case illustrated in Fig. 3, where there are two equilibria E_1 and E_2 in addition to the origin. As will be shown later, if the origin is a stable equilibrium, then E_1 and E_2 are saddles. This means that in such a case the state portrait has the shape associated with regions C or D of Fig. 2.

It should be noted that if, as usual, $b_2 > 0$ and $\Phi(y,0)$ has the shape shown in Fig. 3, then the condition for having multiple equilibria is $a_2 < 0$. This means that the plant is unstable. Notice that the case when a_2 is positive and large enough to avoid the intersection of $-a_2y$ with $b_2\Phi$, is not of interest because, in this case, the origin is unstable.

Moreover, it should be determined how the existence of saddle points is related to the saturation characteristic presented by the FC. The main reason for determining the saddle points is that their existence breaks the global stability of the closed loop system, giving rise to a bounded attraction basin with the global behavior problems mentioned in Sect. 1.

2.1 Local stability of the operating point

Besides the search for other equilibrium points, it is important to check if the desired operating point is stable. The system under study is

$$\begin{aligned} \dot{x} &= Ax - B\Phi(y, \dot{y}), \\ y &= Cx. \end{aligned} \tag{6}$$

The linearization in the neighborhood of the origin of the fuzzy controller under study turns out to be

$$\Phi(y, \dot{y}) = \Phi_y y + \Phi_{\dot{y}} \dot{y} + O(||(y, \dot{y})||^2),$$

with

$$\Phi_y = \left.\frac{\partial \Phi}{\partial y}\right|_{(y,\dot{y})=(0,0)}, \qquad \Phi_{\dot{y}} = \left.\frac{\partial \Phi}{\partial \dot{y}}\right|_{(y,\dot{y})=(0,0)}.$$

Thus sufficiently close to the origin $x = 0$,

$$\dot{x} = Ax - B(\Phi_y y + \Phi_{\dot{y}} \dot{y}),$$

$$[I + \Phi_{\dot{y}} BC]\dot{x} = [A - \Phi_y BC]x,$$

$$\dot{x} = [I + \Phi_{\dot{y}} BC]^{-1}[A - \Phi_y BC]x.$$

Therefore, the stability of the origin is determined by the eigenvalues of

$$[I + \Phi_{\dot{y}} BC]^{-1}[A - \Phi_y BC]. \tag{7}$$

A similar analysis can be developed for the other equilibria E_1 and E_2.

3 Analysis of limit cycles

Once the equilibrium points have been analyzed, the next step is to look for limit cycles. For this purpose, the harmonic balance will be used.

3.1 The describing function method

The existence of limit cycles (periodic orbits) in systems with nonlinearities has traditionally been studied with the well-known describing function method [6], [12], [17], [18]. In order to apply this technique, the linear and nonlinear parts of the system must be arranged as indicated in Fig. 4. Furthermore, the following conditions are normally assumed:

1. There is only one nonlinearity.
2. The nonlinearity is time invariant.
3. The nonlinearity has one input and one output.
4. The linear element behaves as a low-pass filter. This requirement is necessary because the method approximates the response of the nonlinear element by its first harmonic, and higher order harmonics are neglected. One of the causes of the success of the describing function method in control theory is that almost every real plant to be controlled fulfills this requirement.

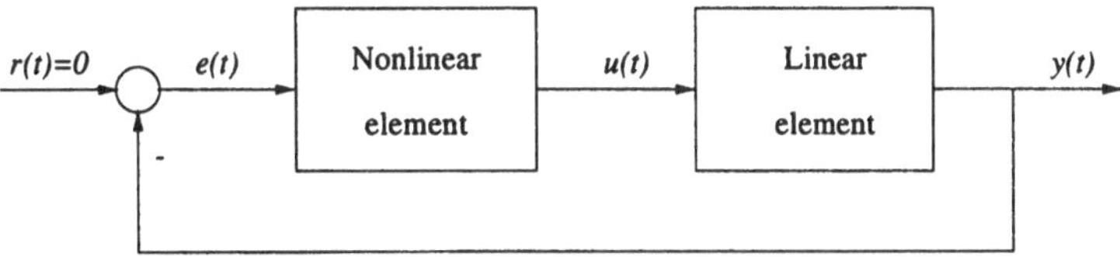

Figure 4: Nonlinear system to which the describing function method is applicable.

In the original method, there was an additional condition by which the nonlinearity had to be odd, that is, $\varphi(x) = -\varphi(-x)$, so that the zero order harmonic of its response equals zero when the input is sinusoidal. This restriction can be relaxed, and a more elaborated theory of harmonic balance is available [6], [8]. Nevertheless, in the sequel and for the sake of simplicity, only the odd case will be considered, but the results are easily generalized.

The describing function of a nonlinear element is obtained calculating the amplitude b and the phase shift ϕ of the first order harmonic $b\sin(\omega t + \phi)$ of the response of the system to a sinusoidal input $a\sin\omega t$. The describing function $N(a,\omega)$ is defined as

$$N(a,\omega) = \frac{b}{a}e^{j\phi},$$

with $j = \sqrt{-1}$. The method provides that, if $G(s)$ is the transfer function of the linear part of Fig. 4, the solutions (a,ω) of

$$1 + N(a,\omega)G(j\omega) = 0, \tag{8}$$

are possible limit cycles of the system. That is, if $a = \bar{a}$ and $\omega = \bar{\omega}$ satisfy (8), a limit cycle will exist with amplitude close to $\bar{a}$ and period close to $2\pi/\bar{\omega}$. The solutions of this equation can be found sketching the curves $G(j\omega)$ for $0 < \omega < \infty$ and $-1/N(a, \omega_i)$ for $0 < a < \infty$ and different values of ω_i. In this way, a family of curves is obtained with a as the varying parameter and ω fixed for each curve. If the curve of $G(j\omega)$ intersects for $\omega = \omega_k$ the curve corresponding to $-1/N(a, \omega_k)$, this intersection is the sign of the existence of a limit cycle. The stability of this limit cycle is studied with the help of the Loeb criterion: the limit cycle will be stable when, for an observer positioned on the $G(j\omega)$ curve and facing in the direction of increasing ω, the $-1/N(a, \omega_k)$ curve crosses from right to left when increasing a. In Fig. 5 the case of a stable limit cycle is presented for $\omega = \omega_4$. The value of the amplitude a will be the one corresponding to the intersection point on the curve $-1/N(a, \omega_4)$.

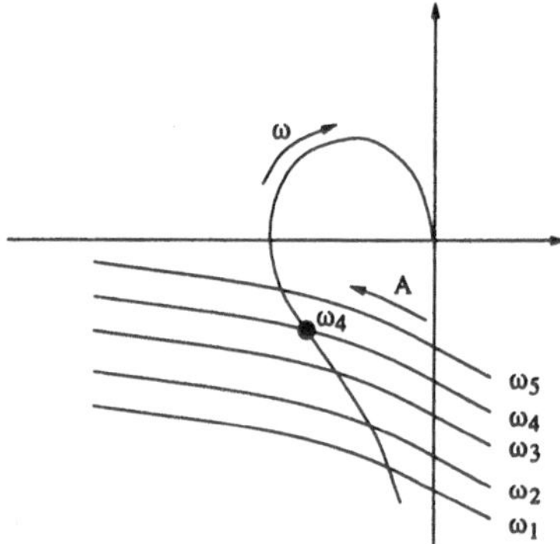

Figure 5: Stable limit cycle for $\omega = \omega_4$.

It must be pointed out that the describing function method is approximate. The amplitude and the frequency of real limit cycles can be slightly different from the predicted values. In some extreme cases, even the existence of limit cycles could be questioned.

3.2 Application to systems with a FC

In order to study the existence of limit cycles in systems that include an FC, the describing function method has been applied. The conditions which must be fulfilled when applying the method recalled in the previous section impose the following restrictions:

- Condition 1 implies that the plant must be reasonably well approximated by a linear model.

- Condition 2 implies that the FC will be time invariant, which is usual if no adaptive policy is adopted.
- Condition 3 means that the FC will have only one input and one output. This happens in the case considered here since the derivative of the error $\dot{e}(t)$ can be obtained from the error $e(t)$. So the only input to the controller is $e(t)$.
- Condition 4 is not a hard restriction since most plants are low-pass filters.

The class of control systems considered in this article can be represented as shown in Fig. 6 and fulfill these restrictions.

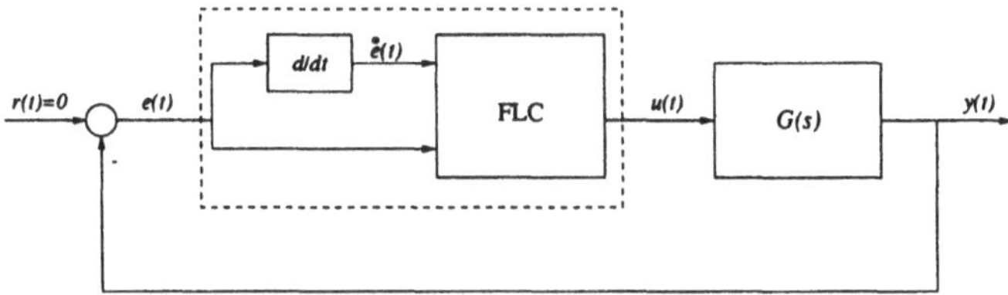

Figure 6: Class of FCs considered.

Another problem is how to evaluate the describing function of the FC. Analytical development of it could be a difficult task. In this article the describing function is obtained by simulation.

4 A family of examples

As mentioned in Sect. 1, this work is inspired by the bifurcation analysis of a second-order linear system with saturation in the feedback path (Fig. 2). The objective of the family of examples of this section is to show that the same kind of qualitative behavior can appear when using a FC to regulate a second-order linear plant. In order to do so, four plants have been chosen, each of them exhibiting one of the different qualitative behaviors displayed in Fig 2.

The four plants will present the same structure (see (1)) with the values for the parameters shown in Table 1. Only plant A is stable, and all of them are minimum phase (nevertheless, the analysis can be carried out with non-minimum phase systems, although, in this case, the stability analysis is more involved). Each plant corresponds to one of the regions of Fig. 2.

Plant	a_1	a_2	b_1	b_2	
A	1	1	1	1	Stable
B	-1	1	1	1	Unstable
C	1	-1	1	1	Unstable
D	-1	-0.3	1	1	Unstable

Table 1: Parameters of the models of the four plants.

For simplicity, the same controller has been applied to the four plants. In a real case, each plant would need a specific design so each one would be controlled by a different FC. Nevertheless, in this example, the behavior of the four plants with the FC designed is satisfactory –at least in the neighborhood of the origin– as will be shown later. The inputs of the FC are the error $e(t) = -y(t)$ and the derivative of this error $\dot{e}(t) = -\dot{y}(t)$. Seven linguistic terms are used for $e(t)$ and three for $\dot{e}(t)$. The membership functions associated with each input variable are shown in Fig. 7. The inference is performed using the

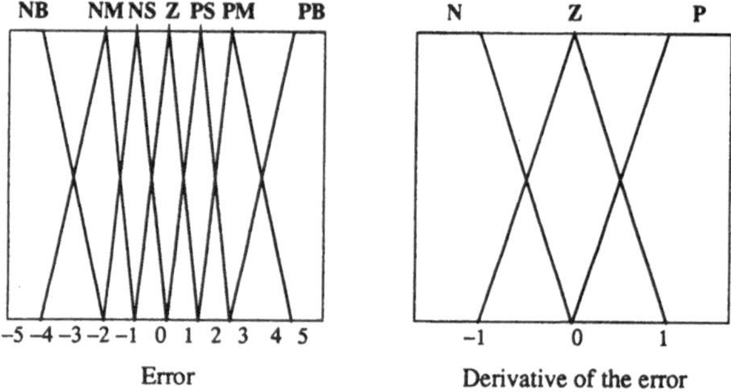

Figure 7: Membership functions of the input variables.

algebraic product. The output of the FC is obtained using the center average defuzzifier. In this case, only the center of the membership functions of the output variable must be defined. Therefore, the rules can be expressed as a numerical array as the one of the present example, represented in Table 2. The resultant control surface is shown in Fig. 8. It is useful to observe the saturation of Φ and that $\Phi(y, \dot{y}) = 0$ is a straight line. This line separates the state space plane into positive and negative regions. It is also called the switching line because when the state trajectory goes across this line, the control switches from positive to negative and vice versa.

This FC seems to be suitable for the four plants considered, as it can be concluded from experiments like the ones illustrated in Fig. 9. In this figure, each plant is simulated with the FC starting with initial conditions close to

		e						
		NB	NM	NS	Z	PS	PM	PB
	N	-5.5	-5.5	-40	-1.0	2.0	2.5	2.5
$\dot{e}$	Z	-4.0	-4.0	-3.0	0.0	3.0	4.0	4.0
	P	-2.5	-2.5	-2.0	1.0	4.0	5.5	5.5

Table 2: Rule base.

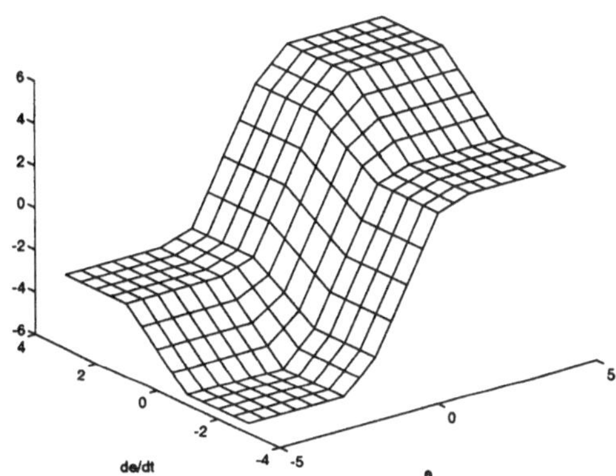

Figure 8: Control surface.

the desired equilibrium. As can be seen, the FC is able to control each plant to zero.

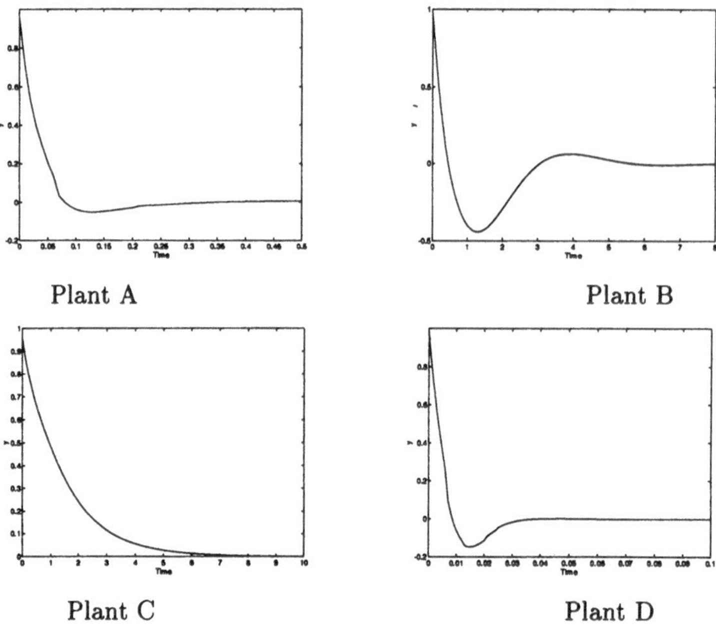

Figure 9: "Local" behavior.

A more rigorous way to test the stability of the origin is using (7). A simple calculation shows that for the system (6) we have

$$[I + \Phi_{\dot{y}}BC]^{-1}[A - \Phi_y BC] = [I + BC]^{-1}[A - 3BC] =$$

$$\frac{1}{2}\begin{bmatrix} 0 & 2 \\ -a_2 - 3 & -a_1 - 4 \end{bmatrix}.$$

The characteristic polynomial of this matrix is $\lambda^2 + (a_1 + 4)\lambda + 2(a_2 + 3) = 0$, so the sufficient conditions for the local stability of system (6) are $a_1 > -4$ and $a_2 > -3$. The four plants of the family fulfill this condition.

In order to study the existence of other equilibrium points, (5) must be solved for each system. Figure 10 shows the graphical solution of this equation. For plants A and B in which $a_2 > 0$ the origin is the only equilibrium point. Plants C and D, with $a_2 < 0$, present two other equilibria. These two equilibria turn out to be saddle points.

In order to predict the existence of limit cycles, the describing function of the FC has been evaluated experimentally. The input $e(t)$ was set equal

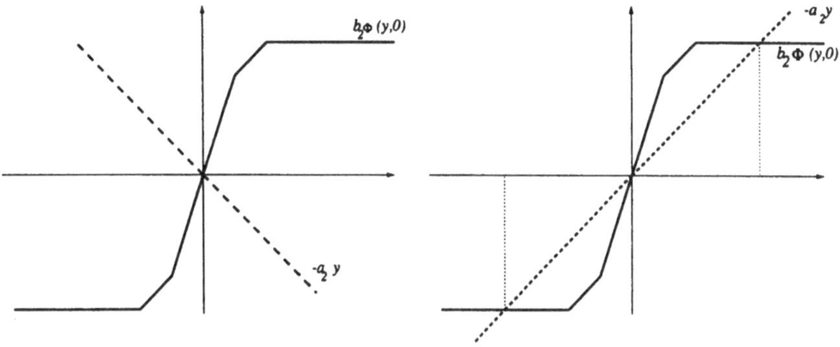

Figure 10: Equilibrium points.

to $a \sin \omega t$ with $0.1 \leq a \leq 100$ and $0.01 \leq \omega \leq 10$. For each value of a and ω, the first order harmonic of the output of the FC was evaluated. The ratio between the amplitude of this harmonic and a is the module of the value of the describing function corresponding to a and ω, while its phase is the phase shift between the harmonic and $a \sin \omega t$. The resultant describing function is shown in Fig. 11 where each curve stands for a value of ω, and is parameterized by a.

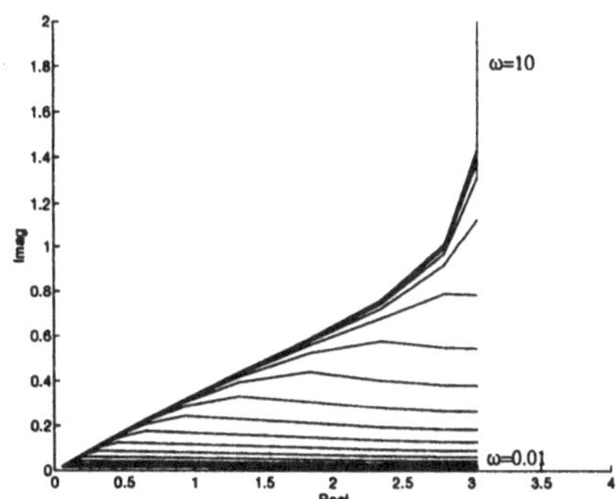

Figure 11: Family of curves representing the describing function $N(a, \omega)$.

In order to solve the harmonic balance (8) the graphs of $G(j\omega)$ and $-1/N(a, \omega)$ are displayed in Fig. 12 for plants A and B, and in Fig. 13 for plants C and D. As can be seen, $G(j\omega)$ only crosses the curves of $-1/N(a, \omega)$ with plants B and D. This fact means that, according to the describing function method, the system corresponding to plants A and C will not present limit cycles, while cases B and D need further study to check if there is any intersection corresponding to the same value of ω. Figure 14 displays the

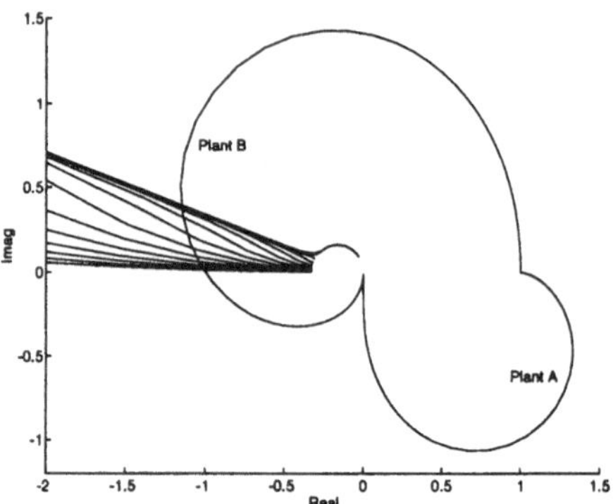

Figure 12: Graphs of $-1/N(a,\omega)$ and $G(j\omega)$ for plants A and B.

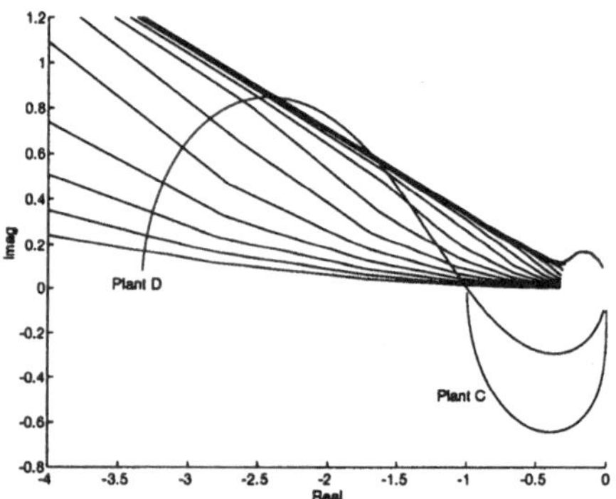

Figure 13: Graphs of $-1/N(a,\omega)$ and $G(j\omega)$ for plants C and D.

sketch of $N(a,\omega)G(j\omega)$ for plants B and D. Solutions of (8), that is, predicted limit cycles, will be reflected in this figure by points where one of the curves of the family $N(a,\omega)G(j\omega)$ crosses the point $(-1,0)$. In the present cases, only a limit cycle is predicted for plant B and and another one for plant D.

Using the Loeb criterion in Figs. 12 and 13, the conclusion is that both limit cycles are unstable.

With this analysis, the expected behavior of the four systems corresponds to the four state portraits presented in Fig. 2. The origin is locally stable

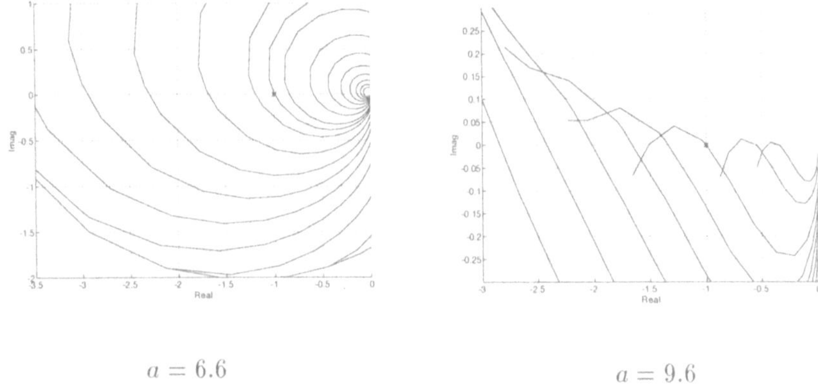

Figure 14: Sketch of $N(a,\omega)G(j\omega)$.

in the four cases. The FC with plant A will not present other equilibria nor limit cycles, this implying that system A is globally stable; system B will show an unstable limit cycle; plant C will present two unstable equilibrium points, and plant D will display both kinds of behavior: equilibrium points and limit cycle.

In order to check the validity of the former study, each system has been simulated for a complete set of initial conditions. Figure 15 shows the results which agree with the analysis. All the trajectories followed by plant A with the FC tend to the origin in spite of using initial conditions quite far from the origin. For plant B, the behavior is qualitatively different depending on whether the initial conditions are inside or outside the unstable limit cycle: in the former case, trajectories tend to infinity while in the second, they tend to the origin as it is shown in Fig. 15. This figure also shows the existence of two saddle points in plant C. They also limit the attraction basin of the origin. Finally, the qualitative behavior of plant D is a combination on the behaviors of plants B and C.

The changes between the state portraits shown in the left column of Fig. 15 are due to the occurrence of the bifurcations mentioned in the introduction. First consider the pitchfork bifurcation at infinity P_∞. This bifurcation happens at the transition between plants A and C, or between plants B and D. In both cases, what happens is that parameter a_2 takes value zero at the transition. That is, parameter a_2 changes from being slightly positive to slightly negative (or vice versa). Keeping in mind Fig. 10, equilibrium points that give rise to the saddles are at the points where the straight line $-a_2 y$ cuts the the curve $b_2\Phi(y,0)$. The slope of the straight line is just $-a_2$, so when $a_2 = 0$, this line is horizontal. A slight change in a_2 around this

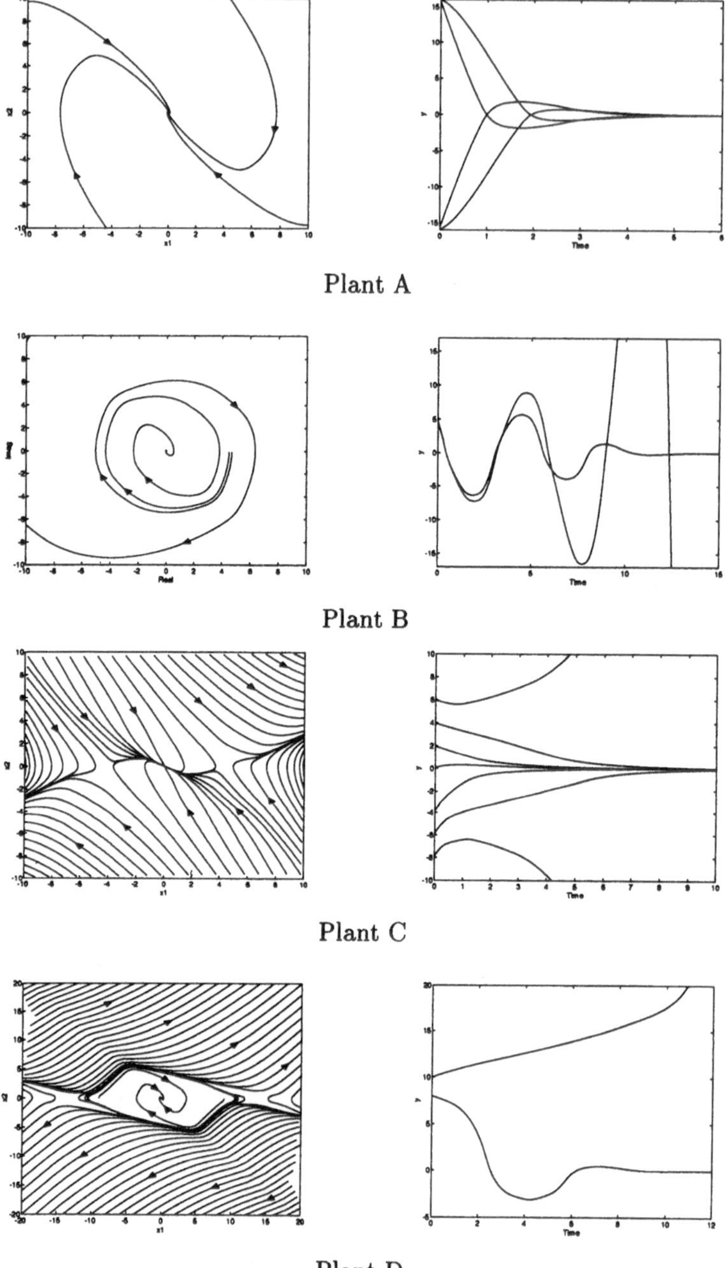

Figure 15: Simulated behavior.

value will mean that the line changes from not cutting the curve $b_2\Phi(y,0)$ to cutting it at infinity. This is the pitchfork bifurcation at infinity P_∞.

The transition between plants A and B is related to a Hopf bifurcation at infinity B_∞. That happens for $a_1 = 0$, being $a_2 > 0$. The intuitive explanation of this bifurcation is not as easy as for the case of the pitchfork bifurcation. However, some insight can be gained. To that end, it should be noted that the unstable limit cycle is produced for the values of a and ω that satisfy (8). For small values of $a_1 > 0$, it is easy to see that $G(j\omega)$ has the shape shown in Fig. 16, where b_1/a_1 is the point where $G(j\omega)$ cuts the horizontal axis for some $\omega > 0$. As a_1 tends to zero, the value of this

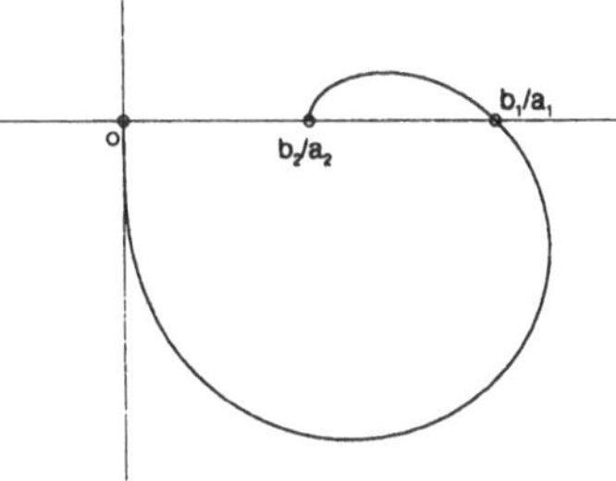

Figure 16: Graph of $G(j\omega)$ for a_1 positive small.

intersection point grows tending to infinity. But as a_1 reaches the value zero and becomes negative, then the shape of $G(j\omega)$ of Fig. 16 changes to the one of Fig. 17, being b_1/a_1 now negative and "coming" from infinity as the absolute value of a_1 grows from $a_1 = 0$. With the situation depicted in Fig.

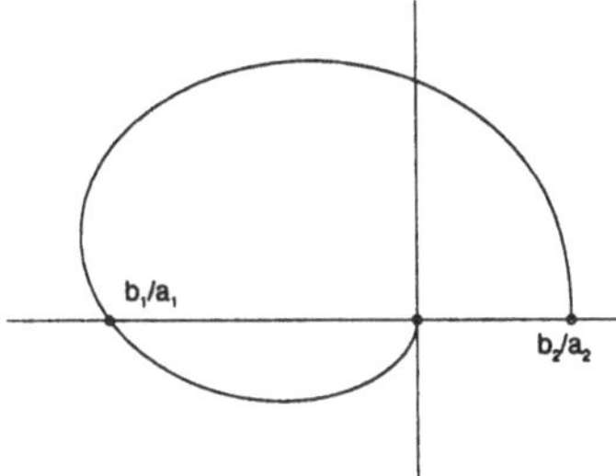

Figure 17: Graph of $G(j\omega)$ for a_1 negative small.

17, (8) has a solution that shows that an unstable limit cycle appears starting with amplitude infinity when a_1 is negative close to zero. That is the Hopf bifurcation at infinity B_∞. For a complete description of this phenomenon, see [5], [9], [15], [16].

Regarding the boundary between plants C and D, the complex phenomenon known as a double saddle connection bifurcation is produced. The reader is referred to the above references for the study of that bifurcation.

In any case, it must be observed that the bifurcations associated with the boundaries are due to the saturation characteristics of the FC. In this way, the global phenomena displayed by a fuzzy control system are inherently related to the saturation every FC includes, and cannot be avoided if the saturation persists. However, it is evident that equilibria other than the origin only appear if the second-order plant is unstable. So only in this last case the system will not be globally stable.

A final experiment has been performed, in order to illustrate the problem which exists beyond the limited local stability. A disturbance is introduced in the simulation of the FC controlling plant C, as it is shown in Fig. 18.

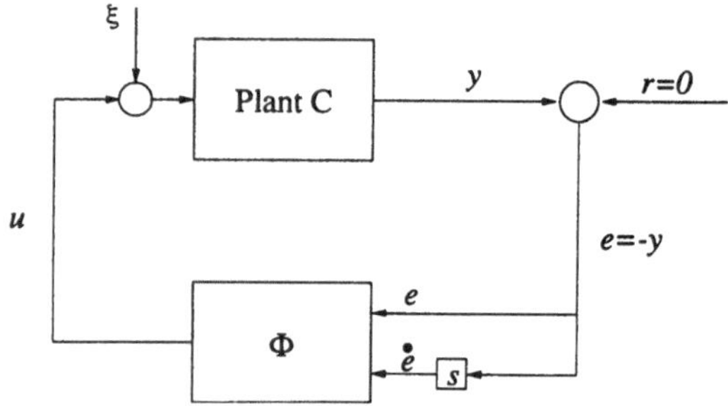

Figure 18: Fuzzy control system with disturbances.

The experiment consists in applying a disturbance $\xi(t) = \xi_0$ from $t = 2$ to $t = 5$ when the system starts from the desired operating point. Fig. 19 shows the simulations corresponding to $\xi_0 = 3$ (curve a) and $\xi_0 = 5$ (curve b). It can be seen that the system is able to recover from the first disturbance. The reason is that when the disturbance finishes, the system has not been led out of the attraction basin of the desired equilibrium. The opposite occurs with the $\xi_0 = 5$. Fig. 20 shows both evolutions in the phase portrait.

5 Conclusions

In this article, qualitative methods of nonlinear, dynamical systems have been applied to a fuzzy control system. With these methods, the global behavior of the system has been analyzed. It has been shown that some of the classical control techniques used to analyze nonlinear systems in the frequency domain may be useful for predicting the global behavior of control systems that in-

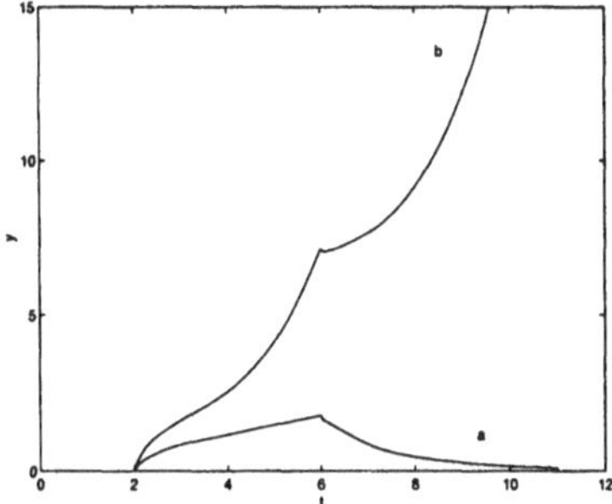

Figure 19: Simulation of the system with disturbances $\xi_0 = 3$ (curve a) and $\xi_0 = 5$ (curve b).

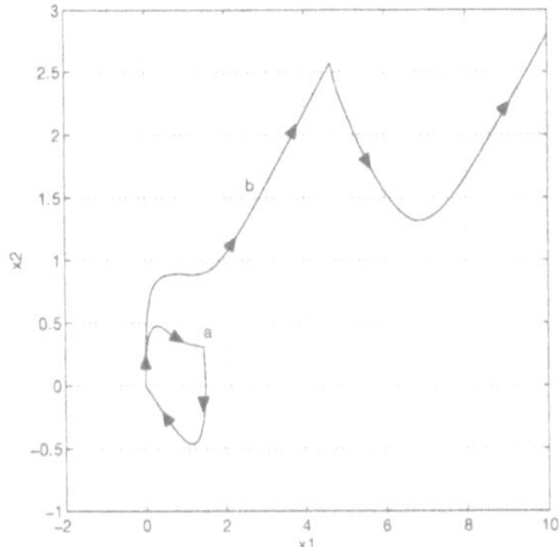

Figure 20: Trajectories of Fig. 19 in the phase portrait.

clude an FC. Moreover, the ways global instability arises have been clarified. In the same way, the analysis is not only based on simulations and the global stability of the system can be studied in a consistent way. The study has been focused on the conditions that must be fulfilled to predict the appearance of multiple equilibria or the existence of limit cycles. Nevertheless, this study is only an introduction of the application of qualitative analysis to FC systems. The order of the system has been restricted to second-order. Work in progress is trying to extend the method to higher order and MIMO systems. Also, the connection between these studies and robustness issues is to be developed.

Acknowledgments The authors would like to thank CICYT from the Spanish Ministerio de Educación for supporting this work under grant TAP–94–0491.

References

[1] Abdelnour, G., Cheung, J.Y., Chang, C.-H. and Tinetti, G. (1993), "Steady-state analysis of a three-term fuzzy controller", *IEEE Transactions on Systems, Man, and Cybernetics*, Vol. 23, No 2, 607–610.

[2] Abdelnour, G., Cheung, J.Y., Chang, C.-H. and Tinetti, G. (1993), "Application of describing funtions in the transient response analysis of a three-term fuzzy controller", *IEEE Transactions on Systems, Man, and Cybernetics*, Vol. 23, No 2, 603–606.

[3] Abed, E.H., Wang, H.O. and Tesi, A. (1996), "Control of bifurcations and chaos", in W.S. Levine (Ed.) *The Control Handbook*, IEEE Press.

[4] Aracil, J., Ollero A. and García-Cerezo, A. (1989), "Stability indices for the global analysis of expert control systems", *IEEE Transactions on Systems, Man, and Cybernetics*, Vol. 19, 988–1007.

[5] Aracil, J., Ponce, E. and Álamo, T. *A Frequency Domain Approach to Bifurcations in Control Systems with Saturation.* Internal Report GAR 1997–01.

[6] Cook, P.A. (1994), *Nonlinear Dynamical Systems*, 2nd. Ed., Prentice Hall.

[7] Driankov, D., Hellendoorm, H. and Reinfrank, M. (1993), *An Introduction to Fuzzy Control*, Springer-Verlag.

[8] Genesio, R. and Tesi, A. (1992), "Harmonic Balance Methods for the Analysis of Chaotic Dynamics in Nonlinear Systems", *Automatica*, Vol. 28, No. 3, 531–548.

[9] Glover, J.N. (1989), "Hopf bifurcations at infinity", *Nonlinear Analysis*, Vol. 13, 1393–1398.

[10] Guckenheimer, J. and Holmes, P. (1983). *Nonlinear Oscillations, Dynamical Systems and Bifurcations of Vector Fields*. Springer Verlag, New York.

[11] Hale, J.K. and Koçak, H. (1991), *Dynamics and Bifurcations*, Springer-Verlag.

[12] Khalil, H.K. (1996), *Nonlinear Systems*, Macmillan.

[13] Kickert, W.J.M. and Mandani, E.H. (1978), "Analysis of a fuzzy logic controller", *Fuzzy Sets and Systems*, 29–44

[14] Kuznetsov, Y.A. (1995), *Elements of Applied Bifurcation Theory*, Springer-Verlag.

[15] Llibre, J. and Ponce, E. (1996), "Global first harmonic bifurcation diagram for odd piecewise linear control systems", *Dynamics and Stability of Systems*, Vol. 11, No. 1, 49–88.

[16] Llibre, J. and Sotomayor, J. (1996), "Phase portraits of planar control systems", *Nonlinear Analysis, Theory, Methods and Applications.*, Vol. 27, No. 10, 1177–1197.

[17] MacFarlane, A.G.J. (Ed.) (1979), *Frequency-Response Methods in Control Systems*, IEEE Press.

[18] Mees, A.I. (1981), *Dynamics of Feedback Systems*, Wiley.

[19] Nayfeh, A.H. and Balachandran, B. (1995), *Applied Nonlinear Dynamics*, Weley.

[20] Ponce, E., Álamo, T. and Aracil, J. (1996), "Robustness and bifurcations for a class of piecewise linear control systems", *1996 IFAC World Congress*, paper 2b–02 5, Vol. E, 61–66.

[21] Strogatz, S.H. (1995), *Nonlinear Dynamics and Chaos*, Addison-Wesley.

[22] Yager, R.R., and Filev, D.P. (1994), *Essentials of Fuzzy Modeling and Control*, Wiley.

[23] Wang, L-X (1997), *A Course in Fuzzy Systems and Control*, Prentice-Hall.

Quadratic Stability of Continuous-Time Fuzzy Control Systems

S. G. Cao , N. W. Rees and G. Feng
Department of Systems and Control
School of Electrical Engineering
University of New South Wales
Sydney 2052, N.S.W., Australia

1 Introduction

Takagi-Sugeno (TS) fuzzy models [4]-[7] are widely employed in fuzzy model-based control. The basic idea is the use of TS fuzzy rules to construct local linear state space models from which local linear controllers can be derived. The stability of the overall system is then determined by Lyapunov stability analysis. It is required that for all local linear models, a common positive definite matrix P is found which satisfies a system of linear inequalities. The method provides for conservative designs and in many cases even if a common P cannot be found, the overall system may still be stabilized by using other design methods. In fact, in [10], a much stronger result is obtained: *if a fuzzy system is quadratically stabilizable via any nonlinear state feedback control law, then, it is also quadratically stabilizable via a linear state feedback control law.*

In [13]-[15] a new design method for fuzzy control systems is developed. The main idea is that instead of using a common P, a set of $(P_1, P_2, \ldots, P_m)$ is used to construct a piecewise differentiable quadratic (PDQ) Lyapunov function for the purpose of stability analysis. Each P_i only corresponds to one subsystem, thus, it is easy to find such a P_i by simply solving a Riccati equation. It is shown that *a fuzzy control system is globally quadratically stable if a suitable set of Riccati equations has positive-definite symmetric solutions.* Once these solutions are obtained, the stabilizing feedback control law can be constructed from them. This result gives a connection between the stability analysis of fuzzy control systems and robust control theory [8]-[10].

In this article we further develop the analysis and design method proposed in [13]-[15] for a class of fuzzy systems which can be described by the dynamical fuzzy model from [12]. It is shown that *the design of a fuzzy control system is equivalent to the design of a set of linear time-invariant "extreme" systems.* Thus, any design method in linear control system theory can be used to design the fuzzy control system. Therefore, if a complex system, or a nonlinear system can be represented by the above mentioned dynamical fuzzy model, we can use the powerful analysis techniques from linear control theory to obtain a global nonlinear control law for the nonlinear system. The most important difference between quadratic stability of a linear system [8]-[10] and the quadratic stability of a fuzzy system [13]-[15] is to be found in the type of boundary conditions required. Since these boundary conditions are the key to the global stability of the fuzzy control system we will discuss them in detail in this article.

The article is organized as follows. Section 2 discusses the fuzzy dynamic model and introduces some definitions. Section 3 presents the quadratic stability results. Section 4 contains some new stability results. Section 5 shows a numerical example. Finally, the article concludes with some brief remarks in Sect. 6.

Notation: The notation used is fairly standard. We denote by $z^T z$ or $||z||^2$ the square norm of a vector $z \in \Re^k$. Given an $n \times n$ matrix N, $\lambda_{max}(N)$ and $\lambda_{min}(N)$ denote its maximum and minimum eigenvalues, respectively. Let A and B be Hermitian matrices. We write $A \geq B$ if the matrix $A - B$ is positive semidefinite. Similarly, $A > B$ means that $A - B$ is positive definite. $||F||_2$ denotes the spectral norm of the matrix F, i.e., $\max_i \sqrt{|\lambda_i(F^T F)|}$. The identity matrix is denoted by I. For a matrix function $\chi(t), \chi(t) : \Re \to \Re^{n \times n}$, the left limit of $\chi(t)$ at $t \in \Re$ is defined as

$$\chi(t^-) := \lim_{\varepsilon \to 0} \chi(t - \varepsilon),$$

if this limit exists, and the right limit of $\chi(t)$ at $t \in \Re$, if it exists, is defined as

$$\chi(t^+) := \lim_{\varepsilon \to 0} \chi(t + \varepsilon).$$

2 The dynamic continuous-time fuzzy model

In order to extend crisp dynamic systems to their fuzzy counterparts we first need to represent a fuzzy set in the form given by the definition below.

Definition 2.1 [12]: A fuzzy set $F(z)$ is a binary set, $F(z) = (\mu(z), \bar{z})$, where $\bar{z}$ is the point at which $\mu(\bar{z}) = 1$ and $\bar{z}$ is called the crisp point of $F(z)$; $\mu(z)$ is the membership function of $F(z)$.

Furthermore, we use the following fuzzy model to represent a multi-input continuous-time system

$$\begin{array}{lll} R^l: & \text{IF} & x_1 \text{ is } F_1^l \; AND \; \ldots x_n \text{ is} F_n^l \\ & \text{THEN} & \dot{x}(t) = A_l x(t) + B_l u(t), \end{array} \tag{1}$$

where $l = 1, 2, \ldots, m$, R^l denotes the *l-th* fuzzy rule, (A_l, B_l) represents the *l-th* local model of the fuzzy system (1), m is the number of fuzzy rules, $x(t) \in \Re^n$ are the state variables, $u(t) \in \Re^p$ are the input variables of the system. The state vector is given as

$$x(t) = (x_1(t), x_2(t), \cdots, x_n(t))^T. \tag{2}$$

The state of the overall system is computed by taking the weighted average of the states for all local models.

Definition 2.2: The *l-th* fuzzy dynamic local model is a binary set,

$$FDLM_l = (\mu_l(t), (A_l, B_l)),$$

where $\mu_l(t) = \mu_l(x(t))$ is the normalized membership function of the fuzzy set F^l, where $F^l = \prod_{i=1}^{n} F_i^l$ and

$$\sum_{l=1}^{m} \mu_l(t) = 1. \tag{3}$$

The pair (A_l, B_l) represents the crisp input-output relationship for the region F^l. (A_l, B_l) is also called the *l-th* subsystem of system (1). Let M be the set of membership functions satisfying (3). It should be noted here that the membership functions of the fuzzy system (1) do not depend on $u(t)$.

Remark 2.1: The *l-th* fuzzy dynamic local model corresponds to the following fuzzy set in the state space: $F_l(x) = (\mu_l(x(t)), \bar{x}_l(t))$. Thus, the crisp state $\bar{x}_l(t)$ and an input u can normally be represented by the crisp system

$$\dot{\bar{x}}_l(t) = A_l \bar{x}_l(t) + B_l u(t).$$

Since at the crisp point $\bar{x}_l(t)$ we have $\mu_l(\bar{x}_l(t)) = 1$ and $\mu_k(\bar{x}_l(t)) = 0, k \neq l$ by (3), we also have $\bar{x}_l(t) = x(t)$. Thus, the meaning of the subsystem in (1) can be explained as the crisp system part of (1) in the region F^l.

Now we can use the fuzzy computational machinery described in [12], that is, center-average defuzzification, product inference and singleton fuzzification, to obtain the global fuzzy model for (1). That is,

$$\dot{x}(t) = A(\mu(t)x(t) + B(\mu(t))u(t), \tag{4}$$

$$A(\mu(t)) = \sum_{l=1}^{m} u_l(t)A_l, \quad B(\mu(t)) = \sum_{l=1}^{m} \mu_l(t)B_l,$$

$$\mu(t) = (\mu_1(t), \mu_2(t), \cdots, \mu_n(t)).$$

Using the membership functions, we can divide the state space into m regions

$$S_l = \{x | \mu_l(x(t)) \geq \mu_i(x(t)) \quad i = 1, \ldots, m, \ i \neq l\}, \quad l = 1, \ldots, m. \tag{5}$$

The boundary of S_l is denoted by

$$\partial S_l = \{x | \mu_l(x(t)) = \mu_i(x(t)) \quad i = 1, 2, \ldots, m, \ i \neq l\}. \tag{6}$$

Then, on every subspace, the fuzzy system (4) can be expressed by

$$\begin{aligned} \dot{x}(t) &= \tilde{A}_l(\mu)x(t) + \tilde{B}_l(\mu)u(t) \\ &= (A_l + \Delta A_l(\mu))x(t) + (B_l + \Delta B_l(\mu))u(t), \quad x(t) \in S_l, \end{aligned} \tag{7}$$

$$\Delta A_l(\mu) = \sum_{i=1, i\neq l}^{m_l} \bar{\mu}_i(t)\Delta A_{li}, \quad \Delta B_l(\mu) = \sum_{i=1, i\neq l}^{m_l} \bar{\mu}_i(t)\Delta B_{li}, \tag{8}$$

where $\bar{\mu}_i \in M$, $\bar{\mu}_i \neq 0$, $\Delta A_{li} = A_i - A_l$, $\Delta B_{li} = B_i - B_l$, and $l = 1, 2, \ldots, m$. Here $\bar{\mu}_1, \ldots \bar{\mu}_{m_l}$ are the m_l membership functions that are not equal to zero when the *l-th* subsystem plays a dominant role, that is, $x(t) \in S_l$.

Let $\tau_i, i = 1, 2, \ldots, \bar{T}$ be $\bar{T}$ time instants at which the state of the system (4) meets the boundary of one of S_l. Since the system (3) is decomposed into several subsystems in (7) defined on several different regions, we will use the following piecewise differentiable quadratic (PDQ) Lyapunov function for the purpose of stability analysis,

$$V(t) = V(x(t)) \hat{=} x^T(t)P(t)x(t), \tag{9}$$

$$P(t) = P_l, \quad x(t) \in S_l, \tag{10}$$

where $(P_1, P_2, \ldots, P_m)$ is a set of $n \times n$ fixed positive-definite symmetric matrices, and $P(t)$ is called the symmetric piecewise differentiable matrix function. All discontinuities of $P(t)$ will occur at the time instants $\tau_i, i = 1, 2, \ldots, \bar{T}$. We assume that $x(t)$ does not jump at the time instants $\tau_i, i = 1, 2, \ldots, \bar{T}$, that is, $x(\tau_i) = x(\tau_i^-) = x(\tau_i^+)$.

For the control of the fuzzy system (7), we will consider a nonlinear controller described by

$$u(t) = K(x(t)) = (\sum_{l=1}^{m} \mu_l(t)K_l)x(t), \tag{11}$$

or the switching control law

$$u(t) = K(x(t)) = K_l x(t), \quad x(t) \in S_l. \tag{12}$$

Remark 2.1: The control law (12) has the following explanation. First, $x(t) \in S_l$, $K_l x(t)$ is used until the state $x(t)$ meets the boundary of S_l. Then $K(x(t))$ will be changed to $K_j x(t)$ when $x(t)$ is in the $S_j, j \neq l$. Similarly, the value of (10) is $P(t) = P_l, x(t) \in S_l$ until the state $x(t)$ leaves S_l. Thus, at the boundary of S_j the value of (2) is well defined.

Using the control law (2) the closed-loop system can be described by the equation

$$\dot{x}(t) = A_c(t)x(t), \tag{13}$$

$$A_c(t) = A(\mu(t)) + B(\mu(t))K(\mu(t)).$$

Definition 2.3: The system (4) is said to be quadratically stabilizable if there exists a feedback control law (2), a set of $n \times n$ fixed positive-definite symmetric matrices $(P_1, P_2, \ldots, P_m)$ and a constant α such that the following conditions are satisfied

$$1) \quad \bar{T} < \infty, \tag{14}$$

$$2) \quad \frac{dV(t)}{dt} = L(x,t) \tag{15}$$

where

$$\begin{aligned} L(x,t) &= x^T(A(\mu(t))^T P(t) + P(t)A(\mu(t))x \\ &\quad + 2x^T P(t)B(\mu(t))K(\mu(t))x \\ &\leq -\alpha\|x\|^2, \end{aligned}$$

$$P(t) = P_l, \quad x(t) \in S_l, \quad l = 1, 2, \ldots, m, \tag{16}$$

for all non-zero $x(t) \in \Re^n, \forall \mu \in M$ and $t \in [0, \infty)$.

Remark 2.2: It should be noted that the PDQ Lyapunov function is a discontinuous function. The following result shows that even if $\frac{dV(t)}{dt} = L(x,t) < 0$, that is, condition (15) holds, we cannot guarantee the global stability of the closed-loop system (13). However, if the boundary condition (14) holds together with the condition (15), we can guarantee that the closed-loop system (13) is globally asymptotically stable.

Theorem 2.1: The closed-loop fuzzy system (13) is globally asymptotically stable, if it is quadratically stabilizable, that is $\bar{T} < \infty$, and there exist a set of feedback gains $(K_1, K_2, \ldots, K_m)$ in (2), a set of positive-definite symmetric matrices $(P_1, P_2, \ldots, P_m)$ and a set of positive-definite symmetric matrices $(Q_1, Q_2, \ldots, Q_m)$ such that

$$A_{cl}(\mu)^T P_l + P_l A_{cl}(\mu) + Q_l \leq 0, \tag{17}$$

$$A_{cl}(\mu) = \tilde{A}_l(\mu) + \tilde{B}_l(\mu)K(\mu),$$

$$x(t) \in S_l, \quad l = 1, 2, \ldots, m,$$

for any $\mu \in M$.

Proof: Let the PDQ Lyapunov function be of the form (2). We have that

$$\begin{aligned} \frac{dV(t)}{dt} &= \dot{V}(t) = \lim_{\Delta \to 0} \frac{1}{\Delta}(V(t+\Delta) - V(t)), \\ &= x^T(A_{cl}^T(\mu)P_l + P_l A_{cl}(\mu))x, \quad x \in S_l, \quad x \notin \partial S_l, \quad t \neq \tau_i. \end{aligned}$$

Using (17) we have $\frac{\dot{V}(t)}{V(t)} \leq -\frac{x^T Q_l x}{x^T P_l x} \hat{=} -\beta_l(x)$ where $x \in S_l$, $x \notin \partial S_l$, and $t \neq \tau_i$. Since $P_l > 0$ and $Q_l > 0$, it then follows that

$$-\beta_l(x) \leq -\bar{\alpha}, \quad \bar{\alpha} \hat{=} \min_l \frac{\lambda_{min}(Q_l)}{\lambda_{max(P_l)}}, \quad x \in S_l, \quad x \notin \partial S_l, \quad t \neq \tau_i.$$

Clearly, $\bar{\alpha} > 0$. Thus

$$\dot{V}(t) \leq -\bar{\alpha} V(t), \quad t \neq \tau_i.$$

When $t = \tau_i$ by (16) and *Remark 2.1* we still have

$$\dot{V}(\tau_i) \leq -\bar{\alpha} V(\tau_i), \quad t = \tau_i.$$

Thus, by the Gronwell-Bellman lemma, we have

$$V(x(t)) \leq exp[-\bar{\alpha}(t - \bar{t})]V(x(\bar{t})), \quad \tau_{i-1} \leq \bar{t} \leq t \leq \tau_i, \quad \tau_0 = 0. \tag{18}$$

Suppose that $x_0 \subset S_k$. Then it follows from (18) that

$$V(x(t)) \leq exp[-\bar{\alpha}t]V(x_0), \quad x(t) \in S_k, \quad 0 \leq t \leq \tau_1. \tag{19}$$

Since

$$\lambda_{min}(P_l)||x||^2 \leq V(t) \leq \lambda_{max}(P_l)||x||^2,$$

$$x(t) \in S_l, \quad l = 1, 2, \ldots, m, \tag{20}$$

and using (19) we have

$$||x(t)|| \le C||x_0||exp[-\alpha(t-0)], \quad 0 \le t \le \tau_1, \tag{21}$$

where $C = \max_i (C_l), C_l = \frac{\lambda_{max}(P_l)^{1/2}}{\lambda_{min}(P_l)^{1/2}}, \alpha = \frac{1}{2}\bar{\alpha}$. For the discontinuous point τ_1 suppose that $x(t)$ enters $S_j, j \neq k$. From (18) and (21) we still have

$$||x(t)|| \le C||x(\tau_1)||exp[-\alpha(t-\tau_1)] \le C^2||x_0||exp[-\alpha(t-0)], \quad \tau_1 \le t.$$

Using a similar line of reasoning, we obtain

$$||x(t)|| \le C^i||x_0||exp[-\alpha(t-0)], \quad \tau_i \le t. \tag{22}$$

Since $\bar{T} < \infty$ it follows from (22) that

$$||x(t)|| \le \bar{C}||x_0||exp[-\alpha t], \quad t \ge 0, \tag{23}$$

$$\bar{C} = C^{\bar{T}} < \infty.$$

Thus, the closed-loop fuzzy system (13) is globally asymptotically stable. Q.E.D.

3 Quadratic stability

In this section, we will present quadratic stability results for the fuzzy system (4). From the previous section, it can be seen that the fuzzy system (4) can be decomposed into m subsystems. However, because each subsystem in (7) is a linear time-varying system, it is still very difficult to design a control law. Later on, it will be seen that the design problem for (4) can be divided into m much easier design problems. We will use the "extreme" system idea to transform (7) into a set of extreme subsystems.

In order to achieve this, first we need to define some important bounds for the fuzzy system (4) The upper bounds are defined by

$$[\Delta\bar{A}_l(\mu)]^T [\Delta\bar{A}_l(\mu)] \le E_l^T E_l = [E_{l1} \;\; E_{l2}]^T [E_{l1} \;\; E_{l2}], \tag{24}$$

$$\Delta\bar{A}_l(\mu) = [\Delta A_l(\mu) \quad \Delta B_l(\mu)] .$$

In the following discussion we assume that there exists a set of membership functions denoted by μ^*'s such that the upper bound (24) can be achieved. These upper bounds are called accurate upper bounds. Otherwise, we can choose a conservative upper bounds called, approximate upper bounds. In this case, all the necessary conditions for quadratic stabilization, which will be discussed in the next section, do not hold. Because the upper bounds are defined by $\left[\Delta\bar{A}_l(\mu)\right]^T \left[\Delta\bar{A}_l(\mu)\right] \le E_l^T E_l$ the choice of E_l in (2) is not unique. Thus, the upper bounds in (24) must be chosen as the worst upper bounds in the stability sense, that is, the upper bounds in (24) will reduce the stability margin of the system.

There are several methods to obtain the approximate upper bounds. The simplest one is through the following procedure. First, for an arbitrary, but fixed $x \in \Re^n$, define the function $g(A) = x^T A^T Ax$. Since the function is convex it follows that

$$g(\sum_{l=1}^{p} \mu_l A_l) \le \sum_{l=1}^{p} \mu_l g(A_l).$$

Then, using the above, we obtain

$$\begin{aligned} x^T(\sum_{j=1}^{m_l} \bar{\mu}_j \Delta\bar{A}_{lj})^T(\sum_{j=1}^{m_l} \bar{\mu}_j \Delta\bar{A}_{lj})x &\le x^T(\sum_{j=1}^{m_l} \bar{\mu}_j \Delta\bar{A}_{lj}^T \Delta\bar{A}_{lj})x \\ &\le (\sum_{j=1}^{m_l} \bar{\mu}_j \|\Delta\bar{A}_{lj}\|_2^2)x^T x \le x^T \bar{\lambda}_l I x, \end{aligned}$$

where $\bar{\lambda}_l = \max(\lambda_{max}(\Delta\bar{A}_{lj}^T \Delta\bar{A}_{lj}), j = 1, 2, \ldots, m_l)$.

Thus, the approximate upper bounds are

$$E_l^T E_l = \bar{\lambda}_l I, \quad l = 1, 2, \ldots, m. \tag{25}$$

In most cases, the approximate upper bounds from (25) are too conservative, and we need other approximate upper bounds. Let

$$g_l(\bar{\mu}_1, \cdots, \bar{\mu}_{m_l}) = tr(\sum_{j=1}^{m_l} \bar{\mu}_j \Delta \bar{A}_{lj})^T (\sum_{j=1}^{m_l} \bar{\mu}_j \Delta \bar{A}_{lj}). \tag{26}$$

The best approximate upper bounds can be found by solving the following nonlinear programming problem

$$max(g_l(\bar{\mu}_1, \cdots, \bar{\mu}_{m_l})), \tag{27}$$

$$\bar{\mu}_1, \bar{\mu}_2, \cdots, \bar{\mu}_{m_l} \in M$$

We can also choose the approximate upper bounds by simple intuition. For example, in some cases when $\mu_l = 0.5$, the g_l is largest because the system is on the boundary of several subsystems. Let $\bar{x}$ be the state such that $\mu_l(\bar{x}) = 0.5$. The approximate upper bounds in (24) are then

$$(\sum_{j=1}^{m_l} \bar{\mu}_j(\bar{x}) \Delta \bar{A}_{lj})^T (\sum_{j=1}^{m_l} \bar{\mu}_j(\bar{x}) \Delta \bar{A}_{lj}) = E_l^T E_l. \tag{28}$$

Using the upper bounds in (24) we can define the following m extreme subsystems,

$$\dot{x}(t) = (A_l + E_{l1})x(t) + (B_l + E_{l2})u(t), \quad x(t) \in S_l. \tag{29}$$

The main purpose of this section is to show that if we can find a fuzzy control law to stabilize the m extreme subsystems in (28), we can also stabilize the original fuzzy system (4) by this same control law. It should be noted here that m extreme subsystems are linear time-invariant systems. Thus, any design methods from linear control theory can be used to design the control law. In this context, the main result of this article can be described by the following theorem.

Theorem 3.1: (Equivalent principle) The fuzzy system (4) is quadratically stabilizable, if and only if there exists a fuzzy control law (2) which can quadratically stabilize the m extreme subsystems in (29) with the accurate upper bounds and $\bar{T} < \infty$.

In the proof, the following results are used.

Lemma 3.1[3]: For any constant $\varepsilon > 0$ and any matrices X and Y with appropriate dimensions, we have

$$X^T Y + Y^T X \leq \varepsilon X^T X + \frac{1}{\varepsilon} Y^T Y.$$

Lemma 3.2: Given the matrices $\Delta A(\mu), \Delta B(\mu)$ that satisfy

$$[\Delta A(\mu)]^T [\Delta A(\mu)] \leq E^T E, \quad \forall \mu \in M, \tag{30}$$

$$[\Delta A(\mu^*)]^T [\Delta A(\mu^*)] = E^T E, \quad \mu^* \in M, \tag{31}$$

and A_0, there exists a positive-definite symmetric matrix P such that

$$[A_0 + \Delta A(\mu)]^T P [A_0 + \Delta A(\mu)] < 0, \tag{32}$$

if and only if there exists a scalar $\varepsilon > 0$ such that the following condition holds

$$A_0^T P + P A_0 + \varepsilon P P + \frac{1}{\varepsilon} E^T E < 0. \tag{33}$$

Proof: Suppose that (33) holds. Then using (30) and *Lemma 3.1* we have $[A_0 + \Delta A(\mu)]^T P + P [A_0 + \Delta A_l(\mu)]$ is equal to $A_0^T P + P A_0 + \Delta A(\mu)^T P + P \Delta A(\mu)$. The latter is less or equal $A_0^T P + P A_0 + \varepsilon P P + \frac{1}{\varepsilon} E^T E$ which in turn is less or equal 0.

In order to prove necessity, suppose that (32) holds. Then we have a positive-definite symmetric matrix P such that

$$x^T \left[[A_0 + \Delta A(\mu)]^T P [A_0 + \Delta A(\mu)] \right] x < 0.$$

Let

$$\Pi = A_0^T P + P A_0.$$

We have

$$x^T \Pi x < -x^T [\Delta A(\mu)^T P + P \Delta A(\mu)] x,$$

for any $\mu \in M$ and $x \in \Re^n, x \neq 0$. It follows that given any $x \in \Re^n, x \neq 0$,

$$(x^T \Pi x)^2 > 4(\max_{\mu \in M}(x^T P \Delta A(\mu) x)^2).$$

By (30), it follows that

$$\begin{aligned}(x^T P \Delta A(\mu) x)^2 &= x^T P P x x^T \Delta A(\mu)^T \Delta A(\mu) x \\ &\leq x^T P P x x^T E^T E x.\end{aligned}$$

Because there exists a μ^* such that (31) holds, it follows that

$$(x^T \Pi x)^2 > 4(\max_{\mu \in M}(x^T P \Delta A(\mu) x)^2) = 4x^T P P x x^T E^T E x.$$

Then the arguments used in the proof are entirely similar to those used in [8]. Hence, the proof is completed. Q.E.D.

Proof of Theorem 3.1: Suppose there exists a fuzzy control law (2) such that it can quadratically stabilize the m extreme subsystems in (29) and there exists a set of the membership functions denoted by $(\mu_1^*, \mu_2^*, \ldots, \mu_m^*)$ such that the upper bound (24) can be achieved, Then, there exists a PDQ Lyapunov function of the form (2) for the extreme systems and $\varepsilon_l > 0, l = 1, 2, \ldots, m$, such that

$$\begin{aligned}L(x,t) &= x^T[(A_l + E_{l1} + (B_l + E_{l2})K(\mu))^T P_l \\ &\quad + P_l(A_l + E_{l1} + (B_l + E_{l2})K(\mu))]x \\ &= x^T\{[(A_l + B_l K(\mu))^T P_l + P_l(A_l + B_l K(\mu)) \\ &\quad + (E_{l1} + E_{l2}K(\mu))^T P_l + P_l(E_{l1} + E_{l2}K(\mu))]\}x < 0, \\ &\quad x \in S_l\end{aligned}$$

Let,

$$A_0 = A_l + B_l K(\mu), \quad \Delta A(\mu) = E_{l1} + E_{l2}K(\mu).$$

Using *Lemma 3.2*, we have

$$\begin{aligned} L(x,t) \le\ & x^T[(A_l + B_l K(\mu))^T P_l + P_l(A_l + B_l K(\mu)) + \\ & \frac{1}{\varepsilon_l}[I \;\; K(\mu)] \begin{bmatrix} E_{l1}^T \\ E_{l2}^T \end{bmatrix} \begin{bmatrix} E_{l1} & E_{l2} \end{bmatrix} \begin{bmatrix} I \\ K(\mu) \end{bmatrix} \\ & + \varepsilon_l P_l P_l] x < 0, \quad x \in S_l. \end{aligned} \tag{34}$$

Using the similar PDQ Lyapunov function, we have

$$\begin{aligned} L(x,t) :=\ & x^T[(A_l + \Delta A_l + (B_l + \Delta B_l)K)^T P_l \\ & + P_l((A_l + \Delta A_l + (B_l + \Delta B_l)K)]x \\ =\ & x^T[(A_l + B_l K)^T P_l + P_l(A_l + B_l K) \\ & + (\Delta A_l + \Delta B_l K)^T P_l + P_l(\Delta A_l + \Delta B_l K)], \quad x \in S_l, \end{aligned}$$

where $K = K(\mu)$.

Using *Lemma 3.1* and (24), we obtain

$$\begin{aligned} L(x,t) \le\ & x^T[(A_l + B_l K)^T P_l + P_l(A_l + B_l K) + \\ & \frac{1}{\varepsilon_l}[I \;\; K] \begin{bmatrix} \Delta A_l^T \\ \Delta B_l^T \end{bmatrix} \begin{bmatrix} \Delta A_l & \Delta B_l \end{bmatrix} \begin{bmatrix} I \\ K \end{bmatrix} + \varepsilon_l P_l P_l] x \\ \le\ & x^T[(A_l + B_l K)^T P_l + P_l(A_l + B_l K) + \\ & \frac{1}{\varepsilon_l}[I \;\; K] \begin{bmatrix} E_{l1}^T \\ E_{l2}^T \end{bmatrix} \begin{bmatrix} E_{l1} & E_{l2} \end{bmatrix} \begin{bmatrix} I \\ K \end{bmatrix} + \varepsilon_l P_l P_l] x, \end{aligned}$$

$x \in S_l$

By (34) it follows that

$$L(x,t) \le -\beta x^T x, \quad \beta > 0, quad x \in S_l, \quad l = 1, 2, \ldots, m. \tag{35}$$

From (35) and $\bar{T} < \infty$ the fuzzy system (4) is quadratically stabilizable, via the fuzzy state feedback control law (2).

Conversely, if the fuzzy system (4) is quadratically stabilizable, let $\mu = \mu_l^*$, $l = 1, 2, \ldots, m$ in the m subsystems (24). Then the m extreme subsystems in (2) are quadratically stabilizable. Hence, the proof is completed. Q.E.D.

From *Theorem 2.1* we know that if the fuzzy system (4) is quadratically stabilizable it is also globally asymptotically stabilizable. If we use the switching control law (12), we have the following important corollary.

Corollary 3.1: The fuzzy system (4) is quadratically stabilizable, if and only if there exists a set of feedback gains $(K_1, K_2, \ldots, K_m)$ such that the following closed-loop extreme subsystems with the accurate upper bounds are quadratically stable and $\bar{T} < \infty$

$$\dot{x}(t) = (A_l + E_{l1}) + (B_l + E_{l2})K_l x(t) \quad x(t) \in S_l, \quad l = 1, 2, \ldots, m. \tag{36}$$

The design problem of (36) is a typical design problem in linear control. We can easily obtain the solution using most standard commercial linear control design software packages.

The above equivalent theorems require accurate upper bounds, but in many cases the accurate upper bounds are very difficult to find, so we have to use approximate upper bounds. Thus, we use the following stability theorem to determine whether the closed-loop fuzzy system (13) is stable. In addition, if a fuzzy control law can be obtained by other methods, then by using the following theorem we can also determine whether this fuzzy control law is a stabilizing control law.

Theorem 3.2: (Stability condition) Given a control law $K(\mu)$ and the approximate upper bound in (24), if $\bar{T} < \infty$ and there exist constants $\varepsilon_l > 0$, $l = 1, 2, \ldots, m$ and a set of positive-definite symmetric matrices $(Q_1, Q_2, \ldots, Q_m)$ such that a set of Riccati equations

$$\begin{aligned} &[A_l + B_l K(\mu)]^T P_l + P_l [A_l + B_l K(\mu)] \\ &\quad + \varepsilon_l P_l P_l + \frac{1}{\varepsilon_l}(E_{l1}^T + E_{l2}^T K(\mu))^T (E_{l1} + E_{l2} K(\mu)) + Q_l = 0, \end{aligned} \tag{37}$$

has a set of positive-definite symmetric solutions $(P_1, P_2, \ldots, P_m)$, then the fuzzy control law $K(\mu)$ is a stabilizing control law. That is, the fuzzy system (4) is quadratically stabilizable via the control law $K(\mu)$. Conversely, if the fuzzy system (4) is quadratically stabilizable and accurate upper bounds can be found, then there exist constants $\varepsilon_l^* > 0$ such that for all ε_l's in

$(0, \varepsilon_l^*)$, the set of Riccati equations (37) with accurate upper bounds has a set of positive-definite symmetric solutions $(P_1, P_2, \ldots, P_m)$.

Proof of Theorem 3.2: Let the PDQ Lyapunov function be of the form (2). If (37) holds, from (35) we know that $L(x,t) < 0, x(t) \in S_l, l = 1, 2, \ldots, m$. Thus the fuzzy system (4) is quadratically stabilizable via the control law $K(\mu)$. Conversely, if the fuzzy system (4) is quadratically stabilizable and μ^* can be found, we have

$$\begin{aligned} L(x,t) &:= x^T[(A_l + \Delta A_l + (B_l + \Delta B_l)K)^T P_l \\ &\quad + P_l((A_l + \Delta A_l + (B_l + \Delta B_l)K)]x \\ &= x^T[(A_l + B_l K)^T P_l + P_l(A_l + B_l K) \\ &\quad + (\Delta A_l + \Delta B_l K)^T P_l + P_l(\Delta A_l + \Delta B_l K)], \quad x \in S_l. \end{aligned}$$

Let,

$$A_0 = A_l + B_l K(\mu), \quad \Delta A(\mu) = \Delta A_l + \Delta B_l K.$$

Using *Lemma 3.2*, we have (37). The proof is completed.

Now, by using the control law (12), we have the following corollary.

Corollary 3.2: (Stability condition) If $\bar{T} < \infty$ and there exist constants $\varepsilon_l > 0, l = 1, 2, \ldots, m$ and a set of positive-definite symmetric matrices $(Q_1, Q_2, \ldots, Q_m)$ such that a set of Riccati equations

$$\begin{aligned} &(A_l + B_l K_l)^T P_l + P_l(A_l + B_l K_l) \\ &\quad + \varepsilon_l P_l P_l + \frac{1}{\varepsilon_l}(E_{l1}^T + E_{l2}^T K_l)^T (E_{l1} + E_{l2} K_l) + Q_l = 0, \end{aligned} \tag{38}$$

has a set of positive-definite symmetric solutions $(P_1, P_2, \ldots, P_m)$, then the fuzzy system (4) is quadratically stabilizable via the control law (2). Conversely, if the fuzzy system (4) is quadratically stabilizable and μ^* can be found, then there exist constants $\varepsilon_l^* > 0, l = 1, 2, \ldots, m$ such that for all ε_l's in $(0, \varepsilon_l^*)$, the set of Riccati equations (38) with accurate upper bounds has a set of positive-definite symmetric solutions $(P_1, P_2, \ldots, P_m)$.

From the *Theorem 3.2* it can be observed that only in the case of accurate upper bounds the stability conditions (37) and (38) are also necessary conditions.

By *Theorem 3.2* we have yet another corollary.

Corollary 3.3: (Equivalent principle) The fuzzy system (4) is quadratically stabilizable, if there exists a fuzzy control law (11) or (12) such that it can quadratically stabilize the m extreme subsystems with the approximate upper bounds in (24) and the stability condition (37) or (38) holds.

By this equivalent theorem, we can see that when using the approximate upper bounds we actually require that the control law (2) not only stabilizes the m extreme subsystems in (24), but also satisfies the stability condition (37) or (38).

The stability condition (37) or (38) is a sufficient condition for quadratic stabilizability of the fuzzy system (4). Thus it gives a conservative result. But if the approximate upper bounds are estimated more accurately, the stability condition becomes less conservative. If the accurate upper bounds can be found, the stability condition (37) or (38) becomes a necessary condition.

Let us note here that the equations in (37) or (38) depend on the parameters ε_l, but by *Theorem 3.2* we can deduce that the positive-definite solutions decrease monotonically as $\varepsilon_l \to 0$. Therefore, we can propose the following stability checking algorithm.

The stability checking algorithm:

Step 1. Suppose that the upper bounds (24) are given and set $\varepsilon_l, l = 1, 2, \ldots, m$ to a set of positive values, for example,

$$\varepsilon_l = 1, \quad \bar{m} = m.$$

Step 2. Determine whether the set of Riccati equations (37) or (38) has a set of positive-definite symmetric solutions. If a set of positive-definite symmetric solutions $(P_1, P_2, \ldots, P_{\bar{m}})$ exists, then the fuzzy system (4) can be quadratically stabilized by the fuzzy control law (11) or (12), Otherwise, proceed to Step 3.

Step 3. Set $\varepsilon_{\bar{l}} = \varepsilon_{\bar{l}}/2, \ \bar{l} = 1, 2, \ldots, \bar{m}$, where $\bar{m}$ is the number of the Riccati equations (37) or (38) which do not have solutions in *Step 2.* If some $\varepsilon_{\bar{l}}$'s are less than the computational accuracy, then stop: the system (4) cannot be stabilized by the fuzzy control law, Otherwise go to *Step 2.*

The iteration over $\varepsilon_{\bar{l}}$ is a binary search. It is already known that if the fuzzy control law is a quadratically stabilizing control law we can always find a suitable $\varepsilon_{\bar{l}}$ such that the $\bar{l}-th$ Riccati equation (37) or (38) has a positive-definite symmetric solution $P_{\bar{l}}$.

The above stability checking algorithm depends on the feedback gains. If we choose the feedback gains from below, we can obtain a necessary and sufficient stabilization condition which only depends on the system matrices and the upper bounds.

First, let the orthogonal-triangular decomposition of $E_{12}^T E_{12}$ be

$$E_{l2}^T E_{l2} = \Sigma_l^T U_l^T U_l \Sigma_l = \Sigma_l^T \Sigma_l, \quad U_l^T U_l = I.$$

If $rank(E_{l2}) < p$, define $\Xi_l = {\Sigma_l^+}^T \Sigma_l^+$, $\Sigma_l^+ = (\Sigma_l \Sigma_l^T)^{-1}\Sigma_l$. Choose Φ_l such that

$$\Phi_l \Sigma_l^T = 0, \quad \Phi_l^T \Phi_l = I.$$

If $rank(E_{l2}) = p$ then Σ_l is square and nonsingular, and $\Xi_l = (\Sigma_l^T \Sigma_l)^{-1} = (E_{l2}^T E_{l2})^{-1}$ and $\Phi_l = 0$.

If $E_{l2} = 0$, then $\Sigma_l = 0, \Phi_l = I$, and $\Xi_l = 0$. If the fuzzy control laws are chosen as

$$u(t) = K_l x(t) \quad if\ x(t) \in S_l,$$

$$K_l = -[\frac{1}{2\varepsilon_l}\Phi_l \Phi_l^T + \Xi_l] B_l^T P_l - \Xi_l E_{l2}^T E_{l1}, \tag{39}$$

then we have the following theorem.

Theorem 3.4: If $\bar{T} < \infty$ and there exist constants $\varepsilon_l > 0, l = 1, 2, \ldots, m$, and a set of positive-definite symmetric matrices $(Q_1, Q_2, \ldots, Q_m)$ such that a set of Riccati equations

$$\begin{aligned}
&(A_l - B_l \Xi_l E_{l2}^T E_{l1})^T P_l + P_l (A_l - B_l \Xi_l E_{l2}^T E_{l1}) \\
&+ P_l (I - B_l \Xi B_l^T - \frac{1}{\varepsilon_l} B_l \Phi_l^T \Phi_l B_l^T) P_l \\
&+ E_{l1}^T (I - E_{l2} \Xi_l E_{l2}^T) E_{l1} + \varepsilon_l I = 0,
\end{aligned} \tag{40}$$

has a set of positive-definite symmetric solutions $(P_1, P_2, \ldots, P_m)$, then the fuzzy system (4) is asymptotically stabilizable. Conversely, if the fuzzy system (4) is quadratically stabilizable and μ^* can be found, then there exist constants $\varepsilon_l^* > 0, l = 1, 2, \ldots, m$ such that for all ε_l's in $(0, \varepsilon_l^*)$ the set of Riccati equations (40) with accurate upper bounds has a set of positive-definite symmetric solutions $(P_1, P_2, \ldots, P_m)$.

Proof of Theorem 3.4: Substituting the control law (39) into (40), we have the Riccati equation (38) [9]. Using the same procedure as in the proof of *Theorem 3.2*, we can obtain the required result.

By using *Theorem 3.3* we can propose the following algorithm to estimate the upper bounds in (24).

Approximate upper bounds search algorithm:

Step 1: Choose approximate upper bounds (24) of the following form

$$E_{l1}^T E_{l1} = \lambda_{l1} \Delta \bar{A}_l(\tilde{\mu})^T, \quad E_{l2}^T E_{l2} = \lambda_{l2} \Delta \bar{B}_l(\tilde{\mu})^T \Delta \bar{B}_l(\tilde{\mu}), \tag{41}$$

$$\tilde{\mu} = [0.5 \ \ 0.5 \ \cdots \ 0.5], \quad l = 1, 2, \ldots, m,$$

or

$$E_{l1}^T E_{l1} = \lambda_{l1} I, \ E_{l2}^T E_{l2} = \lambda_{l2} I, \quad l = 1, 2, \ldots, m. \tag{42}$$

Step 2: Initially, set $\lambda_{l1} = 0, \lambda_{l2} = 0, l = 1, 2, \ldots, m$, and determine whether the set of Riccati equations (38) has a set of positive-definite symmetric solutions. If a set of positive-definite symmetric solutions $(P_1, \ldots, P_m)$ does not exist, the algorithm ends and a stabilization control law cannot be found. Otherwise, go to *Step 3*.

Step 3: Increase the values of $\lambda_{l1} = \lambda_{l1} + \sigma_{l1}, \lambda_{l2} = \lambda_{l2} + \sigma_{l2}, \sigma_{l1}, \sigma_{l2} > 0, l = 1, 2, \ldots, m$. Then, determine whether the set of Riccati equations (38) has a set of positive-definite symmetric solutions $(P_1, P_2, \ldots, P_m)$. If a set of positive-definite symmetric solutions exists, then repeat *Step 3* until the maximum values of $\lambda_{l1}, \lambda_{l2}, l = 1, 2, \ldots, m$, are found, which will enable (38) to have the desired solutions. These values are described by $\bar{\lambda}_{l1}$ and $\bar{\lambda}_{l2}$. Substituting them into (3) we obtain the approximate upper bounds. It should be noted here that the above search algorithm gives the upper bounds which represent the largest allowable interaction among the subsystems.

In light of the above theorems, we can follow the procedure below to evaluate a stabilizing state-feedback control law. This procedure can be easily implemented on any commercial control system software package.

The fuzzy controller design algorithm 1

Step 1. Find the accurate upper bound (24).

Step 2. Design the fuzzy control law (2) such that the m closed-loop extreme subsystems have the required performance. Using the control law (12) we can design m independent subsystems one by one such that every closed-loop extreme subsystem has the required performance.

The above fuzzy controller design algorithm is an accurate algorithm because the accurate upper bounds (24) are used. In some cases the accurate upper bounds are difficult to find. The only information which can be used is the approximate upper bounds. Thus, we revise the above controller design algorithm by using the approximate upper bounds.

The fuzzy controller design algorithm 2:

Step 1: Use the approximate upper bound search algorithm and obtain the largest allowable upper bounds.

Step 2: Substitute these bounds into the m extreme subsystems in (3). Second, design a fuzzy control law (11) or (12) such that the m closed-loop extreme subsystems have the required performance. Then determine whether the stability condition (37) or (38) holds by the stability checking algorithm or simulation. If the stability condition does not hold, then, go to *Step 1* and choose another set of largest allowable upper bounds.

For practical purposes, the most convenient method is to choose a set of approximate upper bounds which are less than the worst-case upper bounds and then design a fuzzy control law. This results in the following algorithm.

The fuzzy controller design algorithm 3:

Step 1: Choose the upper bound of the form (24).

Step 2: Initially, set $\lambda_{l1} = 0, \lambda_{l2} = 0 \quad l = 1, 2, \ldots, m$. Then design a fuzzy control law (2) and examine the closed-loop performance using the stability checking algorithm or simulation. If the performance is satisfactory the algorithm ends and the control law is found. Otherwise, go to *Step 3.*

Step 3: Increase the values of

$$\lambda_{l1} = \lambda_{l1} + \sigma_{l1},\ \lambda_{l2} = \lambda_{l2} + \sigma_{l2},\ \sigma_{l1},\ \sigma_{l2} > 0,\ \lambda_{l1} \le \bar{\lambda}_{l1},\ \lambda_{l2} \le \bar{\lambda}_{l2},$$

$l = 1, 2, \ldots, m.$

Use the new upper bounds in (24) and then design the fuzzy control law (2) and also examine the closed-loop performance by the stability checking algorithm or simulation. If the performance is satisfactory the algorithm ends and the control law is found. Otherwise, repeat *Step 3* until the largest allowable upper bounds are reached.

4 Further results about quadratic stability

All the results in Sect. 3 require the condition $\bar{T} < \infty$. However, this condition is difficult to test. One way to test it is to use a simulation method. In this section we will develop methods to change this condition into one that is easier to check. In the theorem that follows we will use PDQ Lyapunov function and fuzzy control law given as

$$V(t) = V(x(t)) \hat{=} x^T P(t) x(t), \tag{43}$$

$$P(t) = P_l, \quad x(t) \in S_l, \quad t \neq \tau_i, \tag{44}$$

$$P(\tau_i^-) = P_l, \quad x(\tau_i^-) \in \partial S_l, \quad P(\tau_i^+) = P_j, \quad j \neq l, \quad t = \tau_i, \tag{45}$$

$$u(t) = K(x(t)) = \begin{cases} K_l x(t) & x(t) \in S_l, \quad t \neq \tau_i \\ K_j x(\tau_i) & x(\tau_i) \in \partial S_j, \quad t = \tau_i \end{cases}. \tag{46}$$

Remark 4.1: The control law (46) has the following explanation. When $x(t) \in S_i, K_l x(t)$ is used until the state $x(t)$ meets the boundary of S_l. Then $K(x(t))$ will be changed to $K_j x(\tau_i)$ at the boundary of S_j. $S_j, j \neq l$ is one of the regions which satisfies $\partial S_l \cap \partial S_j \neq 0$. Since there are several regions to intersect with S_l we need to choose one in which the boundary condition $P(\tau_i^+) \leq P(\tau_i^-)$ holds.

Theorem 4.1: Given a state feedback fuzzy control law (12) and the approximate upper bound in (24), if there exist constants $\varepsilon_l > 0, l = 1, 2, \ldots, m$, and a set of positive-definite symmetric matrices $(Q_1, Q_2, \ldots, Q_m)$ such that:
(i) a set of Riccati equations

$$[A_l + B_l K_l]^T P_l + P_l [A_l + B_l K_l] + \varepsilon_l P_l P_l$$

$$+\frac{1}{\varepsilon_l}(E_{l1}^T + E_{l2}K_l)^T(E_{l1} + E_{l2}K_l) + Q_l = 0, \tag{47}$$

$$x(t) \in S_l, \quad l = 1, 2, \ldots, m,$$

has a set of positive-definite symmetric solutions $(P_1, P_2, \ldots, P_m)$, and (ii)

$$P(\tau_i^+) \leq P(\tau_i^-), \quad \forall i, \tag{48}$$

then, the closed-loop fuzzy system (13) is globally asymptotically stable.

Proof: From the proof of *Theorem 3.2* we know that the condition (47) means that

$$A_{cl}(\mu)^T P_l + P_l A_{cl}(\mu) + Q_l = 0, \quad x(t) \in S_l, \quad l = 1, 2, \ldots, m.$$

Using the same proof procedure as in *Theorem 2.1* we have

$$V(x(t)) \leq exp[-\bar{\alpha}(t - \bar{t})]V(x(\bar{t})),$$

$$\tau_{i-1} < \bar{t} \leq t < \tau_i, \quad \tau_0 = 0.$$

At the discontinuous point τ_i, by the boundary condition (48), we have

$$V(x(\tau_i^+) \leq V(x(\tau_i^-).$$

Thus

$$V(x(t)) \leq exp[-\bar{\alpha}(t - 0)]V(x(0)), \quad \forall t. \tag{49}$$

From (49) it follows that

$$x(t)^T x(t) \leq c\ exp(-\bar{\alpha}t)x_0^T x_0,$$

$$c = \min_t \frac{\lambda_{max}(P(0))}{\lambda_{min}(P(t))}.$$

Thus

$$\begin{aligned}\int_0^{\tilde{T}} x^T(t)x(t)dt &\le (\int_0^{\tilde{T}} exp(-\alpha t)dt)cx_0^T x_0 \\ &= -\frac{1}{\alpha}[exp(-\alpha\tilde{T}) - 1]cx_0^T x_0.\end{aligned}$$

Taking the limit as $\tilde{T} \to \infty$, it follows that

$$\lim_{\tilde{T}\to\infty} \int_0^{\tilde{T}} x^T(t)x(t)dt \le \frac{c}{\alpha}x_0^T x_0.$$

Thus, the fuzzy system (13) is asymptotically stable in the large. Q.E.D.

In the theorems that follow we still use the PDQ Lyapunov function (2) and the fuzzy control law (2).

Theorem 4.2: Given a state feedback fuzzy control law (12) and the approximate upper bound in (24), if there exist constants $\varepsilon_l > 0, l = 1, 2, \ldots, m$, and a set of positive-definite symmetric matrices $(Q_1, Q_2, \ldots, Q_m)$ such that: (i) a set of Riccati equations

$$\begin{aligned}&[A_l + B_l K_l]^T P_l + P_l[A_l + B_l K_l] + \varepsilon_l P_l P_l \\ &\quad + \frac{1}{\varepsilon_l}(E_{l1}^T + E_{l2}K_l)^T(E_{l1} + E_{l2}K_l) + Q_l = 0,\end{aligned} \tag{50}$$

$$x(t) \in S_l, \quad l = 1, 2, \ldots, m,$$

has a set of positive-definite symmetric solutions $(P_1, P_2, \ldots, P_m)$, and (ii)

$$C \le e^{(\alpha-\xi)\sigma}, \tag{51}$$

for a sufficiently small $\xi > 0$, where $\alpha \hat{=} \min_l \frac{\lambda_{min}(Q_l)}{2\lambda_{max}(P_l)}$, $C = \max_l (C_l)$, $C_l = \frac{\lambda_{max}(P_l)^{1/2}}{\lambda_{min}(P_l)^{1/2}}$, $\sigma = \min_i (\tau_i - \tau_{i-1}), \tau_0 = 0$, then, the closed-loop fuzzy system (13) is globally asymptotically stable.

Proof: From the proof of *Theorem 3.2* we know that the condition (50) means that

$$A_{cl}(\mu)^T P_l + P_l A_{cl}(\mu) + Q_l = 0,$$

$$x(t) \in S_l, \quad l = 1, 2, \ldots, m,$$

for any $\mu \in M$. Using the same proof procedure as in *Theorem 2.1* we have

$$\begin{aligned} \|x(t)\| &\leq Cexp[-\alpha(t-\tau_i)](\prod_{j=0}^{i-1} C\ exp[-\alpha(\tau_{j+1}-\tau_j)])\|x_0\| \quad \tau_0 = 0 \\ &= Ce^{-\xi t}(\prod_{j=0}^{i-1} C\ exp[-(\alpha-\xi)\sigma])\|x_0\|. \end{aligned} \tag{52}$$

From (51) we know that

$$Ce^{-(\alpha-\xi)\sigma} \leq 1. \tag{53}$$

Thus, from (52) and (53) it follows that

$$\|x(t)\| \leq Ce^{-\xi t}\|x_0\|, \quad t \geq 0.$$

Hence the closed-loop fuzzy system (13) is globally asymptotically stable. Q.E.D.

Remark 4.1: The condition (51) in *Theorem 4.2* means that if the nominal closed-loop systems $A_l + B_l K_l$, $l = 1, 2, \ldots, m$ are stable enough in the sense that α is large enough and if the local subsystems are stable in the sense that (50) holds, we can guarantee that the closed-loop fuzzy system (13) is globally asymptotically stable. Thus, condition (46) can be considered as another kind of boundary condition.

However, condition (51) still includes an unknown factor σ. We need to further improve the boundary condition such that it can be checked by only using the closed-loop nominal system matrices $A_l + B_l K_l$, $l = 1, 2, \ldots, m$. The following theorem gives such a result.

Theorem 4.3: Given a state feedback fuzzy control law (12) and the approximate upper bound in (24), if there exist constants $\varepsilon_l > 0, l = 1, 2, \ldots, m$, and a set of positive-definite symmetric matrices $(Q_1, Q_2, \ldots, Q_m)$ such that: (i) a set of Riccati equations

$$[A_l + B_l K_l]^T P_l + P_l[A_l + B_l K_l] + \varepsilon_l P_l P_l$$
$$+\frac{1}{\varepsilon_l}(E_{l1}^T + E_{l2}K_l)^T(E_{l1} + E_{l2}K_l) + Q_l = 0, \qquad (54)$$

$$x(t) \in S_l, \quad l = 1, 2, \ldots, m,$$

has a set of positive-definite symmetric solutions $(P_1, P_2, \ldots, P_m)$, and (ii) given any $\xi \in (0, \alpha)$

$$\|\Delta \bar{A}_c\| < \frac{(\alpha - \xi)^2}{4M_\mu C \ ln \ C}, \qquad (55)$$

where ,

$$\alpha \hat{=} \min_l \frac{\lambda_{min}(Q_l)}{2\lambda_{max}(P_l)}, \quad C = \max_l(C_l), \quad C_l = \frac{\lambda_{max}(P_l)^{1/2}}{\lambda_{min}(P_l)^{1/2}},$$

$$A_{cl} = A_l + B_l K_l, \quad A_{ci} = A_i + B_i K_l,$$

and

$$\Delta \bar{A}_{cl}(t) = \sum_{i=1, i \neq l}^{m_l} \bar{\mu}_i(t)[A_{ci} - A_{cl},]$$

and M_μ satisfies

$$\|\Delta \dot{\bar{A}}_{cl}(t)\| \;=\; \| \sum_{i=1, i \neq l}^{m_i} \dot{\bar{\mu}}_i(t)[A_{ci} - A_{cl}]\|$$

$$\leq \sum_{i=1,i\neq l}^{m_i} \| \max_{nT\leq\xi\leq t} \dot{\mu}_i(\xi)\|\|\|[A_{ci} - A_{cl}]\| \leq M_\mu \|\Delta \bar{A}_c\|, \quad (56)$$

$$\|\Delta \bar{A}_c\| = \max_{l,i}(\|A_{ci} - A_{cl}\|, \ i = 1, 2, \ldots, m_i, \ l = 1, 2, \ldots, m,$$

$$T = \frac{2 \ ln \ C}{(\alpha - \xi)},$$

then the closed-loop fuzzy system (13) is globally asymptotically stable.

Proof: Since (55) holds the solution of (13) for any fixed μ, we have

$$\|x(t)\| \leq C\|x_0\| exp(-\alpha t), \quad t \geq 0. \quad (57)$$

The closed-loop system (13) can be considered as the following time-varying linear system

$$\dot{x}(t) = A_c(t)x(t) = (A_{cl} + \Delta \bar{A}_{cl}(t))x(t), \quad x(t) \in S_l. \quad (58)$$

Now, consider approximating $A_c(t)$ in (58) by the piecewise constant matrix

$$A_{pc}(t) = A_c(nT), \quad nT \leq t \leq (n+1)T, \quad n = 0, 1, 2, \ldots,$$

where T is chosen as

$$\frac{2 \ ln \ C}{(\alpha - \xi)}. \quad (59)$$

Now the system (13) can be rewritten as

$$\dot{x}(t) = A_{pc}(t)x(t) + (A_c(t) - A_{pc}(t))x(t). \quad (60)$$

Since

$$\sum_{l=1}^{m}(\mu_l(t) - \mu_l(nT)) = 0,$$

and

$$||\bar{\mu}_l(t) - \bar{\mu}_l(nT)|| \leq (t - nT) \max_{nT \leq \xi \leq t} \dot{\bar{\mu}}_i(\xi) \hat{=} (t - nT)\dot{\bar{\mu}}_{max}(t),$$

we have

$$\begin{aligned} ||A_m(t) - A_{pc}(t)|| &= || \sum_{i \neq l, i=1}^{m_i} (\bar{\mu}_i(t) - \bar{\mu}_i(nT))(A_{ci} - A_{cl})|| \\ &\leq (t - nT) \sum_{i \neq l, i=1}^{m_i} ||\dot{\bar{\mu}}_{max}(t)|| ||(A_{ci} - A_{cl})||, \end{aligned} \quad (61)$$

$l = 1, 2, \ldots, m.$

It should be noted here that in the case of trapezoidal and triangular membership functions, the derivative of $\mu_l(t)$ does not exist in the conventional sense at some points. In this case we divide the region $[t, nT]$ into several subregions, for example, $[t, nT] = [t, t1] \cup [t1, nT]$, where t_1 is the discontinuous point of $\dot{\mu}_l(t)$. Then

$$\begin{aligned} &|| \sum_{i \neq l, i=1}^{m_i} (\bar{\mu}_i(t) - \bar{\mu}_i(nT))(A_{ci} - A_{cl})|| \\ &\leq || \sum_{i \neq l, i=1}^{m_i} (\bar{\mu}_i(t) - \bar{\mu}_i(t_1) + \bar{\mu}_i(t_1) - \bar{\mu}_i(nT))(A_{ci} - A_{cl})|| \\ &\leq || \sum_{i \neq l, i=1}^{m_i} (\bar{\mu}_i(t) - \bar{\mu}_i(t_1))(A_{ci} - A_{cl})|| \\ &\quad + || \sum_{i \neq l, i=1}^{m_i} (\bar{\mu}_i(t) - \bar{\mu}_i(nT))(A_{ci} - A_{cl})||. \end{aligned}$$

Thus, (61) still holds. From (55) (56) and (61) we have

$$||A_m(t) - A_{pc}(t)|| \leq \frac{(\alpha - \xi)^2}{4C \; ln \; C} T = \frac{(\alpha - \xi)}{2C}. \quad (62)$$

Fact 3.1: Consider the linear system

$$\dot{x}(t) = A_0 x(t) + \Delta A(t)x(t). \tag{63}$$

Suppose that for some $\bar{C}, \lambda$ *and* $\Delta \geq 0$

$$||e^{A_0 t}|| \leq \bar{C}e^{-\lambda t}, \quad ||\Delta A(t)|| \leq \Delta, \quad \forall t \geq 0.$$

Under these conditions,

$$||\varphi(t,x)|| \leq \bar{C}||x||e^{-(\lambda - \bar{c}\Delta)t},$$

where $\varphi(t,x)$ is the solution of (57).

The proof of *fact 3.1* can be found in [11]. It follows from *fact 3.1*, (62) and (57) that on $nT \leq t \leq (n+1)T$,

$$\begin{aligned} ||x(t)|| &\leq Ce^{-\frac{\alpha+\xi}{2}(t-nT)}||x(nT)|| \leq Ce^{-\frac{\alpha+\xi}{2}(t-nT)}(Ce^{-\frac{\alpha+\xi}{2}T})^n||x_0|| \\ &\leq Ce^{-\xi t}e^{-\frac{\alpha-\xi}{2}(t-nT)}(Ce^{-\frac{\alpha-\xi}{2}T})^n||x_0|| \leq Ce^{-\xi t}||x_0||, \end{aligned}$$

which completes the proof. Q.E.D.

Theorem 4.3 requires that the $\Delta\bar{A}_{cl}$'s are sufficiently small. We can choose a large enough m in the fuzzy reference model such that $\Delta\bar{A}_{cl}$'s are indeed sufficiently small. In fact, we can also further decompose the region S_l into K_l subspaces as

$$\begin{aligned} S_{li} = \{x | \alpha_{li} \leq \mu_l(x) < \alpha_{li+1}, \mu_j(x) < \alpha_{l1} < \cdots < \alpha_{lk_l} < 1 \\ j \neq l, \; i = 1,2,\ldots,k_l, \; l = 1,2,\ldots,\bar{m}\}. \end{aligned} \tag{64}$$

$$S_l = \partial S_{l1} \cup \partial S_{l2} \cup \cdots \cup \partial S_{lk_l}.$$

Now we rearrange the regions S_l

$$S_{11} \cup S_{12} \cup \cdots \cup S_{lk_1} \cdots S_{m1} \cup S_{m2} \cup \cdots \cup S_{mk_m} = S_1 \cup S_2 \cup \cdots \cup S_{\bar{M}}, \tag{65}$$

$$\bar{M} = \sum_{l=1}^{m} k_l.$$

In each subregion S_l, the fuzzy system (4) can be expressed by the following subsystems

$$\begin{aligned}\dot{x}(t) &= \tilde{A}_l(\alpha_{li})x(t) + \tilde{B}_l(\alpha_{li})u(t) \\ &= (A_l + \Delta A_l(\alpha_{li}))x(t) + (B_l + \Delta B_l(\alpha_{li}))u(t), \quad x(t) \in S_l \quad (66)\end{aligned}$$

$l = 1, 2, \ldots, \bar{M}.$

Thus, we can choose a large enough $\bar{M}$ in the fuzzy model (66) such that $\Delta\bar{A}_{cl}$ are sufficiently small and guarantee that the closed-loop fuzzy system (13) is globally asymptotically stable. The condition (55) can also be considered as another kind of boundary condition.

5 Example

To illustrate the stabilization algorithm, we consider the problem of balancing an inverted pendulum on a cart. The equations of the motion for the pendulum are

$$\dot{x}_1 = x_2,$$

$$\dot{x}_2 = \frac{g\sin(x_1) - amlx_2^2\sin(2x_1)/2 - a\cos(x_1)u}{4l/3 - aml\cos^2(x_1)}, \quad (67)$$

where x_1 denotes the angle between the pendulum and the vertical, and x_2 is the angular velocity. $g = 9.8m/s^2$ is the gravity constant, m is the mass of the pendulum, M is the mass of the cart, $2l$ is the length of the pendulum, and u is the force applied to the cart. $a = 1/(m+M)$. For the purpose of simulation we choose $m = 2.0kg, M = 8.0kg, 2l = 1.0m$.

The following fuzzy model is then used to design a fuzzy controller.

$$\begin{array}{lll} R^1: & \text{IF} & x_1(t)\text{is about } 0 \\ & \text{THEN} & \dot{x}(t)\ A_1x(t) + B_1u(t), \\ R^2: & \text{IF} & x_1(t)\text{is about } \pm\pi/2 \\ & \text{THEN} & \dot{x}(t)\ A_2x(t) + B_2u(t), \end{array} \quad (68)$$

where

$$A_1 = \begin{bmatrix} 0 & 1 \\ \frac{g}{4l/3-aml} & 0 \end{bmatrix}, \quad B_1 = \begin{bmatrix} 0 \\ -\frac{a}{4l/3-aml} \end{bmatrix},$$

$$A_2 = \begin{bmatrix} 0 & 1 \\ \frac{2g}{\pi(4l/3-aml\beta^2)} & 0 \end{bmatrix}, \quad B_2 = \begin{bmatrix} 0 \\ -\frac{a\beta}{4l/3-aml\beta^2} \end{bmatrix},$$

and $\beta = cos(88^o)$.

We use the following membership functions

$$\begin{aligned} \mu_1(x_1(t)) &= (1 - 1/(1 + exp(-7(x_1(t) - \pi/4)))) \times \\ &\quad (1/(1 + exp(-7(x_1(t) + \pi/4)))), \\ \mu_2(x_1(t)) &= 1 - \mu_1(x_1(t)). \end{aligned}$$

Using the approximate upper bound search algorithm we find the approximation upper bounds as

$$E_1 = E_{11} = E_{21} = 0.1(A_1 - A_2), \quad E_2 = E_{12} = E_{22} = 0.01(B_1 - B_2).$$

Now we use the pole placement method to design the feedback control law. The closed-loop eigenvalues of each local subsystem are chosen as $[-3, -3]$. Then the following feedback gains can be obtained based on the pole placement method

$$K_1 = [149 \quad 34], \qquad K_2 = [3430.7 \quad 1120.9]. \tag{69}$$

When $\varepsilon = 0.01$ and $Q_1 = Q_2 = I$, the Riccati equations (54) have two positive-definite symmetric solutions. Since the two state space models in (68) are in controllable companion form and the closed-loop eigenvalues of two local subsystems are the same we have $||\Delta \bar{A}_C|| = ||\Delta \bar{A}_{C1} - \Delta \bar{A}_{C2}|| = 0$. Thus, the conditions (54) and (55) of *Theorem 4.3* hold. By *Theorem 4.3* the closed-loop fuzzy system is globally asymptotically stable when using this feedback gain. Figure 1 shows the angle response of the pendulum system (67) for initial conditions $x_1 = 65^o,\ x_2 = 0$ and $x_1 = 88^o,\ x_2 = 0$.

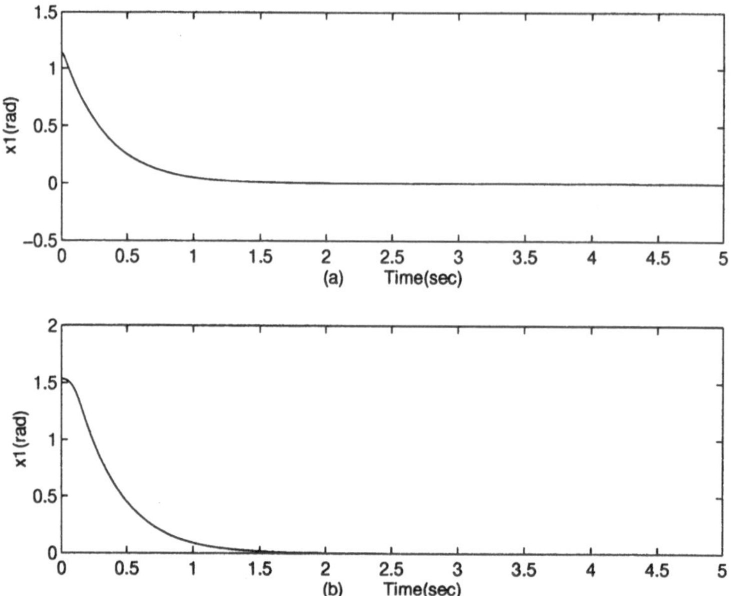

Figure 1: Fuzzy control with (69): (a) $x(0) = [65^o 0]'$, (b) $x(0) = [88^o 0]'$

It can be observed from the above simulation that with the method proposed in this article the pendulum can be well balanced for various initial conditions and the transient performance is also satisfactory. In fact, the transient performance of the system can be improved if the closed-loop eigenvalues of each local subsystem are chosen as $[-10, -10]$. In this case the following feedback gains can be obtained based on the pole placement design method

$$K_1 = [664.7 \;\; 113.3], \qquad K_2 = [20432 \;\; 3736]. \tag{70}$$

When $\varepsilon = 0.01$ and $Q_1 = Q_2 = I$, the Riccati equations (54) have two positive-definite symmetric solutions. We still have $||\Delta \bar{A}_C|| = ||\Delta \bar{A}_{C1} - \Delta \bar{A}_{C2}|| = 0$. Thus, the conditions (54) and (55) of *Theorem 4.3* hold. By *Theorem 4.3* the closed-loop fuzzy system is globally asymptotically stable when using this feedback gain. Figure 2 shows the angle response of the pendulum system (67) for initial conditions $x_1 = 65^o$, $x_2 = 0$ and $x_1 = 88^o$, $x_2 = 0$.

It can be observed from the above simulation results that the closed-loop response becomes faster.

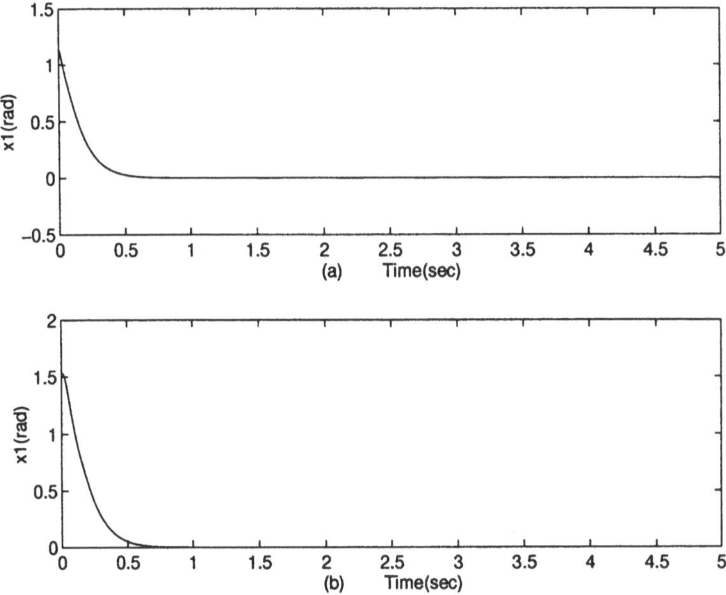

Figure 2: Fuzzy control with (70): (a) $x(0) = [65^o 0]'$, (b) $x(0) = [88^o 0]'$

6 Conclusions

Quadratic stability results for continuous-time fuzzy control systems are presented in this article. The main contribution is the development of the equivalent principle, that is, *the design of a fuzzy control system is equivalent to the design of a set of linear time-invariant "extreme" systems.* We point out that the boundary conditions are the key to the global stability of the fuzzy control system. Thus, a number of results on the boundary conditions are presented. We prove that *the existence of the positive-definite symmetric solutions to a suitable set of Riccati equations together with the boundary conditions guarantees the quadratic stability of the closed-loop fuzzy system. This in turn implies their global asymptotic stability.* More results about boundary conditions can be found in [16]-[18].

References

[1] L. A. Zadeh. Outline of a new approach to the analysis of complex systems and decision processes. *IEEE Trans. Syst. Man, Cybern*, 3:28–44, 1973.

[2] E. H. Mamdani and S. Assilian. Applications of fuzzy algorithms for control of simple dynamic plant. In *Proc. Inst. Elec. Eng.*, volume 121, pages 1585–1588, 1974.

[3] H. Nakanishi, I.B. Turksen, and M. Sugeno. A review and comparison of six reasoning methods. *Fuzzy Sets and Systems*, 57:257–294, 1993.

[4] T. Takagi and M. Sugeno. Fuzzy identification of systems and its application to modeling and control. *IEEE Trans. Sys., Man, Cybern.*, 15:116–132, 1985.

[5] K. Tanaka and M. Sugeno. Stability analysis and design of fuzzy control systems. *Fuzzy Sets and Systems*, 45:135–156, 1992.

[6] K. Tanaka and M. Sano. A robust stabilization problem of fuzzy control systems and its application to backing up control of a truck-trailer. *IEEE Trans. Fuzzy Systems*, 2(2):119–134, 1994.

[7] C. L. Chen, P. C. Chen, and C. K. Chen. Analysis and design of fuzzy control systems. *Fuzzy Sets and Systems*, 57:125–140, 1993.

[8] I. R. Peterson. A stabilization algorithm for a class of uncertain linear system. *System & Control Letters*, 8:351–357, 1987.

[9] K.Zhou and P. P. Khargonekar. Stability robustness bounds for linear state-space models with structured uncertainty. *IEEE Trans. Automat. Contr.*, AC-32:621–623, July 1987.

[10] M. A. Rotea and P. P. Khargoneker. Stabilization of uncertain systems with norm bounded uncertainty - a control Lyapunov function approch. *SIAM J. Control Optim.*, 27:1462–1476, 1989.

[11] W. A. Coppel. Dichotomies in stability theory. In *Lecture Notes in Mathematics 629*. Springer-Verlag, New York, 1978.

[12] S. G. Cao and N. W. Rees. Identification of dynamic fuzzy systems. *Fuzzy Sets and Systems*, 74:307–320, 1995.

[13] S. G. Cao, N. W. Rees, and G. Feng. Quadratic stability analysis and design of continuous-time fuzzy control system. *Int. J. System Science*, 27(2):193–203, 1996.

[14] S. G. Cao, N. W. Rees, and G. Feng. Analysis and design for a class of complex control systems - part I fuzzy modeling and identification. *Automatica*, June 1997.

[15] S. G. Cao, N. W. Rees, and G. Feng. Analysis and design for a class of complex control systems - part II fuzzy controller design. *Automatica*, June 1997.

[16] S. G. Cao, N. W. Rees, and G. Feng. Analysis and design of uncertain fuzzy control systems - part I fuzzy modeling and identification. In *Proc. of IEEE International Conference on Fuzzy Systems*, pages 640–646, New Orleans, USA, 1996.

[17] S. G. Cao, N. W. Rees, and G. Feng. Analysis and design of uncertain fuzzy control systems - part II fuzzy controller design. In *Proc. of IEEE International Conference on Fuzzy Systems*, pages 647–653, New Orleans, USA, 1996.

[18] S. G. Cao, N. W. Rees, and G. Feng. Fuzzy control of nonlinear continuous-time systems. In *Proc. of 35th IEEE Conference on Decision and Control*, pages 592–597, Kobe, Japan, 1996.

Fuzzy Stability Analysis of Fuzzy Systems: A Lyapunov Approach

J.P. Marin
LAAS-CNRS
7, Avenue du Colonel Roche
31077 Toulouse Cedex - FRANCE

1 Introduction

The modelling phase in the analysis of the dynamic behavior of complex systems is crucial. The formalism of the model must satisfy a number of requirements. First, it must be flexible enough to capture all system properties, e.g., nonlinearities and uncertainties). Then it must allow the use of all available sources of information about the system. Finally, it must allow the analysis of the system behavior in a convenient, systematic way.

Because quantitative models (differential equations based) of complex systems do not satisfy the second requirement, linguistic models based on fuzzy rules and fuzzy inference have been proposed as a possible alternative [Zadeh, 73]-[Mamdani, 75]. The advantages offered by linguistic models are the possibility of integrating expert's experience based on heuristic knowledge for representing nonlinear maps as well as dealing with system uncertainties.

In practice however, linguistic models suffer from a combinatorial explosion of the number of fuzzy rules. This problem occurs independently of the nonlinearities of the system and depends on the desired accuracy of the model. To avoid such a drawback, Takagi-Sugeno (TS) models can be used [Takagi, 85]. Such models can be considered as hybrid models between quantitative and linguistic ones. As a matter of fact, uncertainties management is lost by the use of a deffuzzification procedure. Furthermore, formal logical aspects are lost by the use of t-norm operators as implication operators since this type of implications violate the properties of boolean implication.

The first objective of this article is to present a new type of a fuzzy model which bridges the gap between linguistic and Tagaki-Sugeno models. Such hybrid systems are obtained by the use of input-dependent fuzzy sets in the conclusion part (consequent) of the fuzzy rules. They allow the representation of uncertainties and nonlinearities. Moreover, the expert's knowledge can be integrated in the modelling phase. Such models are based on various classes of approximate reasoning. Using a quadratic parameterization of a fuzzy set, we obtain an exact and compact analytical characterization for these hybrid systems.

When a model of the dynamic system is available, we have to decide what is an "acceptable" dynamic behavior. Stability and robustness of dynamic systems is probably one of the most important requirements to characterize the steady-state behavior of the system. Although, such properties have a clear practical significance, there exists various mathematical ways to characterize them. However, classical concepts such as global asymptotic stability of an equilibrium point or input/output stability are not satisfactory since they are not necessary nor sufficient for an acceptable behavior. Consequently, new stability criteria must be derived to formalise the intuitive idea of "acceptable behavior".

According to the formalism by which a model is represented, there exists two main approaches to analyze (robust) stability of dynamic systems. The first one is quantitative" analysis: assuming that a quantitative model of the closed-loop system is available (inclusive Tagaki-Sugeno models), some methodologies have been developed to provide stability proof in a classical sense, i.e., global asymptotic stability of an equilibrium point. This approach suffers from the following drawback: in the case of fuzzy controllers, mathematical properties of the control law must be deduced from the fuzzy formalism and are generally poor. Consequently, stability analysis is conservative [Aracil, 91]-[Wang, 94]. The same approach has been extended to the H^{∞} performance analysis [Marin, 96a].

If both the plant and the controller models are given in Takagi-Sugeno form, stability of the closed-loop system can be studied using Linear Matrix Inequalities [Boyd, 94]-[Tanaka, 92] and quadratic Lyapunov function. Note that the premise parts (antecedents) of the fuzzy rules rule are ignored in this approach. Consequently, the stability of each rule is necessary to conclude the stability of the overall system. Using similar approach, it has been shown that the stability of each rule is not necessary for global stability (see : [Marin, 95a]-[Marin, 95b] and [Marin, 96b] for the extension to the H_{∞} performance). These results are achieved taking into account the premise part of the fuzzy rules.

On another hand, assuming that a linguistic model of the plant is available, few approaches propose a stability analysis in a fuzzy sense [Kiska, 85]-[Jianquin, 93]. Such approaches allow to remove the drastic distinction between stability and instability and seems able to characterize an "acceptable" behavior of the system. However, these methods are based on fuzzy relations, discrete-time fuzzy dynamic systems, or cell-to-cell mappings. Consequently, they suffer from a combinatorial explosion with respect to the dimension of the system and the precision required during the fuzzy modelling phase.

The second objective of this article is to use the new type of a fuzzy model for deriving fuzzy stability criteria. A fuzzy version of Lyapunov Stability theory developed in [Deglas, 82] [Deglas, 84], is extended to define the fuzzy attraction domain of a fuzzy equilibrium set. This "fuzzification" of classical Lyapunov stability theory allows to remove the binary distinction between stability and instability. Then, the use of Matrix Inequalities allows to derive sufficient conditions to compute fuzzy attraction domain of a fuzzy equilibrium set based on quadratic Lyapunov function.

The article is organized as follows. In Sect. 2, we give some basic notation. In Sect. 3, the parameterization of a fuzzy systems is discussed (approximate reasoning methods, parameterization of fuzzy sets and fuzzy relations). In Sect. 4, the so-called α-Q-Stability is introduced in the context of Lyapunov Stability theory applied to a fuzzy dynamic system. The convex relaxation and the connection between the stability of fuzzy systems and real (crisp) systems is studied with respect to some classes of fuzzifiers and defuzzifiers. In Sect. 5, computationally tractable sufficient condition to check the α-Q-Stability of a fuzzy dynamic system is given. Section 6 is dedicated to practical examples. In Sect. 7, some concluding remarks and perspectives for future work are presented.

2 Notation

2.1 Conventional mathematical notation

$\mathcal{N}^{(*)}$ is the set of (strictly) positive natural numbers. $\Re^{+(*)}$ is the set of (strictly) positive real numbers. $\Re^n, \Re^{m\times n}$ are the sets of real k-vectors (column structure), real $m \times n$ matrices. M' is the transpose of a matrix. log(.) stands for the Neperian logarithms. exp (.) stands for the scalar exponential function. $\Lambda_i(P)$ stands for the i-th eigenvalue of the square matrix P. The Euclidean norm is denoted $\|\ .\ \|$. $\circ$ denotes the composition of mappings: given two compatible mappings A, and B, the mapping $C = A \circ B$ is defined as $\forall X, C(X) = A(B(X))$. Given a real vector $x \in \Re^n$, we denote $x_e = (x', 1)' \in \Re^{n+1}$. I_n is the identity matrix of rank n, $n \in \mathcal{N}^*$

2.2 Fuzzy and crisp sets

ϕ is the empty set. We will assume that $\sup_\phi(.) = 0$. The convex hull of a crisp set $\bar{\Omega}$ is denoted $\bar{Co}(\bar{\Omega})$

$\tilde{L}$ (resp. $\bar{L}$) is a multidimensional fuzzy (resp. crisp) set defined by its membership function $\mu_{\tilde{L}} : U_{\tilde{L}} \longrightarrow [0,1]$ (resp. $\mu_{\bar{L}} : U_{\bar{L}} \longrightarrow \{0,1\}$) where $U_{\tilde{L}}$ (resp. $U_{\bar{L}}$) is the universe of discourse of $\tilde{L}$ (resp. $\bar{L}$). $\mathcal{F}(U)$ (resp. $\mathcal{C}(U)$) denotes the set of fuzzy (resp. crisp) sets defined on the set U. The α-cut $\bar{L}^\alpha$ of $\tilde{L}$ is a crisp set defined on $U_{\tilde{L}}$ and specified by $\bar{L}^\alpha = \{x \in U_{\tilde{X}} : \mu_{\tilde{L}}(x) \geq \alpha\}, \alpha \in]0,1]$. We have, $\bar{L}^{\alpha_2} \subseteq \bar{L}^{\alpha_1}, \forall \alpha_1 \leq \alpha_2$.

A fuzzy relation $\tilde{R}$ is a multidimensional fuzzy set defined on the Cartesian product $U_{\tilde{R}} = U_{\tilde{Y}} \times U_{\tilde{X}}$.

$\underline{\tilde{L}}$ is multidimensional anti-fuzzy set defined by its membership function $\mu_{\underline{\tilde{L}}} : U_{\underline{\tilde{L}}} \longrightarrow [0,1]$ with α-cut satisfying : $\underline{\bar{L}}^{\alpha_2} \subseteq \underline{\bar{L}}^{\alpha_1}, \forall \alpha_1 \geq \alpha_2$ where the α-cut of anti-fuzzy sets are defined in the same way as fuzzy sets.

2.3 Operations on fuzzy sets

T denotes a T-conorm. S denotes an S-norm. $\neg$ stands for the complementary operator : $\mu_{\neg\tilde{A}}(u) + \mu_{\tilde{A}}(u) = 1$. $\cup$ (resp. $\cap$) stands for the union (resp. intersection) of crisp sets. $\cup^S$ (resp. $\cap^T$) stands for the union (resp. intersection) of (anti)-fuzzy sets : $\mu_{\tilde{A}\cup^S\tilde{B}}(u) = S(\mu_{\tilde{A}}(u), \mu_{\tilde{B}}(u))$ (resp. $\mu_{\tilde{A}\cap^T\tilde{B}}(u) = T(\mu_{\tilde{A}}(u), \mu_{\tilde{B}}(u))$). $\tilde{L}^e$ is the cylindric extension of $\tilde{L}$ on $U_{\tilde{Y}}$: $\mu_{\tilde{L}_i^e}(v,u) = \mu_{\tilde{L}_i}(u)$ with $U_{\tilde{L}_i^e} = U_{\tilde{Y}} \times U_{\tilde{L}}$.

The convex hull $\widetilde{\text{Co}}(\tilde{L})$ of the fuzzy set $\tilde{L}$ is a fuzzy set and is defined by $\forall \alpha \in]0,1], \bar{Co}^\alpha(\tilde{\Omega}) = \bar{Co}(\bar{\Omega}^\alpha)$. $\partial \bar{L}$ denotes the boundary of the set $\bar{L}$. We define the distance between a point x in $\Re^n$ and a set $\bar{S} \subset \Re^n$ by $d(x, \bar{S}) = \inf_{s\in\bar{S}}(\| x - s \|)$ and the separation of $\bar{X} \subset \Re^n$ from $\bar{S} \subset \Re^n$ by $d^*(\bar{X}, \bar{S}) = \sup_{x\in\bar{X}}(d(x, \bar{S}))$. X or $x \in \Re^n$ stands for the state of a dynamic system, $\dot{X}$ (resp. $\dot{x}$) stands for the time derivative of X (resp. x).

3 Quadratic parameterization of fuzzy systems

In this section, we present a new type of a fuzzy system which bridges the gap between linguistic and Takagi-Sugeno ones. First, the general structure of the model is presented. Various classes of approximate reasoning are defined according to the choice of implication and aggregation operators. The parameterization of the premise and conclusion parts of the fuzzy rules are given. Finally, the quadratic characterization of this type of a fuzzy system is described.

3.1 Generalities

A Fuzzy-Fuzzy mapping (or fuzzy system) $\Sigma_{FF} : \mathcal{F}(U_{\widetilde{X}}) \longrightarrow \mathcal{F}(U_{\widetilde{Y}})$ is defined by the 2-uplet $\{\widetilde{R}, \text{ T}\}$ where $\widetilde{R}$ is a fuzzy relation defined on $U_{\widetilde{Y}} \times U_{\widetilde{X}}$, T is a T-norm and the following inference mechanism

$$\widetilde{Y} = \widetilde{\Sigma_{FF}}(\widetilde{X}) \Longleftrightarrow \mu_{\widetilde{Y}}(y) = \sup_{x \in U_{\widetilde{X}}} T(\mu_{\widetilde{X}}(x), \mu_{\widetilde{R}}(y, x)).$$

In what follows, we will assume T=min and that the fuzzy relation $\widetilde{R}$ is defined by a linguistic rule base. A linguistic rule (denoted L-rule) is written in the form

$$\text{L-rule } i,\ i \in [1..n]\text{: If X is } \widetilde{L}_i \text{ then } \widetilde{Y} = \widetilde{K}_i(X), \tag{1}$$

where $\widetilde{L}_i$ is a multidimensional fuzzy set defined on $U_{\widetilde{X}}$ and $\widetilde{K}_i(X)$ is a multidimensional fuzzy set defined on $U_{\widetilde{Y}}$ and depends on the input X. The fuzzy relation $\widetilde{R}$ is given by

$$\mu_{\widetilde{R}}(y, x) = \text{agr}_{i=1..n}(\mu_{\widetilde{R}_i}(y, x)),$$

where

$$\mu_{\widetilde{R}_i}(y, x) = \text{imp}(\mu_{\widetilde{L}_i}(x), \mu_{\widetilde{K}_i}(y, x)),\ i \in [1..n].$$

In the above, *agr* stands for the aggregation operator (or "also" operator), *imp* denotes the fuzzy implication operator, and $\widetilde{R}_i$ is the fuzzy relation associated with the L-rule i. Consequently, the parameterization of Σ_{FF} is equivalent to : (i) the choice of *agr* and *imp* operators, (ii) the parameterization of the premise part of the L-rule i, i.e., the fuzzy sets $\widetilde{L}_i$) and (iii) the conclusion part of the L-rule i, i.e., the input-dependent fuzzy set $\widetilde{K}_i(X)$). Since we assume $T = min$, the parameterization of Σ_{FF} is achieved by the parameterization of $\widetilde{R}$ or, equivalently, $\bar{R}^\alpha, \alpha \in]0,1]$.

3.2 The implication and aggregation operators

The implication and aggregation operators are mutually connected and cannot be chosen independently. According, to the choice of these operators, we say that the nature of the fuzzy system is disjunctive or conjunctive. In this section, we define disjunctive and conjunctive fuzzy systems. Moreover, a result is given to handle complex structures of premise/conclusion part of the L-rules.

3.2.1 Disjunctive fuzzy systems

The disjunctive fuzzy systems are characterized by the use of a disjunctive aggregation operator and, consequently, a conjunctive implication operator. We choose

$$imp(a,b) = min(a,b), \quad agr(a,b) = max(a,b), \forall (a,b) \in [0,1] \times [0,1].$$

The min implication is known also as the Mamdani's implication.

3.2.2 Conjunctive fuzzy systems

The conjunctive fuzzy systems are characterized by the use of a conjunctive aggregation operatior and, consequently, a logical implication[1]. We choose

$$imp(a,b) = max(1-a,b), \quad agr(a,b) = min(a,b), \forall (a,b) \in [0,1] \times [0,1].$$

The latter implication is known as the Kleene's implication.

3.2.3 Complex premise/conclusion parts of rules

The following result provides a useful tool to simplify the premise and/or conclusion part of the L-rules.

Theorem 1 *Let the fuzzy systems* $\Sigma^i_{FF} = \{\widetilde{R}_{\Sigma^i_{FF}}, min\}$ *and* $\widetilde{R}_{\Sigma^i_{FF}}$ *(*$i = 1, 2$*) be defined by sets of L-rules as*

[1]A logical fuzzy implication is the generalization of the boolean implication.

- $\widetilde{R}_{\Sigma^1_{FF}}$: *L-rule i*, $i \in [1..n]$
  If X *is* $\cup^{max}_{j=1..n_{i,j}}\{\widetilde{L}_{i,j}\}$ *then* $\widetilde{Y} = \cap^{min}(resp.\cup^{max})_{k=[1..n_{i,k}]}\widetilde{K}_{i,k}(X)$.

- $\widetilde{R}_{\Sigma^2_{FF}}$: *L-rule (i,j,k)*, $i \in [1..n]$, $j \in [1..n_{i,j}]$, $k \in [1..n_{i,k}]$:
  If X *is* $\widetilde{L}_{i,j}$ *then* $\widetilde{Y} = \widetilde{K}_{i,k}(X)$.

If Σ^i_{FF}, $i = 1,2$ *have the same nature and are disjunctive (resp. conjunctive) then* $\widetilde{R}_{\Sigma^1_{FF}} = \widetilde{R}_{\Sigma^2_{FF}}$ *(equivalently* $\Sigma^1_{FF} = \Sigma^2_{FF}$*)* .

Remark 1 *Using the distributivity property of min – max operators, it can be shown that any "complex" combination* $\widetilde{F} = C^{max,min}_{l=1..m}\{\widetilde{F}_l\}$ *of union-intersection of fuzzy sets (possibly recursively defined) can be written as the union of intersection of a family* $\widetilde{F}_{p,q}$, $p \in [1..n_p]$, $q \in [1..n_q]$ *of fuzzy sets. The family* $\{\widetilde{F}_{p,q}\}$ *is constructed from the family* $\{\widetilde{F}_l\}$. *We have that* $\widetilde{F} = \cup^{max}_{p=1..n_p}\{\cap^{min}_{q=1..n_q}\{\widetilde{F}_{p,q}\}\}$. *As a consequence of theorem (1), we can assume, without loss of generality, that the fuzzy sets* $\widetilde{L}_i$ *in the premise part of the L-rule i are defined by the intersection of the fuzzy sets* $\widetilde{L}_{i,j}$, *i.e.,* $\widetilde{L}_i = \cap^{min}_{j=1..n_{i,j}}\{\widetilde{L}_{i,j}\}$. *This last form is the standard form of the premises for conjunctive and disjunctive systems. In the same way, we can assume, without loss of generality, that the fuzzy sets* $\widetilde{K}_i(X)$ *in the conclusion part of the L-rule i of a disjunctive (resp. conjunctive) system is defined by the intersection (resp. union) of the fuzzy sets* $\widetilde{K}_{i,k}$: $\widetilde{K}_i(X) = \cap^{min}$ *(resp.* $\cup^{min})_{k=1..n_{i,k}}$ $\{\widetilde{K}_{i,k}(X)\}$. *This last form is the standard form of the conclusions for disjunctive (resp. conjunctive) systems.*

3.3 The premise part of the rules

We saw, in section (3.2.3), that, without loss of generality, we can assume that the fuzzy set $\widetilde{L}_i$, $i \in [1..n]$ can be written as $\widetilde{L}_i = \cap^{\mathrm{T}}_{j=1..n_{i,j}}\{L_{i,j}\}$. In what follows, we assume that $U_{\widetilde{X}} = \Re^n$ and define the so-called Elementary-Quadratic-Fuzzy-Set $\widetilde{L}_{i,j}$, $i \in [1..n]$, $j \in [1..n_{i,j}]$ through the family of α-cuts by a single quadratic constraint. That is

$$\forall a \in]0,1],\ x \in \bar{L}^{\alpha}_{i,j} \Leftrightarrow x'_e F_{i,j}(\alpha) x_e \leq 0,\ i \in [1..n],\ j \in [1..n_{i,j}], \qquad (2)$$

where $F_{i,j}$:$]0,1] \longrightarrow \Re^{(n+1)\times(n+1)}$ and is assumed continuous. In what follows, we assume that $F_{i,j}(\alpha)$ is partitioned as

$$F_{i,j}(\alpha) = \begin{bmatrix} F^x_{i,j}(\alpha) & f_{i,j}(\alpha) \\ f_{i,j}(\alpha)' & f^0_{i,j}(\alpha) \end{bmatrix} \text{ and } F^x_{i,j}(\alpha) \in \Re^{n \times n}.$$

Remark 2 *It can be shown that $F_{i,j}(.)$ defines a fuzzy set $\widetilde{L}_{i,j}$ if and only if $F_{i,j}(.)$ is increasing on $]\alpha_1, \alpha_2]$ where*

$$\alpha_1 = inf_{\alpha \in]0,1], F_{i,j}(\alpha) \not< 0}(\alpha), \text{ and } \alpha_2 = sup_{\alpha \in]0,1], F_{i,j}(\alpha) \not> 0}(\alpha).$$

Moreover, if we impose $\widetilde{L}_{i,j}$ to be normalized, i.e., $\cup_{x \in U_{\widetilde{X}}} \{\mu_{\widetilde{L}_{i,j}}(x)\} =]0,1]$), we obtain $\alpha_1 = 0^+$, and $\alpha_2 = 1$.

Remark 3 *Common practical situation is when the membership function of a fuzzy set $\widetilde{L}_{i,j}$ is defined as*

$$\mu_{\widetilde{L}_{i,j}}(x) = f(-(x - x_{i,j})' P_{i,j}(x - x_{i,j})),$$

where

$$P_{i,j} = P'_{i,j} \in \Re^{n \times n}, x_{i,j} \in \Re^n, \; f : \Re \longrightarrow [0,1]$$

Then, we have, $F_{i,j}(\alpha) = F_{i,j} + F(\alpha)$ *with* $F_{i,j} = \begin{bmatrix} P_{i,j} & -P_{i,j}x_{i,j} \\ -x'_{i,j}P_{i,j} & x'_{i,j}P_{i,j}x_{i,j} \end{bmatrix}$ *and* $F(\alpha) = \begin{bmatrix} 0 & 0 \\ 0 & f^{-1}(\alpha) \end{bmatrix}$, *where* $f^{-1}(\alpha) = \{y \in \Re \text{ such that } f(y) = \alpha\}$.

Remark 4 *If $f = \exp$, $f^{-1} = \log$, $P_{i,j} = P'_{i,j} > 0$,, the latter parameterization defines radial basis function. In this case, the T-norm operator used in the relation $\widetilde{L}_i = \cap^T_{j=1..n_{i,j}} \{\widetilde{L}_{i,j}\}$ can also be chosen as the algebraic product, T=*. Then, $\bar{L}^\alpha_i$ is defined by a single quadratic constraint*

$$F_i(\alpha) = F_i + F(\alpha) \text{ with } F_i = \sum_{j=1..n_{i,j}} F_{i,j} \text{ and } F(\alpha) = \begin{bmatrix} 0 & 0 \\ 0 & \log(\alpha) \end{bmatrix}.$$

Remark 5 *If $f(y) = 1 - \sqrt{-y}$ if $x \geq -1$, 0 elsewhere. $f^{-1}(\alpha) = -(1-\alpha)^2$, $p > 0$, $P_{i,j} = P'_{i,j} > 0$, the latter parameterization defines conic membership function, e.g., triangular membership function in the one-dimensional case.*

Example 1 *Assume that the fuzzy sets $\widetilde{L}_i$ are defined, respectively, by the matrices $L_i(.)$, $i = 1, 2$ where*

$$L_1(\alpha) = \begin{bmatrix} P_1 & -x_1P_1 \\ -P_1x_1 & x_1P_1x_1 + \log(\alpha) \end{bmatrix},$$

and

$$L_2(\alpha) = \begin{bmatrix} P_2^2 & -x_2P_2 \\ -P_2x_2 & x_2P_2^2x_2 - (1-\alpha)^2 \end{bmatrix}.$$

The graphical representation of $\widetilde{L}_1$ *and* $\widetilde{L}_2$ *are shown in Fig. 1.*

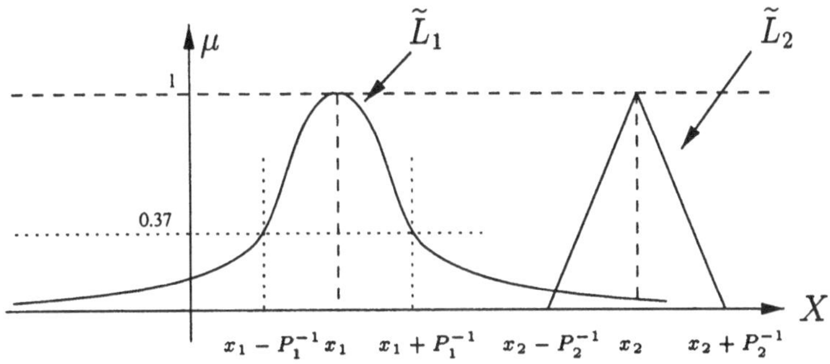

Figure 1: Graphical representation of standard fuzzy sets.

Remark 6 *The complementary fuzzy sets* $\neg\widetilde{L}_{i,j}$ *of* $\widetilde{L}_{i,j}$ *are defined as*

$$x \in \neg\bar{L}^{\alpha}_{i,j} \Leftrightarrow x'_e(-F_{i,j}(1-\alpha))x_e \leq 0.$$

Moreover, we have that $x \in \neg\bar{L}^{\alpha}_i \Leftrightarrow x \in \cup_{j=1..n_{i,j}}\{\neg\bar{L}^{\alpha}_{i,j}\}$. *Consequently, the proposed parameterization of fuzzy sets allows the treatment of the complementary operator.*

In what follows, we need to characterize the cylindric extension $\widetilde{L}^e_i$ on $U_{\widetilde{Y}}$ of a fuzzy set $\widetilde{L}_i$ defined on $U_{\widetilde{X}}$. We have

$$\forall a \in]0,1],\ u \in \bar{L}^{\alpha,e}_{i,j} \Leftrightarrow w'_eP'_eF_{i,j}(\alpha)P_ew_e \leq 0,\ i = 1..n,\ j = 1..n_{i,j}, \quad (3)$$

where $w = (y', x')'$ and the projection operator matrix P_e is defined by $P_ew_e = (n'_y, x', 1)'$ and n_y is the null vector defined on $U_{\widetilde{Y}} = \Re^m$.

3.4 The conclusion part of the rules

We saw that the conclusion part of the L-rule i is characterized by an input-dependent fuzzy set $\widetilde{K}_i(X)$ defined on $U_{\widetilde{Y}}$. In what follows, we assume $U_{\widetilde{Y}} = \Re^m$. Consequently, the problem of the parameterization of the conclusions has the same nature as the parameterization of the fuzzy system Σ_{FF}. This fact proves the recursive nature of fuzzy systems. According to theorem (1) and Remark (1), we assume that the conclusion part $\widetilde{K}_i(X)$ of the L-rule i is disjunctive (resp. conjunctive) and as such is defined by an intersection (resp. union) of input-dependent fuzzy sets $\widetilde{K}_{i,k}(X)$. Let us define the input-dependent fuzzy set $\widetilde{K}_{i,k}(X)$ by the relation

$$\mu_{\widetilde{K}_{i,k}(x)}(y) = \mu_{\widetilde{K}_{i,k}}(y,x), \forall (y,x) \in U_{\widetilde{Y}} \times U_{\widetilde{Y}}$$

Then it follows that the parameterization of the input-dependent fuzzy set $\widetilde{K}_{i,k}(x)$ is equivalent to the parameterization of a fuzzy relation $\widetilde{K}_{i,k}$ defined on $U_{\widetilde{Y}} \times U_{\widetilde{X}}$: $\{\widetilde{K}_{i,k}(X), X \in U_{\widetilde{X}}\} \Leftrightarrow \widetilde{K}_{i,k}$. In what follows, we use quadratic constraints to characterize the fuzzy relation $\widetilde{K}_{i,k}$. We will note that both TS conclusion and Mamdani's conclusions are a particular case of the proposed fuzzy relation parameterization.

A fuzzy relation $\widetilde{K}_{i,k}$ is defined in a similar way as the multidimensional fuzzy sets $\widetilde{L}_i$. However, the parameterization of such a fuzzy relation is a more complex and uses extensively the so-called extension principle [Zadeh, 75]. The construction of $\widetilde{K}_{i,k}(X)$ can be done as follows:

- We define $\widetilde{K}_{i,k,l}, l \in [1..n_{i,k,l}]$ with $U_{\widetilde{K}_{i,k,l}} = \Re^{q_{i,k,l}} \times \Re^{p_{i,k,l}}, p_{i,k,l} \in \mathcal{N}^*, q_{i,k,l} \in \mathcal{N}^*$ as

  $$(u_{i,k,l}, y_{i,k,l}) \in \bar{K}^{\alpha}_{i,k,l} \Leftrightarrow w_e' H_{i,k,l}(\alpha) w_e \leq 0,$$

  where $w = (u'_{i,k,l}, y'_{i,k,l})'$ and

  $$H_{i,k,l} :]0,1] \longrightarrow \Re^{(q_{i,k,l}+p_{i,k,l}+1)\times(q_{i,k,l}+p_{i,k,l}+1)},$$

  is an increasing function (see Remark (2), the generalization to fuzzy relation is straightforward). $\widetilde{K}_{i,k,l}$ is called an Elementary-Quadratic-Fuzzy-Relation. Expressed in a symbolic way, we have

  $$\widetilde{u}_{i,k,l} = \widetilde{K}_{i,k,l}(y_{i,k,l}).$$

- Then, we have

  $$\mu_{\widetilde{K}_{i,k}}(y,x) = \sup_{\substack{y = G_{i,k}x_e + \sum_{l=1..n_{i,k,l}} E_{i,k,l}u_{i,k,l} \\ y_{i,k,l} = D_{i,k,l}x \\ u_{i,k,l}}} T_{l=1..n_{i,k,l}}\{\mu_{\widetilde{K}_{i,k,l}}(u_{i,k,l}, y_{i,k,l}),\}$$

  with $G_{i,k} \in \Re^{m\times(n+1)}$, T =min, $E_{i,k,l} \in \Re^{m\times q_{i,k,l}}$, $D_{i,k,l} \in \Re^{p_{i,k,l}\times n}$.

In symbolic terms we obtain

$$\widetilde{K}_{i,k}(X) = G_{i,k}X_e + \sum_{l=1..n_{i,k}} E_{i,k,l}\widetilde{K}_{i,k,l}(D_{i,k,l}X).$$

Remark 7 *The latter form of conclusion is inspired from the representation of uncertain time-varying linear systems with structured uncertainties used in linear robust control community [Rotea, 93].* $\widetilde{K}_{i,k}(X)$ *can be interpreted as an Affine Time Invariant system with input X and output Y associated with an uncertain feedback connection between some fictitious outputs* $y_{i,k,l}$ *and fictitious inputs* $u_{i,k,l}$.

In what follows, we assume that $G_{i,k}$ and $H_{i,k,l}(\alpha)$ are partitioned as

$$G_{i,k} = \begin{bmatrix} G^x_{i,k} & g_{i,k} \end{bmatrix}, \; H_{i,k,l}(\alpha) = \begin{bmatrix} H^u_{i,k,l}(\alpha) & H^{uy}_{i,k,l}(\alpha) & h^u_{i,k,l}(\alpha) \\ H^{uy}_{i,k,l}(\alpha)' & H^y_{i,k,l}(\alpha) & h^y_{i,k,l}(\alpha) \\ h^u_{i,k,l}(\alpha)' & h^y_{i,k,l}(\alpha)' & h^0_{i,k,l}(\alpha) \end{bmatrix},$$

with

$$G^x_{i,k} \in \Re^{m\times n}, \; H^u_{i,k,l} :]0,1] \longrightarrow \Re^{q_{i,k,l}\times q_{i,k,l}}, \; H^y_{i,k,l} :]0,1] \longrightarrow \Re^{p_{i,k,l}\times p_{i,k,l}}.$$

Remark 8 *Special and common cases of matrices* $H_{i,k,l}(\alpha)$ *are given by*

$$H^b_{i,k,l}(\alpha) = \begin{bmatrix} I_{q_{i,k,l}} & 0 & 0 \\ 0 & -I_{p_{i,k,l}} & 0 \\ 0 & 0 & 0 \end{bmatrix} \text{ and } H^p_{i,k,l}(\alpha) = \begin{bmatrix} 0 & I_d & 0 \\ I_d & 0 & 0 \\ 0 & 0 & 0 \end{bmatrix}.$$

$H^b_{i,k,l}$ *defines bounded-real uncertainties and* $H^p_{i,k,l}$ *defines the positive real uncertainties (with* $d = q_{i,k,l} = p_{i,k,l}$*).*

Remark 4 With respect to the extension to a radial basis function, the use of algebraic product as T-norm and the complementary operator is still valid.

Remark 9 *The term* $G_{i,k}x_e$ *is a redundant term since it can be recasted as a singular quadratic fuzzy relation with*

$$D_{i,k,l} = I_n, \; E_{i,k,l} = I_m \text{ and } H_{i,k,l}(\alpha) = H_{i,k,,l} = \begin{bmatrix} I_m & -G_{i,k} \\ -G'_{i,k} & G'_{i,k}G_{i,k} \end{bmatrix}.$$

However, the use of singular quadratic constraint to model affine relation is not "natural". The use of the explicit affine relation $G_i x_e$ *seems to be preferable and increases the readability of the L-rule i.*

Remark 10 *If* $P_s' H_{i,k,l}(\alpha) P_s = H_{i,k,l}(\alpha), l \in [1..n_{i,k,k}],\ \alpha \in]0,1]$ *where the projection operator matrix* P_s *is defined by* $P_s w_e = (y', \underset{\sim}{n_x'}, 1)'$ *and* n_x *is the null vector defined on* $U_{\widetilde{X}} = \Re^n$, *then the fuzzy relation* $\widetilde{A}_{i,k}$ *degenerates into the cylindric extension of a fuzzy set* $\widetilde{B}_{i,k}$ *defined on* $U_{\widetilde{B}_{i,k}} = U_{\widetilde{Y}} = \Re^m$. *We have* $\mu_{\widetilde{A}_{i,k}}(y, x) = \mu_{\widetilde{B}_{i,k}}(y)$. *In this case, the L-rule i can be rewritten in the symbolic form*

$$\text{L-rule } i,\ i \in [1..n]\text{: If } X \text{ is } \widetilde{L}_i \text{ then } \widetilde{Y} = \widetilde{K}_{i,k}(X) = \widetilde{B}_{i,k}.$$

The above form of L-rule's conclusion is closely related to the Mamdani's conclusion rule and/or Takagi-Sugeno conclusion rule (where $\widetilde{B}_i$ *is a fuzzy singleton).*

If $n_{i,k} = 1$ *and* $n_{i,k,l} = 0$, *the conclusion part of the rule can be written*

$$\text{L-rule } i,\ i \in [1..n]\text{: If } X \text{ is } \widetilde{L}_i \text{ then } Y = G_{i,k}X + g_{i,k}.$$

The above form of L-rule's conclusion is closely related to the Takagi-Sugeno rule with an affine conclusion.

Example 2 *The friction, in a mechanical systems, is generally imperfectly known. However, for small positive velocity, it can be approximated by an affine relation such as*

$$\widetilde{f} = -\widetilde{f}_d - \widetilde{f}_v \dot{x}$$

where $\widetilde{f}$ *is the fuzzy uncertain friction,* $\widetilde{f}_d$ *the fuzzy uncertain dry friction,* f_v *the fuzzy uncertain viscosity coefficient friction, and* $\dot{x}$ *the velocity. The (fuzzy) uncertainties on* f_d *and* f_v *should be considered as correlated. A possible fuzzy model of* $\widetilde{f}$ *is given by*

$$\widetilde{f} = [-f_v \ \ -f_d] \begin{pmatrix} \dot{x} \\ 1 \end{pmatrix} + \widetilde{K}(\dot{x}).$$

with

$$(f, \dot{x})' \in \bar{K}^\alpha \Leftrightarrow$$

$$\begin{pmatrix} f \\ \dot{x} \\ 1 \end{pmatrix}' \begin{bmatrix} 1 & 0 & 0 \\ 0 & -(1-\alpha)^2 \Delta f_v^2 & 0 \\ 0 & 0 & -(1-\alpha)^2 \Delta f_d^2 \end{bmatrix} \begin{pmatrix} f \\ \dot{x} \\ 1 \end{pmatrix} \leq 0.$$

Δf_d *and* Δf_v *are, respectively, the maximum (uncorrelated) uncertainty on the dry friction and the viscosity coefficient friction. The graphical representation of* $\widetilde{f}$ *is given in Fig. 2.*

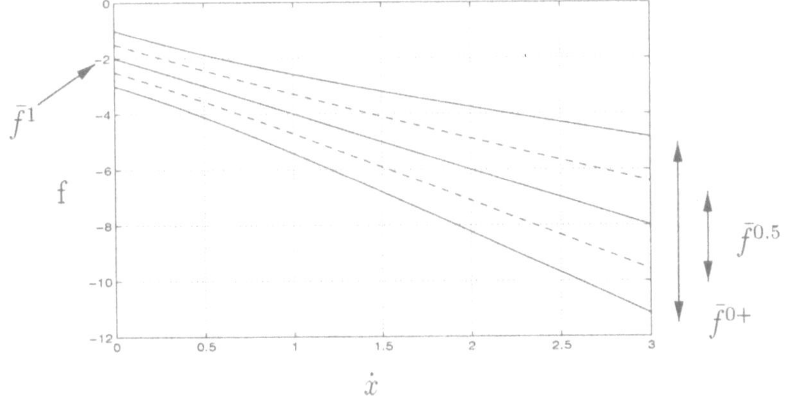

Figure 2: Graphical representation of $\widetilde{f}$.

3.5 The standard form of $\widetilde{R}$

It can be shown that the fuzzy relation $\widetilde{R}$, defined by a set of L-rules, of disjunctive and conjunctive fuzzy systems $\Sigma_{FF} : \mathcal{F}(U_{\widetilde{X}}) \longrightarrow \mathcal{F}(U_{\widetilde{Y}})$ can be written in a unified form called the standard form (SF). The standard form of $\widetilde{R}$ (or equivalently, $\bar{R}^\alpha, \alpha \in]0,1]$) for disjunctive and conjunctive fuzzy systems is given by

$$\text{SF} \; : \bar{R}^\alpha = \cup_{i=1..n}\{\bar{R}_i^\alpha\}, \tag{4}$$

with $\bar{R}_i^\alpha = \cap_{k=1..n_{i,k}}\{\bar{R}_{i,k}^\alpha\}$ and $\bar{R}_{i,k}^\alpha = \cap_{j=1..n_{i,j}}\{\bar{L}_{i,j}^{\alpha,e}\} \cap \bar{K}_{i,k}^\alpha$. $\widetilde{R}_i$ is the fuzzy relation associated with the so-called generalized mathematical rule i (denoted M-rule i). $\widetilde{R}_{i,k}$ is the fuzzy relation associated with the M-rule (i,k).

Remark 11 *For disjunctive systems, the generalized M-rule $\widetilde{R}_i$ is the fuzzy relation associated with the L-rule i. Since, generally, $\forall i \in [1..n]$, $n_{i,k} = 1$, disjunctive systems do not exhibit the generalized M-rule but only M-rules. For conjunctive systems, obtaining the standard form of $\widetilde{R}$ is not straightforward and the use of the distributivity properties of the min-max operators is necessary. Consequently, there is no longer correspondence between L-rules and M-rules and, generally, conjunctive systems exhibit generalized M-rules.*

3.6 Quadratic parameterization of $\widetilde{R}^\alpha$

In this section, we develop an exact analytical parameterization of $\bar{R}^\alpha$ in a compact and general form, using quadratic constraints.

In what follows, we denote $t^i = (x', u'_{i,1}, \dots u'_{i,k} \dots u'_{i,n_{i,k}})'$, $i \in [1..n]$ where
$u_{i,k} = (u'_{i,k,1}, \dots u'_{i,k,l} \dots u'_{i,k,n_{i,k,l}})'$, $k \in [1..n_{i,k}]$, $l \in [1..n_{i,k,l}]$.

We have $t^i \in \Re^{d_i}$ with $d_i = n + \sum_{k=1..n_{i,k}} \sum_{l=1..n_{i,k,l}} p_{i,k,l}$.

From (4), one obtains

$$(y,x) \in \bar{R}^{\alpha} \Leftrightarrow \exists\, i \in [1..n] \text{ such that } \forall k \in [1..n_{i,k}], \forall j \in [1..n_{i,j}],$$
$$(y,x) \in \cap_{j=1..n_{i,j}} \{\bar{L}^{\alpha,e}_{i,j}\} \cap_{k=1..n_{i,k}} \{\bar{K}^{\alpha}_{i,k}\}.$$

Then, a number of equivalences can be shown. First we have that

$$(y,x) \in \bar{L}^{\alpha,e}_{i,j} \Leftrightarrow t_i \in \Omega^{\alpha}_{i,j} \quad \text{with} \quad \Omega^{\alpha}_{i,j} = \{t_i \in \Re^{d_i} : t^{i\,\prime}_e C_{i,j}(\alpha) t^i_e \leq 0\}$$

where

$$C_{i,j} :]0,1] \to \Re^{(d_i+1)\times(d_i+1)}, \quad C_{i,j}(\alpha) = \begin{bmatrix} F^{x}_{i,j}(\alpha) & 0 & f_{i,j}(\alpha) \\ 0 & 0 & 0 \\ f'_{i,j}(\alpha) & 0 & f^{0}_{i,j}(\alpha) \end{bmatrix}.$$

Second, $(y,x) \in \bar{K}^{\alpha}_{i,k} \Leftrightarrow t_i \in \Omega^{\alpha}_{i,k}$ where

$$\Omega^{\alpha}_{i,k} = \{t_i : t^{i\,\prime}_e C_{i,k,l}(\alpha) t^i_e \leq 0, y = G_{i,k} x_e + \sum_{l=1..n_{i,k,l}} E_{i,k,l} u_{i,k,l},\ l \in [1..n_{i,k,l}]\}$$

and $C_{i,k,l}(\alpha) =$

$$\begin{bmatrix} D'_{i,k,l} H^{y}_{i,k,l}(\alpha) D_{i,k,l} & \begin{bmatrix} 0 & \dots & C^{y}_{i,k,l}(\alpha) & \dots & 0 \end{bmatrix} & D'_{i,k,l} h^{y}_{i,k,l} \\ \begin{bmatrix} 0 \\ \vdots \\ C^{y\,\prime}_{i,k,l}(\alpha) \\ \vdots \\ 0 \end{bmatrix} & \begin{bmatrix} 0 & \dots & \dots & \dots & 0 \\ \vdots & \ddots & & & \vdots \\ 0 & \dots & C^{u}_{i,k,l}(\alpha) & \dots & 0 \\ \vdots & & & \ddots & \vdots \\ 0 & \dots & \dots & \dots & 0 \end{bmatrix} & \begin{bmatrix} 0 \\ \vdots \\ c^{u}_{i,k,l}(\alpha) \\ \vdots \\ 0 \end{bmatrix} \\ h^{y\,\prime}_{i,k,l}(\alpha) D_{i,k,l} & \begin{bmatrix} 0 & \dots & c^{u\,\prime}_{i,k,l}(\alpha) & \dots & 0 \end{bmatrix} & h^{0}_{i,k,l}(\alpha) \end{bmatrix}$$

$$C^{u}_{i,k,l}(\alpha) = \begin{bmatrix} 0 & \dots & \dots & \dots & 0 \\ \vdots & \ddots & & & \vdots \\ 0 & \dots & H^{u}_{i,k,l}(\alpha) & \dots & 0 \\ \vdots & & & \ddots & \vdots \\ 0 & \dots & \dots & \dots & 0 \end{bmatrix} \text{ and } c^{u}_{i,k,l} = \begin{bmatrix} 0 \\ \vdots \\ h^{u}_{i,k,l}(\alpha) \\ \vdots \\ 0 \end{bmatrix}$$

and $C^{y}_{i,k,l}(\alpha) = \begin{bmatrix} 0 & \dots H^{y}_{i,k,l}(\alpha) D_{i,k,l} & \dots & 0 \end{bmatrix}$.

Using the above notation, the quadratic parameterization of $\bar{R}^\alpha$ can be written as

$$(y,x) \in \bar{R}^\alpha \Leftrightarrow \exists\ i \in [1..n], t_i \in \Omega_i^\alpha, \tag{5}$$

where $\Omega_i^\alpha = \Omega_{i,p}^\alpha \cap \Omega_{i,c}^\alpha$, $\Omega_{i,p}^\alpha = \cap_{j=1..n_{i,j}}\{\Omega_{i,j}^\alpha\}$ and $\Omega_{i,c}^\alpha = \cap_{k=1..n_{i,k}}\{\Omega_{i,k}^\alpha\}$.

Remark 12 *The expression (5) is called the quadratic parameterization of $\widetilde{R}$ since the sets Ω_i^α are the conjunction of quadratic constraints on the family of vectors $t^i, i \in [1..n]$.*

Remark 13 *It can be shown that a fuzzy system*

$$\Sigma_{FF} = \{\widetilde{R}, min\} : \mathcal{F}(U_{\widetilde{X}}) \to \mathcal{F}(U_{\widetilde{Y}}),$$

is "equivalent" to a family of crisp systems

$$\Sigma_{CC}^\alpha : \mathcal{C}(U_{\widetilde{X}}) \to \mathcal{C}(U_{\widetilde{Y}}), \alpha \in]0,1],$$

(i.e., $\widetilde{Y} = \widetilde{\Sigma_{FF}}(\widetilde{X}) \Leftrightarrow \bar{Y}^\alpha = \bar{\Sigma}_{CC}^\alpha(\bar{Y}^\alpha), \forall \widetilde{X} \in \mathcal{F}(U_{\widetilde{X}})$. It can also be shown that Σ_{CC}^α is defined by the singleton $\{\bar{R}^\alpha\}$),

4 α-Q-Stability of fuzzy dynamic systems

In Sect. 3, we saw that a wide family of fuzzy systems can be parameterized as a set of quadratic constraints. If a given property of a fuzzy system can be expressed as a set of quadratic constraints on the input and output vectors, the use of the so-called S-procedure [Boyd, 94] provides a systematic way to derive (generally sufficient) conditions to check the desired property. Applying this idea to fuzzy dynamic systems, we define the so-called α-Q-Stability. This property allows to unify various concepts of Local/Global, Robust/Nonrobust stability. This in turn allows to evaluate in a flexible way whether a given fuzzy dynamic system exhibits an acceptable dynamic behavior. First, fuzzy dynamic systems are presented. Then, the definition of a fuzzy Lyapunov function and its variation is introduced. Finally, α-Q-Stability of fuzzy dynamic systems is defined. It is shown that this property can be expressed as a set of quadratic constraints on the input and output vectors. Finally, the corresponding sufficient conditions are presented.

4.1 Fuzzy dynamic systems

We start with the following definition.

Definition 1 *A Fuzzy System* $\Sigma_{FF}, \mathcal{F}(I) \rightarrow \mathcal{F}(O) = \{\widetilde{R}, T\}$, *with* $U_{\widetilde{R}} = U_{\widetilde{O}} \times U_{\widetilde{I}}$ *is said to be a Fuzzy Autonomous Dynamical System (FADS) if and only if* $U_{\widetilde{X}} \subseteq U_{\widetilde{I}}$, $U_{\dot{\widetilde{X}}} \subseteq U_{\widetilde{O}}$ *and* $\{U_{\widetilde{I}} - U_{\widetilde{X}}\} = \{U_{\widetilde{O}} - U_{\dot{\widetilde{X}}}\}$. *Moreover,* $\widetilde{R}$ *is a Minimal fuzzy relation of the FADS* Σ_{FF} *if and only if* $\{U_{\widetilde{I}} - U_{\widetilde{X}}\} = \{U_{\widetilde{O}} - U_{\dot{\widetilde{X}}}\} = \Phi$.

In what follows, we assume Σ_{FF} is FADS and $\widetilde{R}$ is a Minimal fuzzy relation. Consequently, $U_{\widetilde{R}} = \{U_{\dot{\widetilde{X}}}\} \times \{U_{\widetilde{X}}\}$. Equivalently, $I = x$ and $O = \dot{x}$. Moreover, we assume $T = \min$.

The FADS Σ_{FF} is governed by the differential fuzzy equation

$$\forall t \in \Re^+, \forall \widetilde{X} \in \mathcal{F}(U_{\widetilde{X}}),\ \dot{\widetilde{X}}(t) = \widetilde{\Sigma}_{FF}(\widetilde{X}(t)),\ \widetilde{X}(0) = \widetilde{X}_0. \tag{6}$$

According to Remark (13), the FADS (6) is governed by a family of differential crisp equations

$$\forall t \in \Re^+,\ \forall \bar{X} \in \mathcal{C}(U_{\widetilde{X}}),\ \forall \alpha \in]0,1],\ \dot{\bar{X}}^{\alpha}(t) = \bar{\Sigma}^{\alpha}_{CC}(\bar{X}^{\alpha}(t)),\ \bar{X}^{\alpha}(0) = \bar{X}^{\alpha}_0. \tag{7}$$

Remark 14 *We call* α*-trajectories a possible solution of the equation*

$$\dot{X}(t) = \Sigma^{\alpha}_{RR}(X(t)), X(0) = X_0,$$

where $\Sigma^{\alpha}_{RR} : U_{\widetilde{X}} \longrightarrow U_{\dot{\widetilde{X}}}$ *is the real-to-real mapping with*

$$\Sigma^{\alpha}_{RR} = Def^{\alpha} \circ \bar{\Sigma}^{\alpha}_{CC} \circ Sing,$$

where $Sing : U_{\widetilde{X}} \rightarrow \mathcal{F}(U_{\widetilde{X}})$ *is the singleton fuzzifier and* $Def^{\alpha} : \mathcal{F}(U_{\dot{\widetilde{X}}}) \rightarrow U_{\dot{\widetilde{X}}}$ *is an unknown defuzzifier satisfying (P1) :* $Def^{\alpha}(\dot{\widetilde{X}}) \in \dot{\bar{X}}^{\alpha}, \forall \dot{\widetilde{X}} \in \mathcal{F}(U_{\dot{\widetilde{X}}})$.

4.2 Fuzzy Lyapunov function and its time derivative

In what follows, $\bar{E}^{\alpha} \subset U_{\widetilde{X}}$ and $\bar{A}^{\alpha} \subset U_{\widetilde{X}}$ are assumed to be compact and convex such that $\bar{N}^{\alpha} = \bar{A}^{\alpha} \cap \neg \bar{E}^{\alpha}$ is nonempty.

We denote $\overline{\partial}\bar{N}^{\alpha} = \partial \bar{A}^{\alpha}$ and $\underline{\partial}\bar{N}^{\alpha} = \partial \bar{I}^{\alpha}$. The convexity assumption with respect to $\bar{A}^{\alpha}$ and $\bar{I}^{\alpha}$ is sometimes too conservative and can be relaxed. However, it fits our requirements for the α-Q-Stability purpose, without any loss of generality.

Definition 2 *A real-valued function* $V : U_{\widetilde{X}} \longrightarrow R^+$, *belongs to the class* $\mathcal{ES}_{\bar{N}^\alpha}$ *if and only if : (L1) : V is defined and continuously differentiable in* $\bar{N}^\alpha$, *(L2) :* $V(x) = V_1^\alpha$, $V_1^\alpha > 0$, $\forall x \in \overline{\partial}\bar{N}^\alpha$, *(L3) : there exists* $p \geq 1$, $\underline{v}, \overline{v} > 0$ *and* $V_0^\alpha \geq 0$ *such that* $\underline{v} * (d(x, \bar{E}^\alpha))^p \leq V(x) - V_0^\alpha \leq \overline{v} * (d(x, \bar{E}^\alpha))^p, \forall x \in \bar{N}^\alpha$.

We define the fuzzy variation of $V \in \mathcal{ES}_{\bar{N}^\alpha}$ denoted $\widetilde{D}V : U_{\widetilde{X}} \longrightarrow \mathcal{F}(\Re)$ along the FADS (6), using the extension principle, as

$$\widetilde{D}V(x) = [\langle \nabla V(x), \dot{x} \rangle \ : \dot{x} = \widetilde{R}(x)].$$

The α-cuts $\bar{D}^\alpha V$ of $\widetilde{D}V$ are defined in a similar way replacing $\widetilde{R}(.)$ by $\bar{R}^\alpha(.)$. In what follows, we denote $D^\alpha V(x) = \sup_{v \in \bar{D}^\alpha V(x)}(v)$.

Remark 15 *If a dynamic system is defined by a real-to-real mapping* Σ_{RR}, *we can also associate with it a Lyapunov function* $V \in \mathcal{ES}_{\bar{N}^\alpha}$. *In this case, the Lyapunov variation along* Σ_{RR} *is a real-valued function and is defined by*

$$DV(x) = [\langle \nabla V(x), \dot{x} \rangle \ : \dot{x} = \Sigma_{RR}(x)]$$

4.3 α-Q-Stability of Fuzzy Dynamical Systems

Definition 3 *The FADS (6) is* α*-Q-Stable in* $\bar{N}^\alpha$ *if and only if* $\exists\ V \in \mathcal{ES}_{\bar{N}^\alpha}$, $\epsilon_\alpha \geq 0$ *such that*

$$\forall x \in \bar{N}^\alpha,\ D^\alpha V(x) + \epsilon_\alpha (V(x) - V_0^\alpha) \leq 0 \ \textit{where}\ V(x) = x'Px. \quad (8)$$

The physical interpretation of definition (3) is given by the corollary below.

Corollary 1 *If the FADS (6) is* α*-Q-Stable in* $\bar{N}^\alpha$ *then*

$$\forall \bar{x}_0 \subseteq \bar{N}^\alpha, sup_{x \in \bar{x}_0}(d^*(\bar{x}^\alpha(t), \bar{E}^\alpha)) \leq ad^*(\bar{x}_0, \bar{E}^\alpha)\exp(-b_\alpha t), \forall t \in \Re^+, \quad (9)$$

with $a = [\frac{\overline{v}}{\underline{v}}]^{\frac{1}{2}}$, $b_\alpha = \frac{\epsilon_\alpha}{2\overline{v}}$, $\underline{v} = min_i(\Lambda_i(P))$, $\overline{v} = max_i(\Lambda_i(P))$.

The above corollary shows that $\bar{E}^\alpha$ is classically stable [Vidyasagar, 92] along the α-trajectories of the FADS (6). Also if, $\epsilon_\alpha > 0$, then $\bar{E}^\alpha$ is exponentially-stable for any uncertain initial condition $\bar{x}_0^\alpha \in \bar{A}^\alpha$.

Note that the quadratic Lyapunov function $V(x) = x'Px, P = P' > 0$ belongs to $\mathcal{ES}_{\bar{N}^\alpha}, \forall \bar{N}^\alpha \subset \Re^n$. (L3) is satisfied with $\| . \|$: being the Euclidean norm. The set $\bar{E}^\alpha$ (resp. $\bar{A}^\alpha$) is defined by $x \in \bar{E}^\alpha \Leftrightarrow x'Px \leq V_0^\alpha$ (resp. $x \in \bar{A}^\alpha \Leftrightarrow x'Px \leq V_1^\alpha$).

Remark 16 *It can be shown that the family $\bar{E}^\alpha, \alpha \in]0;1]$ defines the fuzzy equilibrium set $\widetilde{E}$. The family $\bar{A}^\alpha$ defines the anti-fuzzy attractive set $\underset{\sim}{A}$.*

The concept of α-Q-Stability of $\bar{N}^\alpha$ removes the clear and drastic distinction between classical local/global stability of an equilibrium point and its instability counterpart for several reasons.

- Evaluation of the so-called Space-performances:
  - The knowledge of $\bar{A}^\alpha$ addresses the local/global property of the invariant set $\bar{E}^\alpha$. Recall that for practical purposes, local stability is not sufficient and global stability is not necessary. A practicing control engineer can evaluate if $\bar{A}^\alpha$ is sufficiently large or not. Moreover, for any initial condition in $\bar{A}^\alpha$, $x(t)$ remains in $\bar{A}^\alpha$.
  - The knowledge of $\bar{E}^\alpha$ addresses the problem of output (or state) precision of the system as $t \longrightarrow \infty$. Recall that for practical purposes, the asymptotic stability of an equilibrium point is not necessary and the asymptotic stability of a too large set $\bar{E}^\alpha$ is unacceptable (e.g., large oscillations). A practicing control engineer can evaluate if $\bar{E}^\alpha$ is sufficiently small or not.
  - The linguistic meaning of $\widetilde{E}$ (resp. $\underset{\sim}{A}$) is the following: $\mu_{\neg\widetilde{E}}(x)$ (resp. $\mu_{\neg\underset{\sim}{A}}(x)$) represents the possibility of x to belong to the equilibrium set (resp. to the attraction region of equilibrium set) of the fuzzy dynamic system (6).
- Evaluation of the so-called Time-performance: the knowledge of the convergent rate ϵ_α allows a practicing control engineer to know if the system is fast enough or not.
- Evaluation of the so-called Safety-Performance: all previous types of performance depend on α which can be considered as the credibility of the model. That is, $\alpha = 1$ gives an accurate, but not reliable model of the system and $\alpha = 0^+$ gives a reliable, but imprecise model. Thus, α addresses the trade-off Precision vs. Reliability of the model and, consequently, the trade-off Robustness vs. Performance of the corresponding dynamic system.

Remark 17 *Definition (3) and (1) admit the real Quadratic-Stability counterpart when the FDS (6) is replaced by a dynamic system defined by the differential equation $\dot{X}(t) = \Sigma_{RR}(X(t))$, $X(0) = X_0$ where Σ_{RR} is a real-real mapping. In that case, the α-trajectories, the sets $\bar{E}^\alpha$ and $\bar{A}^\alpha$, the reals $b_\alpha, \epsilon_\alpha$ are independent of α. The Lyapunov function variation $D^\alpha V$ must be replaced by the expression given in Remark (15).*

4.4 The convex relaxation and its application

An important result is given by the following theorem.

Theorem 2 *Consider as given FADS*

$$\Sigma^1_{FF} = \{\widetilde{R}, min\},$$

and FADS

$$\Sigma^2_{FF} = \{\widetilde{Co}(\widetilde{R}), min\}.$$

Then, Σ^1_{FF} is α-Q-Stable in $\bar{N}^\alpha$ if and only if Σ^2_{FF} is α-Q-Stable in $\bar{N}^\alpha$.

The first consequence of theorem (2) is given by the corollary below.

Corollary 2 *Define the FADS*

$$\Sigma^1_{FF} = \{\widetilde{R}_{\Sigma^1_{FF}}, min\},$$

and

$$\Sigma^2_{FF} = \{\widetilde{R}_{\Sigma^2_{FF}}, min\},$$

where $\widetilde{R}_{\Sigma^1_{FF}}$ and $\widetilde{R}_{\Sigma^2_{FF}}$ are minimal and defined by the sets of L-rules

- $\widetilde{R}_{\Sigma^1_{FF}}$ *: L-rulep, $p = 1..n$,*

$$\text{If } \widetilde{X} \text{ is } \widetilde{L}_p \text{ then } \dot{\widetilde{X}} = \widetilde{Co}(\cup_{q\in[1..n_p]}\{\widetilde{K}_{p,q}(X)\}),$$

- $\widetilde{R}_{\Sigma^2_{FF}}$ *: L-rule(p, q), $p = 1..n$, $q = 1..n_q$,*

$$\text{If } \widetilde{X} \text{ is } \widetilde{L}_p \text{ then } \dot{\widetilde{X}} = \widetilde{K}_{p,q}(X).$$

Assume that $\Sigma^i_{FF}, i = 1, 2$ have the same nature. Then, Σ^1_{FF} is α-Q-Stable in $\bar{N}^\alpha$ if and only if Σ^2_{FF} is α-Q-Stable in $\bar{N}^\alpha$.

Corollary (2) is relevant with respect to studying the stability ofa ystem of the form Σ^1_{FF}. As a matter of fact, it is shown that such systems can be transformed into an augmented (in terms of the number of L-rule) fuzzy dynamic system Σ^2_{FF}. Although, Σ^1_{FF} and Σ^2_{FF} are not equivalent (i.e., $\widetilde{R}_{\Sigma^1_{FF}} \neq \widetilde{R}_{\Sigma^2_{FF}}$), their α-Q-Stability properties are equivalent.

4.5 α-Q-Stability of fuzzy systems vs. Quadratic-Stability of real systems

Sometimes, a fuzzy-to-fuzzy mapping $\Sigma_{FF} = \{\widetilde{R}, \mathrm{T}\}$ is associated with a fuzzifier Fuz and a defuzzifier Defuz used to parameterize real-to-real mappings $\Sigma_{RR} = \mathrm{Def} \circ \Sigma_{FF} \circ \mathrm{Fuz}$. The question is: Can we deduce a stability property of the trajectories generated by the dynamic system associated with Σ_{RR} from the α-Q-Stability of the fuzzy dynamic system associated with Σ_{FF} ? The correspondence between α-Q-Stability of Σ_{FF} and Quadratic-Stability of Σ_{RR} is given in Remark (17). A partial answer is given by the following corollary.

Corollary 3 *Consider a fuzzy system $\Sigma_{FF} = \{\widetilde{R}, min\}$ and a real system $\Sigma_{RR} = Def \circ \Sigma_{FF} \circ Fuz$. Then, Σ^1_{FF} is α-Q-Stable in $\bar{N}^\alpha$ if and only if Σ_{RR} is Quadratically-Stable in $\bar{N}^\alpha$ where Fuz is the Singleton fuzzifier and Defuz satisfies (P2): $Def^\alpha(\dot{\tilde{X}}) \in \bar{Co}(\dot{\tilde{X}}^\alpha)$, $\forall \dot{\tilde{X}} \in \mathcal{F}(U_{\dot{\tilde{X}}})$.*

Remark 18 *In corollary (3), the defuzzifier can be any defuzzifier satisfying (P2). Consequently, there may exist a defuzzifier satisfying (P2) such that Σ_{RR} is Quadratically-Stable in $\bar{N}^{alpha}$ whereas the fuzzy dynamic system Σ_{FF} is not α-Q-Stable in $\bar{N}^\alpha$. In this case, there exists α-trajectories of Σ_{FF} which do not satisfy all conditions of definition (3).*

5 Necessary and sufficient condition for α-Q-Stability

In this section, we give a necessary and sufficient condition for the α-Q-Stability of the system (6). First, the Pseudo-α-Q-Stability of a generalized M-rule is defined and a sufficient condition to check Pseudo-α-Q-Stability of a generalized M-rule is presented. Finally, we use the Pseudo-α-Q-Stability of a generalized M-rule to derive a necessary and sufficient condition for α-Q-Stability of the system (6). In what follows, we assume that the fuzzy relation $\widetilde{R}$ of the system (6) is written in the standard form (SF) and is minimal.

5.1 Pseudo-α-Q-Stability of a generalized M-rule

We define the so-called pseudo-α-Quadratic-Stability of a generalized M-rule i in the following way.

Definition 4 *The generalized M-rule i is Pseudo-α-Q-Stable in $\bar{N}^\alpha$ with respect the matrix*

$$P = P' \in \Re^{n\times n},\ P > 0,$$

and

$$\bar{N}^\alpha \subset \Re^n = \{x : -x'Px + V_0^\alpha \leq 0,\ x'Px - V_1^\alpha \leq 0\},$$

if and only if $\exists\ \epsilon_\alpha \geq 0$ *such that*

$$\forall x \in \bar{N}^\alpha,\ (\dot{x}, x) \in \bar{R}_i^\alpha,\ \langle \nabla V(x), \dot{x}\rangle + \epsilon_{\alpha,\alpha}(V(x) - V_0^\alpha) \leq 0, \tag{10}$$

where $V(x) = x'Px$ and $\widetilde{R}_i$ is the fuzzy relation associated with the generalized M-rule i.

Remark 19 *Note that, necessarily, $V_1^\alpha > V_0^\alpha \geq 0$ and $x \in \bar{N}^\alpha \Leftrightarrow V_1^\alpha \geq V(x) \geq V_0^\alpha$. This condition is deduced from assumption (L3).*

5.2 Sufficient condition for Pseudo-α-Q-Stability via Matrix Inequalities

In this section, we use Matrix Inequalities to derive a sufficient condition to check the Pseudo-α-Q-Stability of a generalized M-rule. Let us introduce first some notation.

- $DV_{i,k}(\dot{x}, x, P, \epsilon_\alpha) = 2x'P\dot{x} + \epsilon_\alpha(x'Px - V_0^\alpha)$.

  Using matrix notation and the quadratic parameterization of fuzzy systems presented in Sect. 3.6, we have

  $$DV_{i,k}(\dot{x}, x, P, , \epsilon_\alpha) = t_e^{i\prime} LY_{i,k}(P, \epsilon_\alpha) t_e^i,$$

  with

  $$DV_{i,k}(P, \epsilon_\alpha) =$$

  $$\begin{bmatrix} A_{i,k}'P + PA_{i,k} + \epsilon_\alpha P & \begin{bmatrix} 0 & \dots & M_{i,k} & \dots & 0 \end{bmatrix} & Pg_{i,k} \\ \begin{bmatrix} 0 \\ \vdots \\ M_{i,k} \\ \vdots \\ 0 \end{bmatrix} & \begin{bmatrix} 0 & \dots & \dots & \dots & 0 \\ \vdots & \dots & \dots & \dots & \vdots \\ 0 & \dots & 0 & \dots & 0 \\ \vdots & \dots & \dots & \dots & \vdots \\ 0 & \dots & \dots & \dots & 0 \end{bmatrix} & \begin{bmatrix} 0 \\ \vdots \\ \vdots \\ 0 \end{bmatrix} \\ g_{i,k}'P & \begin{bmatrix} 0 & \dots & \dots & \dots & 0 \end{bmatrix} & -\epsilon_\alpha V_0^\alpha \end{bmatrix}.$$

where $M_{i,k} = P[E_{i,k,1} \ldots E_{i,k,n_{i,k}}]$ and $LY_{i,k}(P,\epsilon_\alpha) \in \Re^{(d_i+1)\times(d_i+1)}$.

- $x \in \bar{N}^\alpha \Leftrightarrow x \in \bar{E}^\alpha$, $\quad x \in \bar{A}^\alpha \Leftrightarrow t_e^{i\,\prime} C_0(P,V_0^\alpha) t_e^i \leq 0$, $t_e^{i\,\prime} C_1(P,V_1^\alpha) t_i^e \leq 0$ with

$$C_0(P,V_0^\alpha) = \begin{bmatrix} -P & 0 & 0 \\ 0 & 0 & 0 \\ 0 & 0 & V_0^\alpha \end{bmatrix}, \quad C_1(P,V_1^\alpha) = \begin{bmatrix} P & 0 & 0 \\ 0 & 0 & 0 \\ 0 & 0 & -V_1^\alpha \end{bmatrix},$$

and $C_r(P,V_r^\alpha) \in \Re^{(d_i+1)\times(d_i+1)}$, $r = 0,1$.

Now we can state the following theorem.

Theorem 3 *If* $\exists\ \tau_{i,k}^l \geq 0,\ \tau_{i,j}^p \geq 0, \tau_{i,k,l}^c \geq 0,\ \tau_{i,r}^d \geq 0,\ t_i \geq 0,\ j \in [1..n_{i,j}], k \in [1..n_{i,k}],\ l \in [1..n_{i,k,l}],\ r = 0,1,\ P \in \Re^{n\times n}, P = P' > 0, \epsilon_\alpha \geq 0, V_r^\alpha \geq 0, r = 0,1$ *and* $V_1^\alpha > V_0^\alpha$ *such that* $\sum_{k=1..n_{i,k}} \tau_{i,k}^l = 1$ *and*

$$F_i = \sum_{k=1..n_{i,k}} \tau_{i,k}^l LY_{i,k}(P,\epsilon_\alpha) - \sum_{j=1..n_{i,j}} \tau_{i,j}^p C_{i,j}(\alpha) - \sum_{r=0,1} \tau_{i,r}^d C_r(P,V_r^\alpha)$$

$$- \sum_{\substack{k=1..n_{i,k}\\ l=1..n_{i,k,l}}} \tau_{i,k,l}^c C_{i,k,l}(\alpha) \leq -t_i I_{d_i+1}, \tag{11}$$

then, the M-rule i is Pseudo-α-Q-Stable in $\bar{N}^\alpha$ *with respect to the matrix P and the set* $\bar{N}^\alpha \subset \Re^n = \{x : -x'Px + V_0^\alpha \leq 0,\ x'Px - V_1^\alpha \leq 0\}$.

Remark 20 *Although Theorem 3 provides only a sufficient condition, it can be proved that if* $\exists\ j_1, j_2 \in [1..n_{i,j}]$ *such that* $\bar{L}_{i,j_1}^\alpha \cap \bar{A}^\alpha \cap \bar{L}_{i,j_2}^\alpha$ *is empty and* $\bar{L}_{i,j_l}$, $l = 1,2$ *are convex, then the condition (11) is also necessary. We believe condition (11) to be also necessary if* $\bar{A}^\alpha \cap \bar{L}_i^\alpha$ *is empty and* $\bar{L}_{i,j}^\alpha$ $j \in [1..n_{i,j}]$ *are convex.*

Remark 21 *If* $n_{i,k} = 1$ *(i.e., the generalized M-rule i is a M-rule),* $n_{i,k,l} = 0$ *and we impose* $\tau_{i,j} = \tau_{i,r} = 0, j \in [1..n_{i,j}], r = 0,1$, *then condition (11) is equivalent to the Quadratic Stability in the large of the system* Σ_a, *where* Σ_a *is the system described by* $\Sigma_a\ :\ \dot{x} = G_{i,k}x_e$. *Necessarily, we have* $g_{i,k} = 0$ *(i.e.,* Σ_a *is a linear system).*

5.3 Necessary and sufficient condition for α-Q-Stability of fuzzy systems

We can state here the following result.

Theorem 4 *The FADS (6) is α-Q-Stable in $\bar{N}^\alpha$ if and only if $\exists\ P = P' \in \Re^{n\times n} > 0$ such that the generalized M-rules $i, i \in [1..n]$ are Pseudo-α-Q-Stable in $\bar{N}^\alpha$ with respect to the matrix P.*

Remark 22 *The test of α-Q-Stability of FADS reduces to solving a system Σ of Matrix Inequalities composed of the conditions (11) written for each generalized M-rule. These conditions can not be recasted as Linear Matrix Inequalities in P, the set of scaling factors τ_*^* and the parameters V_r^α, $r = 0, 1$. However they can be represented as a Bilinear Matrix Inequality (BMI) with respect to all variables. Thus, Σ is also a BMI. BMI are difficult to solve but global solvers are actually under development [Safonov, 94].*

Remark 23 *Another way to circumvent the numerical difficulty due to the bilinear structure of Σ is the following. $\bar{N}^\alpha$ can be approximated by $\bar{N}_a^\alpha$ defined by a set of a priori known quadratic constraints given as: $x \in N_a^\alpha \Leftrightarrow t_e^{i\prime} C_a^z(\alpha) t_e^i \leq 0,\ C_a^z(\alpha) \in \Re^{(d_i+1)\times(d_i+1)}$, $a = 1 \ldots n_a, i = 1 \ldots n$. It can be shown that, if $n_{i,k} = 1, \forall i \in [1; N]$, Σ is converted in a LMIP in the scaling parameters $\tau *^*$, the Lyapunov Matrix P, and ϵ_α. Then, $\bar{N}^\alpha$ is defined by $x \in \bar{N}^\alpha \Leftrightarrow x \in \{x : x'Px - V_1^\alpha \leq 0,\ -x'Px + V_0^\alpha \leq 0$ and $x \in \bar{N}_a^\alpha\}$. The computation of the parameters V_1^α and V_2^α is then achieved by a trivial LMIP [Boyd, 94].*

5.4 Fuzzy modelling of a closed-loop system

In the previous section, we developed a methodology to study α-Q-Stability in $\bar{N}^\alpha$ of a fuzzy dynamic system assuming that the fuzzy relation of the fuzzy system is given in the standard form and is minimal. For practical purposes, the fuzzy dynamic system Σ_{CL} is obtained from the interconnection of the fuzzy model Σ_P of the plant and the fuzzy model Σ_C of the controller. Thus, the problem is how to characterize, in a minimal and standard form, the fuzzy relation of a closed-loop system given the fuzzy relations describing Σ_P and Σ_C.

Theorem 5 *We assume that the fuzzy model of the plant $\Sigma_P = \{\widetilde{R}_p, min\}$ and the controller $\Sigma_C = \{\widetilde{R}_c^{(e)}, min\}$ are given in standard and minimal form. That is*

- $U_{\widetilde{R}_P} = U_{\dot{\widetilde{X}}} \times U_{\widetilde{X}} \times U_{\widetilde{U}}$ *and* $U_{\widetilde{R}_C}$ *(resp.* $U_{\widetilde{R}_c^e}) = U_{\widetilde{X}} \times U_{\widetilde{U}}$ *(resp.* $U_{\dot{\widetilde{X}}} \times U_{\widetilde{X}} \times U_{\widetilde{U}}$*).*
- $\bar{R}_p^\alpha = \cup_{i\in[1,n_p]}\{\cap_{k\in[1..n_{p,i,k}]}\{\bar{L}_{p,i}^{\alpha,e} \cap \bar{K}_{p,i,k}^\alpha\}\}$ *and* $(\dot{X}, X, U) \in \bar{K}_{p,i,k}^\alpha \Leftrightarrow \dot{X} = \bar{K}_{p,i,k}^\alpha((X', U')')$.

- $\bar{R}_c^{\alpha,(e)} = \cup_{j\in[1,n_c]}\{\cap_{l\in[1..n_{c,j,l}]}\{\bar{L}_{c,j}^{\alpha,(e)} \cap \bar{K}_{c,j,l}^{\alpha,(e)}\}\}$ *and* $(X,U) \in \bar{K}_{c,j,l}^{\alpha,(e)} \Leftrightarrow U = \bar{K}_{c,j,l}^{\alpha,(e)}(X)$.

Then, the FADS Σ_{CL} *resulting from the interconnection of* Σ_P *and* Σ_C *is defined as*

$$\Sigma_{CL} = \{\tilde{R}_{CL}, min, U_{\tilde{R}_{CL}} = U_{\tilde{X}} \times U_{\tilde{X}} \times U_{\tilde{U}}\}.$$

$$\tilde{R}_{CL}^{\alpha} =$$

$$\cup_{i\in[1..n_p],j\in[1..n_c]}\{\cap_{k\in[1..n_{p,i,k}],l\in[1..n_{c,j,l}]}\{\{\bar{L}_{c,j}^{\alpha,e} \cap \bar{L}_{p,i}^{\alpha,e}\} \cap \{\bar{K}_{c,j,l}^{\alpha,e} \cap \bar{K}_{p,i,k}^{\alpha}\}\}\}$$

If, moreover, the following assumptions are satisfied:

- *(A1) : The premise part of the L-rule i of the plant is independent of the control input, i.e.,*

$$\mu_{\tilde{L}_{p,i}}((x', u_1')') = \mu_{\tilde{L}_{p,i}}((x', u_2')'), \ \forall \ u_1, u_2 \in U_{\tilde{U}}$$

- *(A2) : the conclusion of the controller's L-rule are defined by an affine relation,*

$$\forall \alpha \in]0,1], \ (X,U) \in \bar{K}_{c,j,l}^{\alpha} \Leftrightarrow U = G_{c,j,l}X_e, \ G_{c,j,l} \in \Re^{m\times(n+1)}$$

then, a minimal fuzzy relation of Σ_{CL} *is given by*

$$\tilde{R}_{CL}^{\alpha} = \cup_{i\in[1..n_p],j\in[1..n_c]}\{\cap_{k\in[1..n_{p,i,k}],l\in[1..n_{c,j,l}]}\{\{\bar{L}_{c,j}^{\alpha,e} \cap \bar{L}_{p,i}^{\alpha,e}\} \cap \{\bar{B}_{i,j,k,l}^{\alpha,e}\}\}\}$$

where $\forall \alpha \in]0,1], \ (\dot{X}, X) \in \bar{B}_{i,j,k,l}^{\alpha} \Leftrightarrow \dot{X} = \bar{K}_{p,i,k}^{\alpha}((X', (G_{c,j,l}X_e)')')'$.

Remark 24 *If the fuzzy models of the plant and the controller are conjunctive (resp. disjunctive) then, we can deduce in a straightforward way, a conjunctive (resp. disjunctive) set of rules such that the associated fuzzy relation is* $\tilde{R}_{CL}$.

Remark 25 *In Theorem 5, it is assumed that the input of the controller is the state (or partial state) of the plant. This case is frequently encountered in practice. The Theorem 5 can be extended to the case in which the input of the controller is an affine function of the state of the plant. More studies are needed in case the input controller is a nonlinear function of the state. Moreover, it is assumed that the controller is static. The generalization to a dynamic controller is straightforward. Some results are presented in [Marin, 97b].*

Remark 26 *The minimal realization of Σ_{CL} is achieved if the premise part of the plant does not depend on the control input. This is a strong assumption which might be unacceptable in practical situation. The study a FADS with non-minimal representation may allow to remove it. Another possible solution is to deduce a minimal representation of a FADS from its non-minimal representation. Some results are presented in [Marin, 97a] and [Marin, 97b].*

6 Examples

In this section, we will study the fuzzy equilibrium set and the fuzzy attraction set of an electric circuit given in Fig. 3.

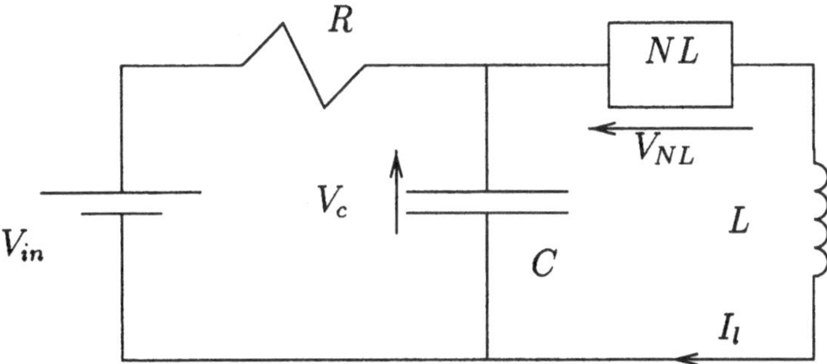

Figure 3: The electric circuit and nonlinear resistance.

Here, R is a linear positive resistance, C is a linear capacitance, L is a self-inductance, and NL is a nonlinear resistance, possibly positive. V_{in} is a continuous source voltage. The state variables are chosen as $x_1 = V_c - V_{in}$ and $x_2 = I_l$, where V_c is the capacitance voltage and I_l the current in the nonlinear resistance NL and self-inductance L. Thus, the state-equation are given by

$$\begin{aligned} \dot{x}_1 &= -\frac{1}{RC}x_1 - \frac{1}{C}x_2, \\ \dot{x}_2 &= \frac{1}{L}x_1 - \frac{1}{L}V_{NL}(x_2). \end{aligned}$$

According to the characteristic of the nonlinear resistance, we will study the fuzzy attraction domain of the system or the fuzzy equilibrium set. We will assume that the fuzzy model of the nonlinear resistance is known.

6.1 The fuzzy model

In this section, we propose a possible fuzzy model for the nonlinear resistance NL and we deduce the fuzzy model of the electric circuit. We assume that the fuzzy model of the nonlinear resistance is disjunctive and given as

- R1 : If x_2 is $-\widetilde{L}_2$ then $V_{NL} = \bar{R}_2 x_2 + \text{Of}$.
- R2 : If x_2 is $\widetilde{L}_1$ then $V_{NL} = -\bar{R}_1 x_2$.
- R3 : If x_2 is $\widetilde{L}_2$ then $V_{NL} = \bar{R}_2 x_2 - \text{Of}$.

The fuzzy sets $\widetilde{L}_1$ and $\widetilde{L}_2$ are represented in Fig. 4.

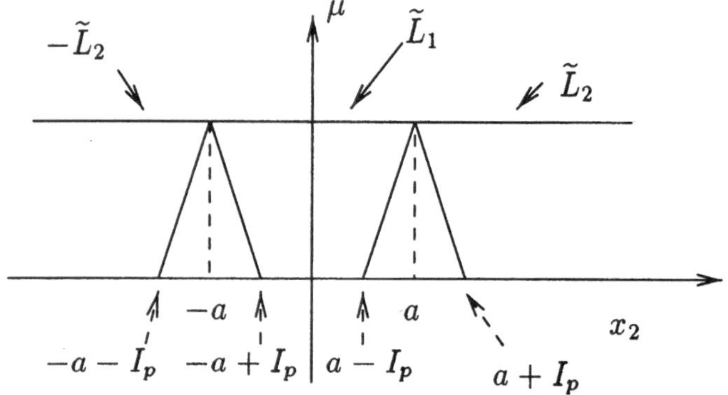

Figure 4: Graphical representation of $\widetilde{L}_1$ and $\widetilde{L}_2$

We have $x = (x_1, x_2)' \in \bar{L}_i^\alpha \Leftrightarrow (x', 1)C_i(\alpha)(x', 1)' \leq 0$ where

$$C_1(\alpha) = \begin{bmatrix} 0 & 0 & 0 \\ 0 & 1 & 0 \\ 0 & 0 & -(a + I_p(1-\alpha))^2 \end{bmatrix},$$

$$C_2(\alpha) = \begin{bmatrix} 0 & 0 & 0 \\ 0 & 0 & -0.5 \\ 0 & -0.5 & a - I_p(1-\alpha) \end{bmatrix}.$$

The crisp sets $\bar{R}_1$ and $\bar{R}_2$ are defined by $\bar{R}_i = [R_i^l; R_i^u]$, $i = 1, 2$. Of is a positive parameter ensuring the continuity of the nonlinear resistance. Consequently, the dynamic system is described by the following FADS

- R1 : If x_2 is $-\widetilde{L}_2$ then $\dot{x} = \bar{A}_2 x - B_2$.

- R2 : If x_2 is $\widetilde{L}_1$ then $V_{NL} = \bar{A}_2 x$.
- R3 : If x_2 is $\widetilde{L}_2$ then $V_{NL} = \bar{A}_2 x + B_2$.

with

$$\bar{A}_1 = \begin{bmatrix} -\frac{1}{RC} & -\frac{1}{C} \\ \frac{1}{L} & -\frac{\bar{R}_1}{L} \end{bmatrix}, \quad \bar{A}_2 = \begin{bmatrix} -\frac{1}{RC} & -\frac{1}{C} \\ \frac{1}{L} & \frac{\bar{R}_2}{L} \end{bmatrix}, \quad \text{and } B_2 = \begin{bmatrix} 0 \\ \text{Of} \end{bmatrix}.$$

By Theorem 2, the α-Quadratic-Stability of the latter FADS is equivalent to the α-Quadratic-Stability of the augmented FADS

- R1 : If x_2 is $-\widetilde{L}_2$ then $\dot{x} = A_2^l x + B_2$.
- R1' : If x_2 is $-\widetilde{L}_2$ then $\dot{x} = A_2^u x + B_2$.
- R2 : If x_2 is $\widetilde{L}_1$ then $V_{NL} = A_1^l x$.
- R2' : If x_2 is $\widetilde{L}_1$ then $V_{NL} = A_1^u x$.
- R3 : If x_2 is $-\widetilde{L}_2$ then $V_{NL} = A_2^l x - B_2$.
- R3' : If x_2 is $-\widetilde{L}_2$ then $V_{NL} = A_2^u x - B_2$.

$$\text{where } A_1^a = \begin{bmatrix} -\frac{1}{RC} & -\frac{1}{C} \\ \frac{1}{L} & -\frac{R_1^a}{L} \end{bmatrix}, \; A_2^a = \begin{bmatrix} -\frac{1}{RC} & -\frac{1}{C} \\ \frac{1}{L} & \frac{R_2^a}{L} \end{bmatrix}, \; a = u, l.$$

Using the symmetric nature of the FADS, we can remove R1 and R1' from the latter rule base. For simplicity, we re-number R1, R1', R2 and R2' to R1, R2, R3 and R4 respectively, and also the matrices A_1^u, A_1^l, A_2^u and A_2^l to A_1, A_2, A_3 and A_4 respectively.

6.2 The fuzzy equilibrium set $\widetilde{E}$.

For our example, we chose $R_1^l = -105\Omega$, $R_1^u = -95\Omega$, $R_2^l = 95\Omega$, $R_2^u = 105\Omega$, $a = 1mA$, $I_p = 0.3mA$. Using the latter parameter, we will briefly describe an iterative algorithm used to approximate the fuzzy invariant set of the system. Due to the negative resistance $\bar{R}_1$ in the fuzzy region $\widetilde{L}_1$, the equilibrium point 0 is unstable in classical sense. Due to the positive resistance $\bar{R}_2$ in the fuzzy region $\widetilde{L}_2$, we can expect the system not to be "unstable" in the whole state-space. Consequently, we will apply the results presented in this article to approximate a possible fuzzy equilibrium set $\widetilde{E}$.

We have $\widetilde{\underline{A}} = \Re^2$. For simplicity, we choose $\epsilon = 0^+$. Applying Theorems 3 and 4, we obtain a set of four Matrix Inequalities

$$\begin{bmatrix} A_1'P + PA_1 & 0 \\ 0 & 0 \end{bmatrix} - \tau_1^p C_1(\alpha) - \tau_{1,0}^d \begin{bmatrix} -P & 0 \\ 0 & V_0^\alpha \end{bmatrix} \leq -t_1 I_3 \quad (12)$$

$$\begin{bmatrix} A_2'P + PA_2 & 0 \\ 0 & 0 \end{bmatrix} - \tau_2^p C_1(\alpha) - \tau_{2,0}^d \begin{bmatrix} -P & 0 \\ 0 & V_0^\alpha \end{bmatrix} \leq -t_2 I_3 \quad (13)$$

$$\begin{bmatrix} A_3'P + PA_3 & PB_2 \\ B_2'P & 0 \end{bmatrix} - \tau_3^p C_2(\alpha) - \tau_{3,0}^d \begin{bmatrix} -P & 0 \\ 0 & V_0^\alpha \end{bmatrix} \leq -t_3 I_3 \quad (14)$$

$$\begin{bmatrix} A_4'P + PA_4 & PB_2 \\ B_2'P & 0 \end{bmatrix} - \tau_4^p C_2(\alpha) - \tau_{4,0}^d \begin{bmatrix} -P & 0 \\ 0 & V_0^\alpha \end{bmatrix} \leq -t_4 I_3 \quad (15)$$

The objective is to solve the latter Matrix Inequalities in order to deduce the fuzzy invariant set $\widetilde{M}$: $\bar{E}^\alpha = \{x \in Re^2 \ : \ x'Px \leq V_0^\alpha\}$. A solution of the latter Matrix Inequalities problem gives a *possible* fuzzy invariant set $\widetilde{M}$. However, we want to compute the *smallest* one in order to deduce more precise and relevant information about the dynamic system. We used the following iterative algorithm to solve *approximately* this optimization problem.

- Step k : For fixed $\tau_{i,0}^d(k)$, $i \in [1;4]$ compute

  $$\min_{P(k),\tau_i^p(k),t_i(k),V_0^\alpha(k)} V_0^\alpha(k),$$

  subject to constraints

  $$P(k) = P'(k) > I_2, \ t_i(k) \geq 0, \tau_i^p(k) \geq 0, \ (12), \ldots, (15).$$

- Step $k+1$: For fixed $P(k), V_0^\alpha(k)$ compute

  $$\max_{t_i(k+1),\tau_i^p(k+1),\tau_{i,0}^d(k+1)} \sum_i t_i(k+1),$$

  subject to constraints

  $$t_i(k+1) \geq 0, \tau_i^p(k+1) \geq 0, \tau_{i,0}^d(k+1), \ (12), \ldots, (15).$$

The presented iterative algorithm is obviously convergent,but does not ensure the global optimum on V_0^α since it might be sensitive to the initial condition on $\tau_{i,0}^d$. The implementation is performed using LMI Lab of the LMI Toolbox. The results of practical experiments are summed up in Tables 1, 2 and Fig. 5.

The tests of Table 1 were performed for $\alpha = 0.5$. The accuracy required on V_0^α to stop the iterative algorithm was 0.1%. Inf denotes the infeasibility of the first LMI problem.

$\tau_{i,0}^d$ initial	1000	3000	5000	7.5000	10000	15000	20000
V_0^α initial	0.3346	0.1376	0.1032	0.0929	0.0977	0.1687	Inf
V_0^α final	0.3346	0.1376	0.1031	0.0929	0.0960	0.1093	
Iterations	2	2	2	2	2	4	

Table 1: Sensitivity to initial conditions.

α	0	0.25	0.5	0.75	1
V_0^α final	0.1187	0.1054	0.0929	0.0812	0.0702

Table 2: "Size" of $\bar{E}^\alpha$ vs α.

The tests in Table 2 were performed with $\tau_{i,0}^d = 7500$ to initialize the algorithm. The Fig. 5 shows the α-cut of the computed fuzzy invariant set $\widetilde{E}$. To evaluate the quality of the computed fuzzy invariant set $\widetilde{E}$, we show possible α-trajectories (in fact, possible α-stable limit-cycle) on the same figure. The result of matrix P was

$$P = \begin{bmatrix} 1.036 & -14.97 \\ -14.97 & 6235.1 \end{bmatrix}.$$

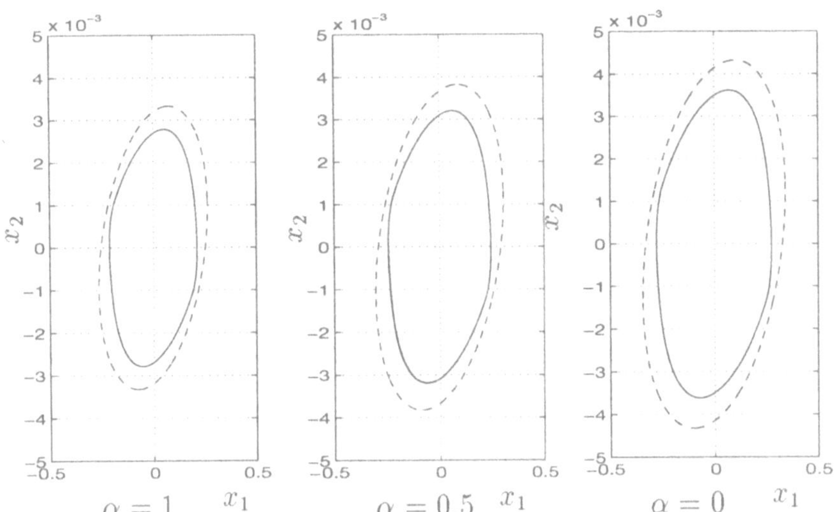

Figure 5: Graphical representation of $\bar{E}^\alpha$ (- -) and possible α-trajectories (-).

6.3 The anti-fuzzy attraction set $\underset{\sim}{A}$.

In this section, all parameters remain the same except for $\bar{R}_1$ and $\bar{R}_2$. We have $R_1^l = 95\Omega$, $R_1^u = 105\Omega$, $R_2^l = -105\Omega$ and $R_2^u = -95\Omega$. Then, the equilibrium point 0 is locally stable, but not globally stable. Thus, we will compute the fuzzy attraction domain $\underset{\sim}{A}$ of the point 0. We have $\widetilde{E} = 0$. The structure of the fuzzy-model is unchanged and using similar arguments to those in the previous section, we obtain a set of four Matrix Inequalities

$$\begin{bmatrix} A_1'P + PA_1 & 0 \\ 0 & 0 \end{bmatrix} - \tau_1^p C_1(\alpha) - \tau_{1,1}^d \begin{bmatrix} P & 0 \\ 0 & -V_1^\alpha \end{bmatrix} \leq -t_1 I_3 \quad (16)$$

$$\begin{bmatrix} A_2'P + PA_2 & 0 \\ 0 & 0 \end{bmatrix} - \tau_2^p C_1(\alpha) - \tau_{2,1}^d \begin{bmatrix} P & 0 \\ 0 & -V_1^\alpha \end{bmatrix} \leq -t_2 I_3 \quad (17)$$

$$\begin{bmatrix} A_3'P + PA_3 & PB_2 \\ B_2'P & 0 \end{bmatrix} - \tau_3^p C_2(\alpha) - \tau_{3,1}^d \begin{bmatrix} P & 0 \\ 0 & -V_1^\alpha \end{bmatrix} \leq -t_3 I_3 \quad (18)$$

$$\begin{bmatrix} A_4'P + PA_4 & PB_2 \\ B_2'P & 0 \end{bmatrix} - \tau_4^p C_2(\alpha) - \tau_{4,1}^d \begin{bmatrix} P & 0 \\ 0 & -V_1^\alpha \end{bmatrix} \leq -t_4 I_3 \quad (19)$$

The objective is to solve the latter Matrix Inequalities in order to deduce the anti-fuzzy attraction set $\underset{\sim}{A}$: $\bar{A}^\alpha = \{x \in \Re^2 \ : \ x'Px \leq V_1^\alpha\}$. A solution of the latter Matrix Inequalities problem gives a *possible* anti-fuzzy attraction set $\underset{\sim}{A}$. However, we desire to compute the *biggest* one in order to deduce more precise and relevant information about the dynamic system. It can be shown that we can fix, at no cost, $\tau_i^p = \tau_{i,1}^d = t_i = 0$, $i = 1, 2$. The iterative algorithm then becomes

- Step k : For fixed $\tau_{i,1}^d(k)$, $i = 3, 4$ compute

  $$\max_{P(k), \tau_i^p(k), t_i(k), V_1^\alpha(k)} V_1^\alpha(k),$$

  subject to constraints

  $$0 < P(k) = P'(k) < I_2, \; t_i(k) \geq 0, \tau_i^p(k) \geq 0, \; i = 3, 4, \; (16), \ldots, (19).$$

- Step $k+1$: For fixed $P(k), V_1^\alpha(k)$ compute

  $$\max_{t_i(k+1), \tau_i^p(k+1), tau_{i,1}^d(k+1)} \sum_i t_i(k+1),$$

  subject to constraints

  $$t_i(k+1) \geq 0, \tau_i^p(k+1) \geq 0, \; \tau_{i,1}^d(k+1), \; (16), \ldots, (19.)$$

τ_i^a initial (10^3)	25	50	75	100	250	500	1000
V_0^α initial(10^{-1})	Inf	1.216	1.052	0.969	0.821	0.772	0.747
V_0^α final (10^{-6})		1.313	1.313	1.313	1.313	1.313	1.313
Iterations		7	10	11	15	20	23

Table 3: Anti-fuzzy attraction set: sensitivity to initial conditions.

The results of practical experiments are summed up in Tables 3, 4 and Fig. 6.

The tests in Table 3 were performed for $\alpha = 0.5$. The accuracy required on V_1^α to stop the iterative algorithm was 0.1%. If denotes the infeasibility of the first LMI problem.

α	0	0.25	0.5	0.75	1
V_0^α final (10^{-5})	0.0890	0.1091	0.1313	0.1441	0.1817

Table 4: Anti-fuzzy attraction set:"size" of $\bar{A}^\alpha$ vs α.

These tests in Table 4 were performed with $\tau_{i,1}^d = 50.10^3$ to initialize the algorithm. The Fig. 6 shows the α-cut of the computed anti-fuzzy attraction set $\underset{\sim}{A}$. To evaluate the quality of the computed anti-fuzzy attraction set $\underset{\sim}{A}$, we show possible α-trajectories (in fact, possible α-unstable limit-cycle) on the same figure. The résult of matrix P was

$$P = \begin{bmatrix} 1.981\ 10^{-4} & 9.895\ 10^{-5} \\ 9.895\ 10^{-5} & 1.000 \end{bmatrix}.$$

6.4 Comments

The previous examples show the applicability of the theory. However, some remarks have to be formulated.

Remark 27 *It seems that the conservatism introduced by the sufficient condition expressed in Theorem 3 is acceptable. The fuzzy invariant set $\widetilde{M}$ is highly capable of approximating the "real" fuzzy convergent set. The results seem a little more conservative in the second example.*

Remark 28 *The numerical properties of the proposed algorithm seem to be highly dependent on the application. In the first case, the number of iterations*

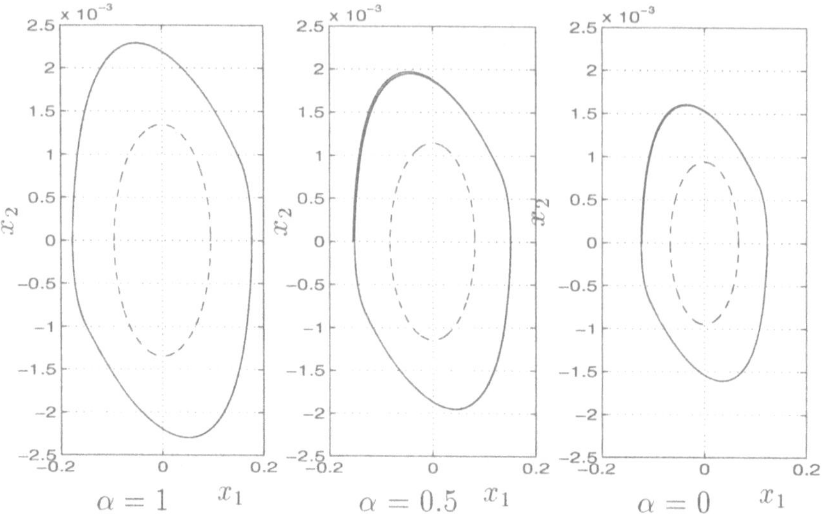

Figure 6: Anti-fuzzy attraction set: graphical representation of $\bar{A}^\alpha$ ('- -') and possible α-trajectories ('-').

is very low,ut the final result depends on the initial conditions. In the second case, the number of iterations is higher, but the final result does not depend on the initial conditions.

Remark 29 *The iterative algorithm used in the example might be improved. For example, the objective of the first step could be the minimization (resp. maximization) of the volume of $\bar{E}^\alpha$ (resp. $\bar{A}^\alpha$). This optimization could be performed using the software described in [S.P. Wu, 96]. Moreover, the optimization criterion used for the second step is more intuitive and almost heuristic. Further studies are required to define it in a more precise way.*

7 Conclusions

In this article, we have presented a new parameterization of fuzzy systems. Its main advantages are the following: (i) the combinatorial explosion of the number of fuzzy rules is avoided by the use of input-dependent fuzzy sets in the conclusion part of the rules, (ii) linguistic model properties are conserved by the use of logical min-max operators and various classes of approximate reasoning techniques can be accomodated, (iii) complex premise/conclusion parts of the fuzzy rules are allowed. Finally, the model is mathematically tractable since a quadratic parameterization (5) is obtained. The quadratic parameterization (5) of the fuzzy relation $\widetilde{R}$ allows a systematic analysis of various properties of fuzzy systems using Matrix Inequalities [Boyd, 94]. The advantage of our stability analysis is that it removes the drastic distinction between stability/instability and sensitivity/insensitivity. This aspect was so far considered as a major drawback in the stability analysis of fuzzy dynamic system. Based on Lyapunov theory, fuzzy stability criteria consist of computing a fuzzy equilibrium set $\widetilde{E}$ and its anti-fuzzy attraction set $\underset{\sim}{A}$. Global/Local and Robust/Nonrobust stability criteria for Linear/Nonlinear systems are particular cases of the presented method. In this sense, the proposed approach can be considered as a unification of various "classical" stability concepts in the context of fuzzy dynamic systems. However, further studies are needed to derive necessary and sufficient conditions for Pseudo-α-Q-Stability of the generalized M-rule. We used iterative algorithms to solve the BMI derived from Theorem 4. The obtained numerical results are promising and show that the theory is applicable. However, the development of more sophisticated solvers of BMI should allow a significant improvement in the numerical results. Note also, that the approach used in this article can be used and extended to analyze the H_∞ performance of a fuzzy dynamic system [Marin, 96a], [Marin, 96b], [Van der Shaft, 96]-[Cao, 96].

Acknowledgments

I would like to thank the students R. Ruiz, A. Juarez, I. Crisostomo, and Dr. A. Herrera of ITESM, Campus Toluca, Mexico for helpful discussions and for providing the technical support for the implementation of the algorithms.

References

[Zadeh, 73] L. A. Zadeh : "Outline of a new approach to the analysis of complex systems and decision processes", IEEE Trans. Syst. Man. Cybern., vol 3, pp 28-44, 1973.

[Zadeh, 75] L. A. Zadeh : "The concept of linguistic variable and its application to approximate reasoning", Information Science, vol 8, pp 199-250, pp 301-358, vol 9, pp 43-80, 1975.

[Mamdani, 75] E. H. Mamdani and S. Assilian : "Application of fuzzy algorithms for control of simple dynamic plant", Proc IEE, vol 121, pp 1585-1588, 1974.

[Takagi, 85] T. Takagi and M. Sugeno : "Fuzzy identification of systems and its application to modelling and control", IEEE Trans. Syst. Man. Cybern., vol 15, pp 116-132, 1985.

[Aracil, 91] J. Aracil & al : "Fuzzy Control of Dynamical Systems. Stability Analysis based on Conicity Criterion", Proc IFSA'91, Vol Engineering, Brussel, pp 5-8, 1991.

[Opitz, 93] H.P. Opitz : " Fuzzy Control and Stability criteria", Proc EUFIT'93, Aachen, pp 130-136,1993.

[Wang, 94] L. Wang and R. Langari : "Fuzzy controller design via hyperstability theory", Proc FUZZ-IEEE'94, pp 178-182, Orlando, 1994.

[Marin, 96a] J.P Marin, A. Titli : "Robust performances of closed-loop fuzzy systems : A global Lyapunov approach", Proc FUZZ-IEEE'96, pp 732-737, New-Orleans, 1996.

[Boyd, 94] S. Boyd, L. El Gahoui, E. Feron, V. Balakrishnan : "Linear Matrix Inequalities in Systems and Control Theory", Volume 15 of SIAM Studies in Applied Mathematics. SIAM, 1994.

[Tanaka, 96] K. Tanaka, K. Ikeda and H.O. Wang : "Robust Stabilization of a class of Uncertain Nonlinear systems via fuzzy control : Quadratic Stabilizability, H^{∞} Control Theory and Linear Matrix Inequalities", IEEE Trans on Fuzzy Systems 4 (1), pp 1-13, 1996.

[Zhao, 96] J. Zhao, V. Wertz and R. Gorez : "Fuzzy gain scheduling based on fuzzy models", Proc FUZZ-IEEE'96, pp 1670-1676, New-Orleans, 1996.

[Tanaka, 94] K. Tanaka, N. Sano : "A robust stabilization problem of fuzzy control systems and its application to backing up truck trailer", IEEE Trans on Fuzzy Systems, pp 29-34, 1994.

[Tanaka, 92] K. Tanaka, M. Sugeno : "Stability analysis and design of fuzzy control system", Fuzzy sets and systems 45, pp 135-156, 1992.

[Marin, 95a] J.P. Marin, A. Titli : "Necessary and Sufficient conditions for Quadratic Stability of a class of Tagaki-Sugeno Fuzzy Systems", Proc EUFIT'95, Aachen, 1995.

[Kim, 95] W.C Kim, C.S. Ahn, W.H. Kwon : "Stability Analysis and stabilization of fuzzy state space models", Fuzzy sets and systems, 71, pp 131-142, 1995.

[Marin, 95b] J.P Marin : "Conditions nécessaires et suffisantes de stabilité quadratique d'une classe de systèmes flous", Proc LFA'95, pp 240-247, Paris, 1995.

[Marin, 96b] J.P Marin : "H_{∞} performance analysis of fuzzy system using quadratic storage function", Proc AADECA'96, pp 7-11, Buenos-Aires,1996.

[Kiska, 85] J.B. Kiska, M.M. Gupta and P.N. Nikiforuk " Energistic Stability of fuzzy dynamic systems", IEEE Trans. Syst. Man. Cybern., 5 (15), pp 783-792, 1985.

[Czogala, 82] E. Czogala, W. Pedrycz : "Control problems in fuzzy systems", Fuzzy sets and systems 7, pp 257-273, 1982.

[Jianquin, 93] C. Jianqin, C. Laijiu : "Study on stability of fuzzy closed-loop systems", Fuzzy sets and systems 57, pp 159-168, 1993.

[Deglas, 82] M. De Glas : "A mathematical theory of fuzzy systems", Fuzzy Information and Decision Process and E. Sanchez (eds.),@North-Holland Publishing Company, 1982.

[Deglas, 84] M. De Glas : "Invariance and Stability of Fuzzy Systems", Journal of Mathematical Analysis and Application, vol 99, pp 299-319, 1984.

[Safonov, 94] M. G. Safonov, K.J. Goh, J.H. Ly : "Control Systems synthesis via Bilinear Matrix Inequalities", Proc ACC, pp 45-49, Baltimore, Maryland, June 1994.

[Rotea, 93] M.A. Rotea, M. Corless, D. Da and I.R. Petersen : "System with structured uncertainty : Relation between Quadratic and Robust Stability", IEEE. Trans. Aut. Cont., vol 38, n 5, pp 799-803, 1993.

[Vidyasagar, 92] M. Vidyasagar : " Nonlinear Systems Analysis", second edition, Prentice Hall, Englewood Cliffs, New Jersey 07632, 1992.

[Van der Shaft, 96] A. Van der Schaft : " L_2-gain and Passivity Techniques in Nonlinear Control", Lecture Notes in Control and information Sciences, Vol 218, Springer, 1996.

[Lu, 95] W.M. Lu, J.C. Doyle : " H_∞ Control of Nonlinear Systems : A Convex Characterization", IEEE. Trans. Aut. Cont., vol 40, n 9, pp 1668-1675, 1995.

[Cao, 96] S.G. Cao, N.W. Rees and G. Feng : " Analysis and Design of Uncertain Fuzzy Control Systems, Part II: Fuzzy Controller Design", Proc FUZZ-IEEE'96, pp 647-653, New-Orleans, 1996.

[Marin, 97a] J.P. Marin, A. Titli : " Robust Quadratic Stabilizability of Non-Homogeneous Sugeno Systems Ensuring Completeness of the Closed-loop systems", in Proc FUZZ-IEEE'97, Barcelona, pp 185-192, 1997.

[Marin, 97b] J.P. Marin, A. Titli : " Robust Quadratic Stabilizability of fuzzy systems using fuzzy dynamic output feedback : a Matrix Inequality approach", in Proc IFSA'97, Prague, vol 4, pp 451-456, 1997.

[S.P. Wu, 96] S.P. Wu, L. Vandenberghe ans S. Boyd : " Software for Determinant Maximization Problems, User's Guide ", Version alpha, Standford University, May 1996

Inverse Fuzzy Process Models for Robust Hybrid Control

M. Fischer and R. Isermann
Darmstadt University of Technology
Institute of Automatic Control
Laboratory for Control Systems and Process Automation
Landgraf-Georg-Str. 4, D-64283 Darmstadt, Germany

1 Introduction

Fuzzy controllers are on their way of becoming a standard tool in industrial automation. [13] gives an overview of possible control concepts involving fuzzy components. It turns out that the application of fuzzy control is particularly effective at the higher levels of automation systems. For this purpose, direct fuzzy controllers are usually designed manually. Experts' knowledge is used to determine the membership functions and the rule base (Fig. 1). This approach allows a fast controller prototyping, but the optimization of the controller usually requires a tedious tuning procedure due to the great number of free parameters and incomplete heuristic knowledge.

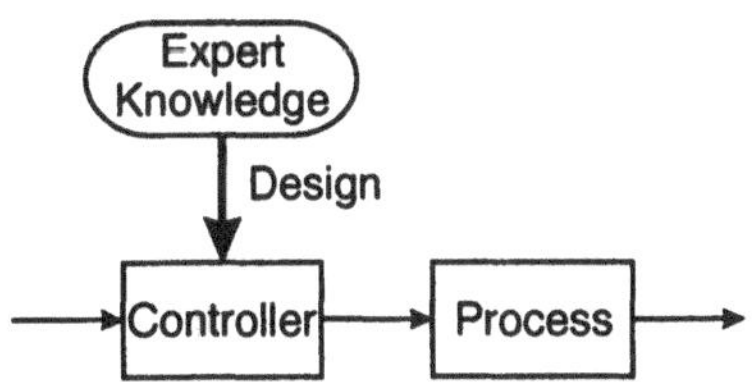

Figure 1: Direct fuzzy controller design.

The idea of fuzzy model-based control [2], [15], [29], illustrated in Fig. 2, attempts to overcome this drawback. Nonlinear dynamic processes can be modeled by means of fuzzy models, i.e., collections of fuzzy if-then rules. This task can be performed either by a human expert who describes the process behavior instead of the control strategy or by means of fuzzy identification from measurement data. These two methods for knowledge acquisition can

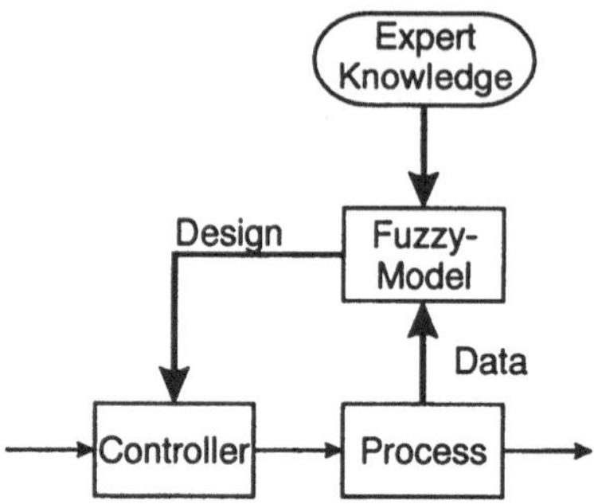

Figure 2: Fuzzy model-based controller design.

be combined, e.g., the expert defines the structure of the fuzzy model and the remaining parameters are determined via identification. Hence, process information from data compensates for deficiencies in the expert knowledge. The controller itself is designed more or less directly from the fuzzy model. Only a few parameters which are also meaningful in terms of control have to be chosen. A variety of well-known design techniques from linear and nonlinar control theory can be used, e.g., feedback linearization [15], predictive control [4], [9], gain scheduling [10], model-based adaptive control [29] etc. Probably, inverse model control is the most intuitive concept [1], [3], [8], [12]. For instance, an inverse neuro-fuzzy plant model (a hybrid fuzzy model) which can be interpreted in the fuzzy sense is used as feedforward controller. In practice, however, it is impossible to obtain the exact inverse of a dynamic, nonlinear process. Since this type of controller does not comprise a feedback of the control variable, it does neither cope with plant-model mismatch nor with external disturbances like changing payload or noise. Hence, the original idea is not suitable for real-world applications. Therefore, the inverse model controller is extended by a feedback loop. The resulting robust hybrid control scheme based on inverse fuzzy process models eliminates the drawbacks mentioned above.

In this article, the identification of nonlinear dynamic processes will be done in terms of Takagi-Sugeno (TS) fuzzy models [26]. Section 2 reviews an appropriate identification algorithm, which builds fuzzy process models from measurement data. Prior knowledge in the form of fuzzy rules can also be considered. Several methods for the special task of inverse model estimation are described in Sect. 3. Besides the analytical inversion of fuzzy models, training from measurement data is also considered. The control scheme is thoroughly analyzed and compared with other approaches in Sect. 4. Finally, in Sect. 5, the control performance is evaluated in controlling the speed of a nonlinear cooling blast with variable dynamics. The controller design procedure is described step by step: a fuzzy process model is identified, the inverse fuzzy model is estimated and interpreted, and the guidelines for controller tuning obtained from theoretical analysis are validated.

2 Fuzzy process models

In this article, nonlinear dynamic single-input/single-output (SISO) systems with input u, output y, dynamic orders n_u and n_y and dead time d

$$y(k) = f(\mathbf{x}(k)), \tag{1}$$

$$\begin{aligned}\mathbf{x}(k) &= [x_1(k)\ x_2(k)\ \ldots\ x_n(k)]^T = [u(k-d)\ u(k-d-1)\ \ldots \\ &\quad u(k-d-n_u)\ y(k-1)\ y(k-2)\ \ldots\ y(k-n_y)]^T,\end{aligned} \tag{2}$$

are approximated by Takagi-Sugeno (TS) fuzzy models [26] at discrete time instants k. The fuzzy rules of the form

$$\begin{aligned} R_j:\ & IF\ \ x_1(k)\ is\ X_{j1}\ \ AND\ \ \ldots\ \ AND\ \ x_n(k)\ is\ X_{jn} \\ & THEN\ \ y(k) = w_{j0} + w_{j1}x_1(k) + \ldots + w_{jn}x_n(k), \\ & \qquad with\ \ n = n_u + n_y,\ j = 1\ldots M,\end{aligned} \tag{3}$$

represent local linear difference equations enhanced by an additional constant w_{j0}. Choosing the product operator as t-norm, the output of the fuzzy system with M rules is obtained as

$$y(k) = \sum_{j=1}^{M} (w_{j0} + w_{j1}x_1(k) + \ldots + w_{jn}x_n(k)) \cdot \Phi_j(\mathbf{x}, \mathbf{c}_j, \sigma_j), \tag{4}$$

where for Gaussian membership functions the weighting functions Φ_j are normalized degrees of fulfillment z_j

$$\begin{aligned}\Phi_j(\mathbf{x}, \mathbf{c}_j, \sigma_j) &= \frac{z_j}{\sum_{i=1}^{M} z_i}, \\ z_i &= \exp\left(-\frac{1}{2}\left(\frac{(x_1-c_{i1})^2}{\sigma_{i1}^2} + v\frac{(x_2-c_{i2})^2}{\sigma_{i2}^2} + \ldots + \frac{(x_n-c_{in})^2}{\sigma_{in}^2}\right)\right).\end{aligned} \tag{5}$$

In [20], this type of fuzzy model is referred to as a local model network which interpolates local linear models by overlapping local basis functions. This interpretation is visualized in Figs. 3 – 6 where a nonlinear process of first order is approximated. The fuzzy model comprises three fuzzy rules

$$\begin{aligned}
R_1:\quad & IF\quad u(k-1)\ is\ small\quad AND\quad y(k-1)\ is\ small \\
& \qquad THEN\quad y(k) = c_1 + b_{11}u(k-1) - a_{11}y(k-1). \\
R_2:\quad & IF\quad u(k-1)\ is\ small\quad AND\quad y(k-1)\ is\ large \\
& \qquad THEN\quad y(k) = c_2 + b_{21}u(k-1) - a_{21}y(k-1). \\
R_3:\quad & IF\quad u(k-1)\ is\ large\quad AND\quad y(k-1)\ is\ don't\ care \\
& \qquad THEN\quad y(k) = c_3 + b_{31}u(k-1) - a_{31}y(k-1).
\end{aligned}$$

Figure 3 shows the Gaussian membership functions on the walls of the box. The three-dimensional Gaussian functions z_j are the firing strengths (degrees of fulfillment of the antecedents) of the three rules. Applying (5) for normalization yields the normalized weighting functions in Fig. 4. The local linear models plotted as planes of different orientation in Fig. 5 are multiplied with the respective normalized weighting functions (compare to (4)). The resulting output of the fuzzy system is plotted in Fig. 6. The surface is a smooth interpolation of the local linear models.

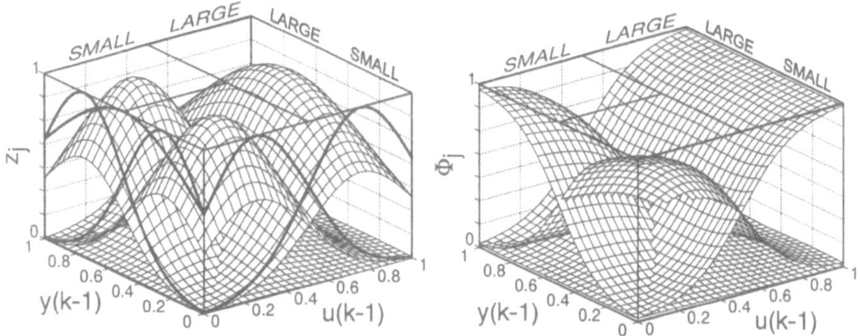

Figure 3: Gaussian membership/weighting functions.

Figure 4: Normalized weighting functions.

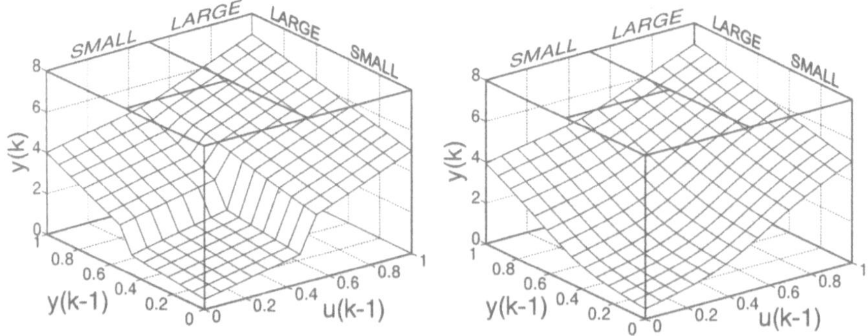

Figure 5: Local linear models.

Figure 6: Fuzzy model output.

Normalized radial basis function networks (RBF) [18] are a special class of the proposed fuzzy model if the linear coefficients are set to $w_{jl} = 0$ for $l \neq 0$ [17]. If this condition is not fulfilled, the fuzzy model is equivalent to the so-called generalized RBF (see [11]). Since RBFs have been shown to be universal approximators, the extended version in (4) obviously has the same property due to the increased number of free parameters.

The task of system identification is to determine all free parameters of the fuzzy model. The following aspects must be considered:

- *Dynamic order:* The dynamic orders of the linear difference equations must be chosen. They might vary for different local models.
- *Dead time:* The dead times must be determined for all local linear models.
- *Membership functions:* The input partitioning, i.e. number, centers c_{ij}, and standard deviations σ_{ij} of the membership functions must be defined.
- *Rule base:* The variables which significantly influence the nonlinear behavior of the process have to be included in the rule antecedent. The antecedents do not necessarily comprise all elements of $\mathbf{x}(k)$ (see (2)). The selection and the logical combination of relevant inputs in the antecedents (if-parts) are crucial tasks.

In general, there are two ways to obtain the above information. Possibly, experts are able to formulate their process knowledge in fuzzy rules. Unfortunately, this usually delivers only a rough idea of the plant behavior, because the expert cannot sense all the details and might not be able to express the observations quantitatively. Therefore, numerous approaches have been proposed which identify nonlinear dynamic fuzzy system models from input-output measurement data, e.g., fuzzy clustering [30], tree construction algorithms [25], or neuro-fuzzy approaches [5].

In the following, the local linear model tree (LOLIMOT) algorithm of Nelles [22] will be used. This method splits the identification into two parts. In an outer loop, the nonlinear parameters for the partitioning of the input space are determined, and in an inner loop the linear parameters of the local models are estimated. For the latter task, a weighted least-squares technique is applied. The optimal parameters $\mathbf{w}_j$ of the j-th model are

$$\mathbf{w}_j = (\mathbf{X}^T \mathbf{Q}_j \mathbf{X})^{-1} \mathbf{X}^T \mathbf{Q}_j \mathbf{y}, \tag{6}$$

where the data matrix $\mathbf{X} = [\mathbf{x}(1)\ \mathbf{x}(2)\ \dots\ \mathbf{x}(N)]^T$ contains measurements over N samples, and $\mathbf{y}$ denotes the desired outputs. The data are weighted with $\mathbf{Q}_j = diag(\Phi_j(\mathbf{x}(1))\ \Phi_j(\mathbf{x}(2))\ \dots\ \Phi_j(\mathbf{x}(N)))$ which contains the firing strength of the respective local model. This local estimation means that all models are estimated independently ignoring the existing overlap. Besides the huge reduction in computational demand (see [20]) the estimation approach has the following appealing property: the local linear models are forced to fit the data locally. Hence, the resulting difference equations are meaningful in terms of gains and time constants. In contrast, global estimation suffers from compensation effects.

The Gaussian basis functions are placed by a tree construction algorithm which proceeds as described hereafter. Initially, only one linear model is estimated for the complete input space, which has the form of a hyper-rectangle. Then, the hyper-rectangle is iteratively divided into two halves in all the dimensions. For each half, membership functions are determined whose standard deviations are proportional to the size of the subspace and the center is placed in the middle of the subspace. For the partitions in each dimension, two local linear models are estimated and the respective global approximation error is calculated. A comparison of the results yields the dimension which is most advantageous for splitting up the input space. Now, the hyper-rectangle (subspace) with the largest local error measure is chosen, and the segmentation procedure is performed for this subspace. The algorithm terminates when the desired overall model quality is reached.

Figure 7 shows the partitioning procedure of LOLIMOT for the example from Figs. 3 – 6. First of all, a global linear model is estimated (1. in Fig. 7). Since the termination tolerance has not been reached yet, it is checked if it

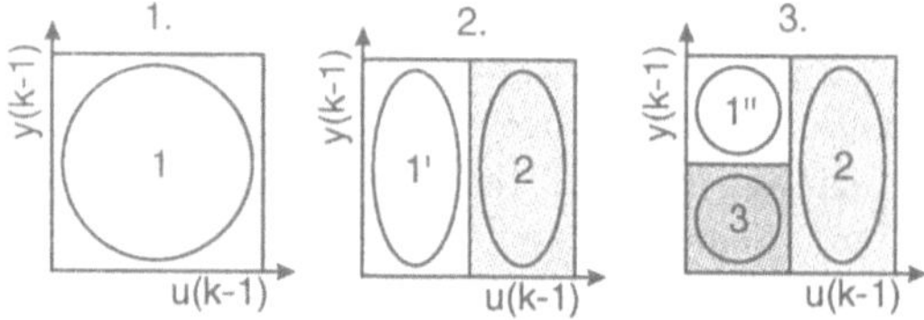

Figure 7: Partitioning procedure of LOLIMOT.

would be more advantageous to split the input space in the $u(k-1)$, or in the $y(k-1)$ dimension. The model resulting for from the split in $u(k-1)$ yields a better performance, and therefore the two local models depicted in 2. in Fig. 7 are arrived at. By means of the local error measure it is determined which local model performs worse. Here, model 1' is selected for further division. In 3. in Fig. 7 it was decided that a split in the $y(k-1)$ dimension results in a better approximation than the alternative split in the $u(k-1)$ dimension.

3 Inverse fuzzy process models

In general, building inverse fuzzy process models is more difficult than constructing forward models. For experts, it is much easier to describe the observed process behavior than to describe a virtual controller that perfectly inverts the process characteristics. Hence, the goal of this section is to derive inverse fuzzy process models either from forward models or from measurement data.

3.1 Analytical inversion of fuzzy models

Some types of fuzzy models can be analytically inverted. Babuska et al. [1] invert fuzzy models with trapezoidal membership functions in the antecedents of the fuzzy if-then rules and singleton membership functions for the consequents (then-parts) of these rules. The analytical inversion of Takagi-Sugeno fuzzy models is also possible if the condition

$$\Phi_j \neq \mathcal{F}(u(k)) \quad \Leftrightarrow \quad \Phi_j \neq \mathcal{F}(x_1(k)) \quad , \quad j = 1 \ldots M, \tag{7}$$

is fulfilled. If the input space is not partitioned in the $u(k)$-dimension, (4) can be solved for $u(k) = x_1(k)$

$$u(k) = \frac{1}{\sum_{j=1}^{M} \Phi_j([x_2 \ldots x_n]^T, \mathbf{c}_j, \sigma_j) \cdot w_{j1}} \cdot \Big(y(k) -$$

$$\sum_{j=1}^{M} (w_{j0} + w_{j2}x_2(k) + \ldots + w_{jn}x_n(k)) \cdot \Phi_j([x_2 \ldots x_n]^T, \mathbf{c}_j, \sigma_j)\Big). \tag{8}$$

The LOLIMOT algorithm from the previous section allows to choose which inputs out of $\mathbf{x}$ are used in the rule antecedent. Hence, if $u(k) = x_1(k)$ is not a candidate for partitioning, the resulting fuzzy model can be analytically inverted.

In the numerator of (7), the inverse local linear models are summed up. In order to obtain a bounded output from each single model, the zeroes of the original local models must be stable. Because this requirement is not met in general the LOLIMOT algorithm has to be modified. When estimating the linear parameters, the zeroes are constrained to be stable. Therefore, the linear least-squares algorithm must be replaced by a nonlinear optimization routine which is capable of handling the constraints. Stability of the polynomial in the numerator can be checked by means of the Schur-Cohn-Jury criterion (see [14]). Of course, the modification also degrades the performance of the forward model, but still the system is more flexible than conventional radial basis function networks (RBF). However, the quality of analytically inverted models cannot be evaluated. Good forward models do not necessarily indicate inverse models of the same performance due to different objective functions.

3.2 Direct inverse learning

A straightforward approach to inverse modeling is to excite the process with a training signal and to identify a fuzzy model from the reverse input/output data, such that the fuzzy system reconstructs the input signal of the process from the given output signal (Fig. 8).

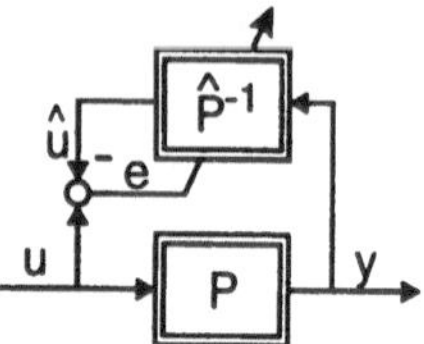

Figure 8: Direct inverse learning.

The standard LOLIMOT algorithm can be employed. Referring to [12], this method suffers from two drawbacks. If the dynamic nonlinear system is not a one-to-one mapping, the identification algorithm receives several desired values u_i for the same input state. If a least-squares approach is used, the fuzzy system will be trained to map y to the mean value of all desired values u_i. Furthermore, an appropriate training signal is difficult to obtain for direct inverse learning. The inverse model is supposed to perform over a wide range of input amplitudes y and for a large bandwidth. In the training process, however, the excitation of the system is introduced as actuation u, so that the persistent excitation of y cannot be guaranteed.

3.3 Specialized inverse learning

Specialized inverse learning [16], [24] as illustrated in Fig. 9 overcomes the drawbacks of direct inverse learning. The inverse model is cascaded with the process or a forward plant model, and the parameters of the former are adjusted in order to minimize the deviation between actual and desired output. Hence, the optimization is a goal-oriented scheme, since the objective of tracking control is pursued in the learning phase. Both the process and the controller being optimized are excited with realistic signals when a typical reference trajectory r is used for training. Actuator saturation can also be considered in the optimization.

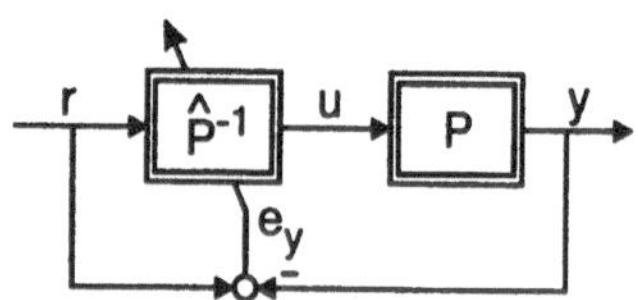

Figure 9: Specialized inverse learning.

The standard LOLIMOT algorithm cannot be directly utilized for specialized inverse learning because the objective function for the estimation of the local linear models is not linear in the parameters anymore. However, the following procedure turns out to be quite effective. Here the partitioning of the input space is performed by direct inverse learning. The scheme is capable of determining the nonlinear characteristics of the plant and yields the structure parameters $\mathbf{c}_j$ and σ_j of the fuzzy model. Then, in the second step, only the parameters w_{ji} of the local models are optimized in the configuration from Fig. 9. The following interactive algorithm is proposed:

1. *User input:* Selection of a local model to be optimized.
2. Levenberg-Marquardt optimization of the cost function

$$J = \sum_k e_y^2(k) \quad , \quad e_y(k) = r(k)\left(1 - P(\hat{P}^{-1}(r(k)))\right) , \tag{9}$$

   by variation of the parameters $\mathbf{w}_j$ of the selected model.
3. Output of the optimized model's parameters and poles; plot of r and $y = P(\hat{P}^{-1}(r))$, which is obtained by simulation with the new parameter set.
4. *User decision:* Based on the information from step 3., it must be decided if the optimized parameters are to be accepted.
5. *User decision:* Continue with step 1. or terminate.

In step 4., the new parameters should only be accepted if they yield a stable local model and a significant reduction of the cost function (9). A rather small improvement of J might indicate that the respective model does not strongly contribute to the output of $\hat{P}^{-1}$. In this case, the parameters tend to be overfitted to the specific training signal r. The inverse model might loose its general validity.

4 Control scheme

Basically, an inverse fuzzy model of the process is used as feedforward controller (Fig. 10, left). As mentioned in Sect. 3, it is impossible to obtain the perfect inverse and the ideal transfer function $P(\hat{P}^{-1}) = 1$. Instead, the output y from Fig. 10 is given by

$$y(s) = (1 + \Delta(s)) \cdot (r(s) + \delta(s)), \tag{10}$$

where s is the Laplace variable, r is the desired output and $1 + \Delta(s)$ is the transfer function obtained from the linearization of $P(\hat{P}^{-1})$ at some operating point. The inner disturbance $\delta(t)$ represents the error introduced by the

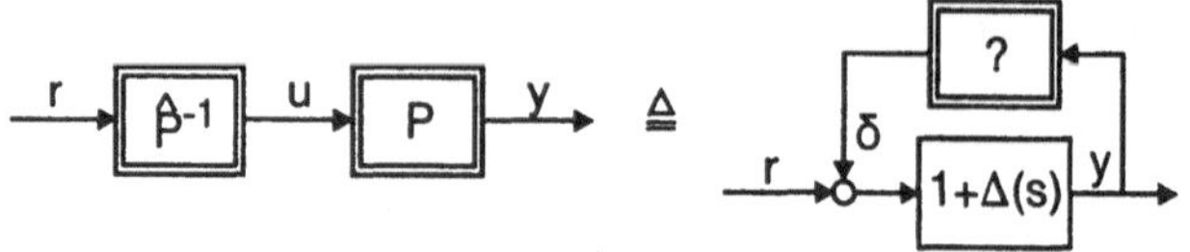

Figure 10: Interpretation of non-ideal inverse model control.

linearization. The functional relation between the output y and this virtual disturbance is not known.

Obviously, the control performance strongly depends on the quality of the inverse model. Therefore, the control scheme has to be expanded by corrective components that can cancel both the disturbance $\delta(t)$ and the dynamic uncertainty $\Delta(s)$. In the field of motion control, Ohnishi [23] introduced a so-called disturbance observer (DISOB) which was refined by Umeno and Hori [27], [28]. It will be shown how this idea can be utilized for inverse fuzzy model-based control.

Figure 11 shows the ideal disturbance observer which exactly reproduces the disturbance $\delta(t)$ in order to cancel $\delta(t)$.

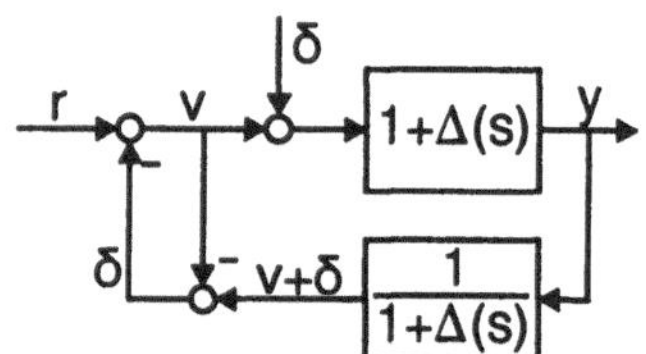

Figure 11: Ideal disturbance observer.

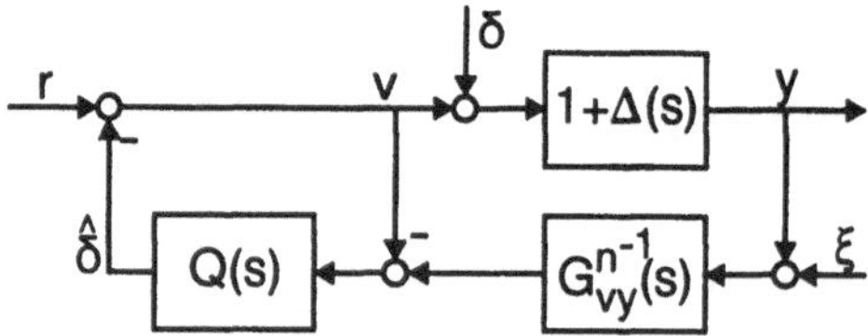

Figure 12: Realizable form of the disturbance observer.

In reality, however, the uncertainty term $\Delta(s)$ is not known. Therefore $1/(1+\Delta(s))$ is replaced by the nominal inverse transfer function $G_{vy}^{n\ -1}(s)$ which is most obviously chosen to be unity. In addition, the realizable form in Fig. 12 considers measurement noise in the output signal y and a filter $Q(s)$ whose purpose will be explained below.

The output is determined to be

$$y(s) = G_{ry}(s)r(s) + G_{\delta y}(s)\delta(s) + G_{\xi y}(s)\xi(s), \tag{11}$$

where

$$G_{ry}(s) = \left.\frac{y(s)}{r(s)}\right|_{\delta,\xi=0} = \frac{(1+\Delta(s))G^n_{vy}(s)}{G^n_{vy}(s) + ((1+\Delta(s)) - G^n_{vy}(s))Q(s)}, \tag{12}$$

$$G_{\delta y}(s) = \left.\frac{y(s)}{\delta(s)}\right|_{r,\xi=0} = \frac{(1+\Delta(s))G^n_{vy}(s)(1-Q(s))}{G^n_{vy}(s) + ((1+\Delta(s)) - G^n_{vy}(s))Q(s)}, \tag{13}$$

$$G_{\xi y}(s) = \left.\frac{y(s)}{\xi(s)}\right|_{r,\delta=0} = -\frac{(1+\Delta(s))Q(s)}{G^n_{vy}(s) + ((1+\Delta(s)) - G^n_{vy}(s))Q(s)}. \tag{14}$$

Choosing ${G^n_{vy}}^{-1}(s) = 1$, the resulting transfer functions from the desired output, the disturbance, and the sensor noise to the output are given as

$$G_{ry}(s) = \frac{1+\Delta(s)}{1+\Delta(s)Q(s)}, \tag{15}$$

$$G_{\delta y}(s) = \frac{(1+\Delta(s))(1-Q(s))}{1+\Delta(s)Q(s)}, \tag{16}$$

$$G_{\xi y} = -\frac{(1+\Delta(s))Q(s)}{1+\Delta(s)Q(s)}. \tag{17}$$

In order to comprehend the purpose of the additional filter $Q(s)$, two special cases are examined. If $Q(s) \approx 1$, (12)–(14) or (15)–(17) become

$$G_{ry}(s) \approx G^n_{vy}(s) = 1 \quad , \quad G_{\delta y}(s) \approx 0 \quad , \quad G_{\xi y}(s) \approx -1. \tag{18}$$

On the other hand, if $Q(s) \approx 0$, the transfer functions become

$$G_{ry}(s) \approx 1 + \Delta(s) \quad , \quad G_{\delta y}(s) \approx 1 + \Delta(s) \quad , \quad G_{\xi y}(s) \approx 0. \tag{19}$$

(18) validates the performance of the ideal disturbance observer in the sense that the transfer function between desired and actual output is unity and the disturbance δ is eliminated. A disadvantage of the ideal observer becomes also visible, namely the direct transmission of the sensor noise to the output. If the feedback path is eliminated in (19), the noise can be rejected. However, the open loop works as a pure inverse model controller, and the model deviations

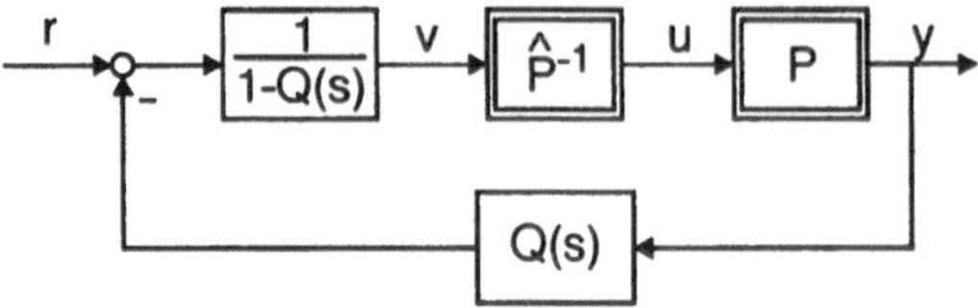

Figure 13: DISOB as standard control loop.

of $\hat{P}^{-1}$are not compensated anymore. This comparison yields a first hint for designing $Q(s)$. If the frequency of the noise is approximately known from experiments, the filter can be shaped so that an appropriate suppression is achieved.

Further guidelines for the design of the filter $Q(s)$ are obtained from a robust stability analysis of the disturbance observer loop with $G^n_{vy} = 1$. Therefore, a block diagram transformation delivers the standard control loop representation in Fig. 13. As proposed by Doyle et. al. [6], sample Nyquist plots of the nominal open-loop transfer function

$$G^n_0(j\omega) = \frac{Q(j\omega)}{1 - Q(j\omega)}, \tag{20}$$

and the actual open-loop transfer function (compare to Fig. 10)

$$G_0(j\omega) = \frac{Q(j\omega)}{1 - Q(j\omega)} \cdot (1 + \Delta(j\omega)), \tag{21}$$

are considered in Fig. 14. A vector from the critical point $(-1, 0)$ to the nominal transfer function $G^n_0(j\omega)$ is defined as

$$h(j\omega) = 1 + G^n_0(j\omega). \tag{22}$$

Additionally, the deviation between actual and nominal transfer function

$$\gamma(j\omega) = G_0(j\omega) - G^n_0(j\omega) = \Delta(j\omega)G^n_0(j\omega), \tag{23}$$

is plotted for an arbitrary frequency ω_0. Assuming that both the nominal closed-loop system and the dynamic uncertainty $\Delta(s)$ are stable, the dashed circle illustrates the robust stability condition

$$\begin{gathered} |\gamma(j\omega)| \le |h(j\omega)| \quad \Leftrightarrow \\ |\Delta(j\omega)G^n_0(j\omega)| \le |1 + G^n_0(j\omega)| \quad \Leftrightarrow \quad |\Delta(j\omega)Q(j\omega)| \le 1. \end{gathered} \tag{24}$$

If the first inequality is satisfied, the curve of $G_0(j\omega)$ cannot be located on the other side of $(-1, 0)$. Hence, instability is excluded.

The information contained in the last inequality in (24) provides the second hint utilized in the design process of the disturbance observer: the larger

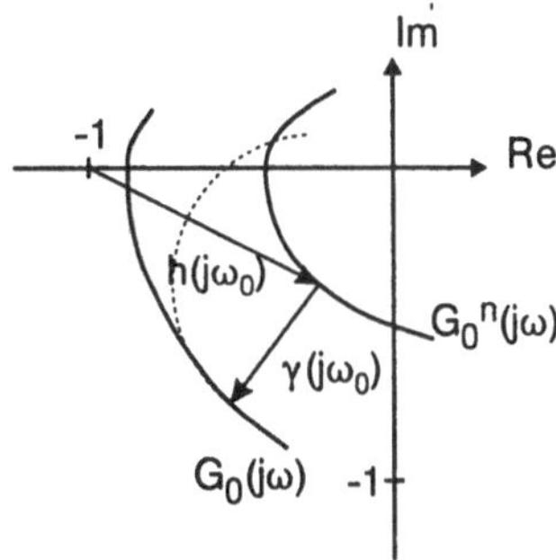

Figure 14: Nyquist plot.

the absolute value of the uncertainty $\Delta(j\omega)$, the smaller the absolute value of $Q(j\omega)$ must be. In general, the proposed inverse fuzzy models are more precise in the low-frequency range. In particular, the static nonlinearity of the process and its inverse are easy to obtain. The dynamics in the high-frequency range might be neglected in the identification due to a reduced model order.

All this suggests a design of $Q(j\omega)$ as a low-pass filter. Because of this, for the experiments described in Sect. 5 the filter

$$Q(s) = \frac{1}{(\tau s)^3 + 3(\tau s)^2 + 3(\tau s) + 1}, \tag{25}$$

will be used which yields the "controller" transfer function

$$\frac{1}{1 - Q(s)} = \frac{(\tau s)^3 + 3(\tau s)^2 + 3(\tau s) + 1}{(\tau s)^3 + 3(\tau s)^2 + 3(\tau s)}. \tag{26}$$

It is to be investigated how the choice of the only design parameter τ influences the control performance.

For implementation, the zero-order-hold equivalents of the continuous transfer functions are derived. Figure 15 illustrates how a dead time of the process can be integrated in the control scheme. For the inverse model controller, the desired output must be known one step in advance. By shifting the predictor z over the summation points, the disturbance observer loop remains realizable because the highest order coefficient of the numerator of $Q(z)$ is zero.

So far, the nominal transfer function was considered as $G^n_{vy}(s) = 1$. From a physical point of view, it might be necessary to change it to a low-pass characteristic

$$G^n_{vy}(s) = \frac{1}{T_n s + 1}. \tag{27}$$

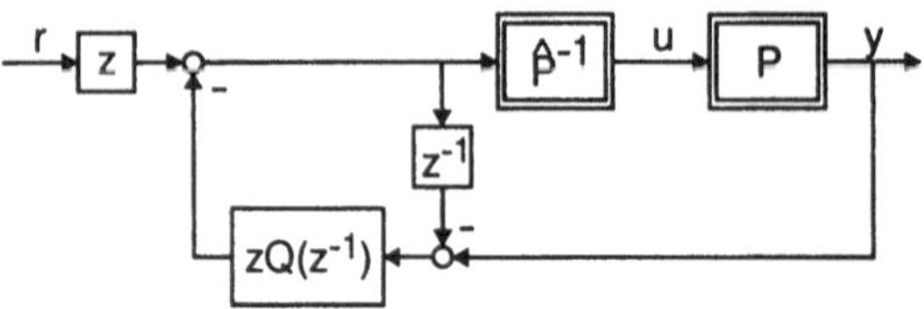

Figure 15: Disturbance observer for a process containing one dead time.

The reason for this is as follows. If the process has some kind of low-pass behavior, the inverse dynamic model has differentiating properties causing large slopes in the actuation u. Since real-world actuators always have a limited rate, the nominal transfer function from (27) appears more realistic. Hence, decreasing the absolute value of the multiplicative uncertainty term

$$\Delta(s) = \frac{G_{vy}(s)}{G_{vy}^n(s)} - 1, \tag{28}$$

increases the stability robustness described by (24). Notice, that the loop still can be realized even if the inverse of (27) is not. Therefore, the order of the low-pass filter $Q(s)$ is chosen higher than the relative order of $G_{vy}^n(s)$ and both transfer blocks are merged in Fig. 12.

The nominal transfer function $G_{vy}^n(s)$ can also be introduced into the cost function (9) for specialized inverse learning in order to create a more realistic objective:

$$J = \sum_k e_y^2(k) \quad , \quad e_y(k) = r\left(G_{vy}^n(z) - P(\hat{P}^{-1}(r(k)))\right). \tag{29}$$

Remarks

- *Similarity to internal model control (IMC)*

The analysis of the controller has yielded some qualitative guidelines for tuning of the linear filters $Q(s)$ and $G_{vy}^n(s)$. Evaluating the overall control scheme, similarities to the well-known internal model control (IMC) [19] can be recognized. Figure 16 shows the standard IMC configuration, where the inverse plant model $\hat{P}^{-1}$ is used as a feedforward controller and the reference value r is corrected by the deviation of the model output y_m from the plant output y. This eliminates steady-state errors due to static errors of the inverse model. The linear feedback filter $G(s)$ serves to reduce the gain and to provide robust controller performance.

Assuming that the inverse model $\hat{P}^{-1}$ perfectly fits the forward model $\hat{P}$, i.e.,

$$\hat{P}^{-1}(\hat{P}) = 1, \tag{30}$$

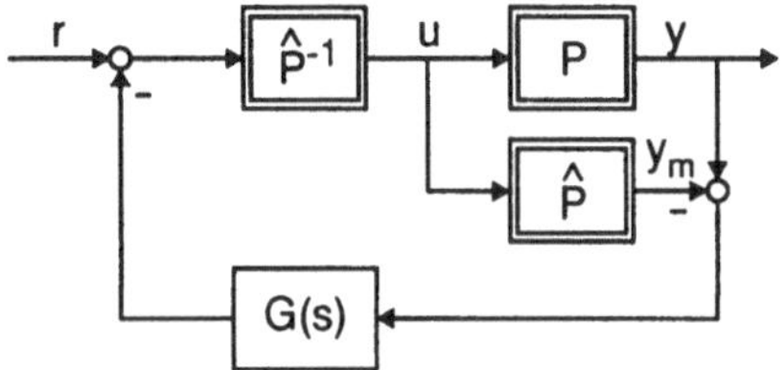

Figure 16: Internal model control (IMC).

the block diagram from Fig. 16 can be transformed into the realizable form of the robust hybrid controller in Fig. 12, if $G^n_{vy}(s) = 1$ and $Q(s) = G(s)$. For the condition given in (30) and $G(s) = 1$, IMC has been proven to be an optimal controller ($y = r$). Hence, this optimality holds also for the presented robust hybrid control scheme.

- *Stability*

As for internal model control, stability of the robust hybrid controller can only be guaranteed for some idealized assumptions. Due to the similarity shown in the previous paragraph the proof itself is nearly the same as for IMC [7]. If inverse fuzzy models are used as controllers, one faces the problem to prove their stability which is not trivial. Consequently, from a practical point of view, it is necessary to provide a supervisory level which observes possible instability and counteracts it either by tuning the filter $Q(s)$ or by switching to a stable backup controller.

5 Experimental results

5.1 Hardware configuration

The applicability of the proposed control scheme is demonstrated by means of a pilot plant. The speed of a radial industrial cooling blast designed for air conditioning of buildings is to be controlled. It is driven by a capacitor motor of 1.44 kW delivering up to 5000 m^3 of air per hour at a maximum speed of 1450 rpm. The power is supplied by phase control, and the speed is measured by a light dependent resistor which senses the light reflected from the rotor wings. The resolution is 4.5 rpm. The signal processing is performed on a PC 486DX2-66 equipped with an i/o board and a real time extension of MATLAB/SIMULINK. Since the original setup rather represents a linear system, which is easy to control, it is equipped with a throttle flap (Fig. 17). This yields a nonlinear dynamic system. Fig. 18 shows the static nonlinearity and Fig. 19 shows a series of step responses. Both plots can be physically interpreted. As long as the throttle is closed, the gain of the fan is large and the system has first-order time-lag behavior with a large time constant. At a speed of about 1000 rpm the throttle opens and the gain promptly decreases. The system's time constant becomes smaller and oscillations indicate complex pairs of poles. Moreover, acceleration and deceleration behaviors are not symmetric. The motor does not actively brake the blast, but only the air drag reduces the speed.

Figure 17: Cooling blast.

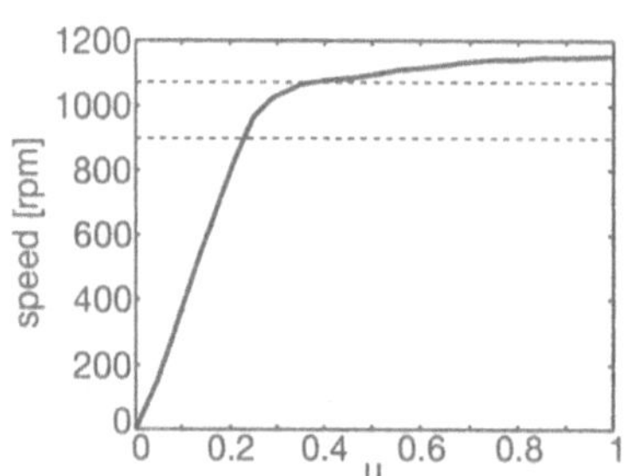

Figure 18: Static I/O mapping.

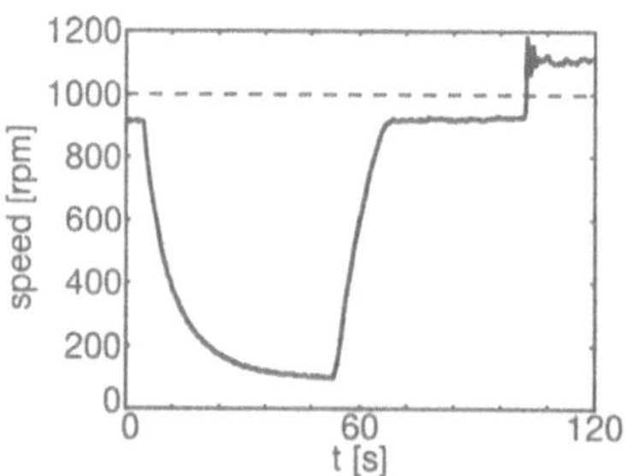

Figure 19: Step responses.

5.2 Nonlinear system identification

First of all, a forward process model is identified from measurement data. As training signal, an amplitude-modulated pseudo-random binary signal (APRBS) [21] depicted in Fig. 20 is utilized. The signal excites the plant dynamics sufficiently and its amplitude covers the whole range of operation. A Takagi-Sugeno fuzzy model with a rather large number of rules is estimated by means of the LOLIMOT algorithm. The dynamic order of the local models is chosen as $n_u = n_y = 2$ in order to allow for an approximation of the oscillatory behavior. The dead time is set to $d = 1$. The normalized mean square error recorded in each iteration is plotted in Fig. 21. It can be seen that the overall model is not significantly improved for more than 12 fuzzy

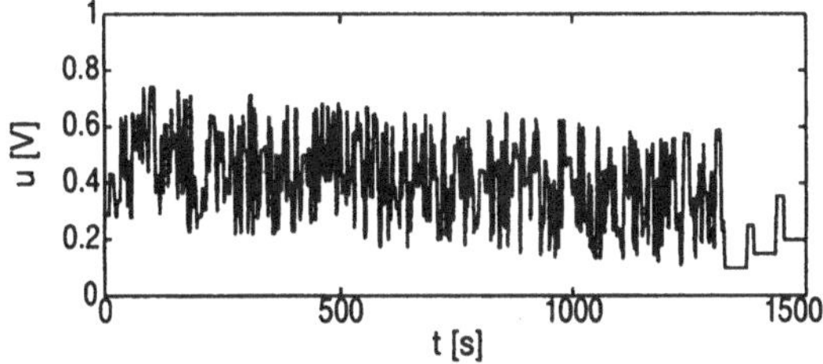

Figure 20: Training data for identification.

rules or 12 local models, respectively. Consequently, this model complexity is selected. The placement of the local linear models is shown in Fig. 22. The input space is partitioned in the $u(k-1)$- and the $y(k-2)$-dimension. A relatively large number of models is placed around the critical speed of 1000 rpm where the process has strongly nonlinear behavior. Figures 23 and 24 show validation data for the fuzzy model. While the approximation quality is very good for low speeds, there are some model deficiencies at high speeds. In fact, this is due to the partly non-deterministic behavior of the cooling blast.

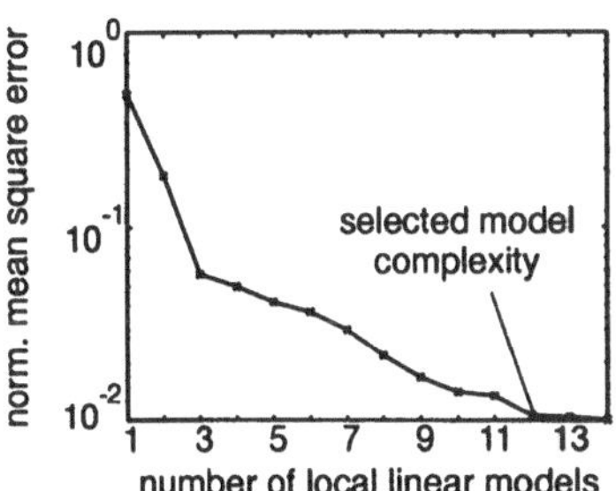

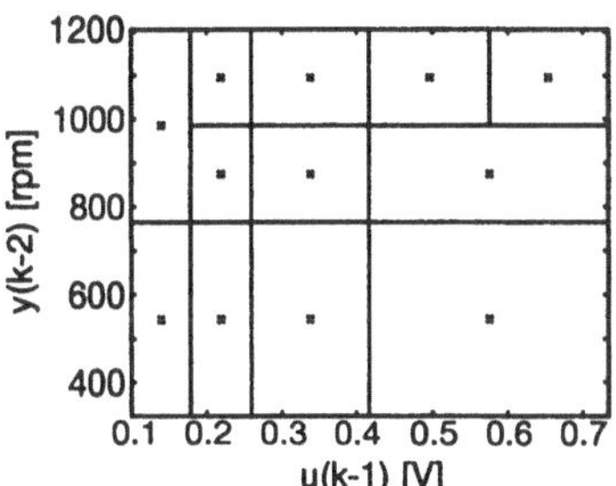

Figure 21: Convergence rate of the LOLIMOT algorithm.

Figure 22: Partitioning of the input space of the forward model.

Subsequently, the inverse model is identified by means of specialized inverse learning described in Sect. 3.3. First, a fuzzy model with 6 rules of the form

$$\begin{aligned} &IF \;\; u(k-1) \; is \; U_j \;\; AND \;\; y(k-2) \; is \; Y_j \\ &THEN \;\; u(k) = c_j + a_{j0}y(k) + a_{j1}y(k-1) + a_{j2}y(k-2) - b_{j1}u(k-1) \end{aligned}$$

is estimated from the training data in Fig. 20 by direct inverse learning (see Fig. 8). For the local linear models in the rule consequents, only one regressor in u is selected, because basically a differential behavior of the inverse model

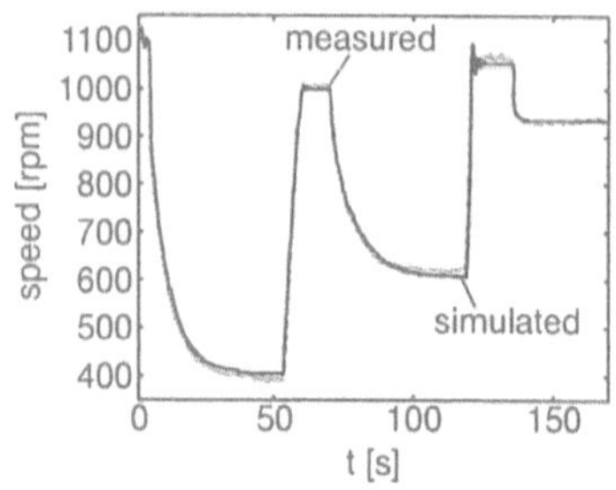

Figure 23: Model validation for low speed.

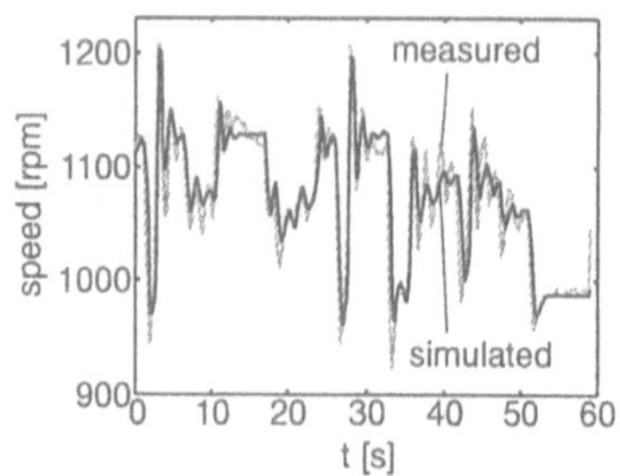

Figure 24: Model validation for high speed.

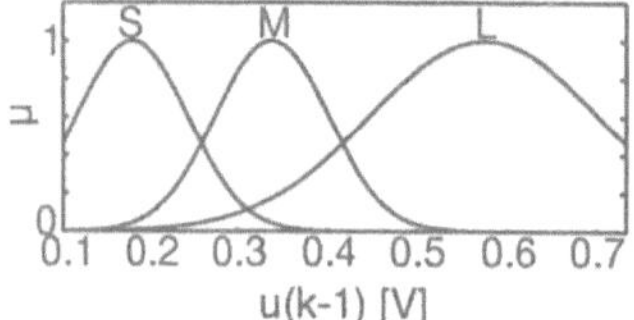

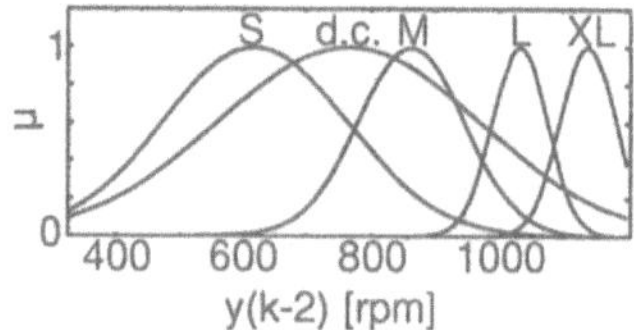

Figure 25: Membership functions of the inverse fuzzy model.

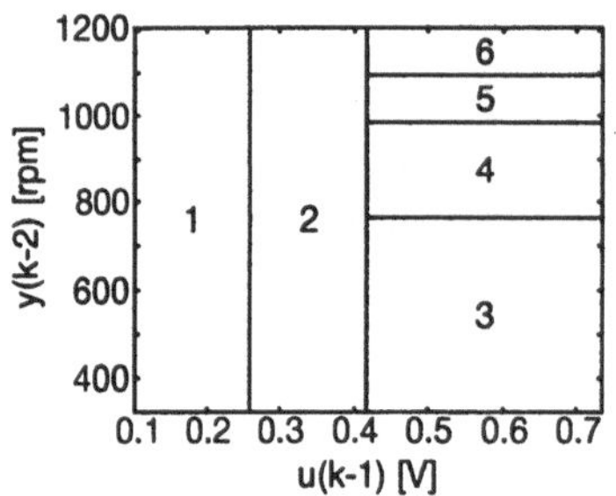

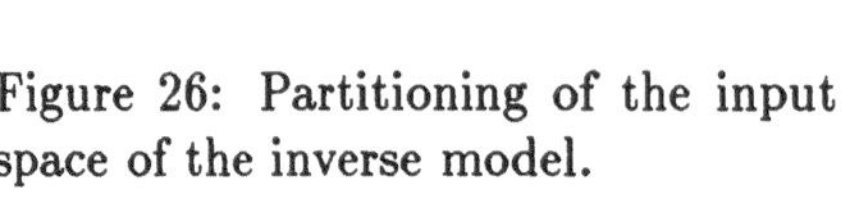

Figure 26: Partitioning of the input space of the inverse model.

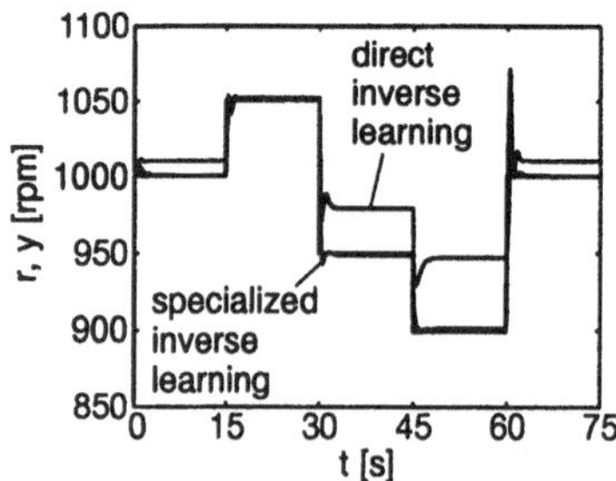

Figure 27: Inverse learning.

is to be achieved. Figures 25 and 26 show the membership functions and the corresponding partitioning of the input space. The consequent parameters of the inverse model are to be improved by specialized inverse learning (see Fig. 9). A step wise changing reference signal r in the range between 700–1100 rpm is applied, and the interactive optimization procedure from Sect. 3.3 is carried out. The simulated validation data in Fig. 27 shows the improvement which is gained when the parameters of only one local model are optimized by the Levenberg-Marquardt algorithm. Both static and dynamic behaviors are significantly improved. Table 1 contains the antecedents of the six fuzzy rules as well as the gains ($K = (a_{j0}+a_{j1}+a_{j2})/b_{j1}$) and the zeroes (obtained by solving the equation: $a_{j0}z_{j_{1,2}}^2 + a_{j1}z_{j_{1,2}} + a_{j2} = 0$) of the corresponding local models after specialized learning. It can be seen, that the gain increases with higher actuation signals u and with higher speed y. At the same time, the zeroes move from real-valued numbers to complex numbers with increasing imaginary parts. Both effects are reasonable in terms of control because the inverse model controller is supposed to compensate for both the plant's gain and the dynamics. As described above, the plant's gain decreases with increasing speed and the dynamic behavior changes from asymptotic to oscillatory. Hence, the overall gain becomes approximately 1, and the zeroes of the controller cancel the plant's poles.

The preceding discussion demonstrates the major advantage of using inverse fuzzy process models for control. In contrast to black-box approximators, the Takagi-Sugeno fuzzy models allow to evaluate the properties of the inverse controller locally, which makes the approach transparent. This is quite important since the interactive procedure of specialized inverse learning is based on numerical computations and does not necessarily guarantee meaningful results for different operating points.

Table 1: Rule base of the inverse model after specialized learning.

	$u(k-1)$	$y(k-2)$	gain K_j	zeroes $z_{j_{1,2}}$
R_1	Small	don't care	$5.3 \cdot 10^{-4}$	$0.96, 0.56$
R_2	Medium	don't care	$2.6 \cdot 10^{-3}$	$0.77 \pm 0.14i$
R_3	Large	Small	$1.4 \cdot 10^{-3}$	$0.81 \pm 0.11i$
R_4	Large	Medium	$3.3 \cdot 10^{-3}$	$0.70 \pm 0.30i$
R_5	Large	Large	$6.7 \cdot 10^{-3}$	$0.66 \pm 0.51i$
R_6	Large	eXtra Large	$9.4 \cdot 10^{-3}$	$0.66 \pm 0.53i$

5.3 Control performance

The horizontal lines in Fig. 18 limit the range of speed which is considered for the evaluation of the controller performance. Since the gain strongly changes, one expects difficulties in controlling the blast with a simple PID controller. For the experiment in Fig. 28, this type of controller was optimized for reference steps at speeds above 1000 rpm. When the throttle is closed, the loop starts oscillating because the controller gain is too large.

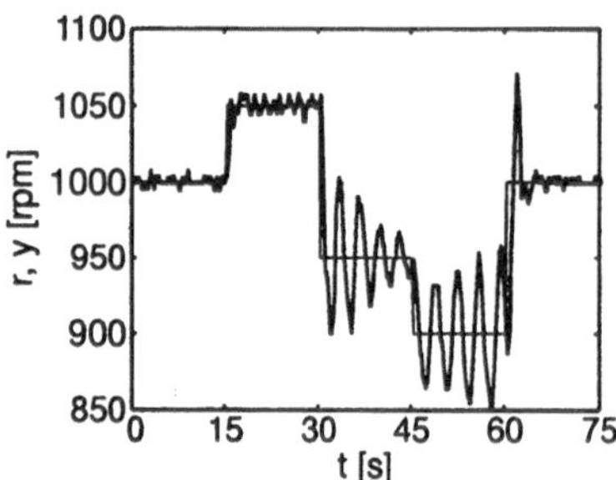

Figure 28: PID control, control signal.

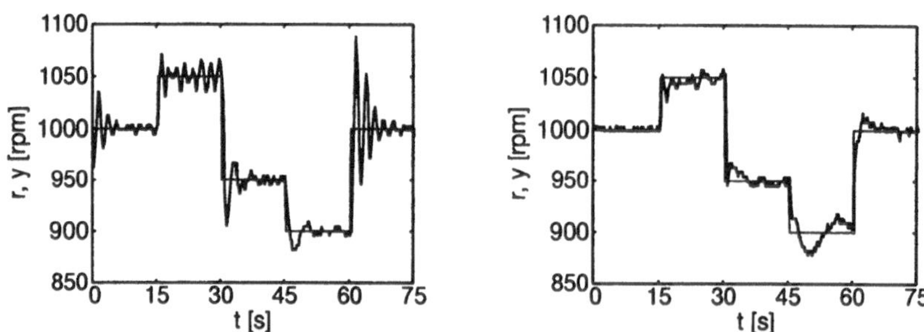

Figure 29: DISOB control, $\tau = 0.2s$, control signal.

Figure 30: DISOB control, $\tau = 0.8s$, control signal.

Indeed, the plant requires a more sophisticated control algorithm. In Figs. 29–31, the robust hybrid control scheme from Fig. 15 is applied with different low-pass filter time constants τ. Choosing $\tau = 0.8s$ yields a reasonable overall performance. If τ is too small (Fig. 29), i.e. the filter cut-off frequency is relatively high, the robust stability condition (24) is endangered. Oscillations are observed, and a further decrease of τ causes an unstable system. On the other hand, if τ is too large (Fig. 31), i.e. the cut-off frequency of $Q(s)$ is relatively low, the disturbance observer compensates for both the dynamic uncertainty $\Delta(s)$ and inner disturbance $\delta(t)$ only in a small frequency band (compare (18),(19)). The static errors of the inverse model are corrected slowly which can also be explained by the controller transfer function (26),

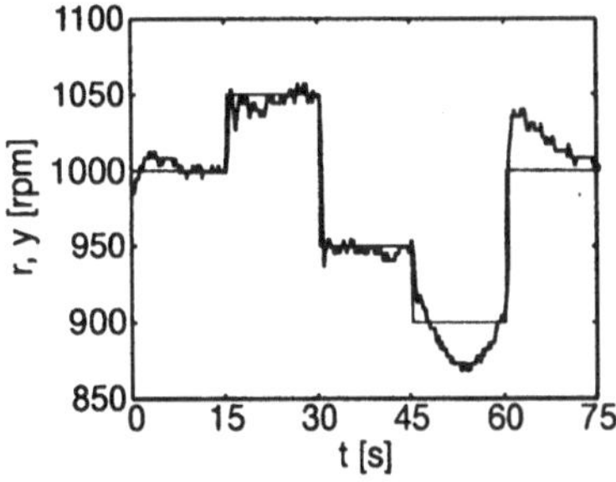

Figure 31: DISOB control, $\tau = 2s$, control signal.

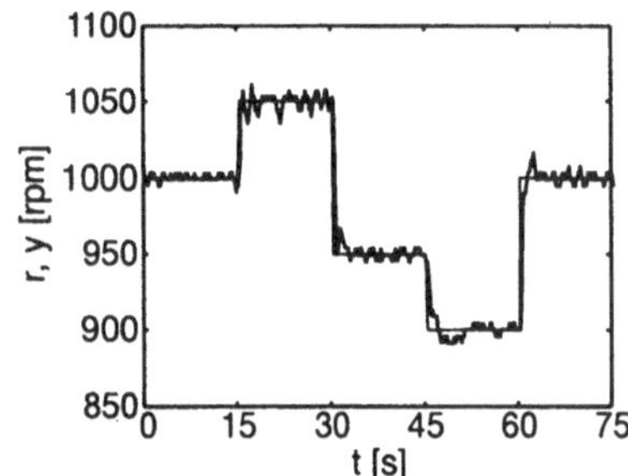

Figure 32: DISOB control, $\tau = 0.3s$, $T_n = 0.3s$, control signal.

where a large value of τ corresponds to a small integral component. So far, the disturbance observer attempts to make the closed loop behave like the nominal transfer function $G^n_{vy}(s) = 1$. This leads to large overshoot for the reference input of 900 rpm. The integrator output of the DISOB sums up the negative control errors without any observable effect. This windup is due to the fact that the process is not actively decelerated. When the sign of the control error switches it takes some time to compensate for the value stored in the integrator. Therefore, in Fig. 32, the nominal transfer function $G^n_{vy}(s)$ is taken from (27) with a time constant $T_n = 0.3s$, which makes the feedback path faster. At the same time, the time constant can be decreased to $\tau = 0.3s$ because the absolute value of the dynamic uncertainty becomes smaller. Considering the large quantization steps of the sensor signal, a fairly good control performance is achieved.

Figures 33 and 34 show the actuation signal corresponding to Fig. 30 and 32 respectively. Changes in the reference signal cause large peaks in the actuation because the inverse fuzzy model has differential behavior. For the rest of the time, the controller output is rather smooth due to the low-pass filter Q. The latter prevents the noise from being amplified by the inverse fuzzy model. In Fig. 34, however, the discretization noise becomes visible since the nominal transfer function has first order time-lag behavior and thereby the inverse transfer function ${G^n_{vy}}^{-1}$ in the feedback path causes the amplification of higher frequencies.

6 Conclusions

A robust hybrid controller comprising inverse fuzzy process models and linear feedback was introduced for SISO processes. The approach can be extended to systems with multiple inputs (MISO) in a straightforward manner if they

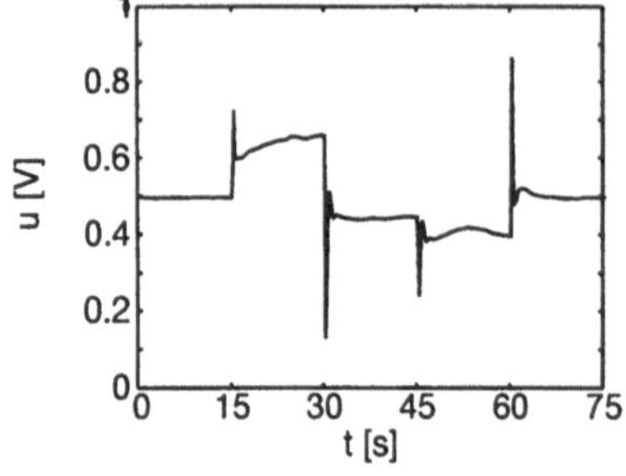

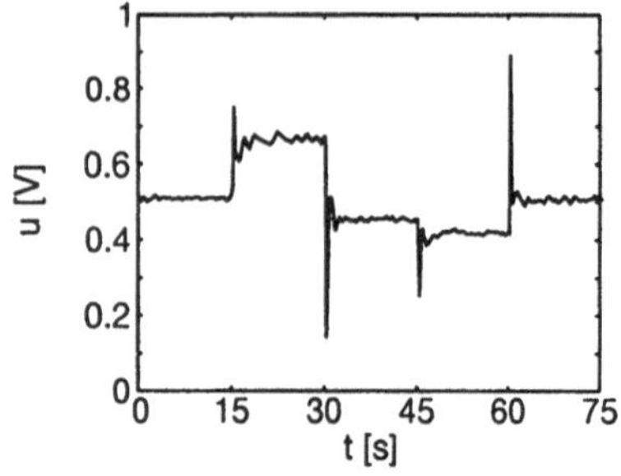

Figure 33: DISOB control, $\tau = 0.8s$, actuation signal.

Figure 34: DISOB control, $\tau = 0.3s$, $T_n = 0.3s$, actuation signal.

have only one manipulated variable and an arbitrary number of measurable disturbances. The inverse fuzzy models can be obtained by means of specialized inverse learning from forward process models. The LOLIMOT algorithm was used to determine the structure of the rule antecedents in the inverse model. The parameters in the rule consequents were optimized by an interactive optimization routine. The design of the feedback controller is relatively simple because only two disturbance observer parameters must be chosen. General hints for tuning were derived from frequency domain analysis. The control of a pilot plant proved the applicability of the approach to real-world processes suffering from disturbances like measurement noise and fuzzy model deviations. The design procedure also demonstrated the interpretability of the inverse fuzzy models. If the gains and poles of the difference equations in the rule consequents are determined they can be used to determine whether the inverse model compensates well for the plant's dynamics in different operating regimes. This is a great advantage of employing fuzzy models in comparison with black-box approximators such as artificial neural networks. In future research, an online-adaptation of the inverse model will be implemented, and the disturbance observer parameters will be tuned by a knowledge-based system, which also supervises stability.

References

[1] R. Babuska, J. Šousa, and H. B. Verbruggen. Model-based design of fuzzy control systems. In *Proceedings of the Third European Congress on Fuzzy and Intelligent Technologies, Aachen, Germany*, pages 837–841, 1995.

[2] R. Babuska and H. B. Verbruggen. An overview of fuzzy modeling for control. *Control Engineering Practice*, 4(11):1593–1606, 1996.

[3] L. Boullart, A. Krijgsman, and R. A. Vingerhoeds, editors. *Application of Artificial Intelligence in Process Control.* Pergamon Press, Oxford, 1992.

[4] H. A. B. te Braake, R. Babuska, and E. v. Can. Fuzzy and neural models in predictive control. *Journal A*, 35(3):44–51, 1994.

[5] M. Brown and C. Harris. *Neurofuzzy Adaptive Modelling and Control.* Prentice Hall, New York, 1994.

[6] J. C. Doyle, B. A. Francis, and A. R. Tannenbaum. *Feedback Control Theory.* Macmillan Publishing Co., New York, 1992.

[7] C. G. Economou, M. Morari, and B. Palsson. Internal model control. 5. extension to nonlinear systems. *Ind. Eng. Chem. Process Des. Dev.*, 21:403–411, 1986.

[8] M. Fischer and R. Isermann. Robust hybrid control based on inverse fuzzy process models. In *Proceedings of the IEEE International Conference on Fuzzy Systems, New Orleans, LA, USA*, pages 1210–1216, 1996.

[9] M. Fischer and O. Nelles. Fuzzy model-based predictive control of nonlinear processes with fast dynamics. In *Proceedings of the Second International ICSC Symposium on Fuzzy Logic and Applications, Zürich, Switzerland*, pages 57–63, 1997.

[10] M. Fischer, O. Nelles, and D. Füssel. Tuning of pid-controllers for nonlinear processes based on local linear fuzzy models. In *Proceedings of the Fourth European Congress on Fuzzy and Intelligent Technologies, Aachen, Germany*, pages 1891–1895, 1996.

[11] K. J. Hunt, R. Haas, and R. Murray-Smith. Extending the functional equivalence of radial basis function networks and fuzzy inference systems. *IEEE Transactions on Neural Networks*, 7(3):776–781, 1996.

[12] K. J. Hunt, D. Sbarbaro, R. Zbikowski, and P.J. Gawthrop. Neural networks for control systems - a survey. *Automatica*, 28(6):1083–1112, 1992.

[13] R. Isermann. On fuzzy logic applications for automatic control, supervision and fault diagnosis. In *Proceedings of the Third European Congress on Fuzzy and Intelligent Technologies, Aachen, Germany*, pages 738–753, 1995.

[14] R. Isermann, K.-H. Lachmann, and D. Matko. *Adaptive Control Systems.* Prentice Hall, New York, 1992.

[15] T.A. Johansen. Fuzzy model based control: Stability, robustness, and performance issues. *IEEE Transactions on Fuzzy Systems*, 2(3):221–234, 1994.

[16] M. I. Jordan and D. E. Rumelhart. Forward models: Supervised learning with a distal teacher. *Cognitive Science*, 16(3):307–354, 1992.

[17] V. Kecman and B.-M. Pfeiffer. Exploiting the structural equivalence of learning fuzzy systems and radial basis function neural networks. In *Proceedings of the Second European Congress on Fuzzy and Intelligent Technologies, Aachen, Germany*, pages 58–66, 1994.

[18] J. Moody and C. Darken. Fast learning in networks of locally-tuned processing units. *Neural Computation*, 1(2):281–294, 1989.

[19] M. Morari and E. Zafiriou. *Robust Process Control.* Prentice Hall, Englewood Cliffs, 1989.

[20] R. Murray-Smith. *A Local Model Network Approach to Nonlinear Modeling.* PhD thesis, University of Strathclyde, UK, 1994.

[21] O. Nelles and R. Isermann. Identification of nonlinear dynamic systems-classical methods versus radial basis function networks. In *Proceedings of the American Control Conference, Seattle, WA, USA*, pages 3786–3790, 1995.

[22] O. Nelles and R. Isermann. Basis function networks for interpolation of local linear models. In *Proceedings of the IEEE Conference on Decision and Control, Kobe, Japan*, 1996.

[23] K. Ohnishi. A new servo method in mechatronics. *Transactions of Japanese Society of Electrical Engineers*, 107(D):83–86, 1987.

[24] D. Psaltis, A. Sideris, and A. A. Yamamura. A multilayered neural network controller. *IEEE Control Systems Magazine*, 8(2):17–21, 1988.

[25] M. Sugeno and G.T. Kang. Structure identification of fuzzy model. *Fuzzy Sets and Systems*, 28(1):15–33, 1988.

[26] T. Takagi and M. Sugeno. Fuzzy identification of systems and its application to modeling and control. *IEEE Transactions on Systems, Man, and Cybernetics*, 15(1):116–132, 1985.

[27] T. Umeno and Y. Hori. Two degrees of freedom controllers for robust servomechanism: Their application to robot manipulators without speed sensors. In *Proceedings of the 1990 IEEE International Workshop on Advanced Motion Control, Yokohama, Japan*, pages 179–188, 1990.

[28] T. Umeno and Y. Hori. Robust speed control of dc servomotors using modern two degrees-of-freedom controller design. *IEEE Transactions on Industrial Electronics*, 38(5):363–368, 1991.

[29] L.-X. Wang. *Adaptive Fuzzy Systems and Control - Design and Stability Analysis.* Prentice Hall, Englewoods Cliffs, 1994.

[30] Y. W. Yoshinari, W. Pedrycz, and K. Hirota. Construction of fuzzy models through clustering techniques. *Fuzzy Sets and Systems*, 54(2):157–165, 1993.

Inverse Fuzzy Model Based Predictive Control

R. Babuška, J.M. Sousa and H.B. Verbruggen
Control Laboratory
Department of Electrical Engineering
Delft University of Technology
PO Box 5031, 2600 GA Delft, The Netherlands

1 Introduction

Conventional fuzzy control is based on expert knowledge in the form of fuzzy if–then rules. Practice shows, however, that it is often not possible to collect sufficient information to design a well-performing fuzzy controller. Human control skills are generally difficult to verbalize, since the operator's control strategy is often based on the simultaneous use of various control principles, combining feedforward, feedback, and predictive strategies in a complex, time-varying fashion. In such a case, the operator may not be able to explain why a particular control action is chosen. Moreover, the rules provided by different operators are often contradictory.

It is thus not surprising that many fuzzy controllers described in the literature are nothing but rule-based representations of classical P(ID) schemes. The desired controller behavior is achieved by tuning the membership functions, scaling factors, and other parameters by a trial-and-error method, using simulation or experiments on the process or its scale model. For complex systems, this tuning may become a tedious and time-consuming trial-and-error procedure. In an industrial environment, on-line trial-and-error controller tuning is often not acceptable for safety, economic and environmental reasons. Hence, methods are needed to alleviate the ad hoc tuning procedures, while preserving the advantages of fuzzy control, such as the capability to control nonlinear systems in a transparent way, and the possibility to include heuristic knowledge.

In this article, we present a method to design a controller based on a fuzzy model of the process. Like in conventional control, the design consists of two steps: 1) system modeling and identification, 2) controller synthesis. However, in contrast to conventional methods, fuzzy models can accurately approximate also highly nonlinear processes and can smoothly integrate prior knowledge with information obtained from process data. If no prior knowledge is available, the rules and membership functions can be extracted from the process measurements, using various techniques, such as fuzzy clustering (Babuška and Verbruggen, 1995; Kaymak and Babuška, 1995; Yoshinari, et al., 1993; Zhao, et al., 1994), neural learning methods or orthogonal least squares (Wang, 1994). Fuzzy modeling and identification are not further discussed in this article, interested readers are referred to the literature cited. An overview of fuzzy modeling for control is given, for instance, by Babuška and Verbruggen (1996).

The control structure used in this article is the standard *nonlinear internal model control* (IMC) scheme (Economou, et al., 1986), depicted in Fig. 1.

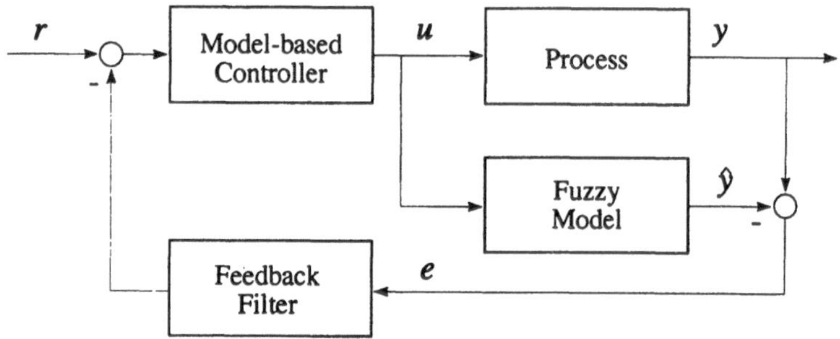

Figure 1: Nonlinear internal model control scheme.

The IMC scheme consists of four blocks: the process to be controlled, a fuzzy model of the process, a controller which is based on this fuzzy model, and a feedback filter. The controller is usually implemented as an inverse of the process' dynamics or as a predictive controller. The purpose of the internal fuzzy model working in parallel with the process is to subtract the effect of the control action from the process output. If the predicted and the measured process outputs are equal, the error e is zero and the controller works in an open-loop configuration. If a disturbance acts on the process output, the feedback signal e is equal to the influence of the disturbance and is not affected by the effects of the control action. This signal is simply subtracted from the reference. With a perfect process model, the IMC scheme is hence able to cancel the effect of unmeasured output-additive disturbances. The feedback filter is introduced in order to filter out the measurement noise and to stabilize the loop by reducing the loop gain.

Two basic properties of the ideal IMC are inherent stability and perfect control. Inherent stability means that if the controller and the process are input–output stable and a perfect model of the process is available, the closed-loop system is input–output stable. If the system is not input–output stable, but it can be stabilized by feedback, IMC can still be applied. Perfect control means that if the controller is an exact inverse of the model, the control is error-free.

In this article, we present two methods to design the controller in the IMC scheme. The first approach uses an analytic inversion of the fuzzy model. Two model structures are considered: the affine Takagi–Sugeno fuzzy model and the linguistic fuzzy model with singleton consequents. The presented methods differ from several other approaches to fuzzy model inversion found in the literature (Braae and Rutherford, 1979; Pedrycz, 1993; Driankov, et al., 1993; Harris, et al., 1993; Raymond, et al., 1995) in that they result in an exact inverse of the model, which guarantees the above-mentioned properties of the nominal IMC controller. However, our approach does not aim at a linguistic interpretation of the controller, as opposed, for instance, to the approach of Braae and Rutherford (1979).

The second approach uses predictive control based on the fuzzy model. The difficulties arising from the inherently non-convex optimization problem in nonlinear predictive control are obviated by using a discrete search technique. With this method, a global minimum is always found in the discretized space of the control variable. Chattering of the input signal is avoided by combining the predictive strategy with the previously described inverse control. An application of this technique to the control of an air-conditioning system is described. The controller based on a fuzzy model is compared with standard linear predictive control, and real-time control results are given.

2 Inversion of an affine Takagi–Sugeno fuzzy model

One possibility to invert a nonlinear fuzzy model is to select a specific structure. Here, we consider the following input–output Takagi–Sugeno (TS) fuzzy model ($i = 1, \ldots, K$)

$$
\begin{aligned}
R_i: \ & \textbf{If } y(k) \textbf{ is } A_{i1} \textbf{ and } \ldots \textbf{ and } y(k-n_y+1) \textbf{ is } A_{in_y} \textbf{ and} \\
& u(k-1) \textbf{ is } B_{i2} \textbf{ and } \ldots \textbf{ and } u(k-n_u+1) \textbf{ is } B_{in_u} \textbf{ then} \\
& y_i(k+1) = \sum_{j=1}^{n_y} a_{ij} y(k-j+1) + \sum_{j=1}^{n_u} b_{ij} u(k-j+1) + c_i,
\end{aligned}
\tag{1}
$$

where A_{il}, B_{il} are fuzzy sets, and a_{ij}, b_{ij}, c_i are crisp consequent (then-part) parameters. The output $y(k+1)$ of the model is computed by the weighted mean formula

$$y(k+1) = \frac{\sum_{i=1}^{K} \beta_i(y(k), \ldots, u(k-n_u+1)) y_i(k+1)}{\sum_{i=1}^{K} \beta_i(y(k), \ldots, u(k-n_u+1))}, \tag{2}$$

where β_i is the degree of fulfillment of the antecedent given by the following product

$$\beta_i(y(k), \ldots, u(k-n_u+1)) = \mu_{A_{i1}}(y(k)) \cdots \mu_{A_{in_y}}(y(k-n_y+1)) \cdot \\ \mu_{B_{i2}}(u(k-1)) \cdots \mu_{B_{in_u}}(u(k-n_u+1)). \tag{3}$$

As the antecedent of (1) does not include the input term $u(k)$, the model output $y(k+1)$ is affine in the input $u(k)$. To see this, denote the normalized degree of fulfillment

$$\lambda_i(y(k), \ldots, u(k-n_u+1)) = \frac{\beta_i(y(k), \ldots, u(k-n_u+1))}{\sum_{j=1}^{K} \beta_j(y(k), \ldots, u(k-n_u+1))}, \tag{4}$$

and substitute the consequent of (1) and the λ_i of (4) into (2)

$$y(k+1) = \sum_{i=1}^{K} \lambda_i(y(k), \ldots, u(k-n_u+1)) \left[\sum_{j=1}^{n_y} a_{ij} y(k-j+1) + \right. \\ \left. + \sum_{j=2}^{n_u} b_{ij} u(k-j+1) + c_i \right] + \sum_{i=1}^{K} \lambda_i(y(k), \ldots, u(k-n_u+1)) b_{i1} u(k). \tag{5}$$

This expression is a special case of a nonlinear affine form

$$y(k+1) = g\left(y(k), \ldots, y(k-n_y+1), u(k-1), \ldots, u(k-n_u+1)\right) + \\ + h(y(k), \ldots, y(k-n_y+1), u(k-1), \ldots, u(k-n_u+1)) u(k). \tag{6}$$

Given the goal that the model output at time step $k+1$ equals the reference output, $y(k+1) = r(k+1)$, the corresponding input, $u(k)$, is computed by a simple algebraic manipulation

$$u(k) = \frac{r(k+1) - g\left(y(k), \ldots, y(k-n_y+1), u(k-1), \ldots, u(k-n_u+1)\right)}{h(y(k), \ldots, y(k-n_y+1), u(k-1), \ldots, u(k-n_u+1))}. \tag{7}$$

In terms of (5) we obtain

$$u(k) = \frac{r(k+1)}{\sum_{i=1}^{K} \lambda_i(y(k), \ldots, u(k-n_u+1)) b_{i1}} \\ - \frac{\sum_{i=1}^{K} \lambda_i(y(k), \ldots, u(k-n_u+1))}{\sum_{i=1}^{K} \lambda_i(y(k), \ldots, u(k-n_u+1)) b_{i1}} \\ \times \left[\sum_{j=1}^{n_y} a_{ij} y(k-j+1) + \sum_{j=2}^{n_u} b_{ij} u(k-j+1) + c_i \right]. \tag{8}$$

3 Inversion of a singleton fuzzy model

In the previous section, an analytical inverse was derived for a TS fuzzy model in the input-affine structure. This section presents a method to invert a fuzzy model of the singleton type

$$
\begin{aligned}
R_i: \ &\textbf{If } y(k) \textbf{ is } A_{i1} \textbf{ and } y(k-1) \textbf{ is } A_{i2} \ldots \textbf{ and } y(k-n_y+1) \textbf{ is } A_{in_y} \\
&\textbf{and } u(k) \textbf{ is } B_{i1} \textbf{ and } u(k-1) \textbf{ is } B_{i2} \ldots \textbf{ and } u(k-n_u+1) \textbf{ is } B_{in_u} \\
&\textbf{then } y_i(k+1) = c_i, \quad i = 1, 2, \ldots, K,
\end{aligned} \tag{9}
$$

where $A_{i1}, \ldots, A_{in_y}$ and $B_{i1}, \ldots, B_{in_u}$ are fuzzy sets, c_i are crisp consequent parameters (fuzzy singletons). Note that this model can also represent systems nonlinear in $u(k)$. To simplify the notation, the rule index i will be omitted in the sequel. The considered fuzzy rule is then given by the following expression

$$
\begin{aligned}
&\textbf{If } y(k) \textbf{ is } A_1 \textbf{ and } y(k-1) \textbf{ is } A_2 \ldots \textbf{ and } y(k-n_y+1) \textbf{ is } A_{n_y} \\
&\textbf{and } u(k) \textbf{ is } B_1 \textbf{ and } u(k-1) \textbf{ is } B_2 \ldots \textbf{ and } u(k-n_u+1) \textbf{ is } B_{n_u} \\
&\textbf{then } y(k+1) \text{ is } c\,.
\end{aligned} \tag{10}
$$

Introduce a state vector $\boldsymbol{x}(k)$ containing the $n_u - 1$ past inputs, the $n_y - 1$ past outputs and the current output, i.e., all the antecedent (if-part) variables in (10) except $u(k)$. That is,

$$\boldsymbol{x}(k) = [y(k), \ldots, y(k-n_y+1), u(k-1), \ldots, u(k-n_u+1)]^T\,. \tag{11}$$

The corresponding fuzzy sets are composed into one multidimensional state fuzzy set X, by applying a t-norm operator on the Cartesian product space of the state variables

$$X = A_1 \times \cdots \times A_{n_y} \times B_2 \times \cdots \times B_{n_u}\,.$$

Denoting B_1 from (10) by B, this fuzzy rule now can be written by

$$\textbf{If } \boldsymbol{x}(k) \text{ is } X \textbf{ and } u(k) \text{ is } B \textbf{ then } y(k+1) \text{ is } c\,. \tag{12}$$

Note that the transformation of (10) into (12) is only a formal simplification of the rule base which does not change the order of the model dynamics, since $\boldsymbol{x}(k)$ is a vector and X is a multidimensional fuzzy set. Let M denote the number of fuzzy sets X_i defined for the state $\boldsymbol{x}(k)$ and N the number of fuzzy sets B_j defined for the input $u(k)$. Assuming that the rule base consists of all possible combinations of X_i and B_j, the total number of rules is $K = MN$.

The entire rule base can be represented as the following table

$$
\begin{array}{c|cccc}
 & & u(k) & & \\
\boldsymbol{x}(k) & B_1 & B_2 & \dots & B_N \\
\hline
X_1 & c_{11} & c_{12} & \dots & c_{1N} \\
X_2 & c_{21} & c_{22} & \dots & c_{2N} \\
\vdots & \vdots & \vdots & \vdots & \vdots \\
X_M & c_{M1} & c_{M2} & \dots & c_{MN}
\end{array}
\tag{13}
$$

By using the product t-norm operator, the degree of fulfillment of the rule antecedent $\beta_{ij}(k)$ is calculated as

$$\beta_{ij}(k) = \mu_{X_i}(\boldsymbol{x}(k))\mu_{B_j}(u(k)) \,. \tag{14}$$

The output $y(k+1)$ of the model is computed as an average of the consequents c_{ij} weighted by the normalized degrees of fulfillment β_{ij}

$$
\begin{aligned}
y(k+1) &= \frac{\sum_{i=1}^{M}\sum_{j=1}^{N}\beta_{ij}(k)c_{ij}}{\sum_{i=1}^{M}\sum_{j=1}^{N}\beta_{ij}(k)} \\
&= \frac{\sum_{i=1}^{M}\sum_{j=1}^{N}\mu_{X_i}(\boldsymbol{x}(k))\mu_{B_j}(u(k))c_{ij}}{\sum_{i=1}^{M}\sum_{j=1}^{N}\mu_{X_i}(\boldsymbol{x}(k))\mu_{B_j}(u(k))} \,.
\end{aligned}
\tag{15}
$$

It is easy to see that when the antecedent membership functions are triangular and form a partition, i.e., $\sum_{i=1}^{M}\mu_{X_i}(\boldsymbol{x}) = 1,\ \forall\boldsymbol{x}$, and $\sum_{j=1}^{N}\mu_{B_j}(u) = 1,\ \forall u$, the above singleton model provides multilinear interpolation between the rule consequents.

The main idea of the inversion method can be outlined as follows. Given a measurement of the state $\boldsymbol{x}(k)$, the multivariate mapping of the fuzzy model, $y(k+1) = f(\boldsymbol{x}(k), u(k))$, can be reduced to the univariate mapping $y(k+1) = f_x(u(k))$ by making use of the model structure. The subscript x denotes that f_x is obtained for the particular state $\boldsymbol{x}$. From this mapping, the inverse mapping $u(k) = f_x^{-1}(r(k+1))$ can be easily found, provided the model is invertible. The concept of invertibility and the respective conditions for the fuzzy model are related to the monotonicity of the model's input–output mapping. A fuzzy model f given by the rule base (12) and the defuzzification method (15) is *invertible* if $\forall x \in \mathcal{X}$ and $\forall y \in \mathcal{Y}$, a unique u exists such that $y = f(x, u)$. It is easy to see that in terms of the model's parameters the monotonicity translates into the following conditions

$$\text{card}(b_j) = 1, \quad j = 1, 2, \dots, N, \text{ and} \tag{16a}$$

$$
\begin{aligned}
&b_1 < b_2 < \dots < b_N \longrightarrow c_{i1} < c_{i2} < \dots < c_{iN}, \quad \text{or} \\
&b_1 < b_2 < \dots < b_N \longrightarrow c_{i1} > c_{i2} > \dots > c_{iN}, \quad i = 1, 2, \dots, M \,.
\end{aligned}
\tag{16b}
$$

Here $b_j = \text{core}(B_j)$ is the core of B_j and card denotes the cardinality of a set.

Example 3.1 Figure 2 presents an example where both of the above conditions are violated. Fuzzy set B_2 does not meet the condition $\text{card}(b_2) = 1$ and furthermore $\text{core}(B_3) > \text{core}(B_1) \to c_{i3} < c_{i1}$ while $\text{core}(B_4) > \text{core}(B_3) \to c_{i4} > c_{i3}$.

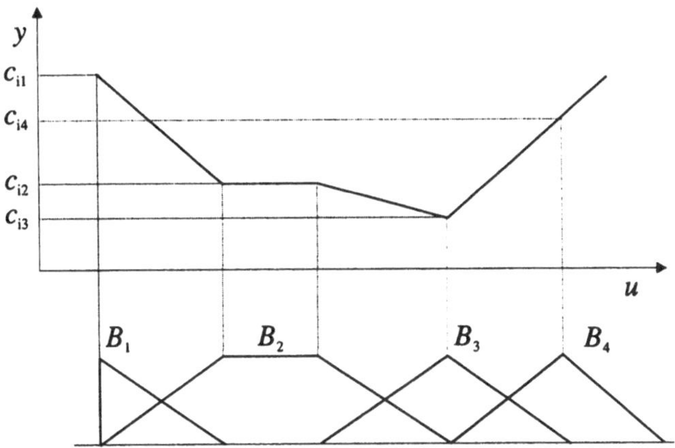

Figure 2: Example of a singleton model which does not satisfy the invertibility conditions.

The inverse of the singleton fuzzy model is formulated in the following theorem.

Theorem 3.1 (Inversion of singleton fuzzy model) *Let the process be represented by an invertible singleton fuzzy model (12) with the defuzzification method given by (15). Further, let the antecedent membership functions form a partition, i.e., let* $\sum_{i=1}^{M} \mu_{X_i}(\boldsymbol{x}) = 1,\ \forall \boldsymbol{x}$, *and* $\sum_{j=1}^{N} \mu_{B_j}(u) = 1,\ \forall u$.

For a given state $\boldsymbol{x}(k)$, *the inverse of this singleton model is given by the following rules*

$$\textbf{If } r(k+1) \textit{ is } C_j(k) \textbf{ then } u(k) \textit{ is } B_j, \quad j = 1, 2, \ldots, N, \tag{17}$$

where C_j *are fuzzy sets with triangular membership functions given by*

$$\mu_{C_1}(y) = \max\left(0, \min(1, \frac{c_2 - y}{c_2 - c_1})\right), \tag{18a}$$

$$\mu_{C_j}(y) = \max\left(0, \min(\frac{y - c_{j-1}}{c_j - c_{j-1}}, \frac{c_{j+1} - y}{c_{j+1} - c_j})\right), \quad 1 < j < N, \tag{18b}$$

$$\mu_{C_N}(y) = \max\left(0, \min(\frac{y - c_{N-1}}{c_N - c_{N-1}}, 1)\right), \tag{18c}$$

and the cores c_j are given by

$$c_j = \sum_{i=1}^{M} \mu_{X_i}(\boldsymbol{x}(k))c_{ij}, \quad j = 1, \ldots, N. \tag{19}$$

The inference (and defuzzification) of the rules (17) is accomplished by the fuzzy-mean method

$$u(k) = \sum_{j=1}^{N} \mu_{C_j}(r(k+1))b_j, \tag{20}$$

where $b_j = \text{core}\, B_j$. *The series connection of the controller and the inverse model, shown in Fig. 3, gives an identity mapping (perfect control) if a* $u(k)$ *exists such that* $r(k+1) = f(\boldsymbol{x}(k), u(k))$. *If no such* $u(k)$ *exists, the difference* $|r(k+1) - f_x(f_x^{-1}(r(k+1)))|$ *is minimal.*

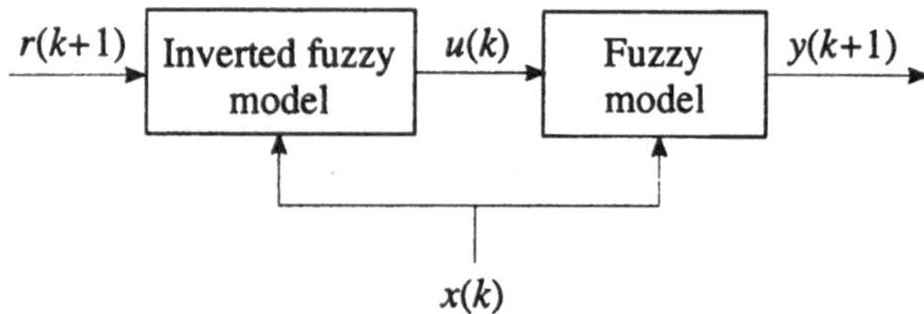

Figure 3: Series connection of the fuzzy model and the controller based on the inverse of this model.

The proof is given in the Appendix.

Remark 1: Note that the membership functions C_j depend on the process state, see equations (18) and (19). The control law is hence dynamically varying according to the state.

Remark 2: For a non-invertible rule base a set of possible control commands can be found by splitting the rule base into two or more invertible parts. For each part, a control action is found by using Theorem 3.1. Among these control actions, only one has to be selected, which requires some additional criteria, such as minimal control effort (minimal $u(k)$ or $|u(k) - u(k-1)|$), etc. Note, however, that this solution is in general different from the solution obtained by predictive control where the control effort is included in the objective function. The invertibility of the fuzzy model can also be checked in run-time, by checking the monotonicity of the aggregated consequents c_j with respect to the cores of the input fuzzy sets b_j, see (19). This is useful, since nonlinear models can be noninvertible only locally, resulting in a kind of exception in the inversion algorithm. Moreover, for models adapted on line, this check is necessary.

Remark 3: For fuzzy models with input delays $y(k+1) = f(\boldsymbol{x}(k), u(k-n_d))$, the inversion cannot be applied directly, as it would give a control action $u(k-n_d)$, n_d steps delayed. In order to generate the appropriate value $u(k)$, the model must be inverted n_d samples ahead, i.e., $u(k) = f^{-1}(r(k+n_d+1), x(k+n_d))$, where

$$\begin{aligned} x(k+n_d) = [&y(k+n_d), \ldots, y(k+1), \ldots \\ &y(k-n_y+1+n_d), u(k-1), \ldots, u(k-n_u+1)] \,. \end{aligned} \tag{21}$$

The unknown values, $y(k+1), \ldots, y(k+n_d)$, are predicted recursively using the fuzzy model

$$\begin{aligned} y(k+i) &= f(x(k+i-1), u(k+i-1)), \\ x(k+i) &= [y(k+i), \ldots, y(k-n_y+1+i), \\ &\quad u(k-1+i), \ldots u(k-n_u+1+i)], \quad \text{for } i = 1, \ldots, n_d \,. \end{aligned} \tag{22}$$

Apart from the computation of the membership degrees, both the model and the controller can be implemented using matrix operations and linear interpolations, which makes the algorithm suitable for real-time implementation.

Example 3.2 Consider a fuzzy model of the form $y(k+1) = f(y(k), y(k-1), u(k))$ where two linguistic terms {*low*, *high*} are used for $y(k)$ and $y(k-1)$ and three terms {*small*, *medium*, *large*} for $u(k)$. The complete rule base consists of $2 \times 2 \times 3 = 12$ rules

If $y(k)$ is *low* and $y(k-1)$ is *low* and $u(k)$ is *small* **then** $y(k+1)$ is c_{11}

If $y(k)$ is *low* and $y(k-1)$ is *low* and $u(k)$ is *medium* **then** $y(k+1)$ is c_{12}

...

If $y(k)$ is *high* and $y(k-1)$ is *high* and $u(k)$ is *large* **then** $y(k+1)$ is c_{43}

In this example, $\boldsymbol{x}(k) = [y(k), y(k-1)]$, $X = \{(low \times low), (low \times high), (high \times low), (high \times high)\}$, $M = 4$ and $N = 3$. The rule base is represented by the table

$\boldsymbol{x}(k)$	$u(k)$ *small*	*medium*	*large*
X_1 (*low* × *low*)	c_{11}	c_{12}	c_{13}
X_2 (*low* × *high*)	c_{21}	c_{22}	c_{23}
X_3 (*high* × *low*)	c_{31}	c_{32}	c_{33}
X_4 (*high* × *high*)	c_{41}	c_{42}	c_{43}

(23)

Given the state $\boldsymbol{x}(k) = [y(k), y(k-1)]$, the degree of fulfillment of the first antecedent proposition "$\boldsymbol{x}(k)$ is X_i", is calculated as $\mu_{X_i}(\boldsymbol{x}(k))$. For X_2, for instance, $\mu_{X_2}(\boldsymbol{x}(k)) = \mu_{\text{low}}(y(k))\mu_{\text{high}}(y(k-1))$. Using (19), one obtains the cores $c_j(k)$

$$c_j(k) = \sum_{i=1}^{4} \mu_{X_i}(\boldsymbol{x}(k))c_{ij}, \quad j = 1, 2, 3 \,. \tag{24}$$

An example of membership functions for fuzzy sets C_j, obtained by (18), is shown in Fig. 4.

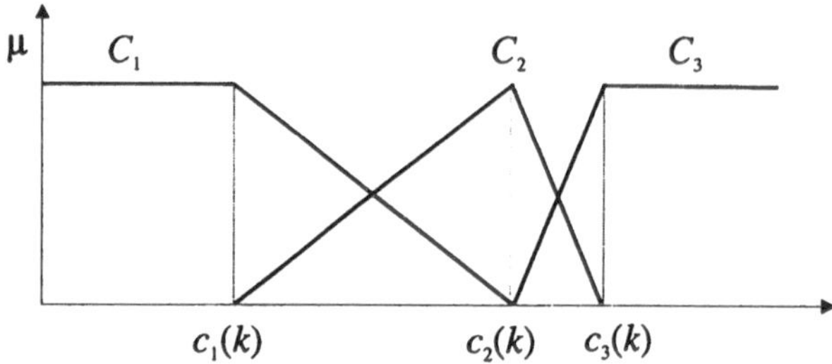

Figure 4: Fuzzy partition created from $c_1(k)$, $c_2(k)$ and $c_3(k)$.

Assuming that $b_1 < b_2 < b_3$, the model is (locally) invertible if $c_1 < c_2 < c_3$ or if $c_1 > c_2 > c_3$. In such a case, the following rules are obtained

$$R_i: \textbf{If } r(k+1) \text{ is } C_i \textbf{ then } u(k) \text{ is } B_i, \quad i = 1, 2, 3\,.$$

Otherwise, if the model is not invertible, e.g., if $c_1 > c_2 < c_3$, the above rule base can be split in two rule bases. The first contains rules 1 and 2 and the second contains rules 2 and 3.

4 Predictive control

Model-based predictive control (MBPC) is a general methodology for solving control problems in the time domain. It is based on three main concepts: 1) Explicit use of a model to predict the process output at future discrete time instants, over a prediction horizon. 2) Computation of a sequence of future control actions over a control horizon by minimizing a given objective function. 3) Receding horizon strategy, so that only the first control action in the sequence is applied, the horizons are moved towards the future and optimization is repeated. Because of the optimization approach and the explicit use of the process model, MBPC can realize multivariable optimal control, deal with nonlinear processes, and can efficiently handle constraints.

Prediction and control horizons. The future process outputs are predicted over the *prediction horizon* H_p using a model of the process: $\hat{y}(k+i)$ for $i = 1, \ldots, H_p$. These values depend on the current process state, and on the future control signals $u(k+i)$ for $i = 0, \ldots, H_c - 1$, where $H_c \leq H_p$ is the *control horizon*. The control variable is manipulated only within the control horizon and remains constant afterwards, i.e., $u(k+i) = u(k+H_c-1)$ for $i = H_c, \ldots, H_p - 1$, see Fig. 5.

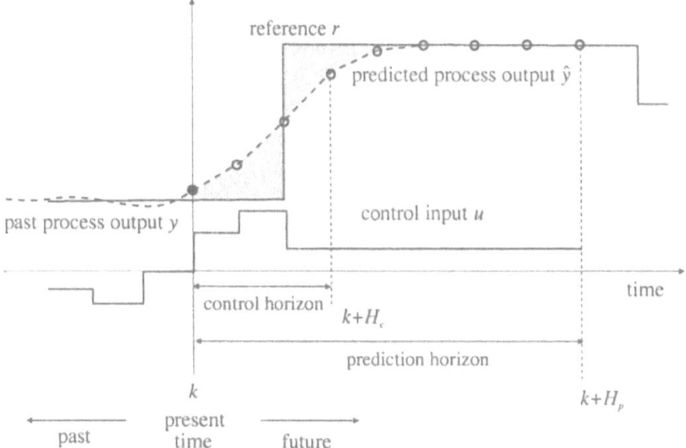

Figure 5: The basic principle of model-based predictive control.

Optimization of an objective function. The sequence of future control signals $u(k+i)$ for $i = 0, \ldots, H_c - 1$ is computed by optimizing a given objective (cost) function, in order to bring and keep the process output as close as possible to the given reference trajectory r, which can be the set-point itself or, more often, some filtered version of it. Most often, the objective functions used are modifications of the following quadratic function (Clarke, et al., 1987)

$$J = \sum_{i=1}^{H_p} \alpha_i \left(r(k+i) - \hat{y}(k+i)\right)^2 + \sum_{i=1}^{H_c} \beta_i \Delta u(k+i-1)^2 . \tag{25}$$

The first term accounts for minimizing the variance of the process output from the reference, while the second term represents a penalty on the control effort (related, for instance, to energy). The latter term can also be expressed by using u itself, or other filtered forms of u, depending on the problem (Soeterboek, 1992). The coefficients α_i and β_i define the weighting of the output error and the control effort with respect to each other, and with respect to the prediction step.[1] For systems with a dead time of n_d samples, only outputs from time $k + n_d$ are considered in the objective function, because outputs before this time cannot be influenced by the control signal $u(k)$. Similar reasoning holds for nonminimum phase systems. Level and rate constraints of the control input, or other process variables can be

[1] Tuning rules were suggested for these parameters (Soeterboek, 1992).

specified as a part of the optimization problem. Generally, any other suitable cost function can be used, but for a quadratic cost function, a linear, time-invariant model, and, in the absence of constraints, an explicit analytic solution of the above optimization problem can be obtained. Otherwise, in the presence of nonlinearities and constraints, a nonconvex optimization problem must be solved iteratively at each sampling period. This hampers the application of nonlinear MBPC to fast systems where iterative optimization techniques cannot be properly used, due to short sampling times and extensive computation times. Iterative optimization algorithms, such as the Nelder-Mead method or sequential quadratic programming, usually converge to local minima, which results in poor solutions of the optimization problem. An alternative approach is presented in the following section. The optimization problem is formulated as a search in a discrete space of control actions. A discrete optimization method is then used to find an optimal control action.

Receding horizon principle. Only the control signal $u(k)$ is applied to the process. At the next sampling instant, the process output $y(k+1)$ is available and the optimization and prediction can be repeated with the updated values. This is called the receding horizon principle.

4.0.1 Branch-and-bound optimization

The branch-and-bound (B&B) method is a structured search technique that belongs to a general class of combinatorial programming methods (Lawler and Wood, 1966; Mitten, 1970). It is useful for solving problems for which direct solution methods either do not exist or are inefficient. The B&B method is based on the fact that, in general, only a small number of the possible solutions need actually to be enumerated, while the remaining solutions are eliminated through the application of bounds. The B&B algorithm involves two basic operations: *branching*, i.e., dividing possible solutions into subsets, and *bounding*, i.e., eliminating those subsets that are known not to contain the solutions. The basic B&B scheme is a recursive application of these two operations.

Figure 6 illustrates the principle of the branch-and-bound method. The fuzzy model is schematically denoted by

$$\boldsymbol{x}(k+1) = f(\boldsymbol{x}(k), u(k)), \tag{26}$$

where $\boldsymbol{x}(k)$ and $u(k)$ are the state and the input at time step k, respectively. The input $u(k)$ takes values from a set of linguistic terms B_j (or other suitable values)

$$\boldsymbol{B} = \{B_1, B_2, \ldots, B_N\}. \tag{27}$$

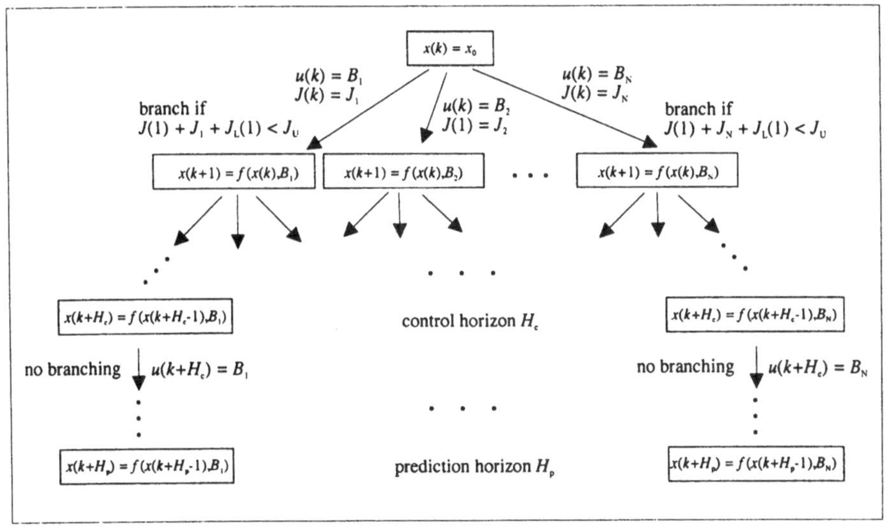

Figure 6: Branch-and-bound optimization.

Let $i = 1, 2, \ldots, H_p$ denote the level of the tree ($i = 0$ at the initial node) and let j denote the branch corresponding to the control alternative B_j. At level i of the tree, N control alternatives are considered, yielding a maximum of N branches. Clearly, application of the branching alone would result in the search of the entire tree, i.e., N^{H_c} possibilities, which is computationally prohibitive, except for very small problems. A substantial part of the search space can be eliminated by employing lower and upper bounds on the cost function. A particular branch j at level i is followed only if the cumulative cost $J(i)$, given by

$$J(i) = \sum_{m=1}^{i} \alpha_m \left(r(k+m) - \hat{y}(k+m)\right)^2 + \sum_{m=1}^{i} \beta_m \Delta u(k+m-1)^2, \qquad (28)$$

plus the cost from the level i to the terminal level H_p is lower than an upper bound on the total cost, denoted J_U. The cost from level i to level H_p is generally unknown, but can be expressed as a sum of two terms. The first one is the cost $J_j(i)$, associated with the transition $\boldsymbol{x}(k+1) = f(\boldsymbol{x}(k), B_j)$, which is computed by evaluating the respective element in the cost function (25). The second one is an estimated lower bound on the cost over the remaining steps $i+1, \ldots, H_p$, denoted $J_L(i+1)$. Hence the condition for branching reads

$$J(i) + J_j(i) + J_L(i+1) < J_U. \qquad (29)$$

Note that no branching takes place for $i > H_c - 1$ (beyond the control horizon), i.e., the last control move $u(k + H_c - 1)$ is applied recursively to the model, until H_p is reached.

The efficiency of the bounding mechanism depends on the quality of the bound estimates. The upper bound should be as low as possible and the lower bound as large as possible, in order to decrease the number of branches. The availability of these estimates depends on the particular problem. If no mechanism for computing the bounds is available, the upper bound is initially set to a sufficiently large value (∞). The first path in the tree search follows the "greedy" strategy of choosing the smallest $J_j(i)$ at each level i. This corresponds to the inverse model solution as described in Section 2, where the control actions $u(i)$ only take values B_j, $j = 1, 2, \ldots, N$. When following a constant or slowly varying references, the terminal cost $J(H_p)$, in most cases, represents the optimum, or a close upper limit of it. The upper bound is set to this value, i.e., $J_U = J(H_p)$. If at a later stage of the tree search, $J(H_p)' < J_U$ is found, J_U is replaced by $J(H_p)'$. In the absence of a better estimate, the lower bound is simply set to $J_L(i) = 0$ for all $i = 1, 2, \ldots, H_c - 1$. Practical experience with this algorithm shows that these 'worst-case' estimates still prevent the algorithm from exploring a large portion of the search space. However, one should not forget that the computational complexity of the algorithm remains exponential, which makes it prohibitively expensive for large H_c and N.

The B&B optimization technique applied to predictive control has three major advantages over other nonlinear optimization methods:

1. The global optimum (minimum in the above formulation) is always found (intrinsic property of the B&B method). This is a significant advantage, as it guarantees the optimality of the controller performance, in the discrete space of control alternatives, without any assumptions as to the form of the cost function. Some issues connected with this discretization are discussed below.
2. The algorithm does not need any initial guess and hence its performance cannot be negatively influenced by a poor initialization, as in the case of iterative optimization methods.
3. The B&B method implicitly deals with constraints. In fact, the presence of constraints improves the efficiency of bounding, as it restricts the search space by eliminating the control alternatives that result in violation of the constraints. Many other optimization techniques perform worse when tight constraints are imposed. The B&B technique also deals with discrete control alternatives in a natural way. In many industrial systems, some of the control variables are restricted to discrete values, which presents problems for numerical techniques based on the computation or estimation of gradients.

Apart from the prohibitive computational complexity for large problems, the most serious drawback of the described approach is the restriction of the possible control actions to a set of discrete alternatives (linguistic terms, or other suitably chosen values). For a continuous control space, this discretization may cause oscillations of the outputs around the reference trajectory. To avoid this undesirable feature, a control scheme is proposed that combines the B&B predictive optimization strategy with the continuous control law by inverting the plant model described previously.

Predictive control can be regarded as a generalization of the inverse model control approach. Assume that $H_p = H_c = 1$ and $\beta_m = 0$, $\forall m$ in (25), and that a control command $u(k)$ exists such that $y(k+1) = r(k+1)$ (i.e., the inverse exists), a global optimum of the functional (25) yields $J^* = 0$ and $u(k)^* = u(k)$. In other words, without constraints and without penalizing the control action, the one-step-ahead predictive control strategy is equivalent to the inverse model control strategy, where the inverse is computed numerically, by means of function minimization. Extending the prediction and control horizons, adding the control signal to the objective function, and including constraints can be regarded as generalization of inverse model control.

This observation led to the idea of combining both control strategies (Babuška, et al., 1995). When a recursive application of the inverse model control law over the entire prediction horizon results in a violation of a constraint (at any step), the B&B scheme is used, since predictive control will result in better performance. However, if no constraints are violated, the first control action computed by the inverse model is applied to the process.

The proposed scheme avoids oscillations that would otherwise occur because of the discretization of the control space. With constant or slowly varying references, the constraints are typically not violated, and the inversion can be applied, yielding a continuous (interpolated) control action.

5 Example: application to an air-conditioning system

The presented control schemes have been applied to temperature control of a fan-coil unit, which is a part of an air conditioning system installed in a test cell at TU Delft (van Paassen and Lute, 1993). Hot water at 65 °C is supplied to the coil which exchanges the heat between the water and the surrounding air. In the fan-coil unit, the air coming from the outside (primary air) is mixed with the return air from the room (recirculated or secondary air). The flows of primary and secondary air are controlled by the outside and return dampers, and by the speed of the fan which forces the air to pass through the coil, heating or cooling the air. A simplified diagram of the system is given in Fig. 7.

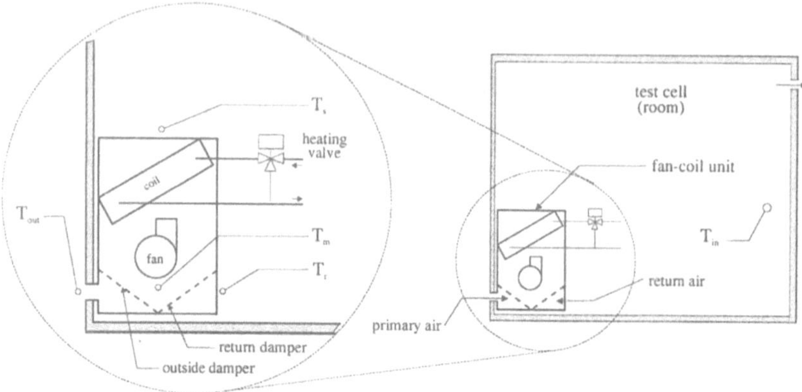

Figure 7: A schematic diagram of the air conditioning system.

The global control goal for this system is to keep the temperature of the test cell T_{in} at a certain reference value, and assures that enough ventilation and renovated air is supplied to the room. For this purpose three different control actions can be used.

1. Speed of the fan. The fan has three different velocities: low, medium and high.
2. Position of the outside and return dampers. The dampers can be open in different positions, controlling the amounts of air coming from outdoors and returned from the test cell.
3. Position of the heating valve. The amount of water entering the heat exchanger is controlled by the heating valve.

In this article, we present results for a part of the above control problem. The fan is kept at low speed and both dampers are half open, allowing ventilation from the outside, and the return of some air from the test cell to the fan-coil. Only the heating valve is used as a control input. As shown in Fig. 7, temperatures can be measured at different places of the installation. For controlling the room temperature T_{in}, the most relevant measurement point is the supply temperature T_s, measured just after the coil, see Fig. 7.

First the TS model with the affine structure (1) was constructed from process measurements. The inputs of the model are the opening of the heating valve ($u(k) \in [0, 1]$, with 0 standing for the valve completely closed), and the mixed air temperature T_m measured just before the fan. The supply temperature T_s is the output of the model. The input variables were selected on the basis of correlation analysis and physical understanding of the process.

By denoting $y_1 = \mathrm{T_s}$ and $y_2 = \mathrm{T_m}$, the state vector $\boldsymbol{x}(k)$ of (11) is given by

$$\boldsymbol{x}(k) = [y_1(k), y_1(k-1), y_2(k), y_2(k-1)] \ .$$

The orders of the inputs and outputs were chosen by comparing several candidate structures of first-order and second-order models in terms of the prediction error criterion.

The antecedent membership functions and the consequent parameters were estimated from a set of input–output measurements by fuzzy clustering and least-squares methods (Babuška and Verbruggen, 1995). The identification data set contains $N = 800$ samples, collected in two different day times (morning and afternoon), using an input signal data shown in Fig. 8a. A sampling period of 30s was used. The excitation signal u consists of a multi-sinusoidal signal with five different frequencies and amplitudes, and of pulses with random amplitude and width. This signal is designed to cover the entire range of the control valve positions and to contain the important frequencies in the expected range of process dynamics. The mean value of this excitation signal is decreasing in order to avoid overheating of the test cell, see Fig. 8a. The measured mixed-air temperature $\mathrm{T_m}$ is given in Fig. 8b.

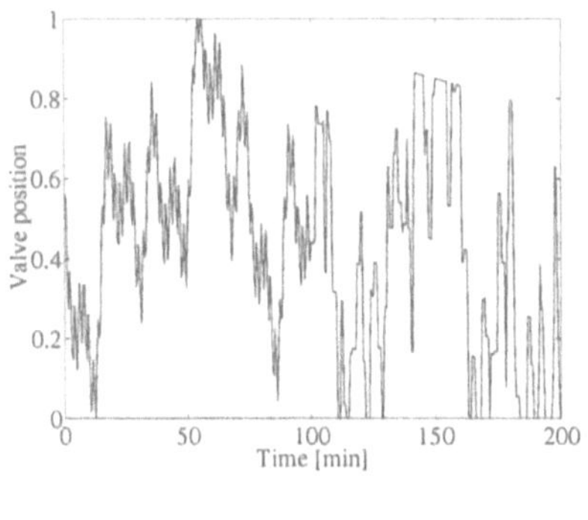

(a) Valve position.

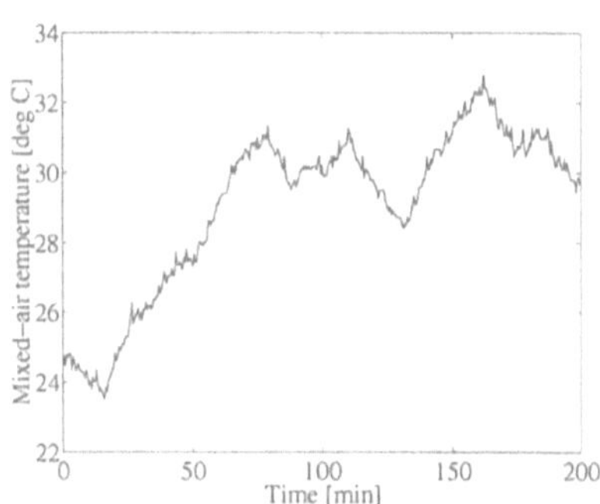

(b) Mixed-air temperature.

Figure 8: Identification data.

A separate data set was recorded on another day for validation purposes. Figure 9a shows the performance of the fuzzy model on this validation data set. To be able to evaluate the results of fuzzy modeling and control, a linear state-space model

$$\begin{aligned} \boldsymbol{x}(k+1) &= \mathbf{A}\boldsymbol{x}(\mathbf{k}) + \mathbf{Bu}(\mathbf{k}), \\ \boldsymbol{y}(k) &= \mathbf{C}\boldsymbol{x}(\mathbf{k}), \end{aligned} \tag{30}$$

was obtained by subspace identification (Verhaegen and Dewilde, 1992). This fifth-order model has one input, the heating valve, and two outputs, the mixed-air temperature T_m and the supply temperature T_s. The matrices **A**, **B** and **C** are:

$$\mathbf{A} = \begin{bmatrix} 0.880 & -0.053 & 0.529 & -0.039 & -0.151 \\ -0.070 & 0.931 & 0.239 & -0.274 & 0.133 \\ -0.091 & -0.294 & 0.104 & -0.977 & 0.107 \\ 0.001 & -0.015 & 0.051 & 0.232 & 0.784 \\ -0.001 & -0.006 & -0.007 & 0.189 & -0.258 \end{bmatrix},$$

$$\mathbf{B} = \begin{bmatrix} -15.065 \\ -8.855 \\ -6.742 \\ -0.311 \\ -0.631 \end{bmatrix},$$

$$\mathbf{C} = \begin{bmatrix} -0.220 & 0.382 & -0.051 & -0.051 & -0.085 \\ -0.518 & -0.196 & 0.816 & 0.145 & -0.064 \end{bmatrix}.$$

Fig. 9b compares the supply temperature measured and predicted by the linear model. One can see that the performance of the linear model is much worse than in the case of the fuzzy model.

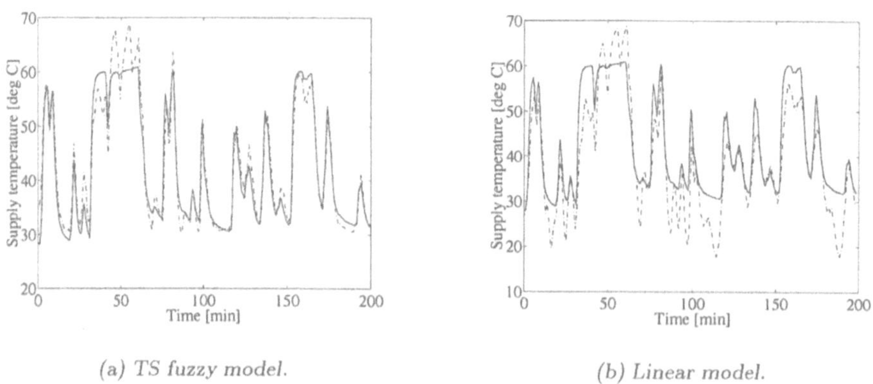

(a) TS fuzzy model.

(b) Linear model.

Figure 9: Validation results. The solid line is measured output, the dashed line is the model output.

The implementation of the IMC scheme to the fan-coil is depicted in Fig. 10. The controller's inputs are the setpoint, the predicted supply temperature $\hat{y}_1$, and the measured mixed-air temperature y_2. The error signal, $e(k) = y_1(k) - \hat{y}_1(k)$, is passed through a first-order low-pass digital Butterworth filter F_1. The filtered error $e_f(k)$ is given by: $e_f(k) =$

$b_1 e(k) + b_2 e(k-1) - a_2 e_f(k-1)$, where $b_1 = b_2 = 0.086$, $a_2 = -0.83$. Another filter F_2 is designed to filter measurement noise of y_2: $y_{2f}(k) = d_1 y_2(k) + d_2 y_2(k-1) - c_2 y_{2f}(k-1)$, with $d_1 = d_2 = 0.245$, $c_2 = -0.51$. Both filter were designed empirically, using simulation.

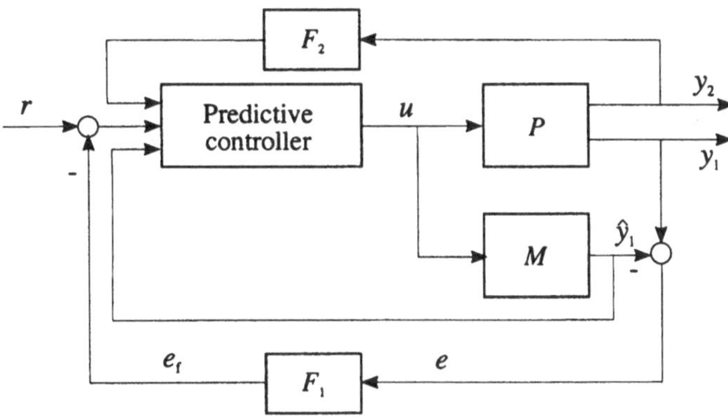

Figure 10: Implementation of the supervisory fuzzy predictive control scheme for the fan-coil using the IMC structure.

The results obtained by using the affine TS fuzzy model and the linear model within the IMC scheme are first compared in simulation. The TS fuzzy model simulates the process. Prediction and control horizons are set to $H_c = H_p = 5$. The upper graph in Fig. 11 presents a response of the supply temperature to several steps in the reference for the fuzzy predictive controller. The corresponding control actions are given in the lower graph. Figure 12 shows the same signals for the predictive controller based on the linear model.

Both controllers are able to follow the reference without steady-state errors, but the response of the fuzzy controller is much smoother than that of the controller based on the linear model. The reason is that in the latter case, the IMC scheme has to compensate severe model-plant mismatches, which causes the oscillations observed in Fig. 12.

The fuzzy controller was also tested in real time in the test cell. Figure 13 shows results obtained for $H_c = H_p = 1$. Contrary to the simulations, this setting gave the best results for the experimental process. This can be attributed to a significant model-plant mismatch which leads to inaccurate predictions for a longer horizon.

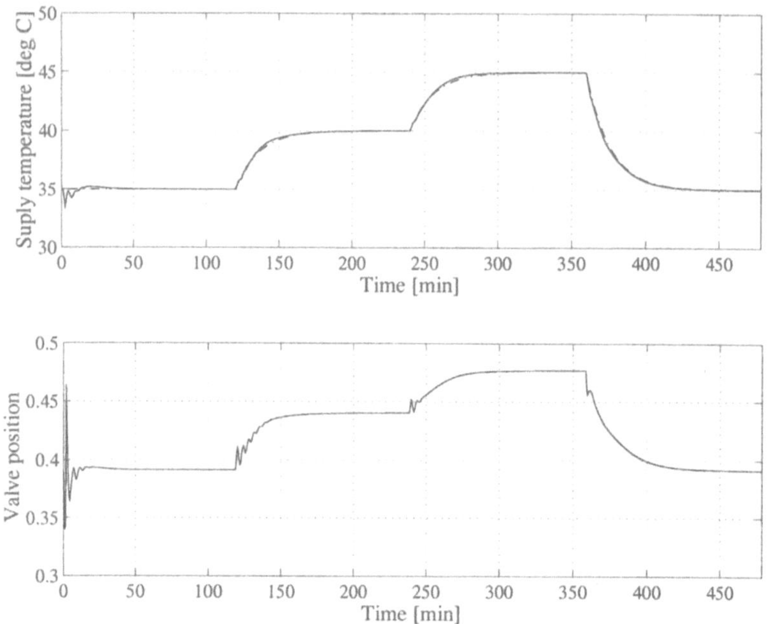

Figure 11: Simulated response of the predictive controller based on the affine TS model. Solid line is the measured output, dashed line is the reference.

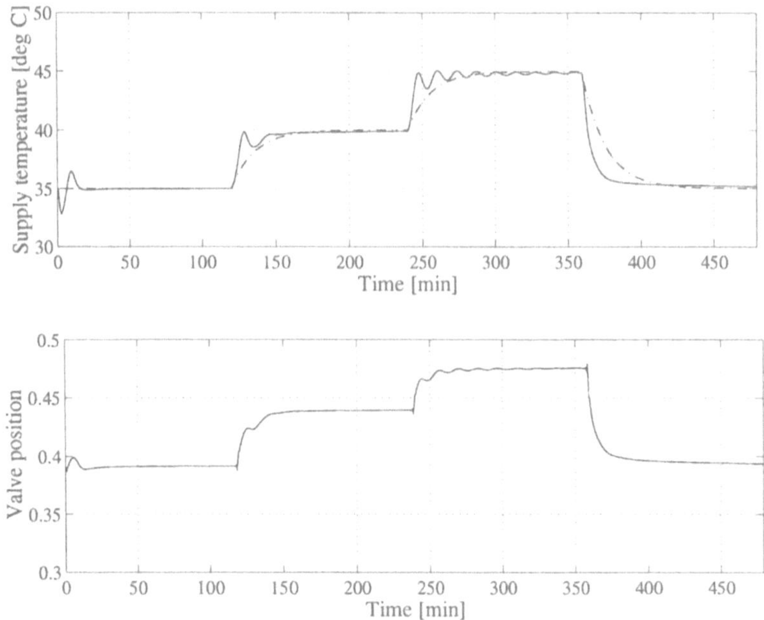

Figure 12: Simulated response of the predictive controller based on a linear model. Solid line is the measured output, dashed line is the reference.

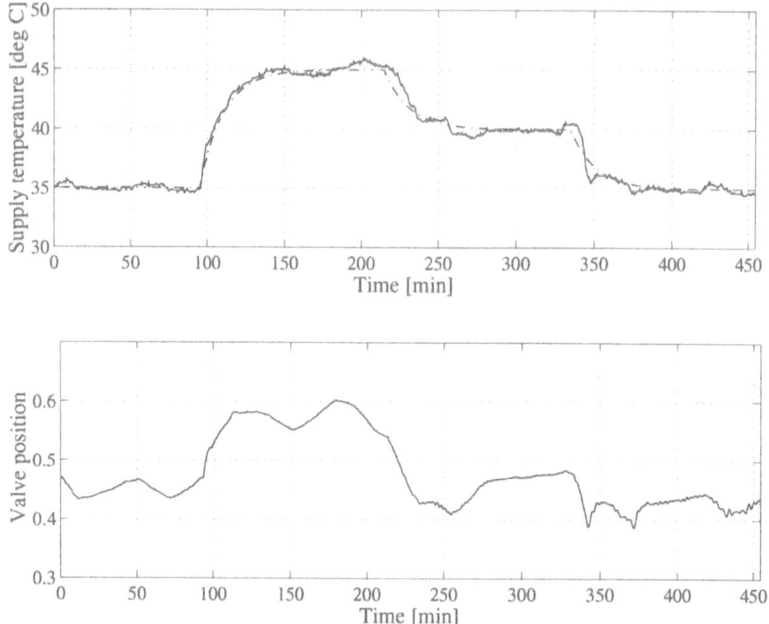

Figure 13: Real-time response using the predictive controller based on the affine TS fuzzy model. Solid line is the measured output, dashed line is the reference.

6 Conclusions

Techniques to develop a nonlinear controller based on a fuzzy model of the process under consideration have been presented. First, methods to analytically invert an affine TS fuzzy model and a singleton model were described. These methods guarantee an exact inversion of the considered fuzzy models. Both inverse models are given by closed analytical formulae, which is an advantage for a fast real-time implementation. The exactness of the inverse guarantees the theoretical properties of the nominal IMC controller, i.e., inherent stability (for stable and invertible systems), and 'perfect' control with zero error. In practice, however, modeling errors and disturbances act on the process, which gives rise to robustness issues. Robustness of the presented schemes will be studied in the future.

A more general approach is to use a fuzzy model as a predictor within a nonlinear model-based predictive control scheme. However, since fuzzy models are generally nonlinear, a nonconvex optimization problem must be solved within each sampling period. Existing iterative techniques for nonconvex optimization often converge to local solutions, which deteriorate the controller performance and hamper the practical use of the method. To circumvent the optimization problems, a discrete optimization scheme based on a branch-and-bound method was proposed. An advantage of this approach is that a global optimum in the space of discrete control actions is always found. This technique, however, is limited to short control horizons, as the computational load of the branch-and-bound technique increases rapidly with the dimension of the optimization problem. The described methods have been applied to temperature control of a fan-coil, which is a part of an air conditioning system. This process is highly nonlinear which hampers the use of linear control techniques. Simulation and real-time results of the inverse and predictive controllers are given.

7 Appendix: Proof of Theorem 3.1

Using the product t-norm for the *and* connective in the rule antecedent, rule (12) can be rewritten in the following form:

$$\textbf{If } \boldsymbol{x}(k) \text{ is } X_i \textbf{ then } (\textbf{if } u(k) \text{ is } B_j \textbf{ then } y(k+1) \text{ is } c_{ij})\,. \tag{31}$$

For a given state $\boldsymbol{x}(k)$, the degree of fulfillment of the first antecedent proposition, "$\boldsymbol{x}(k)$ is X_i", can be calculated as $\mu_{X_i}(\boldsymbol{x}(k))$. Then the consequents of the rules containing a particular B_j, that is, columns in (13), can be aggregated (there are M such rules for each B_j). The aggregated consequents

$c_j(k)$ are given by (19). A set of N rules for the input $u(k)$ is obtained

$$\textbf{If } u(k) \text{ is } B_j \textbf{ then } y(k+1) \text{ is } c_j(k), \qquad j = 1, \ldots, N. \tag{32}$$

Each of the above rules is inverted by exchanging the antecedent and the consequent, which yields the rules

$$\textbf{If } y(k+1) \text{ is } c_j(k) \textbf{ then } u(k) \text{ is } B_j \qquad j = 1, \ldots, N. \tag{33}$$

Since $c_j(k)$ are singletons, it is necessary to interpolate between the consequents $c_j(k)$ in order to obtain $u(k)$. This interpolation is accomplished by means of triangular fuzzy sets C_j defined by equations (18) and (19). To show that the series connection of the controller and of the inverse model gives an identity mapping, or a mapping with a minimal error, two situations must be distinguished

1. $r(k+1) \in [c_1, c_N]$, i.e., the desired output is in the range of $f_x(u(k))$, i.e., can be reached from the current state within one time step.
2. $r(k+1) \notin [c_1, c_N]$, , i.e., the desired output cannot be reached from the current state in one time step.

Consider the first situation. For the sake of brevity, the time argument k is omitted. As the input fuzzy sets B_j form a partition, at most, two adjacent control rules are active simultaneously. Denote these two rules by indices j and $j+1$. Using equations (18) and (20), the control input is given by

$$u = \frac{r - c_j}{c_{j+1} - c_j} b_j + \frac{c_{j+1} - r}{c_{j+1} - c_j} b_{j+1} = \frac{r(b_j - b_{j-1}) + c_{j+1}b_{j+1} - c_j b_j}{c_{j+1} - c_j}. \tag{34}$$

Analogously, the output of the fuzzy model is given by

$$y = \frac{u - b_j}{b_{j+1} - b_j} c_j + \frac{b_{j+1} - r}{b_{j+1} - b_j} c_{j+1}. \tag{35}$$

After substituting u from (34) into (35), and carrying out some elementary algebraic manipulations, the following expression is obtained

$$y = \frac{r(b_j - b_{j+1}) + c_{j+1}b_{j+1} - c_{j+1}b_j}{(c_{j+1} - c_j)(b_{j+1} - b_j)} c_j + \frac{c_j b_j - c_j b_{j+1} - r(b_j - b_{j+1})}{(c_{j+1} - c_j)(b_{j+1} - b_j)} c_{j+1}. \tag{36}$$

Further simplification leads to

$$y = \frac{c_{j+1} - r}{(c_{j+1} - c_j)} c_j + \frac{r - c_j}{(c_{j+1} - c_j)} c_{j+1} = \frac{c_{j+1}c_j - rc_j + rc_{j+1} - c_j c_{j+1}}{(c_{j+1} - c_j)} = r. \tag{37}$$

Now consider the second situation. In the case that $r(k+1) > c_N$, $\mu_{C_N}(r(k+1)) = 1$, which gives the control input $u(k) = b_N$. In the fuzzy model,

$\mu_{B_N}(u(k)) = 1$, which yields the model output $y(k+1) = c_N$. As $c_N > c_j$, $1 \le j \le N-1$, the difference $|r(k+1) - y(k+1)|$ is the least possible. Analogically for $r < c_1$, one obtains $y(k+1) = c_1$ as a result of the control action $u(k) = b_1$, which is the best possible control command, since $c_1 < c_j$, $2 \le j \le N$. □

Acknowledgements

The authors thank Dr. van Paassen of the Laboratory for Refrigerating Engineering and Indoor Climate Control of the Mechanical Engineering faculty at TU Delft for providing the test cell for experimentation. This work was partially supported by the Portuguese foundation Fundação Calouste Gulbenkian.

References

Babuška, R., J. Sousa and H.B. Verbruggen (1995). Model-based design of fuzzy control systems. In *Proceedings Third European Congress on Intelligent Techniques and Soft Computing EUFIT'95*, Aachen, Germany, pp. 837–841.

Babuška, R. and H.B. Verbruggen (1995). Identification of composite linear models via fuzzy clustering. In *Proceedings European Control Conference*, Rome, Italy, pp. 1207–1212.

Babuška, R. and H.B. Verbruggen (1996). An overview of fuzzy modeling for control. *Control Engineering Practice 4*(11), 1593–1606.

Braae, M. and D.A. Rutherford (1979). Theoretical and linguistic aspects of the fuzzy logic controller. *Automatica 15*, 553–577.

Clarke, D.W., C. Mohtadi and P.S. Tuffs (1987). Generalised predictive control. part 1: The basic algorithm. part 2: Extensions and interpretations. *Automatica 23*(2), 137–160.

Driankov, D., H. Hellendoorn and M. Reinfrank (1993). *An Introduction to Fuzzy Control.* Springer, Berlin.

Economou, C.G., M. Morari and B.O. Palsson (1986). Internal model control. 5. Extension to nonlinear systems. *Ind. Eng. Chem. Process Des. Dev. 25*, 403–411.

Harris, C.J., C.G. Moore and M. Brown (1993). *Intelligent Control, Aspects of Fuzzy Logic and Neural Nets.* Singapore: World Scientific.

Kaymak, U. and R. Babuška (1995). Compatible cluster merging for fuzzy modeling. In *Proceedings FUZZ-IEEE/IFES'95*, Yokohama, Japan, pp. 897–904.

Lawler, E.L. and E.D. Wood (1966). Branch-and-bound methods: A survey. *Journal of Operations Research 14*, 699–719.

Mitten, L.G. (1970). Branch-and-bound methods: General formulation and properties. *Journal of Operations Research 18*, 24–34.

Paassen, van, A.H.C. and P.J. Lute (1993). Energy saving through controlled ventilation windows. In *3rd European Conference on Architecture*, Florence, Italy.

Pedrycz, W. (1993). *Fuzzy Control and Fuzzy Systems (second, extended,edition)*. John Willey and Sons, New York.

Raymond, C., S. Boverie and A Titli (1995). Fuzzy multivariable control design from the fuzzy system model. In *Proceedings Sixth IFSA World Congress*, Sao Paulo, Brazil.

Soeterboek, R. (1992). *Predictive Control: A Unified Approach.* New York, USA: Prentice Hall.

Verhaegen, M. and P. Dewilde (1992). Subspace model identification. Part I: the output-error state space model identification class of algorithms. *International Journal of Control 56*, 1187–1210.

Wang, L.-X. (1994). *Adaptive Fuzzy Systems and Control, Design and Stability Analysis.* New Jersey: Prentice Hall.

Yoshinari, Y., W. Pedrycz and K. Hirota (1993). Construction of fuzzy models through clustering techniques. *Fuzzy Sets and Systems 54*, 157–165.

Zhao, J., V. Wertz and R. Gorez (1994). A fuzzy clustering method for the identification of fuzzy models for dynamical systems. In *9th IEEE International Symposium on Intelligent Control*, Columbus, Ohio, USA.

Stable Adaptive Control Using Fuzzy Systems and Neural Networks

Jeffrey T. Spooner and Kevin M. Passino
Dept. Electrical Engineering
The Ohio State University
2015 Neil Ave., Columbus, OH 43210-1272, USA

1 Introduction

Fuzzy controllers have stirred a great deal of excitement in some circles since they allow for the simple inclusion of heuristic knowledge about how to control a plant rather than requiring exact mathematical models. This can sometimes lead to good controller designs in a very short period of time. In situations where heuristics do not provide enough information to specify all the parameters of the fuzzy controller a priori, researchers have introduced adaptive schemes that use data gathered during the on-line operation of the controller, and special adaptation heuristics, to automatically learn these parameters (see e.g. [1] - [12] or the references therein). To date, stability conditions have not been provided for any of the approaches in [1] - [12], but Langari and Tomizuka [13] and others have developed stability conditions for (non-adaptive) fuzzy controllers and recently several innovative stable adaptive fuzzy control schemes have been introduced [14] - [17]. Moreover, closely related neural control approaches have been studied [18] - [23]. In this article, we seek to introduce an adaptive fuzzy or neural control approach which is guaranteed to operate properly under less restrictive assumptions and for more general continuous-time nonlinear systems.

A direct adaptive controller is introduced which attempts to directly adjust the parameters of a fuzzy or neural controller to achieve asymptotic tracking of a reference input. Within [15] a stable direct adaptive control scheme based on standard fuzzy systems was presented for a class of plants with constant input gain and no zero dynamics. Asymptotic tracking convergence was proven for this scheme if a certain approximation error is square integrable. The direct adaptive scheme in [15] thus has the same deficiencies as in [14, 19] by requiring square integrability of the approximation error.

It was shown that radial basis neural networks and standard fuzzy systems may provide asymptotic tracking of a reference signal for a class of nonlinear plants even if the estimation error is not square integrable [20, 16]. Within this article, we present a direct adaptive scheme which uses standard fuzzy systems (mamdani fuzzy systems), Takagi-Sugeno fuzzy systems, or neural networks to achieve stable tracking of a reference input for a class of plants with zero dynamics and a state dependent input gain. If knowledge of how to design the controller is available either in the form of linguistics or mathematical equations, this information may be incorporated into the direct adaptive scheme to accelerate convergence. A control smoothing scheme may be used to reduce the control action, while maintaining closed-loop stability. This article is a condensed version of [24], where more details are provided and stability conditions are derived for an analogous indirect adaptive control approach.

Our direct adaptive scheme has many differences from the existing techniques (i.e. those presented in [15, 20, 16]). Particularly, (i) the stability results presented here may be applied to systems with a state dependent input gain, whereas [15, 20] consider a class of nonlinear plants with constant input gain and [16] only considers the special case of unity gain; (ii) none of the results in [15, 20, 16] considered systems containing zero dynamics; (iii) unlike [15], our direct adaptive algorithm algorithm ensures that even if the representation error is not square integrable, then the tracking error will go to zero (or to an ϵ-boundary layer of zero for the smoothed control version); (iv) our direct adaptive controller allows for Takagi-Sugeno fuzzy systems, standard fuzzy systems, or neural networks; (v) the direct adaptive technique presented here allows for the inclusion of a known controller, u_k, so that it may be used to either enhance the performance of some prespecified controller, or stand alone as a stable adaptive controller; and (vi) furthermore our approach allows for the incorporation of heuristics about the inverse plant dynamics to speed adaptation.

It should be mentioned that other work has been completed in combining conventional stable adaptive control and intelligent control. Within [17] a nonlinear discrete-time plant is represented by a linear regression form using Takagi-Sugeno fuzzy systems to provide global stability. A discrete time adaptive routine is presented in [22] which uses layered neural networks to provide stable adaptive tracking provided some initialization conditions are satisfied. Finally, in [23], a new adaptive routine using dynamic neural networks is presented with stability investigated using a singular perturbation model of the plant [25]. Moreover, the approaches in [24] are extended to the multi-input multi-output case in [26, 27]. Other extensions to prediction, the discrete-time case, and decentralized control are given in [28, 29, 30, 31, 32].

This article is organized as follows: In Sect. 2, we define a class of Takagi-

Sugeno fuzzy systems and show that a large class of fuzzy systems and neural networks may be represented using the same functional form. The plant model used throughout our analysis is defined within Sect. 3, while Section 4 presents the direct adaptive scheme and associated stability proof. In Sect. 5 we illustrate the concepts on the longitudinal control of a vehicle in an automated lane. Section 6 contains the concluding remarks where we discuss both the advantages and disadvantages of the adaptive scheme.

2 Fuzzy systems and neural networks

In this section, we define the Takagi-Sugeno fuzzy system and show that a class of standard fuzzy systems [1] and some neural networks are a special case of this model. Fuzzy systems and neural networks are of particular interest due to their universal approximation capabilities. That is, by adjusting parameters of a sufficiently large fuzzy system or neural network one is able to represent an arbitrary smooth function over a compact region.

2.1 Takagi-Sugeno fuzzy systems

A multiple-input/single-output (MISO) fuzzy system is a nonlinear mapping from an input vector $X = [x_1, x_2, \ldots, x_n]^T \in \mathsf{R}^n$ (T denotes transpose) to an output $y = \mathcal{F}(X) \in \mathsf{R}$ (we will use $\mathcal{F}$ to represent a fuzzy system or neural network throughout our analysis). Using the Takagi-Sugeno model [34], the fuzzy system is characterized by a set of p if-then rules, stored in a rule base, expressed as

$$
\begin{array}{lllll}
R_1: & \textbf{If} & (\tilde{x}_1 \text{ is } \tilde{F}_1^i \textbf{ and } \cdots \textbf{ and } \tilde{x}_n \text{ is } \tilde{F}_n^j) & \textbf{Then} & c_1 = g_1(X). \\
\vdots & & \vdots & & \vdots \\
R_p: & \textbf{If} & (\tilde{x}_1 \text{ is } \tilde{F}_1^k \textbf{ and } \cdots \textbf{ and } \tilde{x}_n \text{ is } \tilde{F}_n^l) & \textbf{Then} & c_p = g_p(X).
\end{array}
$$

Here $\tilde{F}_b^a$ is the a-th linguistic value associated with the linguistic variable $\tilde{x}_b$ that describes input x_b, and $c_q = g_q(X)$ is the consequence of the q-th rule and $g_q : \mathsf{R}^n \to \mathsf{R}$. Using fuzzy set theory, the rule base is expressed as

[1] It is assumed that the reader has some familiarity with fuzzy systems. For an introduction, see [2, 5, 14, 33, see p. 37]

$$\begin{array}{lllll} R_1: & \textbf{If} & (F_1^i \textbf{ and } \cdots \textbf{ and } F_n^j) & \textbf{Then} & c_1 = g_1(X). \\ \vdots & & \vdots & & \vdots \\ R_p: & \textbf{If} & (F_1^k \textbf{ and } \cdots \textbf{ and } F_n^l) & \textbf{Then} & c_p = g_p(X). \end{array}$$

In the above F_b^a is a fuzzy set defined by

$$F_b^a := \left\{(x_b, \mu_{F_b^a}(x_b)) : x_b \in \mathsf{R}\right\}. \tag{1}$$

The membership function, $\mu_{F_b^a} \in [0,\ 1]$, quantifies how well the linguistic variable $\tilde{x}_b$, that represents x_b, is described by the linguistic value $\tilde{F}_b^a$. There are many ways to define membership functions [14]. For instance, Table 1 specifies triangular membership functions with "center" c and "width" w, and it specifies Gaussian membership functions with "center" c and "width" σ (see Figure 5 in Sect. 5 for graphical representation).

	Triangular
Left	$\mu(x) = \begin{cases} 1 & \text{if } x \leq c \\ \max\left(0, 1 + \frac{c-x}{w}\right) & \text{otherwise} \end{cases}$
Centers	$\mu(x) = \begin{cases} \max\left(0, 1 + \frac{x-c}{w}\right) & \text{if } x \leq c \\ \max\left(0, 1 + \frac{c-x}{w}\right) & \text{otherwise} \end{cases}$
Right	$\mu(x) = \begin{cases} \max\left(0, 1 + \frac{x-c}{w}\right) & \text{if } x \leq c \\ 1 & \text{otherwise} \end{cases}$
	Gaussian
Left	$\mu(x) = \begin{cases} 1 & \text{if } x \leq c \\ \exp\left(-\left(\frac{x-c}{\sigma}\right)^2\right) & \text{otherwise} \end{cases}$
Centers	$\mu(x) = \exp\left(-\left(\frac{x-c}{\sigma}\right)^2\right)$
Right	$\mu(x) = \begin{cases} \exp\left(-\left(\frac{x-c}{\sigma}\right)^2\right) & \text{if } x \leq c \\ 1 & \text{otherwise} \end{cases}$

Table 1: Some standard membership functions.

The antecedent fuzzy set, $F_1 \times F_2 \times \cdots \times F_n$ (fuzzy Cartesian product), of each rule is quantified by the "t-norm" [14] which may be defined by, for example, the min-operator, or the product-operator,

$$\mu_{F_1 \times \cdots \times F_n}(x_1, \ldots, x_n) := \min\left\{\mu_{F_1}(x_1), \ldots, \mu_{F_n}(x_n)\right\}, \tag{2}$$

or

$$\mu_{F_1 \times \cdots \times F_n}(x_1, \ldots, x_n) := \mu_{F_1}(x_1) \cdot \ldots \cdot \mu_{F_n}(x_n), \tag{3}$$

respectively (notice that for convenience, we have removed the superscripts from the F_b^a). Using singleton fuzzification, defuzzification may be obtained using

$$\mathcal{F}(X, A) = \frac{\sum_{i=1}^{p} c_i \mu_i}{\sum_{i=1}^{p} \mu_i}, \tag{4}$$

where $\mu_i := \mu_{F_1 \times \cdots \times F_n}(x_1, \ldots, x_n)$ is the value that the membership function (defined via 2 or 3) for the antecedent of the i-th rule takes on at $X = [x_1, \ldots, x_n]^T$. The fuzzy system parameters, A, will be defined below. It is assumed that the fuzzy system is defined so that for all $X \in \mathsf{R}^n$, we have $\sum_{i=1}^{p} \mu_i \neq 0$. We may express (4) equivalently as

$$\mathcal{F}(X, A) = c^T \zeta, \tag{5}$$

where $c^T := [c_1 \ \cdots \ c_p]$ and $\zeta^T := [\mu_1 \ \cdots \ \mu_p] / [\sum_{i=1}^{p} \mu_i]$. We assume that $\mathcal{F}(\cdot)$, the mapping produced by the fuzzy system, is Lipschitz continuous [35].

In this article, the consequents (then-parts) of each rule are taken as a linear combination of a set of Lipschitz continuous functions, $\theta_k(X) \in \mathsf{R}$, $k = 1, 2, \ldots, m-1$, so that

$$c_i = g_i(X) = a_{i,0} + a_{i,1}\theta_1(X) + \cdots + a_{i,m-2}\theta_{m-2}(X) + a_{i,m-1}\theta_{m-1}(X), \tag{6}$$

where $i = 1, \ldots, p$. Define the following:

$$z = \begin{bmatrix} 1 \\ \theta_1(X) \\ \vdots \\ \theta_{m-1}(X) \end{bmatrix} \in \mathsf{R}^m, \tag{7}$$

and

$$A^T = \begin{bmatrix} a_{1,0} & a_{1,1} & \cdots & a_{1,m-1} \\ a_{2,0} & a_{2,1} & \cdots & a_{2,m-1} \\ \vdots & \vdots & \ddots & \vdots \\ a_{p,0} & a_{p,1} & \cdots & a_{p,m-1} \end{bmatrix}. \tag{8}$$

The consequence vector associated with the fuzzy rules is now given by $c = A^T z$, so that the output of the fuzzy system may now be expressed as

$$\mathcal{F}(X, A) = z^T A \zeta. \tag{9}$$

Clearly (9) is a special form of a Takagi-Sugeno fuzzy system.

2.2 Standard fuzzy systems

Standard fuzzy systems naturally allow for the inclusion of heuristics into controller design. In standard fuzzy control, the output of a fuzzy system may be found using the center of gravity operation, which for a wide class of fuzzy systems is expressed as

$$\mathcal{F}(X) = \frac{\sum_{i=1}^{p} c_i \xi_i}{\sum_{i=1}^{p} \xi_i}, \tag{10}$$

where c_i is the center of the output membership function associated with the i-th rule, and ξ_i is the area of the implied membership function associated with the i-th rule (i.e. ξ_i is the area of the output membership function that is modified via the fuzzy implication that represents the i-th rule). This fits the form of (9) with $z = [1]$, $A = [c_1 \ldots c_p]$, and $\zeta_i = \xi_i / \sum_{i=1}^{p} \xi_i$, so that this standard fuzzy system is a special case of the Takagi-Sugeno fuzzy system defined by (9). Other standard fuzzy systems, such as those that use centroid defuzzification, will also fit the form of (9).

2.3 Neural networks

Our framework allows for the use of neural networks in which a single hidden layer of radial basis functions are used, or if a special form of two hidden layers is used. Figure 1 demonstrates these two cases. With a single hidden layer of radial basis functions, the output of neural network is given by

$$\mathcal{F}(X, A) = c^T \zeta, \tag{11}$$

where $\zeta \in \mathsf{R}^p$ are (possibly normalized) radial basis functions (e.g., squashing functions characterized by Gaussian functions [36]), and c^T is a vector of adapting weights. This type of system may be described by (9), with $z = [1]$ and $A = c^T$. As it is well-recognized in the literature, this is exactly the same representation as used with standard fuzzy systems [14].

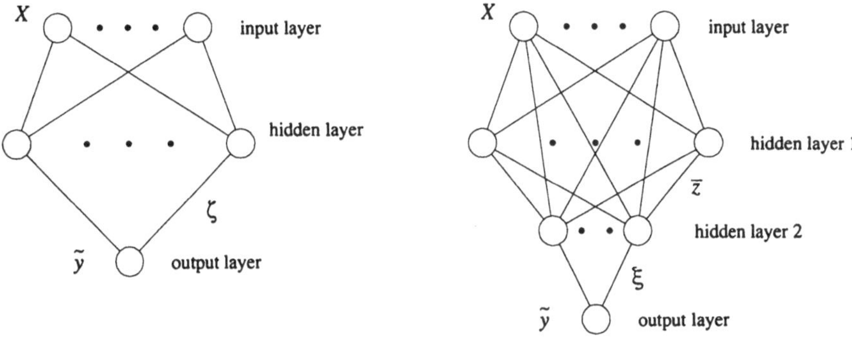

Figure 1: Two types of neural networks which may be used with the adaptive techniques.

A second type of neural network considered in this article is one in which there are two hidden layers, with the second hidden layer of a special form. The output of the first hidden layer produces a vector of functions

$$\bar{z} = [\theta_1 \cdots \theta_m] . \tag{12}$$

The nodes which make up the first hidden layer may be normalized radial basis functions, squashing functions, or any other standard neural basis function [36]. Here we allow both the output of the first hidden layer and the original input to be passed to the second hidden layer (see Fig. 1). The output of the i-th node of the second hidden layer is given by

$$\xi_i = \zeta_i(\bar{z}, X) \left(b_{i,0} + \sum_{j=1}^{m} b_{i,j}\theta_j + \sum_{j=1}^{n} b_{i,j+m}x_j \right) , \tag{13}$$

where $\zeta_i(\bar{z}, X)$ are squashing functions, or radial basis functions which may be normalized, and $b_{i,0}$ is the bias for i-th node. The output of the neural network is taken as a linear combination of the outputs of the second hidden layer, that is

$$\mathcal{F}(X, A) = \sum_{j=1}^{p} c_i \xi_i . \tag{14}$$

We may combine (13) and (14) to obtain

$$\mathcal{F}(X, A) = \sum_{j=1}^{p} \zeta_i(z, X) \left(a_{i,0} + \sum_{j=1}^{m} a_{i,j}\theta_j + \sum_{j=1}^{n} a_{i,j+m}x_j \right). \tag{15}$$

which may be expressed in the form of (9) with $z = [1\ \theta_1\ \cdots\ \theta_m x_1\ \cdots\ x_n]^T$, and $A = [a_{i,j}]$ with $a_{i,j} = c_i b_{i,j}$. Note that, z may, or may not, include any θ_i or x_i.

A key requirement on the fuzzy systems and neural networks presented above is that their adjustable parameters appear linearly in the output expression. Within the adaptive framework to follow, we shall typically refer to Takagi-Sugeno fuzzy systems within our discussion, however, any of the above fuzzy or neural network systems apply.

3 Plant model

Our objective is to design an adaptive control system which will cause the output of a relative degree r plant, y_p, to track a desired output trajectory, y_m (a relative degree r plant is one in which the plant input appears in the output dynamics after r differentiations of the output). The desired output trajectory may be defined by a signal external to the control system so that the first r derivatives of y_m may be measured, or by a reference model, with relative degree greater than or equal to r which characterizes the desired performance. With these considerations, we make the following assumption about the reference signal (let $y_m^{(r)}$ denote the r-th derivative of y_m with respect to time):

(R1) Reference Input Assumption

The desired output trajectory and its derivatives $y_m, y_m^{(1)}, \ldots, y_m^{(r)}$ are measurable and bounded.

Here we consider the SISO plant

$$\dot{X} = f(X) + g(X)u_p, \tag{16}$$

$$y_p = h(X), \tag{17}$$

where $X \in \mathsf{R}^n$ is the state vector, $u_p \in \mathsf{R}$ is the input, $y_p \in \mathsf{R}$ is the output of the plant and functions $f(X), g(X) \in \mathsf{R}^n$, and $h(X) \in \mathsf{R}$ are smooth. If the system has "strong relative degree" r, then

$$\begin{aligned} \dot{\xi}_1 &= \xi_2 = L_f h(X), \\ &\vdots \\ \dot{\xi}_{r-1} &= \xi_r = L_f^{r-1} h(X), \\ \dot{\xi}_r &= L_f^r h(X) + L_g L_f^{r-1} h(X) u_p, \end{aligned} \tag{18}$$

with $\xi_1 = y_p$, which may be rewritten as

$$y_p^{(r)} = (\alpha_k(t) + \alpha(X)) + \beta(X) u_p, \tag{19}$$

where $L_g^r h(X)$ is the r-th Lie derivative of $h(X)$ with respect to g ($L_g h(X) = \frac{\partial h}{\partial X} g(X)$, and e.g. $L_g^2 h(X) = L_g(L_g h(X))$), and it is assumed that for some $\beta_0 > 0$, we have $|\beta_k(t) + \beta(X)| \geq \beta_0$ so that it is bounded away from zero (for convenience we assume that $\beta_k(t) + \beta(X) > 0$, however, the following analysis may easily be modified for systems which are defined with $\beta_k(t) + \beta(X) < 0$). We will assume that $\alpha_k(t)$ is a known component of the dynamics of the plant (that may depend on the state) or known exogenous time dependent signals and that $\alpha(X)$ represents nonlinear dynamics of the plant that are unknown. Throughout the analysis to follow, $\alpha_k(t)$ may be set to zero for all $t \geq 0$.

The dynamics for a relative degree r plant described by (16) may be written in normal form as

$$\dot{\xi}_1 = \xi_2, \tag{20}$$

$$\vdots \quad \vdots \tag{21}$$

$$\dot{\xi}_{r-1} = \xi_r, \tag{22}$$

$$\dot{\xi}_r = \alpha(\xi, \pi) + \beta(\xi, \pi) u_p, \tag{23}$$

$$\dot{\pi} = \Psi(\xi, \pi), \tag{24}$$

with $\pi \in \mathsf{R}^{n-r}$, and $y_p = \xi_1$. The zero-dynamics of the system are given as

$$\dot{\pi} = \Psi(0, \pi). \tag{25}$$

We will consider the adaptive control of plants with no zero dynamics, or plants which have exponential attractive zero dynamics (i.e., plants where (25) is exponentially stable when the states π move outside a ball $|\pi| > B$). The two plant types are characterized by the following assumptions.

(P1) Plant Assumption

The plant is of relative degree $r = n$ (i.e. no zero dynamics), or the plant is of relative degree $r < n$ with exponentially attractive zero dynamics such that

$$\begin{aligned} \tfrac{d}{dt}x_i &= x_{i+1}, \quad i = 1, \ldots, n-1, \\ \tfrac{d}{dt}x_r &= \alpha(X) + \alpha_k(t) + (\beta(X) + \beta_k(t))\, u_p, \end{aligned}$$

where $y_p = x_1$, with $\alpha_k(t)$ a known function. It is also assumed that there exists some $\beta_0 > 0$ such that $\beta(X) \geq \beta_0$, and that $x_1, \ldots, x_r$ are measurable.

If $r < n$, we may use Lipschitz properties of $\Psi(\xi, \pi)$ to see that plants containing zero dynamics have bounded states when the output and corresponding derivatives are bounded [35]. For some positive constants γ_1, γ_2, γ_3, γ_4, and B and function v_1 we have

$$\gamma_1|\pi|^2 \leq v_1(\pi) \leq \gamma_2|\pi|^2, \tag{26}$$

$$\frac{dv_1}{d\pi}\Psi(0,\pi) \leq -\gamma_3|\pi|^2, \text{ if } |\pi| > B, \tag{27}$$

$$\left|\frac{dv_1}{d\pi}\right| \leq \gamma_4|\pi|, \tag{28}$$

if the zero dynamics are exponentially attractive. If the output dynamics are bounded such that $|\xi| \leq k_1$, where k_1 is some positive constant, then using (27), we find

$$\begin{aligned} \dot{v}_1 &= \frac{dv_1}{d\pi}\Psi(\xi,\pi) \\ &\leq -\gamma_3|\pi|^2 + \frac{dv_1}{d\pi}\left(\Psi(\xi,\pi) - \Psi(0,\pi)\right), \text{ if } |\pi| > B. \end{aligned} \tag{29}$$

If $\Psi(\xi, \pi)$ is Lipschitz in ξ, then $|\Psi(\xi,\pi) - \Psi(0,\pi)| \leq k_2|\xi|$ for some positive k_2. Using this, if $|\pi| > B$ we now have

$$\begin{aligned} \dot{v}_1 &\leq -\gamma_3|\pi|^2 + \left|\frac{dv_1}{d\pi}\right| |(\Psi(\xi,\pi) - \Psi(0,\pi))| \\ &\leq -\gamma_3|\pi|^2 + \gamma_4 k_2|\xi||\pi| \\ &\leq -\gamma_3|\pi|^2 + \gamma_4 k_1 k_2|\pi|. \end{aligned} \tag{30}$$

Therefore, $\dot{v}_1 \leq 0$ if $|\pi| \geq \max(B, \gamma_4 k_1 k_2/\gamma_3)$. This ensures boundedness of ξ and π, therefore all the system states are bounded if the output and corresponding derivatives are bounded.

4 Direct adaptive control

Within this section, we define an "output error direct adaptive controller" (using the terminology from [35]), as shown in Fig. 2. While an indirect adaptive controller attempts to identify the plant dynamics and then develop a controller based on the current best guess at the plant dynamics, a direct adaptive controller directly adjusts the parameters of a controller to meet some performance specifications [24].

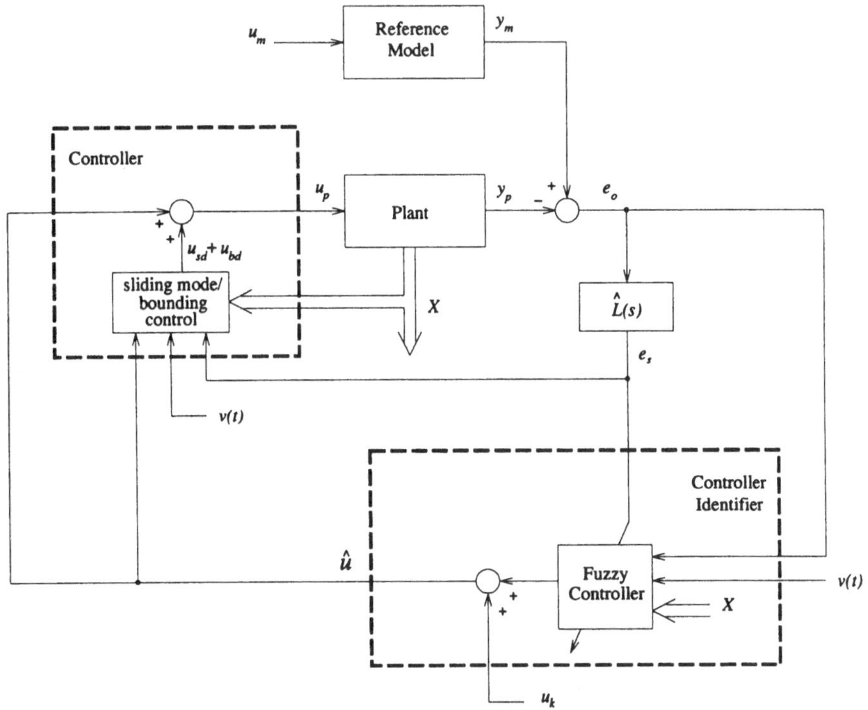

Figure 2: A direct adaptive control system with a reference model.

In addition to the plant assumption **P1**, we require the following plant assumption when using the direct adaptive controller:

(P2) Plant Assumption

Given $y_p^{(r)} = (\alpha(X) + \alpha_k(t)) + \beta(X)u_p$, we require that there exists positive constants β_0 and β_1 such that $0 < \beta_0 \leq \beta(X) \leq \beta_1 < \infty$ and some function $B(X) \geq 0$ such that $|\dot{\beta}(X)| = \left|\frac{\partial\beta}{\partial X}\dot{X}\right| \leq B(X)$ for all $X \in S_x$. Here, as earlier, $\alpha_k(t)$ is a known time dependent signal.

The first part of **P2** introduces a new requirement that the controller gain $\beta(X)$ be bounded from above by a constant β_1. In general, this will not pose a large restriction upon the class of plants since situations in which a finite input will cause an infinitely large effect upon $y_p^{(r)}$ rarely occur in physical plants. The second restriction within **P2** requires that $|\dot{\beta}(X)| \leq B(X)$ for some $B(X) > 0$. We know that $|\dot{\beta}(X)| \leq \|\partial\beta(X)/\partial X\|\|\dot{X}\|$ thus if $\|\partial\beta(X)/\partial X\|$ and $\|\dot{X}\|$ are bounded, then some $B(X)$ may be found. Once again if we consider physical plants with finite controller gain, then $\|\partial\beta(X)/\partial X\|$ will be bounded. If $y_p^{(i)}, i = 0, \ldots, r$ is bounded, then plants with no zero dynamics are ensured that $\|\dot{X}\|$ is bounded since the states may be written in terms of the outputs, $y_p^{(i)}, i = 0, \ldots, r-1$. If a plant has zero dynamics, but $\beta(X)$ is not dependent upon the zero dynamics, then once again we have $|\dot{\beta}(X)|$ bounded.

Consider the control law

$$u_p = \hat{u} + u_{sd} + u_{bd}. \tag{31}$$

The direct adaptive control law is comprised of a bounding control term, u_{bd}, a sliding mode control term u_{sd}, and an adaptive control term, $\hat{u}$. Here we define $\nu(t) = y_m^{(r)} + \eta e_s + \bar{e}_s - \alpha_k(t)$, with $\bar{e}_s = \dot{e}_s - e_o^{(r)}$, and $\eta > 0$. The tracking error is defined as $e_s = k^T e$ where $e = \left[e_o \; \dot{e}_o \; \cdots \; e_o^{(r-1)}\right]^T$, $k = [k_0 \; \cdots \; k_{r-2} \; 1]^T$, and $e_o = y_m - y_p$, thus $\bar{e}_s = [\dot{e}_o \cdots e_o^{(r-1)}][k_0 \cdots k_{r-2}]^T$. We pick the elements of k such that $\hat{L}(s) = s^{r-1} + k_{r-2}s^{r-2} + \cdots + k_1 s + k_0$ has its roots in the open left half plane. The goal of the adaptive algorithm is to "learn" how to control the plant to drive e_s to zero. Thus e_s is a measure of the tracking error. Each of the control terms in (31) will now be described.

4.1 Direct adaptive term

Using feedback linearization [35], we know that there exists some ideal controller

$$u^* = \frac{1}{\beta(X)}(-\alpha(X) + \nu(t)), \tag{32}$$

where $\nu(t)$ is a free parameter. We may express u^* in terms of a Takagi-Sugeno fuzzy model, so that

$$u^* = \mathcal{F}_u(X, \nu, A_u^*) + u_k + d_u(X), \tag{33}$$

where u_k is a known part of the controller (possibly a fuzzy, PID, or some other type of controller). Since the indirect adaptive controller attempted to determine a feedback linearizing controller based on a best guess of the plant dynamics, we allowed for the inclusion of $\alpha_k(t)$ and $\beta_k(t)$ so that known parts of the plant may be included. The direct adaptive controller, however, attempts to directly determine a controller, so within this section we allow for a known part of the controller that is perhaps specified via heuristics or past experience with the application of conventional direct control. We also define the ideal direct control control parameters

$$A_u^* \in \mathsf{R}^{m_u \times p_u}, A_u^* = \arg \min_{A_u \in \Omega_u} \left[\sup_{X \in S_x, \nu \in S_\nu} |\mathcal{F}_u(X, \nu, A_u) - (u^* - u_k)| \right], \tag{34}$$

so that $d_u(X)$ is the error which arises when u^* is represented by a fuzzy system or neural network with ideal parameters, and ν is within the space $S_\nu \subset \mathsf{R}$. We assume that $D_u(X) \geq |d_u(X)|$, where $D_u(X)$ is a known bound on the error in representing the ideal controller with a fuzzy system. We see that if $|d_u(X)|$ is to be small, then our fuzzy controller will require X and ν to be available, either through the input membership functions or through z_u^T. The fuzzy approximation of the desired control is

$$\hat{u} = \mathcal{F}_u(X, \nu, A_u) + u_k, \tag{35}$$

where the matrix A_u is updated on line. The parameter error matrix for the direct adaptive controller

$$\Phi_u(t) = A_u(t) - A_u^*, \tag{36}$$

is used to define the difference between the parameters of the current controller and the desired controller.

Using the control (31), the r-th derivative of the output error becomes

$$e_o^{(r)} = y_m^{(r)} - \alpha(X) - \alpha_k(t) - \beta(X)\left(\hat{u} + u_{sd} + u_{bd}\right). \tag{37}$$

Using the definition of u^*, (32), we may rearrange (37) so that

$$\begin{aligned} e_o^{(r)} &= y_m^{(r)} - \alpha(X) - \alpha_k(t) - \beta(X)u^* \\ &\quad -\beta(X)\left(\hat{u} - u^*\right) - \beta(X)(u_{sd} + u_{bd}) \\ &= -\eta e_s - \bar{e}_s - \beta(X)\left(\hat{u} - u^*\right) - \beta(X)(u_{sd} + u_{bd}). \end{aligned} \tag{38}$$

We may alternatively express (38)

$$\dot{e}_s + \eta e_s = -\beta(X)\left(\hat{u} - u^*\right) - \beta(X)(u_{sd} + u_{bd}). \tag{39}$$

We now define the bounding control term, u_{bd}, for the direct adaptive controller.

4.2 Bounding control

The bounding term for the direct adaptive controller is determined by considering

$$v_{bd} = \frac{1}{2}e_s^2. \tag{40}$$

We differentiate (40) and use (39) to obtain

$$\begin{aligned} \dot{v}_{bd} &= -\eta e_s^2 - e_s\left[\beta(X)(\hat{u} - u^*) + \beta(X)(u_{sd} + u_{bd})\right] \\ &\leq -\eta e_s^2 + |e_s|\left[\beta(X)(|\hat{u}| + |u^*|)\right] - \beta(X)(u_{sd} + u_{bd})e_s, \end{aligned} \tag{41}$$

where we take $\text{sgn}(u_{si}) = \text{sgn}(u_{bi})$ which will be shown later. We do not explicitly know u^*, however, so the bounding controller will be implemented using $\alpha_1(X) \geq |\alpha(X)|$ as defined for the indirect adaptive controller. We choose

$$u_{bd} = \Pi(t)k_{bd}(t)\text{sgn}(e_s). \tag{42}$$

Let ϵ_M and M_e be fixed parameters such that $0 < \epsilon_M \leq M_e$ with

$$\Pi(t) = \begin{cases} 1 & \text{if } M_e \leq |e_s| \\ (|e_s| + \epsilon_M - M_e)/\epsilon_M & \text{if } M_e - \epsilon_M \leq |e_s| < M_e \\ 0 & \text{otherwise} \end{cases}, \tag{43}$$

and

$$\text{sgn}(x) := \begin{cases} 1 & x > 0 \\ -1 & x < 0 \end{cases}. \tag{44}$$

The sliding mode gain is now chosen as

$$k_{bd}(t) = |\hat{u}| + (\alpha_1(X) + |\nu|)/\beta_0. \tag{45}$$

We note that $|u^*| \leq (\alpha_1(X) + |\nu|)/\beta_0$. With this definition of u_{bd}, we are guaranteed that $|e_s| \leq M_e$ if the initial conditions are such that $|e_s(0)| \leq M_e$.

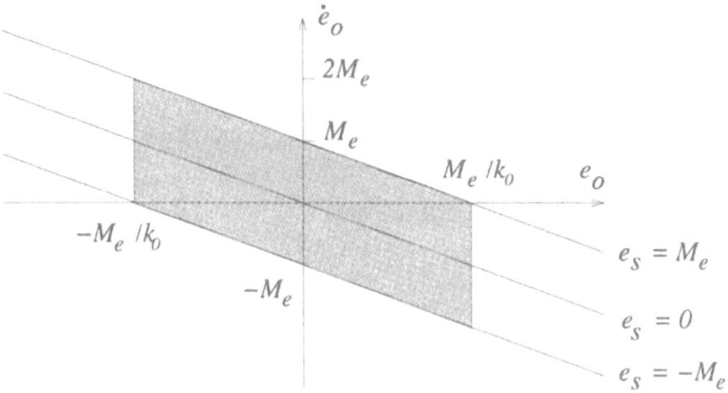

Figure 3: Boundedness around the manifold $e_s = \dot{e}_o + k_o e_o = 0$.

At this point, it is convenient to define transfer functions

$$\hat{G}_i(s) := \frac{s^i}{\hat{L}(s)}, \; i = 0, \ldots, r-1, \tag{46}$$

which are stable since $\hat{L}(s)$ has its poles in the open left half plane. Since $e_o^{(i)} = \hat{G}_i(s)e_s$, with e_s bounded, then $e_o^{(i)} \in \mathcal{L}_\infty$ ($\mathcal{L}_\infty = \{z(t) : \sup_t |z(t)| <$

$\infty\}$). This is shown for the case $e_s = \dot{e}_o + k_o e_o$ in Figure 3 where if $|e_s| \leq M_e$ then e_o and $\dot{e}_o$ stay in the shaded region (i.e. $|e_o| \leq M_e/k_o$ and $|\dot{e}_o| \leq 2M_e$). This may be extended to higher dimensional systems as

$$|e_o^{(i)}| \leq M_e \|\hat{G}_i(s)\|_1, i = 0, \ldots, r-2, \tag{47}$$

and since $e_o^{(r-1)} = e_s - \sum_{i=0}^{r-2} k_i e_o^{(i)}$ the triangular inequality may be used to show that

$$|e_o^{(r-1)}| \leq M_e + M_e \sum_{i=0}^{r-2} k_i \|\hat{G}_i(s)\|_1, \tag{48}$$

for all time if $|e_s| \leq M_e$ and the initial conditions are such that $|e_o^{(i)}(0)| \leq M_e \|\hat{G}_i(s)\|_1, i = 0, \ldots, r-2$. The transfer function 1-norm is defined as $\|\hat{G}_i(s)\|_1 := \int_{-\infty}^{\infty} |g_i(\tau)| d\tau$, where $g_i(t)$ is the impulse response of $\hat{G}_i(s)$. Using the example of $e_s = \dot{e}_o + k_0 e_o$, we obtain $\hat{G}_0(s) = 1/(s + k_0)$, which has an impulse response function of $g_0(t) = e^{-k_0 t}$ with the 1-norm $\|\hat{G}_0(s)\|_1 = 1/k_0$. Using (47) and (48), we obtain the bounds $|e_o| \leq M_e/k_0$ and $|\dot{e}_o| \leq 2M_e$, as shown in Fig. 3. Overall, we see that (47) and (48) provide explicit bounds on the output error when the bounding control u_{bi} is used.

4.3 Adaptation algorithm

Now consider the parameter update law

$$\dot{A}_u(t) = Q_u^{-1} z_u \zeta_u^T e_s, \tag{49}$$

where Q_u is diagonal with positive elements. Using the fuzzy adaptation law defined by (49) we are not guaranteed that $A_u \in \Omega_u$. To ensure this, a projection algorithm may be used. The parameter space is defined so that the parameters are bounded by $A_u \in \left[A_u^{min}, A_u^{max}\right]$. Define $\bar{a}_{u_{i,j}}$ to be the i,j-th element of $z_u \zeta_u^T e_s$. The parameter matrix is updated according to

$$\dot{A}_u(t) = Q_u^{-1} \hat{A}, \tag{50}$$

where the elements of $\hat{A}_u(t)$ are defined by

$$\hat{a}_{u_{i,j}} = \begin{cases} 0 & \text{if } a_{u_{i,j}} \notin (A_u^{min}, A_u^{max}) \text{ and } \bar{a}_{u_{i,j}}(a_{u_{i,j}} - a^c_{u_{i,j}}) > 0 \\ \bar{a}_{u_{i,j}} & \text{otherwise} \end{cases} \tag{51}$$

with $A_u^c \in \left(A_u^{min}, A_u^{max}\right)$. This modified update law will ensure that the parameter matrices will stay within the feasible parameter space.

4.4 Sliding mode control term

A sliding mode control term is used to compensate for the representation error which arises due to expressing u^* in terms of an ideally tuned fuzzy system or neural network.

We now define the sliding mode control term for the direct adaptive controller as

$$u_{sd} = k_{sd}(t)\text{sgn}(e_s), \tag{52}$$

where

$$k_{sd}(t) = \frac{B(X)|e_s|}{2\beta_0^2} + D_u(X). \tag{53}$$

4.5 Stability properties

The controller assumption for the direct adaptive control scheme is stated below.

(C2) Control Assumption

The fuzzy systems (neural networks) are defined such that $D_u(X) \in \mathcal{L}_\infty$, for $X \in S_x \subseteq \mathsf{R}^n$ and there is some known continuous function such that $\alpha_1(X) \geq |\alpha(X)|$.

The properties of the closed-loop system using the direct adaptive controller are now summarized.

Theorem 1: *If a unique closed-loop state trajectory exists for a plant satisfying* **P1** *and* **P2**, *with control law (31) satisfying* **C2** *and parameter adaptation law (49), then the following hold:*

(1) $X \in \mathcal{L}_\infty$,

(2) $A_u \in \mathcal{L}_\infty$, *and*

(3) $\lim_{t\to\infty} |e_o| = 0$.

Proof: Consider the following Lyapunov equation candidate

$$V_d = \frac{1}{2\beta(X)} e_s^2 + \frac{1}{2}\mathbf{tr}(\Phi_u^T Q_u \Phi_u), \tag{54}$$

where $Q_u \in \mathrm{R}^{m_u \times m_u}$ is positive definite and diagonal. Since $0 < \beta_0 \leq \beta(X) \leq \beta_1 < \infty$, V_d is radially unbounded. If $V_d \to 0$, then both the tracking and learning objectives have been fulfilled. Taking the derivative of (54) yields

$$\dot{V}_d = \frac{e_s}{\beta(X)}\left[\dot{e}_s\right] + \mathbf{tr}(\Phi_u^T Q_u \dot{\Phi}_u) - \frac{\dot{\beta}(X) e_s^2}{2\beta^2(X)}. \tag{55}$$

Substituting $\dot{e}_s$, as defined in (39), yields

$$\begin{aligned} \dot{V}_d &= \frac{e_s}{\beta(X)}\left[-\eta e_s - \beta(X)\left(\hat{u} - u^*\right) - \beta(X)(u_{sd} + u_{bd})\right] \\ &\quad + \mathbf{tr}(\Phi_u^T Q_u \dot{\Phi}_u) - \frac{\dot{\beta}(X) e_s^2}{2\beta^2(X)}. \end{aligned} \tag{56}$$

Since $\dot{\Phi}_u = \dot{A}_u$, the parameter update law may be used to show

$$\begin{aligned} \dot{V}_d &\leq -\frac{\eta}{\beta(X)} e_s^2 - \left[z_u^T \Phi_u \zeta_u + d_u + u_{sd} + u_{bd}\right] e_s \\ &\quad + \mathbf{tr}(z_u^T \Phi_u \zeta_u)(e_s - q(t)) - \frac{\dot{\beta}(X) e_s^2}{2\beta^2(X)}. \end{aligned} \tag{57}$$

Equation (57) may equivalently be expressed as

$$\dot{V}_d \leq -\frac{\eta}{\beta(X)}e_s^2 - q(t)z_u^T\Phi_u\zeta_u - \left(\frac{\dot{\beta}(X)e_s}{2\beta^2(X)} + d_u\right)e_s - e_s(u_{sd}+u_{bd}). \quad (58)$$

If u_{bd} is as defined in (42), then

$$\begin{aligned}\dot{V}_d &\leq -\frac{\eta}{\beta(X)}e_s^2 - \left(\frac{\dot{\beta}(X)e_s}{2\beta^2(X)} + d_u\right)e_s - e_s u_{sd} \\ &\leq -\frac{\eta}{\beta_1}e_s^2 + \left(\frac{|\dot{\beta}(X)||e_s|}{2\beta^2(X)} + |d_u|\right)|e_s| - e_s u_{sd}. \qquad (59)\end{aligned}$$

The sliding mode term is then used to ensure that $\dot{V}_d \leq -\eta e_s^2/\beta_1$. $V_d \in \mathcal{L}_\infty$ also implies that $A_u \in \mathcal{L}_\infty$.

We thus have

$$\begin{aligned}\int_0^\infty \eta e_s^2 dt &\leq -\int_0^\infty \beta_1 \dot{V}_d dt \\ &= -\beta_1 V_d(\infty) + \beta_1 V_d(0) \qquad (60)\end{aligned}$$

This establishes that $e_s \in \mathcal{L}_2$ ($\mathcal{L}_2 = \{z(t) : \int_0^\infty z^2(t)dt < \infty\}$) since $V_d \in \mathcal{L}_\infty$. In addition, we know that $e_o^{(i)} \in \mathcal{L}_\infty, i = 0, \ldots, r-1$ since $e_s \in \mathcal{L}_\infty$ and $e_o^{(i)} = G_i(s)e_s$, with all the poles of $G_i(s), i = 0, \ldots, r-1$ in the open left half plane. Using, (39) one may show that $\dot{e}_s \in \mathcal{L}_\infty$. Since $e_s \in \mathcal{L}_2, \mathcal{L}_\infty$ and $\dot{e}_s \in \mathcal{L}_\infty$, by Barbalat's Lemma we have asymptotic stability of e_s (i.e. $\lim_{t\to\infty} e_s = 0$), which implies asymptotic stability of e_o. □

Remark 1 The bounding control term, u_{bd}, for the direct adaptive controller is used to restrict the output trajectory so that a smaller fuzzy controller or neural network may be used to approximate the ideal feedback controller, u^*. The sliding mode control term, u_{sd}, is required due to the modeling errors between the ideal feedback controller and the fuzzy controller or neural network with optimal parameters. The adaptive control term, $\hat{u}$, is then used to ensure asymptotic convergence of the tracking error.

Remark 2 The direct adaptive scheme allows for the inclusion of u_k so that a control engineer may use conventional techniques to develop an initial control design and then use the above adaptive technique to neural networks in parallel to meet the tracking requirements. For example, some PID controller design may provide moderate performance, however, the above direct adaptive technique may be used to meet tracking requirements. Even if u_k produces an unstable closed loop system by itself, the use of the above direct adaptive scheme will result in asymptotically stable tracking.

Remark 3 The direct adaptive scheme does not require that the parameter set Ω_u be defined so that $\mathcal{F}(\cdot)$ is bounded away from a particular region. This is often not the case with indirect adaptive schemes [24].

Remark 4 We may also use an adaptive estimate for some constant such that $|d_u| \leq D_u$ in a similar fashion as described in Remark 1.5. The sliding mode gain is modified to be $k_{sd}(t) = \frac{B(X)|e_s|}{2\beta_0^2} + \hat{D}_u$, where $\dot{\hat{D}}_u = |e_s|/q_u$, with $q_u^{-1} > 0$. To show stability using this adaptive algorithm for the sliding mode gain, the term $q_u(\hat{D}_u - D_u)^2/2$ may be added to the Lyapunov candidate for the direct adaptive controller defined in (54). This may also be modified for the smoothed version of the direct adaptive controller.

Remark 5 As with the indirect adaptive scheme, our direct adaptive scheme has many differences from the existing techniques (i.e. those presented in [15, 20, 16]). Particularly, (i) the results of Theorem 2 may be applied to systems with a state dependent input gain $\beta(X)$, whereas [15, 20] consider a class of nonlinear plants with constant input gain (i.e. $\beta(X) = \beta$, a constant) and [16] only considers the special case of $\beta = 1$; (ii) none of the results in [15, 20, 16] considered systems containing zero dynamics; (iii) unlike [15], our direct adaptive algorithm algorithm ensures that even if the approximation error d_u is not square integrable, then the tracking error will go to zero (or to an ϵ-boundary layer of zero for the smoothed control version); (iv) our direct adaptive controller allows for Takagi-Sugeno fuzzy systems, standard fuzzy systems, or neural networks; (v) the above direct adaptive technique allows for the inclusion of a known controller, u_k, so that it may be used to either enhance the performance of some prespecified controller, or stand alone as a stable adaptive controller; and (vi) furthermore, as we show in the next section, our approach allows for the incorporation of heuristics about the inverse plant dynamics to speed adaptation.

4.6 Smoothing the control action

The sliding mode control term, u_{sd}, may introduce a high frequency signal to the plant which may excite unmodeled dynamics. To avoid this, we now consider a "smoothed" version of the previous adaptive controller in which the tracking error, e_s, is driven to an ϵ-neighborhood of $e_s = 0$. The error is now defined in terms of a dead zone so that

$$e_\epsilon := e_s - \epsilon \mathrm{sat}\left(e_s/\epsilon\right), \tag{61}$$

where $\epsilon > 0$ is the size of the dead zone, and

$$\mathrm{sat}(x) = \begin{cases} 1 & \text{if } 1 \leq x \\ x & \text{if } -1 < x < 1 \\ -1 & \text{if } x \leq -1 \end{cases} . \tag{62}$$

From the above definition, we see that $e_\epsilon = 0$ when e_s is within the dead zone. The smoothed control law is defined as

$$u_p = \hat{u} + u_{sd} + u_{bd}, \tag{63}$$

where each of the terms in (63) is modified as described next.

We now let

$$\hat{u} = \mathcal{F}(X, \nu_\epsilon, A_u) + u_k, \tag{64}$$

where

$$\nu_\epsilon = y_m^{(r)} + \eta e_\epsilon + \bar{e}_s - \alpha_k(t). \tag{65}$$

The bounding controller is now defined as

$$u_{bd} = \Pi(t) k_{bd}(t) \mathrm{sgn}(e_\epsilon), \tag{66}$$

with $\epsilon < M_e$ and $\Pi(t)$ as defined in (43). The parameter update law when using a dead zone is modified to

$$\dot{A}_u(t) = Q_u^{-1} z_u \zeta_u^T e_\epsilon, \tag{67}$$

so that the parameters are not modified when e_σ is within the dead zone. Also, as $\epsilon \to 0$, the smoothed control law approaches the previous sliding mode control law. Using the smoothed control law, the error dynamics become

$$\dot{e}_s = -\eta e_\epsilon - \beta(X)\,(\hat{u} - u^*) - \beta(X)(u_{sd} + u_{bd}). \tag{68}$$

The closed-loop properties are now summarized in the theorem below.

Theorem 2: *If a unique closed-loop state trajectory exists for a plant satisfying* **P1** *and* **P2**, *with smoothed control law (31) satisfying* **C2** *and parameter adaptation law (67), then the following hold:*

(1) $X \in \mathcal{L}_\infty$,

(2) $A_u \in \mathcal{L}_\infty$, *and*

(3) $\lim_{t\to\infty} e_\epsilon = 0$.

Proof: Consider the Lyapunov candidate

$$V_d = \frac{1}{2\beta(X)} e_\epsilon^2 + \frac{1}{2}\mathbf{tr}(\Phi_u^T Q_u \Phi_u). \tag{69}$$

Even though e_ϵ is not continuously differentiable, we still have $\frac{d}{dt}e_\epsilon^2 = 2e_\epsilon \dot{e}_s$. Taking the derivative of (69) one obtains

$$\begin{aligned} \dot{V}_d &= \frac{e_\epsilon}{\beta(X)}\,[\dot{e}_s] + \mathbf{tr}(\Phi_u^T Q_u \dot{\Phi}_u) - \frac{\dot{\beta}(X) e_\epsilon^2}{2\beta^2(X)} \\ &\leq -\frac{\eta}{\beta_1} e_\epsilon^2 + \left(\frac{|\dot{\beta}(X)||e_\epsilon|}{2\beta^2(X)} + |d_u| \right) |e_\epsilon| - e_\epsilon u_{sd}. \end{aligned} \tag{70}$$

Since $e_\epsilon \mathrm{sat}(e_s/\epsilon) = |e_\epsilon|$, this simplifies to

$$\dot{V}_d \leq -\eta e_\epsilon^2/\beta_1. \tag{71}$$

Since $\dot{V}_d \leq 0$, we have $e_\epsilon, \Phi_u \in \mathcal{L}_\infty$ which implies that $e_s, A \in \mathcal{L}_\infty$. In addition, (71) may be used to establish $e_\epsilon \in \mathcal{L}_2$. Since e_ϵ is not continuously differentiable, however, we will consider the signal e_ϵ^2. We know that $e_\epsilon \in \mathcal{L}_2$ implies that $e_\epsilon^2 \in \mathcal{L}_1$ and $e_\epsilon \in \mathcal{L}_\infty$ implies that $e_\epsilon^2 \in \mathcal{L}_\infty$. In addition, $\frac{d}{dt}e_\epsilon^2 \in \mathcal{L}_\infty$ so from Barbalat's lemma $e_\epsilon^2 \to 0$ as $t \to \infty$, and thus $|e_\epsilon| \to 0$ as $t \to \infty$. □

5 Example: An automated highway system

Due to increasing traffic congestion, there has been an renewed interest in the development of an Automated Highway System (AHS) in which high traffic flow rates may be safely achieved. Since many of today's automobile accidents are caused by human error, automating the driving process may actually increase the safety of the highway. Vehicles will be driven automatically with onboard lateral and longitudinal controllers. The lateral controllers will be used to steer the vehicles around corners, make lane changes, and perform additional steering tasks. The longitudinal controllers will be used to maintain a steady velocity if a vehicle is traveling alone (conventional cruise control), follow a lead vehicle at a safe distance (car following, see Fig. 4), or perform other speed/tracking tasks. For more details on intelligent vehicle highway systems see [37] and [38]. Within this section, we will apply the above adaptive techniques to the car following problem, or longitudinal control of a vehicle within an AHS.

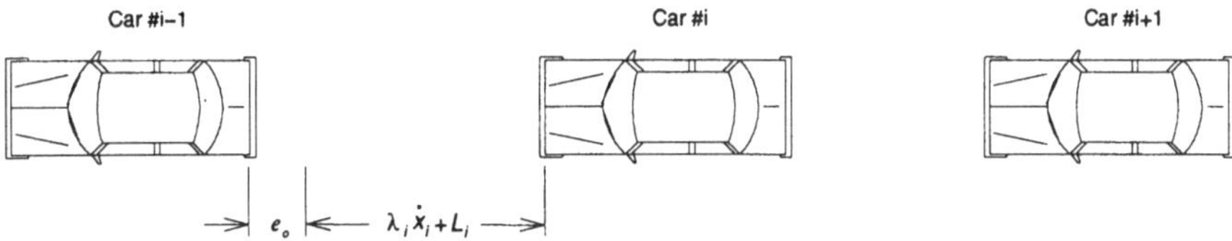

Figure 4: Car following within an Automated Lane.

The dynamics of the car following system for the i-th vehicle may be described by the state vector $X_i = [\delta_i, v_i, f_i]^T$, where $\delta_i = x_i - x_{i-1}$ is the intervehicle spacing between the i-th and $i-1$-st vehicles, v_i is the i-th vehicle's velocity, and f_i is the driving/braking force applied to the longitudinal dynamics of the i-th vehicle. The longitudinal dynamics may be expressed as

$$\dot{\delta} = v - v_{i-1}, \tag{72}$$

$$\dot{v} = \frac{1}{m}\left(-A_\rho v^2 - d + f\right), \tag{73}$$

$$\dot{f} = \frac{1}{\tau}(-f + u_p), \tag{74}$$

where u_p is the control input (if $u_p > 0$, then it represents a throttle input and if $u_p < 0$, it represents a brake input), and the vehicle variables and parameters are summarized in Table 2 (we assume that the variables and parameters are associated with the i-th vehicle, unless subscripts indicate otherwise).

x	vehicle position
v	vehicle velocity
f	applied force in longitudinal direction
$m = 1300kg$	mass of the vehicle
$A_\rho = 0.3Ns^2/m^2$	aerodynamic drag
$d = 100N$	constant frictional force
$\tau = 0.2s$	engine/brake time constant

Table 2: Automobile variables and parameters.

The plant output is $y_p = \delta + \lambda v, \lambda > 0$. This measurement allows for a velocity dependent intervehicle spacing. As the velocity of the i-th vehicle increases, the distance between the i-th and $i-1$-st vehicles should increase. A standard good driving rule for humans is to allow an intervehicle spacing of one vehicle length per 10mph (this roughly corresponds to $\lambda = 0.9$). With $\lambda \neq 0$, the plant is of relative degree 2 since

$$\begin{aligned} y_p^{(2)} &= \dot{v} + \lambda \ddot{v} - \dot{v}_{i-1} \\ &= \frac{1}{m}\left[-A_\rho v^2 - d + f\right] + \frac{\lambda}{m}\left[-2A_\rho v\dot{v} - \frac{1}{\tau}f\right] + \frac{\lambda}{m\tau}u_p - \dot{v}_{i-1} \quad (75) \end{aligned}$$

This is clearly of the form required by both the indirect and direct adaptive schemes (i.e., (19)) with

$$\alpha(X) = \frac{1}{m}\left[-A_\rho v^2 - d + f\right] + \frac{\lambda}{m}\left[-2A_\rho v\dot{v} - \frac{1}{\tau}f\right], \tag{76}$$

$$\beta(X) = \frac{\lambda}{m\tau}, \tag{77}$$

where $\alpha_k(t) = -\dot{v}_{i-1}$ for all $t \geq 0$ and for any $X \in \mathsf{R}^3$. We see that $\beta(X) \geq \beta_0 > 0$ for $\beta_0 = \frac{\lambda}{m_1\tau_1}$, where the vehicle parameters are defined within the intervals $m \in [m_0, m_1]$ and $\tau \in [\tau_0, \tau_1]$, where $m_0, \tau_0 > 0$.

The zero dynamics are found by setting $y_p = 0$, which results in $\lambda\dot{\delta} = -\delta - \lambda v_{i-1}$. The zero dynamics are thus exponentially attractive since if we let $v_1 = \delta^2$, we obtain

$$\dot{v}_1 = -\frac{2}{\lambda}\delta(\delta + \lambda v_{i-1}). \tag{78}$$

If we assume that $|v_{i-1}| \leq V_m$, some bound on achievable velocities for the vehicles, then

$$\dot{v}_1 \leq -\frac{2a}{\lambda}\delta^2, \text{ if } |\delta| \geq |\tfrac{\lambda V_m}{1-a}|, \tag{79}$$

where $0 < a < 1$. Thus as long as $\lambda > 0$, we are ensured exponential attractivity of the zero dynamics.

5.1 Direct adaptive control

Since we desire that $y_p \to 0$, here we simply select $y_m = 0$ so that

$$e_o = -y_p. \tag{80}$$

Since the plant is of relative degree 2, the error metric is defined as

$$e_s = \dot{e}_o + k_0 e_o. \tag{81}$$

For this example, we simply choose $k_0 = 1$ (i.e., the desired tracking eigenvalue is at -1). If e_s is to be measured, then sensors will need to obtain $\delta, \dot{\delta}, v, \dot{v}$, and $\dot{v}_{i-1}$. With such sensors, assumption **P2** is satisfied. Using the definition of $\hat{G}_i(s)$, from (46), we see that $||\hat{G}_0(s)||_1 = 1/k_0 = 1$. Thus if we want the bounding control term to be defined such that $|e_o| \leq 1$ meter, we use (47) to pick $M_e = 1$ and $\epsilon_M = 0.1$. Ideally, we will not need to use the bounding controller unless the initial conditions are such that $e_o \geq 1$. If the following vehicle becomes too close to the lead vehicle, however, the bounding controller may be used to ensure that the two vehicles do not collide. The upper bound on $\alpha(X)$ is found from

$$|\alpha(X)| \leq \frac{A_\rho}{m}\left(|v| + 2\lambda|\dot{v}|\right)|v| + \frac{|d|}{m} + \frac{1}{m}\left(1 + \frac{\lambda}{\tau}\right)|f|. \tag{82}$$

Using bounds on the vehicle parameters, we obtain

$$|\alpha(X)| \leq \frac{A_{\rho_1}}{m_0}\left(|v| + 2\lambda|\dot{v}|\right)|v| + \frac{|d_1|}{m_0} + \frac{1}{m_0}\left(1 + \frac{\lambda}{\tau_0}\right)|f| = \alpha_1(X), \tag{83}$$

where $A_\rho \in [A_{\rho_0}, A_{\rho_1}]$ and $d \in [d_0, d_1]$, with $A_{\rho_0}, d_0 \geq 0$. If vehicle variable bounds are known, these may be used within (83) rather than the instantaneous variable values.

$M_e = 1$	$k_0 = 1$	$\eta = 1$
$\epsilon_M = 0.1$	$\epsilon = 0.05$	$Q_u^{-1} = \begin{bmatrix} 500 & 0 \\ 0 & 500 \end{bmatrix}$

Table 3: Control parameters for the direct algorithm.

Plant assumption **P2** is clearly satisfied since $\beta(X)$ is a constant so that $\dot{\beta}(X) = 0$. We define the known controller to be a simple proportional controller

$$u_k = k_p e_o, \tag{84}$$

with $k_p = 100$. The rule base for the fuzzy controller that generates $\hat{u}$ in (35) is

R_1 :	**If**	vel is slow **and** e_s is neg	**Then**	$c_1 = a_{1,0} + a_{1,1}\nu(t)$
R_2 :	**If**	vel is med **and** e_s is neg	**Then**	$c_2 = a_{2,0} + a_{2,1}\nu(t)$
R_3 :	**If**	vel is fast **and** e_s is neg	**Then**	$c_3 = a_{3,0} + a_{3,1}\nu(t)$
R_4 :	**If**	vel is slow **and** e_s is zero	**Then**	$c_4 = a_{4,0} + a_{4,1}\nu(t)$
R_5 :	**If**	vel is med **and** e_s is zero	**Then**	$c_5 = a_{5,0} + a_{5,1}\nu(t)$
R_6 :	**If**	vel is fast **and** e_s is zero	**Then**	$c_6 = a_{6,0} + a_{6,1}\nu(t)$
R_7 :	**If**	vel is slow **and** e_s is pos	**Then**	$c_4 = a_{7,0} + a_{7,1}\nu(t)$
R_8 :	**If**	vel is med **and** e_s is pos	**Then**	$c_5 = a_{8,0} + a_{8,1}\nu(t)$
R_9 :	**If**	vel is fast **and** e_s is pos	**Then**	$c_6 = a_{9,0} + a_{9,1}\nu(t)$

so that $z_u = [1,\ \nu(t)]^T$, $A_u \in \mathsf{R}^{2\times 9}$, and $\zeta_u \in \mathsf{R}^9$, with $\nu(t) = 2\dot{e}_o + e_o + \dot{v}_{i-1}$ (recall that $r = 2$, $k_0 = 1$, $y_m = 0$, and $\alpha_k(t) = -\dot{v}_{i-1}$). Here we picked $\nu(t)$ so that the fuzzy system may approximate u^* with a small number of rules since u^* is directly a function of $\nu(t)$. We use the smoothed version of the adaptive control with $\epsilon = 0.05$. The t-norm was taken as the minimum operator as in (2). The membership functions were defined as Gaussian membership functions, as shown in Figure 5 (the Gaussian membership functions are defined as in Table 1 with $\sigma = 2.5$ and 0.5 for μ_{vel} and μ_{e_s}, respectively).

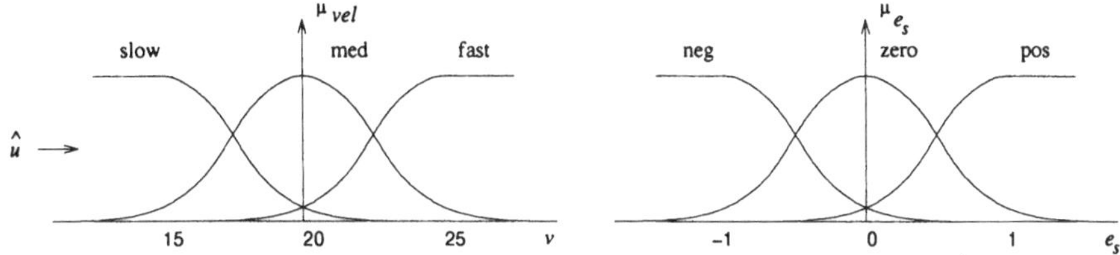

Figure 5: Membership functions for the direct adaptive controller.

Here we assume that our adaptive system, when properly tuned, should be able to approximate the ideal control to within 100 Newtons, so $D_u(X) = 100$. The projection limits were chosen to be $a_{u_{i,j}}^{min} = -2000$ and $a_{u_{i,j}}^{max} = 2000$ for all i, j. The control parameters are summarized in Table 3.

Using the direct adaptive controller, we attempt to control the following vehicle to track the lead vehicle. The velocity profiles for the lead and following vehicles are shown in Figure 6. The output error, e_o, is shown in Figure 7. Notice that the output error is of the order of magnitude for which we designed our adaptive controller. The control input for the lead and following vehicles is shown in Figure 8. We see that although the controller required a few seconds to "learn" how to control the vehicle, the output of the following vehicle quickly matched the output of the lead vehicle. Figure 9 shows how well the known control portion, u_k, is able to track the position of the lead vehicle when no adaptive control portion is included, i.e., $u_p = u_k$. Even though u_k is apparently able to stabilize the system, the control results in very poor performance.

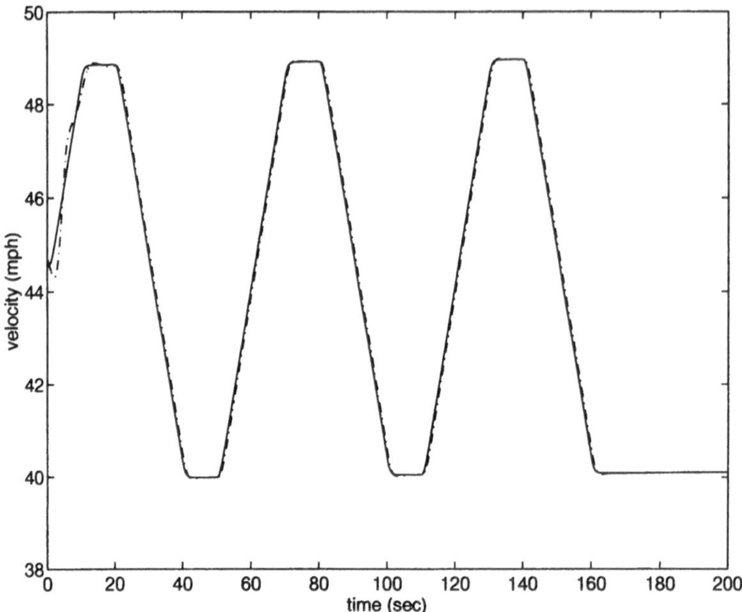

Figure 6: Velocity profiles of the lead car (–) and following car (···).

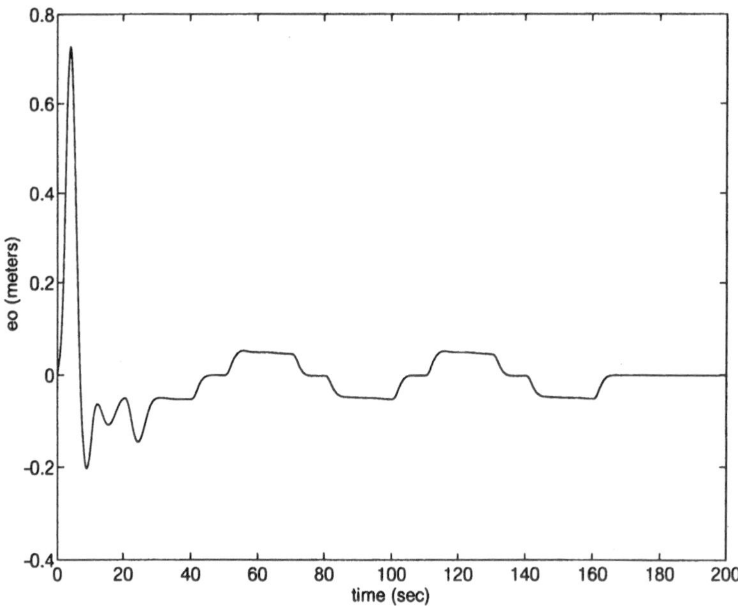

Figure 7: Output error, e_0, for the car following problem using direct adaptive control.

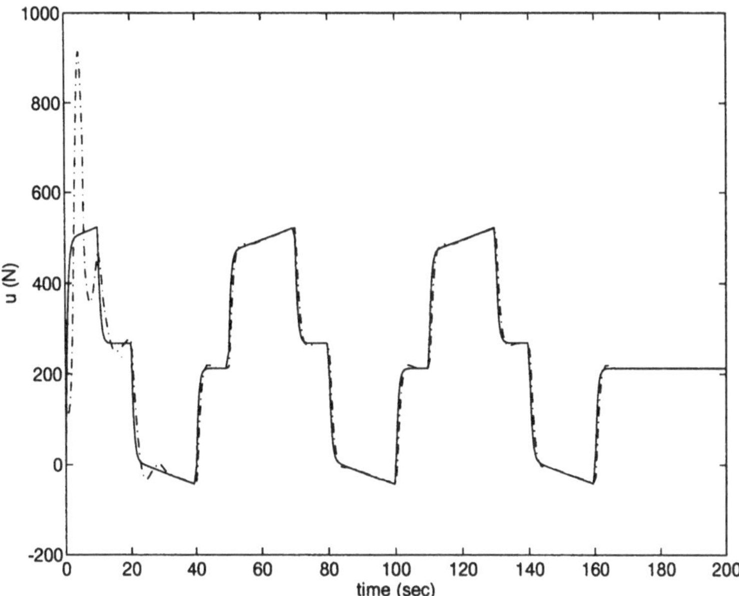

Figure 8: Control input for the lead car (–) and the following car (···).

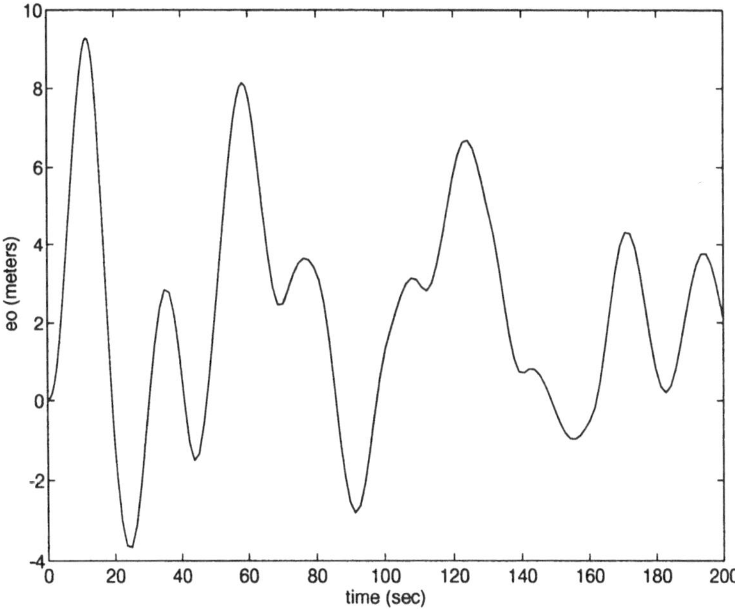

Figure 9: Tracking performance, e_o, when the direct adaptive control is not used, i.e. $u = u_k$.

6 Concluding remarks

Stable closed-loop adaptive systems which take advantage of the universal approximation capabilities of fuzzy systems and neural networks may often be formed using the techniques of robust adaptive control [39]. Combining the advantages of fuzzy systems and neural networks with robust adaptive control often results in an very powerful techniques which may be applied to a wide variety of systems. Within this article, a direct adaptive control scheme was presented for a class of continuous-time nonlinear plants, conditions were provided for its stable operation, and we showed how it could be used for the longitudinal control of a vehicle in an automated lane. The adaptive scheme is particularly powerful since it (i) does not require a complete model of the plant; (ii) uses a general Takagi-Sugeno fuzzy system that includes a certain class of fuzzy systems and neural networks as a special case; and (iii) allows for the inclusion of a priori knowledge in the form of mathematical equations or heuristics to be loaded into the controller. Consider, for example, the case where a linear quadratic regulator (LQR) is designed based on a linear representation of a nonlinear plant. Setting z_u equal to the plant states, it is possible to use the direct adaptive controller as an adaptive LQR, by assigning each column of A_u to the coefficients found for the standard LQR. Each rule may be used to define the space over which each "local LQR" is to adapt. The article [40] shows how this and other ideas can be used in the implementation of the adaptive fuzzy controller presented in this article for a variety of applications. Again, we refer the reader to [24] for more details, particularly a more detailed discussion of the advantages *and* disadvantages of the approach discussed in this article.

References

[1] T. Procyk and E. Mamdani, "A linguistic self-organizing process controller," *Automatica*, vol. 15, no. 1, pp. 15–30, 1979.

[2] W. Pedrycz, *Fuzzy Control and Fuzzy Systems.* New York: John Wiley and Sons, second ed., 1993.

[3] M. Sugeno, ed., *Industrial Applications of Fuzzy Control.* Amsterdam: Elsevier Science Publishers B.V., 1985.

[4] A. Kandel and G. Langholz, *Fuzzy Control Systems.* Boca Raton, FL: CRC Press, 1993.

[5] D. Driankov, H. Hellendoorn, and M. M. Reinfrank, *An Introduction to Fuzzy Control.* Berlin Heidelberg: Springer-Verlag, 1993.

[6] R. J. Marks II, ed., *Fuzzy Logic Technology and Applications.* New York: The Institute of Electrical and Electronics Engineers, 1994.

[7] P. J. Antsaklis and K. M. Passino, eds., *An Introduction to Intelligent and Autonomous Control.* Norwell, MA: Kluwer Academic Publishers, 1993.

[8] J. R. Layne, K. M. Passino, and S. Yurkovich, "Fuzzy learning control for anti-skid braking systems," *IEEE Trans. Control Systems Tech.*, vol. 1, pp. 122–129, June 1993.

[9] J. R. Layne and K. M. Passino, "Fuzzy model reference learning control for cargo ship steering," *IEEE Control Systems Magazine*, vol. 13, pp. 23–34, Dec. 1993.

[10] W. A. Kwong, K. M. Passino, E. G. Lauknonen, and S. Yurkovich, "Expert supervision of fuzzy learning systems for fault tolerant aircraft control," *Proc. of the IEEE, Special Issue on Fuzzy Logic in Engineering Applications*, vol. 83, pp. 466–483, March 1995.

[11] V. G. Moudgal, W. A. Kwong, K. M. Passino, and S. Yurkovich, "Fuzzy learning control for a flexible-link robot," *To appear in IEEE Transactions on Fuzzy Systems*, vol. 3, May 1995.

[12] W. A. Kwong and K. M. Passino, "Dynamically focused fuzzy learning control," *To appear in IEEE Trans. Syst. Man, Cybern.*, 1995.

[13] G. Langari and M. Tomizuka, "Stability of fuzzy linguistic control systems," in *Proc. of 29th Conf. Decision Contr.*, vol. 4, (Honolulu, Hawaii), pp. 2185–2190, 1990.

[14] L.-X. Wang, *Adaptive Fuzzy Systems and Control: Design and Stability Analysis.* Englewood Cliffs, NJ: Prentice Hall, 1994.

[15] L.-X. Wang, "Stable adaptive fuzzy control of nonlinear systems," in *Proc. of 31st Conf. Decision Contr.*, (Tucson, Arizona), pp. 2511–2516, 1992.

[16] C.-Y. Su and Y. Stepanenko, "Adaptive control of a class of nonlinear systems with fuzzy logic," *IEEE Trans. Fuzzy Systems*, vol. 2, pp. 285–294, November 1994.

[17] T. A. Johansen, "Fuzzy model based control: Stability, robustness, and performance issues," *IEEE Trans. Fuzzy Systems*, vol. 2, pp. 221–234, August 1994.

[18] K. S. Narendra and K. Parthasarathy, "Identification and control of dynamical systems using neural networks," *IEEE Trans. Neural Networks*, vol. 1, no. 1, pp. 4–27, 1990.

[19] M. M. Polycarpou and P. A. Ioannou, "Identification and control of nonlinear systems using neural network models: Design and stability analysis," Electrical Engineering – Systems Report 91-09-01, University of Southern California, September 1991.

[20] R. M. Sanner and J.-J. E. Slotine, "Gaussian networks for direct adaptive control," *IEEE Trans. Neural Networks*, vol. 3, pp. 837–863, 1992.

[21] A. Yeşildirek and F. L. Lewis, "A neural network controller for feedback linearization," in *Proc. of 33rd Conf. Decision Contr.*, (Lake Buena Vista, FL), pp. 2494–2499, December 1994.

[22] F.-C. Chen and H. K. Khalil, "Adaptive control of nonlinear systems using neural networks," *Int. J. Control*, vol. 55, pp. 1299–1317, 1992.

[23] G. A. Rovithakis and M. A. Christodoulou, "Adaptive control of unknown plants using dynamical neural networks," *IEEE Trans. Syst. Man, Cybern.*, vol. 24, pp. 400–412, March 1994.

[24] J. T. Spooner and K. M. Passino, "Stable adaptive control using fuzzy systems and neural networks," *IEEE Trans. Fuzzy Systems*, vol. 4, pp. 339–359, August 1996.

[25] D. G. Taylor, P. V. Kokotovic, R. Marino, and I. Kanellakopoulos, "Adaptive regulation of nonlinear systems with unmodeled dyanamics," *IEEE Trans. Automat. Contr.*, vol. 34, pp. 405–412, April 1989.

[26] R. Ordóñez, J. T. Spooner, and K. M. Passino, "Stable multiple input multiple output adaptive fuzzy control," in *Proc. of 35th Conf. Decision Contr.*, (Kobe, Japan), pp. 610–615, December 1996.

[27] R. Ordóñez and K. M. Passino, "Stable multiple input multiple output direct adaptive fuzzy control," in *Proc. 1997 American Control Conf.*, (Albuquerque, NM), 1997.

[28] J. T. Spooner, R. Ordóñez, and K. M. Passino, "Adaptive prediction using fuzzy systems and neural networks," in *Proc. 1997 American Control Conf.*, (Albuquerque, NM), 1997.

[29] J. T. Spooner, R. Ordóñez, and K. M. Passino, "Direct adaptive fuzzy control for a class of nonlinear discrete time systems," in *Proc. 1997 American Control Conf.*, (Albuquerque, NM), 1997.

[30] J. T. Spooner, R. Ordóñez, and K. M. Passino, "Indirect adaptive fuzzy control for a class of nonlinear discrete time systems," in *Proc. 1997 American Control Conf.*, (Albuquerque, NM), 1997.

[31] J. T. Spooner and K. M. Passino, "Direct adaptive fuzzy control for a class of decentralized systems," in *Proc. 1997 American Control Conf.*, (Albuquerque, NM), 1997.

[32] J. T. Spooner and K. M. Passino, "Indirect adaptive fuzzy control for a class of decentralized systems," in *Proc. 1997 American Control Conf.*, (Albuquerque, NM), 1997.

[33] K. M. Passino and S. Yurkovich, "Fuzzy control," in *The Control Handbook* (W. Levine, ed.), Boca Raton, FL: CRC Press, 1996.

[34] T. Takagi and M. Sugeno, "Fuzzy identification of systems and its application to modeling and control," *IEEE Trans. Syst. Man, Cybern.*, vol. 15, pp. 116–132, 1985.

[35] S. Sastry and M. Bodson, *Adaptive Control: Stability, Convergence, and Robustness.* Englewood Cliffs, NJ: Prentice Hall, 1989.

[36] J. Hertz, A. Krogh, and R. G. Palmer, *Introduction to the Theory of Neural Computation.* Addison-Wesley Publishing Company, 1991.

[37] R. E. Fenton and R. J. Mayhan, "Automated highway studies at The Ohio State University—an overview," *IEEE Trans. Vehicular Technology,* vol. 40, pp. 100–113, February 1991.

[38] S. E. Shladover, C. A. Desoer, J. K. Hedrick, M. Tomizuka, J. Walrand, W.-B. Zhang, D. H. McMahon, H. Peng, S. Sheikholeslam, and N. McKeown, "Automatic vehicle control developments in the PATH program," *IEEE Trans. Vehicular Technology,* vol. 40, pp. 114–130, February 1991.

[39] Ioanno and Sun, *Robust Adaptive Control.* Prentice Hall, 1996.

[40] R. Ordóñez, J. Zumberge, J. T. Spooner, and K. M. Passino, "Adaptive fuzzy control: Experiments and comparative analyses." unpublished manuscript, 1995.

An Adaptive Fuzzy Sliding-Mode Controller

R. G. Berstecher and R. Palm
Siemens AG
Corporate Technologies, Dept. ZT IK 4
81730 Munich, Germany

H. Unbehauen
Ruhr-University Bochum
Faculty of Electrical Engineering
Control Engineering Laboratory
44780 Bochum, Germany

1 Introduction

In this article a linguistic, heuristics-based adaptation algorithm for a fuzzy sliding-mode controller (FSMC) is presented. The algorithm relies on linguistic knowledge in form of fuzzy if-then rules which reflect how an experienced operator would adapt the controller in order to obtain desired closed-loop behaviour. In Fig. 1 the structures of the underlying control and adaptation loops are shown.

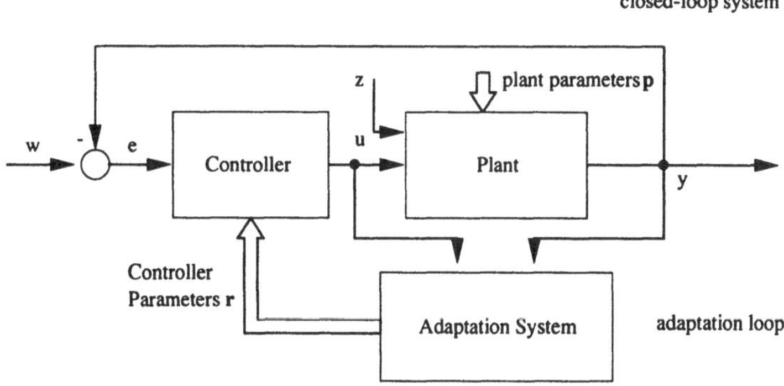

Figure 1: Control and adaptation loop

The adaptation parameters are grouped into the parameter vector $\mathbf{r}$. The linguistically defined adaptation law uses the switching variable s_σ of the FSMC, its time derivative $\dot{s}_\sigma$, and the manipulated variable u in order to compute a resulting adaptation parameter change $\dot{\mathbf{r}}$. The analytic description of the adaptation law reveals similarities to conventional control theory and allows one to relate uncertainties in the system description with the expected adaptation parameter convergence. Then, the heuristics-based, linguistic description of the adaptation law is analysed with conventional methods and stability and convergence properties are studied via Lyapunov's second method.

In Sect. 2, the basic principles of FSMCs are briefly reviewed. In Sect. 3, the heuristics-based motivation for the linguistic adaptation law is presented and an equivalent mathematical description is derived. The latter is used in Sect. 4 to analyse stability properties of the closed-loop system, including the adaptation block, and the convergence properties of the adaptation law. In Sect. 5 additional aspects of the adaptation algorithm are discussed. In Sect. 6 guidelines for the implementation of the adaptive fuzzy sliding-mode controller are presented. In Sect. 7, the adaptation law is applied to the control of a simulated two-link robot arm. Section 8 presents conclusions.

2 The fuzzy sliding-mode controller

In the following, we consider a nonlinear n-th order system of the form

$$x^{(n)} \quad = \quad f(x, \dot{x}, \ldots, x^{(n-1)}) + b(x, \dot{x}, \ldots, x^{(n-1)})u. \tag{1}$$

Let $b(x, \dot{x}, \ldots, x^{(n-1)}) > 0$ and the state $\mathbf{x} = \left(x \; \dot{x} \; \ldots \; x^{(n-1)}\right)^{\mathrm{T}}$ has to follow the desired trajectory $\mathbf{x}_{\mathrm{d}} = \left(x_{\mathrm{d}} \; \dot{x}_{\mathrm{d}} \; \ldots \; x_{\mathrm{d}}^{(n-1)}\right)^{\mathrm{T}}$. Then s_σ is the distance between the state vector $\mathbf{x}$ and the reference state vector. It is called the switching variable and is defined as

$$s_\sigma = \left(\frac{\mathrm{d}}{\mathrm{d}t} + \lambda_{\mathrm{R}}\right)^{n-1} \tilde{e} = \sum_{i=0}^{n-1} \begin{pmatrix} n-1 \\ i \end{pmatrix} {\lambda_{\mathrm{R}}}^{i} \tilde{e}^{(n-1-i)}, \tag{2}$$

with

$$\tilde{\mathbf{e}} = \mathbf{x} - \mathbf{x}_{\mathrm{d}} = \left(\tilde{e} \; \dot{\tilde{e}} \; \ldots \; \tilde{e}^{(n-1)}\right)^{\mathrm{T}}. \tag{3}$$

$s_\sigma = 0$ represents a switching surface in the state space. λ_{R} is a parameter specifying s_σ and therewith the dynamics of the closed-loop behaviour according to [8].

Table 1: Rule base for a FSMC

Rule R 1:	If	s_σ	is	*NB*	Then	u	is	*PB*
Rule R 2:	If	s_σ	is	*NM*	Then	u	is	*PM*
Rule R 3:	If	s_σ	is	*NS*	Then	u	is	*PS*
Rule R 4:	If	s_σ	is	*Z*	Then	u	is	*Z*
Rule R 5:	If	s_σ	is	*PS*	Then	u	is	*NS*
Rule R 6:	If	s_σ	is	*PM*	Then	u	is	*NM*
Rule R 7:	If	s_σ	is	*PB*	Then	u	is	*NB*,

Fuzzy controllers which use the switching variable s_σ to calculate the manipulated variable u are called FSMCs. Their fuzzy rules have the form given in Table 1. In Table 1, N denotes negative, P positive, Z zero, B big, M medium and S small linguistic values whose meaning is defined by corresponding membership functions. The manipulated variable u is the output of the FSMC while the switching variable s_σ is its input.

A boundary layer is introduced for conventional sliding-mode controllers in order to reduce chattering. The output value is linearly interpolated between the positive and negative control value in the boundary layer. However, the robust control properties of sliding-mode controllers are maintained [8]. Fuzzy sliding-mode controllers are described by the choice of the switching variable s_σ and the nonlinear interpolation in the boundary layer.

The shape of the nonlinear transfer characteristic of a FSMC depends not only on the values of u, but also on the membership functions of the rule antecedents (if-parts) and consequents (then-parts) and the defuzzification method. In what follows, g_{fc} represents the transfer characteristic of the FSMC and describes the dependence of the manipulated variable u on the switching variable s_σ and the adaptation parameters $\mathbf{r} = (r_1, \ldots, r_i, \ldots, r_N)$ that are to be defined later.

By choosing the center-of-sums defuzzification method, the output u of the FSMC is computed as

$$u = g_{\mathrm{fc}}(s_\sigma, \mathbf{r}) = \frac{\sum\limits_{i=1}^{N} \mu_i(s_\sigma) M_i}{\sum\limits_{i=1}^{N} \mu_i(s_\sigma) A_i}, \tag{4}$$

where μ_i is the degree to which the antecedent of the i-th fuzzy rule is satisfied by a particular crisp value of s_σ, M_i and A_i are the moment and area of the membership function defining the linguistic value of u in the i-th rule.

The point at which a triangular membership function has value 1 is called a supporting point. Here, the N points $s_{\sigma,j}$, $j = 1, \ldots, N$, with membership 1 to the membership functions defining the linguistic values of s_σ, are the supporting points in the (u, s_σ) diagram (see Fig. 2). At the j-th supporting

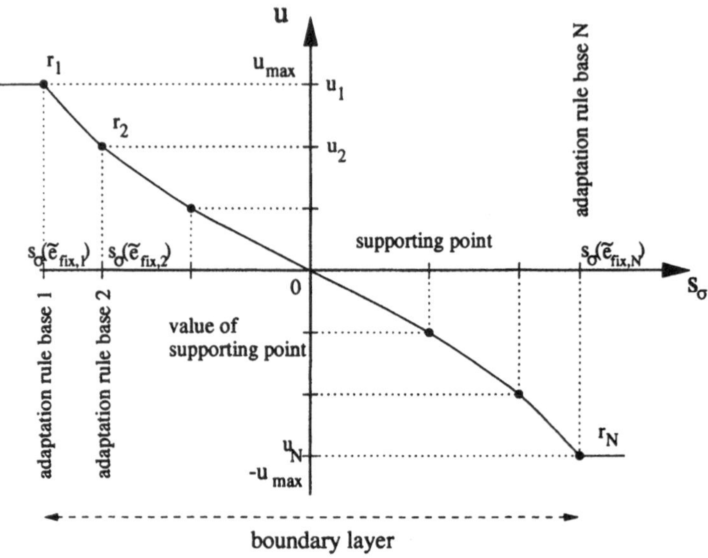

Figure 2: Nonlinear transfer characterstic of the FSMC

point the antecedents of the other $N-1$ rules have degrees of satisfaction equal to zero and therefore these rules do not fire. Thus, the output of the FSMC at the j-th supporting point $s_{\sigma,j}$, according to (4) and the j-th rule R^j in Table 1, is

$$u(s_{\sigma,j}) = u_j = \frac{\mu_j(s_{\sigma,j})M_j}{\mu_j(s_{\sigma,j})A_j} = \frac{M_j}{A_j}. \tag{5}$$

The points $(s_{\sigma,j}, \frac{M_j}{A_j})$ lie on the nonlinear transfer characteristic of the FSMC in the (u, s_σ) diagram.

3 The linguistic adaptation law

3.1 Motivation for the linguistic adaptation strategy

The adaptation strategy is defined by heuristics-based linguistic rules describing how an expert would adapt the FSMC in order to achieve a desired closed-loop behaviour [1]. Such a description by fuzzy rules allows an obvious and easy representation of the adaptation law.

The output u_j of every fuzzy rule is adapted at its corresponding supporting point. At such a point the output of the FSMC can be calculated directly, i.e. without interpolation. The nonlinear transfer characteristic is represented by the parameter values r_i at the supporting points $s_{\sigma,i}$ following Fig. 2. The adaptation parameters r_i for the N supporting points are grouped together in the adaptation vector

$$\mathbf{r} = (r_1 \; \dots \; r_i \; \dots \; r_N)^{\mathrm{T}}. \tag{6}$$

Depending on the adaptation parameters r_i of the FSMC, the resulting parameter adaptation is linear or nonlinear. If the output of the FSMC depends linearly on the vector $\mathbf{r}$, then the output u can be described by

$$u = \boldsymbol{\zeta}(s_\sigma(\tilde{\mathbf{e}}))^{\mathrm{T}}\mathbf{r}, \tag{7}$$

where the vector function $\boldsymbol{\zeta}(\tilde{\mathbf{e}})$ is a nonlinear function of the error vector $\tilde{\mathbf{e}}$.

In order to deal with a linear adaptation problem in r_i, the output value u of the FSMC is only adapted at the supporting points $s_{\sigma,i}$ by changing the moments M_i in (4). A change in the areas A_i, however, would result in a nonlinear adaptation problem. The vector of adaptation parameters then is

$$\mathbf{r} = (r_1 \; \dots \; r_i \; \dots \; r_N)^{\mathrm{T}} = (M_1 \; \dots \; M_i \; \dots \; M_N)^{\mathrm{T}}. \tag{8}$$

By choosing the membership functions of the consequents as singletons, the moments are equal to the location of the respective singletons. If the membership functions of the consequents are described by triangular membership functions such that $\mu_i(s_\sigma) + \mu_{i+1}(s_\sigma) = 1$ for each i in the whole domain, then the shift of a supporting point would result in a change of the moment as shown in Fig. 3. In turn, the areas of the neighbouring membership functions would be altered. Therefore, the output value of the FSMC would depend nonlinearly on the parameters and (7) would then approximate the "correct" output value. If the membership functions of the consequents are realised by triangluar membership functions, then by changing only their location and keeping their area, the output u is given by (7).

The desired closed-loop behaviour is formulated by an expert at the supporting points, where the adaptation rules for every supporting point are put together into a rule base corresponding to this supporting point.

The adaptation strategy changes the i-th output u_i of the FSMC by adapting its corresponding r_i in a manner that depends on the desired closed-loop behaviour. For systems of second order in the $(\tilde{e}, \dot{\tilde{e}})$-phase plane, the

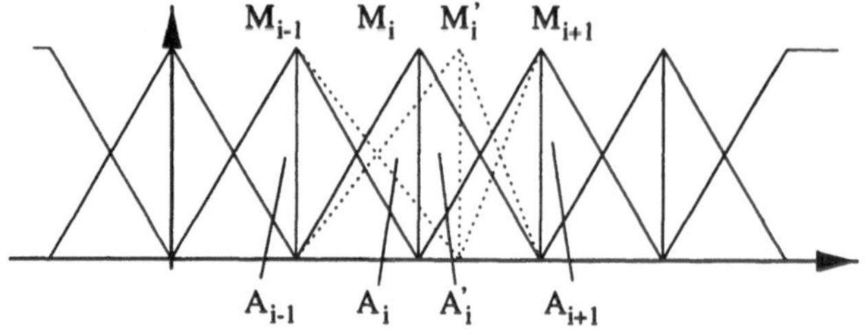

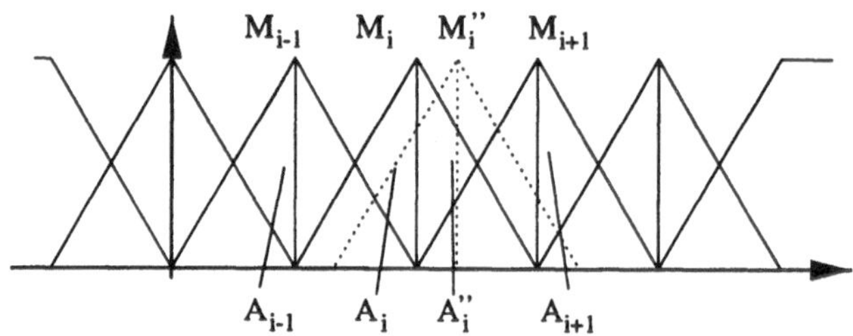

Figure 3: Change of the location of the output membership functions

point $Q = Q(\tilde{e}, \dot{\tilde{e}})$ in Fig. 4 represents the error $\tilde{\mathbf{e}} = \mathbf{x} - \mathbf{x}_\mathrm{d}$. In Fig. 4, the switching line $s_\sigma = 0$, the distance s_σ between $Q(\tilde{e}, \dot{\tilde{e}})$ and the switching line, and its time derivative $\dot{s}_\sigma$ are shown. The adaptation law uses

- the distance s_σ of Q from the switching surface $s_\sigma = 0$ in accordance with (2),
- the approach velocity $\dot{s}_\sigma$ of Q to the switching surface $s_\sigma = 0$, and
- the manipulated variable u

in the antecedents in the fuzzy rules describing the linguistic adaptation strategy. Based on these, each fuzzy rule computes a parameter change $\dot{r}_i$ for the i-th component r_i of the parameter vector $\mathbf{r}$. The L rules R_i^j, defining the adaptation strategy for r_i at $s_{\sigma,i}$, thus have the form

$$\mathrm{R}_i^j : \text{If } s_\sigma \text{ is } LS_\sigma^j \text{ and } \dot{s}_\sigma \text{ is } L\dot{S}_\sigma^j \text{ and } u \text{ is } LU^j \text{ Then } \dot{r}_i \text{ is } L\dot{R}_i^j. \quad (9)$$

LS_σ^j,$L\dot{S}_\sigma^j$, LU^j and $L\dot{R}_i^j$ represent the linguistic values for the variables involved in the rule antecedent and consequent respectively. Each such linguistic value is defined by a membership function. The switching variable s_σ

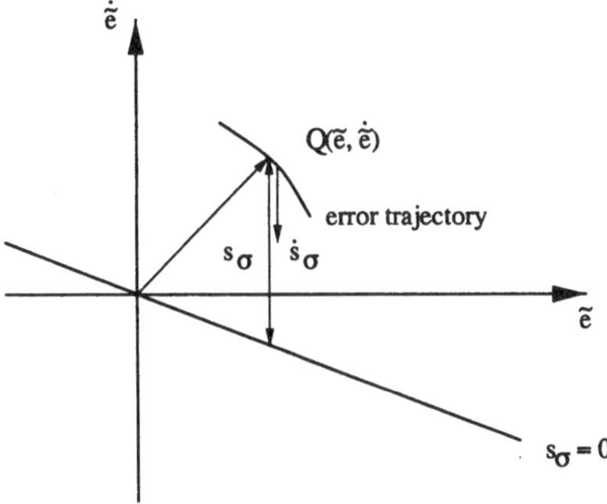

Figure 4: Representation of the linguistically defined adaptation algorithm in the phase plane

is calculated according to (2) and (3). s_σ is the same for all the rules that constitute the i-th rule base. These rules are responsible for the adaptation of the FSMC at the i-th supporting point. The aim of the adaptation rules is to achieve an optimal approach velocity to the switching surface $s_\sigma = 0$: an approach velocity which is too slow or even vanishing must be avoided. On the other hand, at small distances s_σ, the approach velocity to the switching surface $s_\sigma = 0$ should not be too high in order to prevent undesired chattering.

If, for example, 's_σ is positive big', '$\dot{s}_\sigma$ is negative' and 'u is negative', then the negative output value u of the controller should not be changed. In Table 2, this is expressed by rule R_i^7 with '$\dot{r}_i$ is Z'. In Table 2, P represents positive, Z zero, N negative and B big. The adaptation rule uses heuristics-based linguistic knowledge, because the approach velocity $\dot{s}_\sigma$ should be moderate for a medium distance and negative control value u. If, in another example, 's_σ is positive big', '$\dot{s}_\sigma$ is positive' and 'u is negative', then the output value u of the FSMC is not big enough in order to guarantee $\dot{s}_\sigma < 0$. Thus the negative output of the FSMC is to be reduced significantly by $\dot{r}_i < 0$. In Table 2, this situation is represented by R_i^9.

As already mentioned, fuzzy controllers which calculate the manipulated variable u by using the switching variable s_σ are fuzzy sliding-mode controllers. Fuzzy rules that calculate the derivative of the switching variable by using the switching variable s_σ and the manipulated variable u characterise the closed-loop system behaviour. In [3], [9], adaptation rules are derived from the assessment of s_σ and $\dot{s}_\sigma$. In contrast, we derive a parameter change of the adaptation parameters r_i from s_σ, $\dot{s}_\sigma$ and u, according to (9). Such fuzzy rules represent the desired closed-loop behaviour in a heuristics-based linguistic manner since they heuristically relate linguistically expressed system states, s_σ, to linguistic values of the control effort u and the system output $\dot{s}_\sigma$ [2]. Thus, the FSMC performance is enhanced by a linguistic adaptation law. At the same time, the transparency of the controller design is maintained [5]. In Table 2 the rules for the supporing point, characterised by 's_σ is PB', are shown.

Table 2: Rule base of the adaptation rules for the FSMC

	If s_σ is LS_σ^j	and $\dot{s}_\sigma$ is $L\dot{S}_\sigma^j$	and u is LU^j	Then $\dot{r}_i$ is $L\dot{R}_i^j$
Rule R_i^1:	PB	NB	NB	PB
Rule R_i^2:	PB	N	NB	P
Rule R_i^3:	PB	Z	NB	Z
Rule R_i^4:	PB	P	NB	N
Rule R_i^5:	PB	PB	NB	NB
Rule R_i^6:	PB	NB	N	P
Rule R_i^7:	PB	N	N	Z
Rule R_i^8:	PB	Z	N	N
Rule R_i^9:	PB	P	N	NB
⋮	⋮	⋮	⋮	⋮

3.2 Description of the linguistic adaptation algorithm with conventional control methods

A deliberate change of the controller parameters r_i forms the basis of the adaptation algorithm. The values r_i themselves have immediate impact on the manipulated variable u of the FSMC $\dot{s}_\sigma$. The desired closed-loop behaviour depends on the set of fuzzy rules from Table 2. The fuzzy rules calculate, depending on the desired behaviour at the supporing points, a change of r_i. For the adaptation of each parameter r_i, a separate fuzzy rule base and therewith adaptation strategy is defined.

By influencing the parameters r_i only at the supporting points $s_{\sigma,i}$, coupling effects between the different parameters are avoided. Figure 5 illustrates that a change in r_i by Δr_i does not alter the output of the FSMC at the supporting points$s_{\sigma,j}$, $j \neq i$.

The switching variable s_σ in Table 2 characterises *one* supporting point of the transfer characteristic and therefore is identical for all adaptation rules. By representing $\dot{s}_\sigma$, u and $\dot{r}_i$ as in Table 2 and omitting s_σ in the look-up

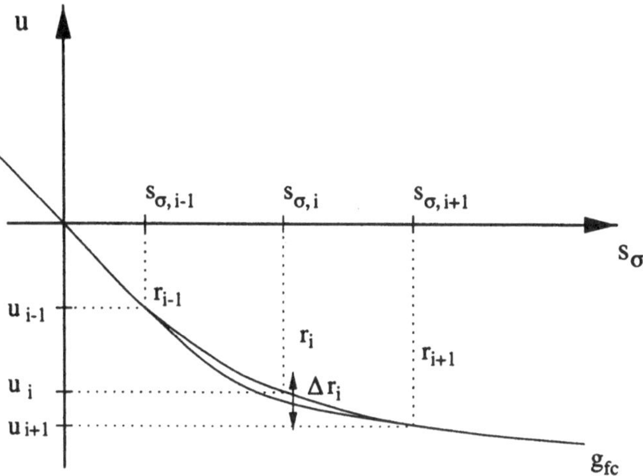

Figure 5: Dependence of the transfer characteristic of the FSMC on the parameter value r_i at $s_{\sigma,i}$

table representation from Fig. 6, the similarity of the adaptation algorithm with the structure of a FSMC becomes apparent. This fuzzy rule base has the structure of a FSMC because, as shown in Fig. 6, in the cells parallel to the main diagonal (cells with values Z), the same output value $\dot{r}_i$ is always chosen. This FSMC in the adaptation loop is used here for the adaptation of the FSMC in the underlying control loop. The FSMC in the adaptation loop is used for the adaptation of the parameter $s_{\sigma,i}$. The index α, i indicates the switching variable characterising the adaptation at the i-th supporting point whereas the index σ characterises the switching variable of the FSMC in the control loop. Since the switching line $s_{\alpha,i} = 0$ comes into play with respect to the adaptation of the parameter r_i, it is called an adaptation line in the following. The adaptation line $s_{\alpha,i} = 0$ can - in analogy to $s_\sigma = 0$ - analytically be described by

$$s_{\alpha,i} = s_{\alpha,i}(\dot{s}_\sigma, u) = \lambda_{\alpha,i}\,(\dot{s}_\sigma - \dot{s}_{\sigma,i_0}) + (u - u_{i_0}) = 0. \tag{10}$$

In the above expression, $\lambda_{\alpha,i} > 0$ indicates the slope of the adaptation line $s_{\alpha,i} = 0$, and $(\dot{s}_{\sigma,i_0}, u_{i_0})$ defines a point on the adaptation line $s_{\alpha,i} = 0$. The point $(\dot{s}_{\sigma,i_0}, u_{i_0})$ on the adaptation line $s_{\alpha,i} = 0$ represents a pair of reference values of $\dot{s}_\sigma$ and u at which the parameter r_i is not adapted. Taking into account the slope $\lambda_{\alpha,i}$, we obtain the set of all points $(\dot{s}_\sigma, u)$ at which no adaptation is performed. Because for points on the adaptation line $\dot{r}_i = 0$ holds, we have

$$u = -\lambda_{\alpha,i}(\dot{s}_\sigma - \dot{s}_{\sigma,i_0}) + u_{i_0}. \tag{11}$$

u \ $\dot{s}_\sigma$	NB	N	Z	P	PB
NB	PB	P	Z	N	NB
N	P	Z	N	NB	NB
Z	Z	N	NB	NB	NB
P	N	NB	NB	NB	NB
PB	NB	NB	NB	NB	NB

Figure 6: Linguistically defined adaptation algorithm for $\dot{r}_i$ where the antecedent 's_σ is PB' from (9) is omitted

Examined analytically, the adaptation rules define a switching line that is given by the parameter $\lambda_{\alpha,i}$ and (local) reference points $(s_{\sigma,i_0}, u_{i_0})$. The adaptation rules following from (9) shows the combination of $\dot{s}_\sigma$ and u at the supporting points at which r_i is to be adapted. The choice of the parameters $\lambda_{\alpha,i}$, $\dot{s}_{\sigma,i_0}$ and u_{i_0} then is driven by the desired closed-loop behaviour at the i-th supporting point as well as the characteristics of the underlying system. The linguistically expressed heuristic knowledge about the process is incorporated here by choosing the membership functions for $\dot{s}_\sigma$ and u. The choice of the membership degrees at the i-th supporting point corresponds to the parameters $\dot{s}_{\sigma,i_0}$, u_{i_0}, and the slope of the adaptation line $\lambda_{\alpha,i}$.

The variable $\dot{s}_\sigma$ is calculated by differentiating (2) and a consequent substitution in (1). That is,

$$\begin{aligned}\dot{s}_\sigma &= \sum_{i=0}^{n-1} \binom{n-1}{i} \lambda_R{}^i \tilde{e}^{(n-i)} \\ &= \sum_{i=1}^{n-1} \binom{n-1}{i} \lambda_R{}^i \tilde{e}^{(n-i)} + \tilde{e}^{(n)} \\ &= \sum_{i=1}^{n-1} \binom{n-1}{i} \lambda_R{}^i \tilde{e}^{(n-i)} + f(\mathbf{x}) + b(\mathbf{x})\, u - x_d^{(n)}. \end{aligned} \tag{12}$$

The unknown nonlinear functions $f(\mathbf{x})$ and $b(\mathbf{x})$ are estimated by $\hat{f}(\mathbf{x})$ and $\hat{b}(\mathbf{x})$. (12) describes a straight line for $\dot{s}_\sigma$ and u in the $(\dot{s}_\sigma, u)$-plane

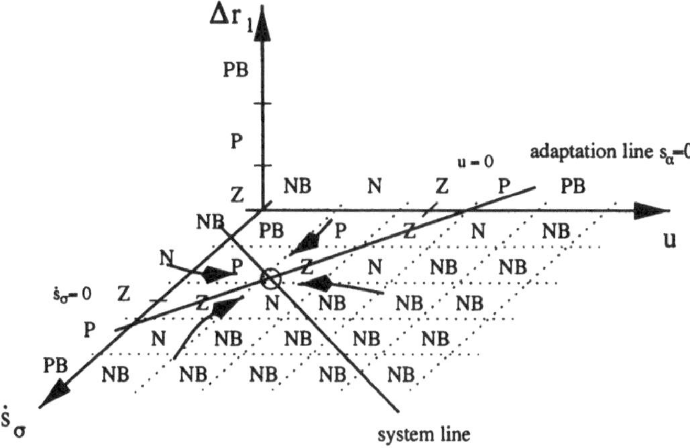

Figure 7: Adaptation line of the i-th fuzzy rule base in the $(\dot{s}_\sigma, u)$-plane

which is also called the system line. Fig. 7 represents a moment t_k at which $s_\sigma(t_k) = s_{\sigma,i}$ for $\mathbf{x}(t_k)$ and $\mathbf{x}_\mathrm{d}(t_k)$ is supposed to hold. At the supporting point, $s_{\alpha,i}(t_k) = 0$ represents the optimal parameter $r_i(t_k)$ with respect to the adaptation criterion, which is given by $\dot{r}_i = 0$. Consider now the intersection point of the adaptation line $s_{\alpha,i}(t_k) = 0$ from (10) with the system line from (12). This intersection point then represents, at supporting point $s_{\sigma,i}$, the optimal combination of $\dot{s}_\sigma$ and u. For the intersection point, the control parameter $r_i(t_k)$ is not adapted because $s_{\alpha,i}(t_k) = 0$ is equal to a parameter change $\dot{r}_i = 0$ (follows from (10)). Then the manipulated variable u at time t_k has its optimal value u_opt. The equation

$$u(t_k) = g_\mathrm{fc}(s_\sigma, \mathbf{r})\big|_{s_{\sigma,i}} = g_\mathrm{fc}(s_{\sigma,i}, \mathbf{r}_\mathrm{opt}) = u_\mathrm{opt}, \tag{13}$$

implicitely expresses $r_i(t_k)$ of the parameter vector $\mathbf{r}_\mathrm{opt}(t_k)$ at supporting point $s_{\sigma,i}$ at time t_k. At this supporting point the parameters r_j, $j \neq i$, according to (4), do not have to be considered. This is illustrated in Fig. 8.

As mentioned before, each of the N components r_i, $i = 1, \ldots, N$, of the parameter vector of the FSMC is adapted by its own fuzzy rule base. For the adaptation of the parameter vector $\mathbf{r}$, its velocity vector, given as

$$\dot{\mathbf{r}} = (\dot{r}_1 \ \dot{r}_2 \ \ldots \ \dot{r}_N)^\mathrm{T}, \tag{14}$$

is used.

In the preceding it was shown how the parameter values $r_i(t_k)$ are adapted for t_k at the supporting points $s_{\sigma,i}$, $i = 1, \ldots, N$. The intersection of system

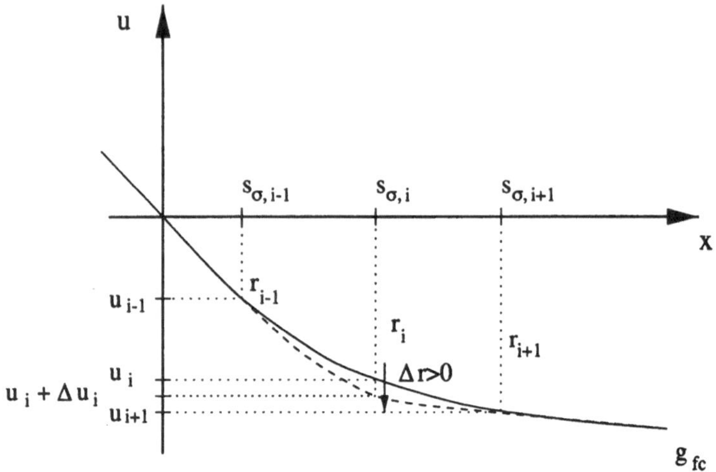

Figure 8: Adaptation of the control parameter r_i with $\dot{r}_i$

and adaptation lines represents the optimal control value u_{opt} which, according to (13), is equivalent to the optimal parameter value $\mathbf{r}_{\text{opt}}$. For variable $\mathbf{x}(t)$ and $\mathbf{x}_\text{d}(t)$, in (12) $f(\mathbf{x}(t))$, $b(\mathbf{x}(t))$, $s_\sigma(t)$, $u(t) = g_{\text{fc}}(s_\sigma(t), \mathbf{r}(t))$ and therewith $\dot{s}_\sigma(t)$ also change. Starting from a variable state vector $\mathbf{x}(t)$ and the desired closed-loop behaviour $\mathbf{x}_\text{d}(t)$, $s_{\alpha,i} = 0$ in (10) means that the parameter value r_i is optimal and does not need to be adapted, because $\dot{r}_i = 0$. When $\dot{s}_\sigma(t)$ and $u(t)$ in (10) change over time, then even for fixed parameters $\mathbf{r}$ the values of the switching variables $s_{\alpha,i}(t)$ also change. A changing $s_{\alpha,i}(t)$ then leads to variable parameter values $r_i(t)$ even if parameter $r_i(t_k)$ is optimal at $s_{\sigma,i}$ with $r_i(t_k) = r_{i,\text{opt}}$. These changes of the parameters, however, are not the result of an intentional parameter change but are due to $\dot{s}_\sigma(t)$ and $u(t)$. In the neighbourhood of $s_{\sigma,i}$ $\dot{s}_\sigma(t)$ and $u(t) = g_{\text{fc}}(s_\sigma(t), \mathbf{r}(t))$ do not vary too much. For this reason the parameter $r_i(t)$ does not vary too much either. The more $s_\sigma(t)$ differs from $s_{\sigma,i}$, the more difficult it is to estimate the effects from changes in $\dot{s}_\sigma(t)$ and $u(t)$. In order to reduce the undesired parameter changes, the membership functions of 's_σ is LS_σ' should only have a small support around the respective supporting point $s_{\sigma,i}$. The effect of the parameter fluctuations then is not completely cancelled but is instead significantly reduced.

In order to eliminate the dependence of r_i on s_σ and u, $\dot{s}_{\sigma,i_0}$ and u_{i_0} are subsequently replaced by state dependent and therewith time-dependent correction terms $\dot{s}_{\sigma,\text{neu}_i}(t)$ and $u_{\text{neu}_i}(t)$. Formally speaking, the adaptation line for the variables $\mathbf{x}$ and $\mathbf{x}_\text{d}$ is described by

$$s_{\alpha,i} = \lambda_{\alpha,i}\left(\dot{s}_\sigma - \dot{s}_{\sigma,\text{neu}_i}\right) + \left(u - u_{\text{neu}_i}\right). \tag{15}$$

The correction term $\dot{s}_{\sigma,\mathrm{neu}_i}$ is defined as

$$\dot{s}_{\sigma,\mathrm{neu}_i} = \dot{s}_{\sigma,i_0} + \dot{s}'_i. \tag{16}$$

Here,

$$\begin{aligned} s'_i &= \left(\lambda_\mathrm{R} + \frac{\mathrm{d}}{\mathrm{d}t}\right)^{n-1} (x - x_{\mathrm{fix},i}) \\ &= \sum_{i=0}^{n-1} \binom{n-1}{i} \lambda_\mathrm{R}^i x^{(n-1-i)} - \sum_{i=0}^{n-1} \binom{n-1}{i} \lambda_\mathrm{R}^i x_{\mathrm{fix},i}{}^{(n-1-i)} \end{aligned} \tag{17}$$

has the time-derivative

$$\begin{aligned} \dot{s}'_i &= \sum_{i=0}^{n-1} \binom{n-1}{i} \lambda_\mathrm{R}^i x^{(n-i)} - \sum_{i=0}^{n-1} \binom{n-1}{i} \lambda_\mathrm{R}^i x_{\mathrm{fix},i}{}^{(n-i)} \\ &= \sum_{i=1}^{n-1} \binom{n-1}{i} \lambda_\mathrm{R}^i x^{(n-i)} + x^{(n)} - \sum_{i=0}^{n-1} \binom{n-1}{i} \lambda_\mathrm{R}^i x_{\mathrm{fix},i}{}^{(n-i)}. \end{aligned} \tag{18}$$

The state vector $\mathbf{x}_{\mathrm{fix},i}$ is a state vector at a distance $s_{\sigma,i}$ from the switching surface $s_\sigma = 0$ and thus corresponds to the i-th supporting point. For constant $\mathbf{x}_\mathrm{d}$, $\mathbf{x}_{\mathrm{fix},i}$ is constant.

For variable $\mathbf{x}_\mathrm{d}(t)$, $\tilde{\mathbf{e}}(t) = \mathbf{x}_{\mathrm{fix},i}(t) - \mathbf{x}_\mathrm{d}(t)$ represents a trajectory of the error vector $\tilde{\mathbf{e}}$ at a distance $s_{\sigma,i}$ from the switching surface $s_\sigma = 0$. In the sequel, time dependencies are not given in order to enhance representational clarity.

$x^{(n)}$ and $x_{\mathrm{fix},i}^{(n)}$ are estimated by a model of the plant. Therefore, for the calculation of the correction terms an estimated model of the plant is included in the form of $\hat{f}(\mathbf{x})$, $\hat{b}(\mathbf{x})$, $\hat{f}(\mathbf{x}_{\mathrm{fix},i})$ and $\hat{b}(\mathbf{x}_{\mathrm{fix},i})$. That is

$$\hat{x}^{(n)} = \hat{f}(\mathbf{x}) + \hat{b}(\mathbf{x})\, g_\mathrm{fc}(s_\sigma, \mathbf{r}), \tag{19}$$

$$\hat{x}_{\mathrm{fix},i}^{(n)} = \hat{f}(\mathbf{x}_{\mathrm{fix},i}) + \hat{b}(\mathbf{x}_{\mathrm{fix},i})\, g_\mathrm{fc}(s_{\sigma,i}, \mathbf{r}). \tag{20}$$

Thus, $\dot{s}'_i$ is obtained from (18) by taking into account (19) and (20).

Namely,

$$
\begin{aligned}
\dot{s}_i' &= \\
&\sum_{i=1}^{n-1}\binom{n-1}{i}\lambda_{\rm R}^i x^{(n-i)} - \sum_{i=1}^{n-1}\binom{n-1}{i}\lambda_{\rm R}^i x_{{\rm fix},i}{}^{(n-i)} \\
&+ \hat{f}(\mathbf{x}) + \hat{b}(\mathbf{x})\, g_{\rm fc}(s_\sigma, \mathbf{r}) - \hat{f}(\mathbf{x}_{{\rm fix},i}) - \hat{b}(\mathbf{x}_{{\rm fix},i})\, g_{\rm fc}(s_{\sigma,i}, \mathbf{r}) \\
&= \sum_{i=1}^{n-1}\binom{n-1}{i}\lambda_{\rm R}^i x^{(n-i)} - \sum_{i=1}^{n-1}\binom{n-1}{i}\lambda_{\rm R}^i x_{{\rm fix},i}{}^{(n-i)} \\
&+ \left(\hat{f}(\mathbf{x}) - \hat{f}(\mathbf{x}_{{\rm fix},i})\right) + \left(\hat{b}(\mathbf{x})\, g_{\rm fc}(s_\sigma, \mathbf{r}) - \hat{b}(\mathbf{x}_{{\rm fix},i})\, g_{\rm fc}(s_{\sigma,i}, \mathbf{r})\right).
\end{aligned}
\tag{21}
$$

Substituting (21) in (16), we obtain $\dot{s}_{\sigma,{\rm neu}_i}$

$$
\begin{aligned}
\dot{s}_{\sigma,{\rm neu}_i} &= \dot{s}_{\sigma,i_0} + \left(\sum_{i=1}^{n-1}\binom{n-1}{i}\lambda_{\rm R}^i x^{(n-i)} + \hat{f}(\mathbf{x}) + \hat{b}(\mathbf{x})\, g_{\rm fc}(s_\sigma, \mathbf{r})\right) \\
&- \left(\sum_{i=1}^{n-1}\binom{n-1}{i}\lambda_{\rm R}^i x^{(n-i)} + \hat{f}(\mathbf{x}_{{\rm fix},i}) + \hat{b}(\mathbf{x}_{{\rm fix},i})\, g_{\rm fc}(s_{\sigma,i}, \mathbf{r})\right)\Bigg|_{\mathbf{x}=\mathbf{x}_{{\rm fix},i}} \\
&= \dot{s}_{\sigma,i_0} + \lambda_{\rm R}^{n-1}(\dot{x} - \dot{x}_{{\rm fix},i}) + (n-1)\lambda_{\rm R}^{n-2}(\ddot{x} - \ddot{x}_{{\rm fix},i}) \\
&+ \hat{f}(\mathbf{x}) - \hat{f}(\mathbf{x}_{{\rm fix},i}) + \hat{b}(\mathbf{x}) g_{\rm fc}(s_\sigma, \mathbf{r}) - \hat{b}(\mathbf{x}_{{\rm fix},i}) g_{\rm fc}(s_{\sigma,i}, \mathbf{r}) - x_{\rm d}^{(n)}.
\end{aligned}
\tag{22}
$$

The correction term $u_{{\rm neu}_i}$ is calculated by

$$
u_{{\rm neu}_i} = u_{i_0} - g_{\rm fc}(s_{\sigma,i}, \mathbf{r}) + g_{\rm fc}(s_\sigma, \mathbf{r}). \tag{23}
$$

(22) and (23) represent the coordinates of the point $A(\dot{s}_{\sigma,{\rm neu}_i}, u_{{\rm neu}_i})$, through which the adaptation line $s_{\alpha,i} = 0$ with slope $\lambda_{\alpha,i}$ is drawn (see Fig. 7). The membership values are also shown for $\dot{s}_\sigma$, u and $\dot{r}_i$ along the axis. The rule consequents of Table 2 are given in the $(\dot{s}_\sigma, u)$-plane for 's_σ is PB'. The adaptation line $s_{\alpha,i} = 0$ and the system line (see (10) and (12)) are drawn in the $(\dot{s}_\sigma, u)$ - plane for $s_\sigma = s_{\sigma,i}$. The arrows show the direction, in which $\dot{s}_\sigma$ and u are adapted through r_i and $\dot{r}_i$ for fixed $\mathbf{x}$ and $\mathbf{x}_{\rm d}$.

The new correction terms are dependent of: (i) the state vector $\mathbf{x}$, (ii) the parameter vector $\mathbf{r}$, and (iii) the estimates $\hat{f}$ and $\hat{b}$. Then the switching variable of the adaptation line $s_{\alpha,i} = 0$ is calculated as

$$
\begin{aligned}
s_{\alpha,i} &= \lambda_{\alpha,i}\left(\dot{s}_\sigma - \dot{s}_{\sigma,\mathrm{neu}_i}\right) + \left(u - u_{\mathrm{neu}_i}\right) \\
&= \lambda_{\alpha,i}\Big[-\dot{s}_{\sigma,i_0} + \Big(\lambda_{\mathrm{R}}^{(n-1)}(\dot{x} - \dot{x}_{\mathrm{d}}) + \ldots + \\
&\qquad\qquad (n-1)\lambda_{\mathrm{R}}(x^{(n-1)} - x_{\mathrm{d}}^{(n-1)})\Big)\Big|_{\mathbf{x}=\mathbf{x}_{\mathrm{fix},i}}\Big] \\
&\quad + \lambda_{\alpha,i}\Big[\Big(f(\mathbf{x}) - \hat{f}(\mathbf{x})\Big) + (b(\mathbf{x}) - \hat{b}(\mathbf{x}))g_{\mathrm{fc}}(s_\sigma, \mathbf{r}) + \\
&\qquad\qquad \hat{f}(\mathbf{x}_{\mathrm{fix},i}) + \hat{b}(\mathbf{x}_{\mathrm{fix},i})g_{\mathrm{fc}}(s_{\sigma,i}, \mathbf{r}) - x_{\mathrm{d}}^{(n)}\Big] \\
&\quad - u_{i_0} + g_{\mathrm{fc}}(s_{\sigma,i}, \mathbf{r}).
\end{aligned} \tag{24}
$$

For the adaptation, the closed-loop behaviour is important. Again, the model of the plant is incorporated in the adaptation process by using $\hat{f}$ and $\hat{b}$ which can be identified through an ordinary identification algorithm.

In order to enhance the representational clarity of (24) and to be able to find upper bounds for the parameter uncertainty, the term

$$
\begin{aligned}
&\lambda_{\alpha,i}\hat{b}(\mathbf{x}_{\mathrm{fix},i})g_{\mathrm{fc}}(s_{\sigma,i}, \mathbf{r}_{\mathrm{opt}}) - \lambda_{\alpha,i}\hat{b}(\mathbf{x}_{\mathrm{fix},i})g_{\mathrm{fc}}(s_{\sigma,i}, \mathbf{r}_{\mathrm{opt}}) + \\
&\qquad g_{\mathrm{fc}}(s_{\sigma,i}, \mathbf{r}_{\mathrm{opt}}) - g_{\mathrm{fc}}(s_{\sigma,i}, \mathbf{r}_{\mathrm{opt}}) = 0
\end{aligned} \tag{25}
$$

is added to the right-hand side of (24). The terms in (24), characterising the behaviour of the controlled plant at the supporting points $s_{\sigma,i}$, can now be grouped together as

$$
\begin{aligned}
&\lambda_{\alpha,i}\Bigg[-\dot{s}_{\sigma,i_0} + \Big(\lambda_{\mathrm{R}}^{n-1}(\dot{x} - x_{\mathrm{d}}) + \ldots + \\
&\quad (n-1)\lambda_{\mathrm{R}}(x^{(n-1)} - x_{\mathrm{d}}^{(n-1)})\Big)\Big|_{\mathbf{x}=\mathbf{x}_{\mathrm{fix},i}} \\
&\quad + \Big(\hat{f}(\mathbf{x}_{\mathrm{fix},i}) + \hat{b}(\mathbf{x}_{\mathrm{fix},i})g_{\mathrm{fc}}(s_{\sigma,i}, \mathbf{r}_{\mathrm{opt}}) - x_{\mathrm{d}}^{(n)}\Big)\Bigg] + \\
&\quad \Big(-u_{i_0} + g_{\mathrm{fc}}(s_{\sigma,i}, \mathbf{r}_{\mathrm{opt}})\Big) \\
= \; &\lambda_{\alpha,i}\left(\hat{\dot{s}}_{\sigma,i} - \dot{s}_{\sigma,i_0}\right) + (u_{\mathrm{opt}} - u_{i_0}),
\end{aligned} \tag{26}
$$

where

$$
\begin{aligned}
\hat{\dot{s}}_{\sigma,i} &= \hat{f}(\mathbf{x}_{\mathrm{fix},i}) + \hat{b}(\mathbf{x}_{\mathrm{fix},i})\, g_{\mathrm{fc}}(s_{\sigma,i}, \mathbf{r}_{\mathrm{opt}}) + \\
&\quad \Big(\lambda_{\mathrm{R}}^{n-1}(\dot{x} - \dot{x}_{\mathrm{d}}) + \ldots + (n-1)\lambda_{\mathrm{R}}(x^{(n-1)} - x_{\mathrm{d}}^{(n-1)})\Big)\Big|_{\mathbf{x}=\mathbf{x}_{\mathrm{fix},i}}
\end{aligned} \tag{27}
$$

is the estimated value $\hat{\dot{s}}_{\sigma,i}$ of $\dot{s}_{\sigma,i}$ at the i-th supporting point and u_{opt} is the optimal manipulated control value at the same supporting point

$$u_{\text{opt}} = g_{\text{fc}}(s_{\sigma,i}, \mathbf{r}_{\text{opt}}). \tag{28}$$

At supporting point $s_{\sigma,i}$ $\mathbf{x} = \mathbf{x}_{\text{fix},i}$ holds. Therefore the optimal parameter vector $\mathbf{r} = \mathbf{r}_{\text{opt}}$ corresponding to (10) is a solution of (26):

$$\lambda_{\alpha,i}\left(\hat{\dot{s}}_{\sigma,i} - \dot{s}_{\sigma,i_0}\right) + (u_{\text{opt}} - u_{i_0}) = 0. \tag{29}$$

When the optimal parameter vector $\mathbf{r} = \mathbf{r}_{\text{opt}}$ is reached, then $\dot{\mathbf{r}} = 0$ holds and in particular, $\dot{r}_i = 0$. Therefore, for $\mathbf{r} = \mathbf{r}_{\text{opt}}$ the parameter adaptation is done. From (24) and (25) and taking (26) into account, we obtain

$$s_{\alpha,i} = \lambda_{\alpha,i}\Big[\Big(f(\mathbf{x}) + b(\mathbf{x})g_{\text{fc}}(s_\sigma, \mathbf{r})\Big) - \Big(\hat{f}(\mathbf{x}) + \hat{b}(\mathbf{x})g_{\text{fc}}(s_\sigma, \mathbf{r})\Big)\Big] + \\ \Big(\lambda_{\alpha,i}\hat{b}(\mathbf{x}_{\text{fix},i}) + 1\Big)\Big[g_{\text{fc}}(s_{\sigma,i}, \mathbf{r}) - g_{\text{fc}}(s_{\sigma,i}, \mathbf{r}_{\text{opt}})\Big]. \tag{30}$$

The model uncertainties with respect to f and b are described by

$$\Delta f(\mathbf{x}) = f(\mathbf{x}) - \hat{f}(\mathbf{x}), \tag{31}$$

$$\Delta b(\mathbf{x}) = b(\mathbf{x}) - \hat{b}(\mathbf{x}). \tag{32}$$

From (30) it then follows

$$s_{\alpha,i} = \lambda_{\alpha,i}\Big[\Delta f(\mathbf{x}) + \Delta b(\mathbf{x})g_{\text{fc}}(s_\sigma, \mathbf{r})\Big] + \\ \Big(\lambda_{\alpha,i}\hat{b}(\mathbf{x}_{\text{fix},i}) + 1\Big)\Big[g_{\text{fc}}(s_{\sigma,i}, \mathbf{r}) - g_{\text{fc}}(s_{\sigma,i}, \mathbf{r}_{\text{opt}})\Big]. \tag{33}$$

In (33) the model uncertainties in the form of $\Delta f(\mathbf{x})$ and $\Delta b(\mathbf{x})$ are related to the deviations of the manipulated variable $u = g_{\text{fc}}(s_{\sigma,i}, \mathbf{r})$ from the optimal manipulated variable value $u_{\text{opt}} = g_{\text{fc}}(s_{\sigma,i}, \mathbf{r}_{\text{opt}})$. Inexact estimates in the form of (31) and (32) lead to an adaptation switching variable $s_{\alpha,i} \neq 0$. The adaptation algorithm adapts the parameter value r_i, such that $s_{\alpha,i} = 0$. This parameter value then is optimal for the given system description. If a system identification block is used in order to estimate $\hat{f}$ and $\hat{b}$, then the adaptation law is an indirect one. If no such block is used, then the adaptation scheme is said to be direct.

4 Stability analysis of the closed-loop system and parameter convergence

Parameter r_i is adapted via $\dot{r}_i$ by using the switching variable $s_{\alpha,i}$ according to (33). All N parameters, r_1 to r_N, of the parameter vector $\mathbf{r}$ are adapted at their corresponding supporting points. The adaptation implicitely refers to the plant model at the supporting points $s_{\sigma,i}$ as given by (27) and (33). The convergence of the adaptation parameters and the stability of the closed loop including the adaptation block are treated with a special Lyapunov-like function V [10], [6], [11]. The positive semi-definite function V is chosen as

$$V(s_\sigma; s_{\alpha,1}, \ldots, s_{\alpha,N}) = \frac{1}{2} s_\sigma^2 + \frac{1}{2} \sum_{i=1}^{N} s_{\alpha,i}^2. \tag{34}$$

The switching surface $s_\sigma = 0$ is always stable refering to (2). According to (33), $s_{\alpha,i} = 0$ ensures that the parameter r_i converges to its optimal value $r_{i,\mathrm{opt}}$. For the latter, $s_\sigma \dot{s}_\sigma|_{s_\sigma = s_{\sigma,i}} < 0$ holds at the i-th supporting point.

In order to deal with a Lyapunov-like function, the time-derivative $\dot{V}$ must obey

$$\frac{\mathrm{d}}{\mathrm{d}t} V(s_\sigma; s_{\alpha,1}, \ldots, s_{\alpha,N}) = s_\sigma \dot{s}_\sigma + \sum_{i=1}^{N} s_{\alpha,i} \dot{s}_{\alpha,i} < 0. \tag{35}$$

If each term in (35) is negative, then the above is guaranteed. The control law is realised by a FSMC with $u = g_{\mathrm{fc}}(s_\sigma, \mathbf{r})$.

4.1 Closed-loop stability properties of the FSMC

For the derivative of the switching variable $\dot{s}_\sigma$ it follows (according to (12)) that

$$\begin{aligned}
\dot{s}_\sigma &= \frac{\mathrm{d}}{\mathrm{d}t}\left(\sum_{i=0}^{n-1} \binom{n-1}{i} {\lambda_\mathrm{R}}^i \tilde{e}^{(n-1-i)} \right) \\
&= \sum_{i=0}^{n-1} \binom{n-1}{i} {\lambda_\mathrm{R}}^i \tilde{e}^{(n-i)} \\
&= \sum_{i=1}^{n-1} \binom{n-1}{i} {\lambda_\mathrm{R}}^i \tilde{e}^{(n-i)} + x^{(n)} - x_\mathrm{d}^{(n)} \\
&= \sum_{i=1}^{n-1} \binom{n-1}{i} {\lambda_\mathrm{R}}^i \tilde{e}^{(n-i)} + f(\mathbf{x}) + b(\mathbf{x}) g_{\mathrm{fc}}(s_\sigma, \mathbf{r}) - x_\mathrm{d}^{(n)}.
\end{aligned}$$

If for the switching variable s_σ we have that either one of the following two inequalities holds

$$s_\sigma \dot{s}_\sigma \quad < \quad -\eta_\sigma |s_\sigma|, \qquad \eta_\sigma \in \Re^+, \tag{36}$$

or

$$\dot{s}_\sigma \operatorname{sgn}(s_\sigma) \quad < \quad \eta_\sigma, \tag{37}$$

then the boundary layer of the FSMC is a region of attraction.

If for the manipulated variable u of the FSMC we have that the inequality from below holds

$$g_{\mathrm{fc}}(s_\sigma, \mathbf{r}) \operatorname{sgn}(s_\sigma) < \\ -\frac{1}{b(\mathbf{x})} \left[\left(\sum_{i=1}^{n-1} \binom{n-1}{i} \lambda_{\mathrm{R}}{}^i \tilde{e}^{(n-i)} + f(\mathbf{x}) - x_{\mathrm{d}}^{(n)} \right) \operatorname{sgn}(s_\sigma) + \eta_\sigma \right] \tag{38}$$

then the boundary layer is a region of attraction.

For the coordinates of point Q in the $(\tilde{e}, \dot{\tilde{e}}, \ldots, \tilde{e}^{(n-1)})$-phase plane, the two differential equations from below hold:

$$\dot{s}_\sigma \quad = \quad b(\mathbf{x}) g_{\mathrm{fc}}(s_\sigma, \mathbf{r}) + \sum_{i=1}^{n-1} \binom{n-1}{i} \lambda_{\mathrm{R}}{}^i \tilde{e}^{(n-i)} + f(\mathbf{x}) - x_{\mathrm{d}}^{(n)}, \tag{39}$$

$$s_\sigma \quad = \quad \left(\lambda_{\mathrm{R}} + \frac{\mathrm{d}}{\mathrm{d}t} \right)^{n-1} \tilde{e}. \tag{40}$$

The manipulated variable $u = g_{\mathrm{fc}}(s_\sigma, \mathbf{r})$ drives the vector $\tilde{\mathbf{e}}$ to the switching surface and makes it then slide to the origin of the $(\tilde{e}, \dot{\tilde{e}}, \ldots, \tilde{e}^{(n-1)})$-phase plane. From $\tilde{\mathbf{e}} = \mathbf{x} - \mathbf{x}_{\mathrm{d}} = \mathbf{0}$ it follows that $\mathbf{x} = \mathbf{x}_{\mathrm{d}}$.

The first differential equation (39) stands for the movement of point Q towards the switching surface $s_\sigma = 0$. The second differential equation (40) describes the movement of Q parallel to the switching surface $s_\sigma = 0$.

The resulting movement of Q is then given by the superposition of the two velocity components in the $(\tilde{e}, \dot{\tilde{e}})$-phase plane as shown in Fig. 9. Outside the boundary layer of the FSMC, the component $\dot{s}_\sigma$ usually dominates the approach velocity to the switching surface. In the boundary layer, the manipulated variable u is reduced thus slowing the approach velocity $\dot{s}_\sigma$. Hence, the importance of the component parallel to $s_\sigma = 0$ increases.

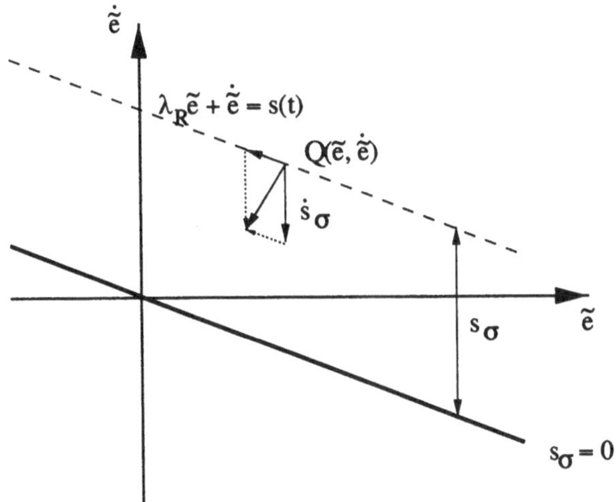

Figure 9: Direction of the error in the phase space

4.2 Convergence of the adaptation parameters

The transfer characteristic of the FSMC is adapted at the N supporting points. The convergence of the adaptation parameters is guaranteed if each term $s_{\alpha,i}\ \dot{s}_{\alpha,i}$ is negative such that the Lyapunov-like function $\dot{V}$ (see (35)) is negative.

Each term is negative if either one of the two inequalities from below holds:

$$s_{\alpha,i}\ \dot{s}_{\alpha,i} \leq -\eta_\alpha |s_{\alpha,i}|, \qquad \eta_\alpha \in \Re^+, \text{or} \tag{41}$$

$$\dot{s}_{\alpha,i}\ \mathrm{sgn}(s_{\alpha,i}) \leq -\eta_\alpha \tag{42}$$

By differentiating (33), we obtain

$$\begin{aligned} \dot{s}_{\alpha,i} \;=\; & \lambda_{\alpha,i}\left[\frac{\partial \Delta f(\mathbf{x})}{\partial \mathbf{x}} + \frac{\partial\left(\Delta b(\mathbf{x}) g_{\mathrm{fc}}(s_\sigma, \mathbf{r})\right)}{\partial \mathbf{x}}\right]^{\mathrm{T}} \dot{\mathbf{x}} + \\ & \lambda_{\alpha,i}\Delta b(\mathbf{x}) \left[\frac{\partial g_{\mathrm{fc}}(s_\sigma, \mathbf{r})}{\partial \mathbf{x}_{\mathrm{d}}}^{\mathrm{T}} \dot{\mathbf{x}}_{\mathrm{d}} + \frac{\partial g_{\mathrm{fc}}(s_\sigma, \mathbf{r})}{\partial \mathbf{r}}^{\mathrm{T}} \dot{\mathbf{r}}\right] + \\ & \lambda_{\alpha,i}\frac{\mathrm{d}\hat{b}(\mathbf{x}_{\mathrm{fix},i})}{\mathrm{d}t}\left[g_{\mathrm{fc}}(s_{\sigma,i}, \mathbf{r}) - g_{\mathrm{fc}}(s_{\sigma,i}, \mathbf{r}_{\mathrm{opt}})\right] + \\ & \left(\lambda_{\alpha,i}\hat{b}(\mathbf{x}_{\mathrm{fix},i}) + 1\right) \frac{\partial g_{\mathrm{fc}}(s_{\sigma,i}, \mathbf{r})}{\partial \mathbf{r}}^{\mathrm{T}} \dot{\mathbf{r}}. \end{aligned} \tag{43}$$

The term

$$\lambda_{\alpha,i}\left[\frac{\partial\Delta f(\mathbf{x})}{\partial\mathbf{x}}+\frac{\partial\left(\Delta b(\mathbf{x})g_{\mathrm{fc}}(s_\sigma,\mathbf{r})\right)}{\partial\mathbf{x}}\right]^{\mathrm{T}}\dot{\mathbf{x}}$$

in (43) represents the influence of the system uncertainties on the switching variable of the adaptation. These uncertainties are the result of inexact or insufficient process knowledge in the form of $\Delta f(\mathbf{x})$ and $\Delta b(\mathbf{x})$. The term

$$\lambda_{\alpha,i}\Delta b(\mathbf{x})\left[\frac{\partial g_{\mathrm{fc}}(s_\sigma,\mathbf{r})}{\partial\mathbf{x}_{\mathrm{d}}}^{\mathrm{T}}\dot{\mathbf{x}}_{\mathrm{d}}+\frac{\partial g_{\mathrm{fc}}(s_\sigma,\mathbf{r})}{\partial\mathbf{r}}^{\mathrm{T}}\dot{\mathbf{r}}\right]$$

in (43) represents the effect which the uncertainty $\Delta b(\mathbf{x})$ has on the parameter adaptation. The term

$$\lambda_{\alpha,i}\frac{\mathrm{d}\hat{b}(\mathbf{x}_{\mathrm{fix},i})}{dt}\left[g_{\mathrm{fc}}(s_{\sigma,i},\mathbf{r})-g_{\mathrm{fc}}(s_{\sigma,i},\mathbf{r}_{\mathrm{opt}})\right]$$

in (43) describes the result of the time-dependent change of the estimated input coefficient $\hat{b}(\mathbf{x}_{\mathrm{fix},i}(t))$. Finally, the term

$$\left(\lambda_{\alpha,i}\hat{b}(\mathbf{x}_{\mathrm{fix},i})+1\right)\frac{\partial g_{\mathrm{fc}}(s_{\sigma,i},\mathbf{r})}{\partial\mathbf{r}}^{\mathrm{T}}\dot{\mathbf{r}} \tag{44}$$

in (43) describes the dependence of the FSMC on the chosen parameters and the adaptation law at the i-th supporting point $s_{\sigma,i}$.

From (42) and taking into account (43) the following convergence condition follows

$$\Bigg(\lambda_{\alpha,i}\left[\frac{\partial\Delta f(\mathbf{x})}{\partial\mathbf{x}}+\frac{\partial\left(\Delta b(\mathbf{x})g_{\mathrm{fc}}(s_\sigma,\mathbf{r})\right)}{\partial\mathbf{x}}\right]^{\mathrm{T}}\dot{\mathbf{x}}+ \tag{45}$$

$$\lambda_{\alpha,i}\Delta b(\mathbf{x})\left[\frac{\partial g_{\mathrm{fc}}(s_\sigma,\mathbf{r})}{\partial\mathbf{x}_{\mathrm{d}}}^{\mathrm{T}}\dot{\mathbf{x}}_{\mathrm{d}}+\frac{\partial g_{\mathrm{fc}}(s_\sigma,\mathbf{r})}{\partial\mathbf{r}}^{\mathrm{T}}\dot{\mathbf{r}}\right]+$$

$$\lambda_{\alpha,i}\frac{\mathrm{d}\hat{b}(\mathbf{x}_{\mathrm{fix},i})}{dt}\left[g_{\mathrm{fc}}(s_{\sigma,i},\mathbf{r})-g_{\mathrm{fc}}(s_{\sigma,i},\mathbf{r}_{\mathrm{opt}})\right]+$$

$$\left(\lambda_{\alpha,i}\hat{b}(\mathbf{x}_{\mathrm{fix},i})+1\right)\frac{\partial g_{\mathrm{fc}}(s_{\sigma,i},\mathbf{r})}{\partial\mathbf{r}}^{\mathrm{T}}\dot{\mathbf{r}}\Bigg)\ \mathrm{sgn}(s_{\alpha,i})\le-\eta_\alpha. \tag{46}$$

4.2.1 Choice of the adaptation law for exact modelling

In order to simplify the representation, first the case of exact modelling of the plant (1) is considered. According to (31) and (32), $\Delta f(\mathbf{x})=\Delta b(\mathbf{x})=0$

holds. The switching variables $s_{\alpha,i}$ are obtained in this case from (33) as

$$s_{\alpha,i} = \left(\lambda_{\alpha,i}\hat{b}(\mathbf{x}_{\text{fix},i}) + 1\right)\left[g_{\text{fc}}(s_{\sigma,i},\mathbf{r}) - g_{\text{fc}}(s_{\sigma,i},\mathbf{r}_{\text{opt}})\right]. \tag{47}$$

For the time-derivative of $\dot{s}_{\alpha,i}$, by the use of (42) the following condition follows

$$\begin{aligned}\dot{s}_{\alpha,i}\ \text{sgn}(s_{\alpha,i}) \quad &= \quad \left(\lambda_{\alpha,i}\frac{\mathrm{d}\hat{b}(\mathbf{x}_{\text{fix},i})}{\mathrm{d}t}\left[g_{\text{fc}}(s_{\sigma,i},\mathbf{r}) - g_{\text{fc}}(s_{\sigma,i},\mathbf{r}_{\text{opt}})\right] + \right.\\ &\left.\left(\lambda_{\alpha,i}\hat{b}(\mathbf{x}_{\text{fix},i}) + 1\right)\frac{\partial g_{\text{fc}}(s_{\sigma,i},\mathbf{r})}{\partial\mathbf{r}}^{\mathrm{T}}\ \dot{\mathbf{r}}\right)\text{sgn}\ (s_{\alpha,i}) < -\eta_\alpha.\end{aligned} \tag{48}$$

For $\mathbf{r}$ an adaptation law $\dot{\mathbf{r}}$ has to be chosen such that (48) holds and the term $s_{\alpha,i}\ \dot{s}_{\alpha,i}$ has a negative value in (35).

At the i-th supporting point $s_{\sigma,i}$, only the i-th component r_i of $\mathbf{r}$ for $g_{\text{fc}}(s_{\sigma,i},\mathbf{r})$ is of interest because all other parameters r_j, $j = 1,\ldots,N, j \neq i$ do not have any influence on $g_{\text{fc}}(s_{\sigma,i},\mathbf{r})$. First, we obtain from (4)

$$\begin{aligned}g_{\text{fc}}(s_{\sigma,i},\mathbf{r}) \quad &= \quad \frac{\sum\limits_{j=1}^{N}\mu_j(s_{\sigma,i})M_j}{\sum\limits_{j=1}^{N}\mu_j(s_{\sigma,i})A_j}\\ &= \quad \frac{\mu_i(s_{\sigma,i})M_i}{\mu_i(s_{\sigma,i})A_i} = \frac{M_i}{A_i}.\end{aligned}$$

The dependence of the FSMC on the parameter vector $\mathbf{r}$

$$\frac{\partial g_{\text{fc}}(s_\sigma,\mathbf{r})}{\partial\mathbf{r}} = \begin{pmatrix}\frac{\partial g_{\text{fc}}(s_\sigma,\mathbf{r})}{\partial r_1}\\ \vdots\\ \frac{\partial g_{\text{fc}}(s_\sigma,\mathbf{r})}{\partial r_i}\\ \vdots\\ \frac{\partial g_{\text{fc}}(s_\sigma,\mathbf{r})}{\partial r_N}\end{pmatrix} \tag{49}$$

at the i-th supporting point $s_{\sigma,i}$ is given by

$$\left.\frac{\partial g_{\mathrm{fc}}(s_\sigma,\mathbf{r})}{\partial \mathbf{r}}\right|_{s_\sigma=s_{\sigma,i}} = \begin{pmatrix} 0 \\ \vdots \\ 0 \\ \frac{\partial g_{\mathrm{fc}}(s_{\sigma,i},\mathbf{r})}{\partial r_i} \\ 0 \\ \vdots \\ 0 \end{pmatrix} = \begin{pmatrix} 0 \\ \vdots \\ 0 \\ \frac{1}{A_i} \\ 0 \\ \vdots \\ 0 \end{pmatrix}. \tag{50}$$

For (44) we then obtain

$$\frac{\partial g_{\mathrm{fc}}(s_{\sigma,i},\mathbf{r})}{\partial \mathbf{r}}^{\mathrm{T}} \dot{\mathbf{r}} = \left(0 \;\ldots\; 0 \;\frac{1}{A_i}\; 0 \;\ldots\; 0\right) \begin{pmatrix} \dot{r}_1 \\ \vdots \\ \dot{r}_{i-1} \\ \dot{r}_i \\ \dot{r}_{i+1} \\ \vdots \\ \dot{r}_N \end{pmatrix} = \frac{\dot{r}_i}{A_i}. \tag{51}$$

Then, at $s_{\sigma,i}$, only the i-th component of the velocity vector $\dot{\mathbf{r}}$ in the adaptation law is of interest. (48) then becomes

$$\begin{aligned} \dot{s}_{\alpha,i}\,\mathrm{sgn}(s_{\alpha,i}) = & \left(\lambda_{\alpha,i}\frac{\mathrm{d}\hat{b}(\mathbf{x}_{\mathrm{fix},i})}{\mathrm{d}t}\left[g_{\mathrm{fc}}(s_{\sigma,i},\mathbf{r}) - g_{\mathrm{fc}}(s_{\sigma,i},\mathbf{r}_{\mathrm{opt}})\right] + \right. \\ & \left.\left(\lambda_{\alpha,i}\hat{b}(\mathbf{x}_{\mathrm{fix},i}) + 1\right)\frac{1}{A_i}\,\dot{r}_i\right)\,\mathrm{sgn}(s_{\alpha,i}) < -\eta_\alpha. \end{aligned} \tag{52}$$

For slowly time-varying reference trajectories, $\hat{b}(\mathbf{x}_{\mathrm{fix},i})$ changes slowly, i. e. $\frac{\mathrm{d}\hat{b}(\mathbf{x}_{\mathrm{fix},i})}{\mathrm{d}t}$ can be neglected for such reference trajectories.

The adaptation law for $\dot{r}_i$ is then chosen with the signum-function $\mathrm{sgn}(s_{\alpha,i})$ as

$$\boxed{\dot{r}_i = -\frac{1}{\lambda_{\alpha,i}\hat{b}(\mathbf{x}_{\mathrm{fix},i}) + 1}\, c_{ii}\, \frac{\partial g_{\mathrm{fc}}(s_{\sigma,i},\mathbf{r})}{\partial r_i}\,\mathrm{sgn}(s_{\alpha,i})} \tag{53}$$

The positive constants c_{ii} are such that $s_{\alpha,i}\dot{s}_{\alpha,i}$ has a negative value in (35) and the adaptation parameters converge.

Using the vector functions $\mathbf{g}_{\mathrm{fc}}(\mathbf{s}_\sigma, \mathbf{r}) = \left(g_{\mathrm{fc}}(s_{\sigma,1}, \mathbf{r}) \ \ \ldots \ \ g_{\mathrm{fc}}(s_{\sigma,N}, \mathbf{r})\right)^{\mathrm{T}}$, $\mathbf{sgn}(\mathbf{s}_\alpha) = \left(\mathrm{sgn}(\mathbf{s}_{\alpha,1}) \ \ \ldots \ \ \mathrm{sgn}(\mathbf{s}_{\alpha,\mathbf{N}})\right)^{\mathrm{T}}$, and with the diagonal matrices $\mathbf{C} = \mathrm{diag}(c_{ii})$ and $\mathbf{\Lambda}_\alpha = \mathrm{diag}(\frac{1}{\lambda_{\alpha,i}\hat{b}(\mathbf{x}_{\mathrm{fix},i})+1})$, $i = 1, \ldots, N$, the N adaptation laws

$$\boxed{\dot{\mathbf{r}} = -\mathbf{\Lambda}_\alpha \ \mathbf{C} \ \frac{\partial \mathbf{g}_{\mathrm{fc}}(\mathbf{s}_\sigma, \mathbf{r})}{\partial \mathbf{r}} \ \mathbf{sgn}(\mathbf{s}_\alpha)} \tag{54}$$

can be grouped together.

4.2.2 Closed-loop stability behaviour in the boundary layer of the FSMC

The adaptation law at $s_{\sigma,i}$ is defined in such a way so that for the switching variable s_σ at the i-th supporting point we have that

$$s_\sigma \ \dot{s}_\sigma \Big|_{s_\sigma = s_{\sigma,i}} < -\eta_\sigma \left|s_\sigma\right| \Big|_{s_\sigma = s_{\sigma,i}} \tag{55}$$

The parameter value r_i is adapted by (53) so that the term of (52) is negative at the supporting points.

Sliding-mode controllers can be used for the control of a stationary operating point as well as for trajectory following. The desired control value $w = x_{\mathrm{d}}$ does not enter the stability considerations explicitely because sliding-mode controllers are supposed to switch instantaneously. However, the upper bound of the manipulated variable u of the sliding-mode controller depends on the n-th derivative of $x_{\mathrm{d}}^{(n)}(t)$ as it can be seen from (38). The faster $x_{\mathrm{d}}(t)$ changes, the bigger the manipulated value has to be in order to guarantee stability. If a quasi-static $x_{\mathrm{d}}(t)$ can be assumed, then a smaller value of the manipulated variable is sufficient for plant stabilisation.

The transfer characteristic of the FSMC is strictly monotonously increasing between the supporting points. If at the same time the manipulated variable at the Nth supporting point obeys

$$\begin{aligned} g_{\mathrm{fc}}(s_{\sigma,N}, \mathbf{r}_{\mathrm{opt}}) \ &< \\ &-\frac{1}{\hat{b}(\mathbf{x})}\left(\left|\sum_{i=1}^{n-1}\binom{n-1}{i}\lambda_{\mathrm{R}}{}^{i}\tilde{e}^{(n-i)} + \hat{f}(\mathbf{x}) - x_{\mathrm{d}}^{(n)}\right| + \left|\eta_\sigma\right|\right) \end{aligned} \tag{56}$$

for $\Phi > s_\sigma > s_{\sigma,N}$, then we have

$$s_\sigma \, \dot{s}_\sigma \quad < \quad -\eta_\sigma |s_\sigma| \tag{57}$$

for $s \geq \Phi$ and $\Phi > s \geq s_{\sigma,N}$. Then the part of the boundary layer described by $-\Phi < s_\sigma < s_{\sigma,N}$ is a region of attraction. The same holds for all the other $N-1$ supporting points. With this approach, the effective width of the boundary layer can be reduced.

If the plant from (1) is known then by (42) the convergence of the adaptation parameters to their linguistically defined optimal values is guaranteed. If the equations (57) hold then the manipulated variable is such that it also guarantees the stability of the boundary layer.

4.2.3 Stability and convergence analysis for approximate modelling

According to (33), the adaptation algorithm for parameter r_i refers to the switching variable $s_{\alpha,i}$: the parameter values r_i are changed such that $s_{\alpha,i} = 0$ is reached. Referring to (33) and for $s_{\alpha,i} = 0$ we obtain

$$\begin{aligned} &\lambda_{\alpha,i}\Big[\Delta f(\mathbf{x}) + \Delta b(\mathbf{x}) g_{\text{fc}}(s_\sigma, \mathbf{r})\Big] + \\ &\quad \Big(\lambda_{\alpha,i}\hat{b}(\mathbf{x}_{\text{fix},i}) + 1\Big)\Big[g_{\text{fc}}(s_{\sigma,i}, \mathbf{r}) - g_{\text{fc}}(s_{\sigma,i}, \mathbf{r}_{\text{opt}})\Big] \;=\; 0. \end{aligned} \tag{58}$$

(58) can be interpreted as a *balance condition.* In this equation, model uncertainties are related to deviations of r_i from the optimal value $r_{i,\text{opt}}$. $s_{\alpha,i} = 0$ means that for a model matching the real plant, the optimal parameter vector $\mathbf{r}_{\text{opt}}$ is reached. According to (58), the model uncertainties $\Delta f(\mathbf{x})$ and $\Delta b(\mathbf{x})$ will necessarily lead for $s_{\alpha,i} = 0$ to a deviation $\Delta r_{i,\text{opt}}$ of the parameter value r_i from the optimal value $r_{i,\text{opt}}$. The design parameter $\lambda_{\alpha,i}$ describes the effect of the model uncertainties in the balance condition. From (58), we obtain

$$\begin{aligned} &-\frac{\lambda_{\alpha,i}}{\lambda_{\alpha,i}\hat{b}(\mathbf{x}_{\text{fix},i}) + 1}\,[\Delta f(\mathbf{x}) \quad + \\ &\qquad \Delta b(\mathbf{x}) g_{\text{fc}}(s_\sigma, \mathbf{r})] = g_{\text{fc}}(s_{\sigma,i}, \mathbf{r}) - g_{\text{fc}}(s_{\sigma,i}, \mathbf{r}_{\text{opt}}). \end{aligned} \tag{59}$$

The relation of $\lambda_{\alpha,i}$ to $\Big(\lambda_{\alpha,i}\hat{b}(\mathbf{x}_{\text{fix},i}) + 1\Big)$ states the importance of the model uncertainties for the deviation of the adaptation parameters from their optimal values $\mathbf{r}_{\text{opt}}$. The bigger $\lambda_{\alpha,i}$ is, the stronger the undesired effect of the model uncertainties feeds.

In case the plant is completely known, the parameter value converges to its optimal value. (58) identifies a relation between the model-plant match quality on the one side and convergence behaviour on the other side. The better the match, the smaller the maximally expected parameter uncertainties $\Delta r_{i,\text{opt}}$ are. The term

$$\left(\lambda_{\alpha,i}\hat{b}(\mathbf{x}_{\text{fix},i}) + 1\right) \left[g_{\text{fc}}(s_{\sigma,i}, \mathbf{r}) - g_{\text{fc}}((s_{\sigma,i}, \mathbf{r}_{\text{opt}})\right] \tag{60}$$

results in the adaptation of r_i at $s_{\sigma,i}$ via a local, linguistically defined adaptation criterion. It enables the adaptation throughout the whole definition space of s_σ. Also, convergence of r_i is guaranteed.

For a different initialisation of the state vector $\mathbf{x}_0$ and the control parameter vector $\mathbf{r}_0$, the correction term in (24) ensures that the always same parameter value is obtained after convergence is reached.

5 Remarks about the adaptation algorithm

5.1 Parameter chattering and implementation of a FSMC in the adaptation loop

During adaptation, a chattering of the adaptation parameters occurs according to (53) and(54), respectively.

The parameter chattering is the result of a too high approach velocity of the parameters and the inability of the FSMC to realize infinitely fast switching. Therefore the sliding-mode controller in the adaptation loop (see (53) or (54)) is substituted by a FSMC. The maximal parameter velocity $\dot{r}_{i,\max}$ is defined by the above adaptation laws. Between these, the parameter adaptation rate $\dot{r}_i$ in the boundary layer of the FSMC is interpolated. This structure was already defined through the linguistic adaptation algorithm. During the analytical examination of the adaptation law, however, first a condition was derived to guarantee convergence of the adaptation parameters, starting from the adaptation line $s_{\alpha,i} = 0$. The resulting sliding-mode controller, however, produces undesired parameter chattering. Therefore, as with classic sliding-mode controllers, the sliding-mode controller in the adaptation loop is replaced with a linguistically defined fuzzy sliding-mode controller.

5.2 Quasi-continuous adaptation rate

For the purpose of implementation, time-discrete controllers are used, and the continuous adaptation rate $\dot{\mathbf{r}}$ (see (54)) is approximated by

$$\dot{\mathbf{r}} \approx \frac{\Delta \mathbf{r}}{T}, \tag{61}$$

where T represents the sampling period. If the controller parameters are only adapted after every m-th sampling interval, $\dot{\mathbf{r}}$ is approximated by

$$\dot{\mathbf{r}} \approx \frac{\Delta \mathbf{r}}{m\,T}. \tag{62}$$

If the adaptation law from (54) is given in a time-discrete form, then by taking into account (61) and (62), we obtain

$$\mathbf{r}\left((m+1)T\right) = \mathbf{r}(mT) + T\Delta\mathbf{r}(mT), \qquad m \in \mathbb{Z}. \tag{63}$$

6 Adaptive controller implementation

6.1 Design steps for the controller and the adaptation algorithm

6.1.1 Design of the FSMC in the control loop

The membership functions of the input of the FSMC are chosen as triangular membership functions, with overlap of 1. The membership functions of the output are also represented by triangular membership functions. The rule-base itself is given in diagonal form. The inference method is realised by the Larsen-operator. The rule consequents of different rules are merged through addition. From this, the crisp output of the FSMC is calculated through defuzzification by the method-of-weighted sums. The implementation steps are the following ones:

1. Definition of ranges for the state variables.
2. Calculation of the error vector $\tilde{\mathbf{e}}$ from the state vector $\mathbf{x}$ and the reference vector $\mathbf{x}_{\mathrm{d}}$ according to (3).
3. Choice of parameter λ_{R} for (2).
4. Definition of the switching hyperplane s_σ according to (2).

5. Characterisation of the range of the switching variable s_σ and thus of the operating range of the fuzzy sliding-mode controller.

6. Determination of the minimal and maximal manipulated variable $u_{\min}$ and $u_{\max}$ for values of the manipulated variable that are outside the boundary layer. Mostly, $u_{\min} = -u_{\max}$. $u_{\max} = K$ has to be chosen greater than the sum of the absolute values of the model uncertainties and disturbances in order to ensure an attracting boundary layer.

7. Without boundary layer, the closed-loop behaviour is given by $s_\sigma = 0$. However, chattering is undesired. Therefore, a small boundary layer is introduced by g_{fc} providing a balance between the desired tracking accuracy and the control activity. It has to be taken into account that the switching variable s_σ for time-discrete control varies around the switching hyperplane. This is due to the fact that the manipulated variable u cannot be switched at any instant. The bigger the variation of the manipulated variable, the bigger value for it is chosen.

8. Choice of the number and position of the supporting points of the FSMC; sometimes symmetry around operating points can be used in order to halve the number of adaptation parameters.

9. Choice of the membership functions of the FSMC's inputs and outputs.

10. Definition of the parameter values of the fuzzy sliding-mode controller in (8) determining the nonlinear transfer characteristic.

11. Design of the FSMC from the rules of Table 1.

6.1.2 Implementation of the FSMC in the adaptation loop

The number and the position of the supporting points are defined heuristically. The bigger the boundary layer of the FSMC is, the more supporting points are chosen. As a rule of thumb, the supporting points are chosen closer around the origin of the transfer characteristic. With increasing s_σ, the distance between the supporting points increases as well. For the definition of the correction terms, a model of the plant is used. According to the balance condition in (58), a more exact knowledge about the plant leads to a better convergence behaviour of the controller parameters. The design can be done in the following steps.

1. Heuristic choice of the optimal approach velocity $\dot{s}_\sigma$ and the corresponding optimal manipulated variable u of the FSMC through $\dot{s}_{\sigma,i_0}$ and u_{i_0} at the i-th supporting point by choosing the membership functions for $\dot{s}_\sigma$ and u in (10).

2. Determination of the slope $\lambda_{\alpha,i}$ of the adaptation line $s_{\alpha,i} = 0$ in (10). According to Fig. 7, $0 < \lambda_{\alpha,i} < 1$ means that the approach velocity $\dot{s}_\sigma$ is reached even for large changes in the manipulated variable u. For $\lambda_{\alpha,i} > 1$, the manipulated variable u changes more slowly relative to the approach velocity $\dot{s}_\sigma$.
3. Inclusion of the description of the model in form of the estimated model $\hat{f}(\mathbf{x})$ and $\hat{b}(\mathbf{x})$ for the correction terms $\dot{s}_{\sigma,\mathrm{neu}_i}$ and u_{neu_i} in (15).
4. Choice of the factors c_{ii} of the adaptation law according to (53), such that convergence of the adaptation parameters is guaranteed.
5. Introduction of a boundary layer in the adaptation loop in order to prevent an exaggerated varying of the parameters.
6. Design of the linguistic adaptation block with the structure of the FSMC.

In an implementation as a time-discrete system, $\dot{s}_\sigma$ is at every kth sampling instant approximated by

$$\dot{s}_\sigma \approx \frac{s_\sigma(kT) - s_\sigma((k-1)T)}{T}, \qquad k \in \mathbb{Z}, \tag{64}$$

where T represents the sampling interval.

7 Two-link robot arm

The adaptation algorithm is tested on an example of a two-link robot arm. As the robot arm is planar, the internal link coordinates $\boldsymbol{\theta}$ can be directly calculated from the coordinates of the work space - possibly cartesian coordinates. This is also known as the solution of the inverse problem. According to the kinematic properties of the robot, i.e.

$$\mathbf{x} = \mathbf{f}(\boldsymbol{\theta}), \tag{65}$$

the cartesian coordinates $\mathbf{x}$ have to be transformed into angular coordinates $\boldsymbol{\theta}$ according to

$$\boldsymbol{\theta} = \mathbf{f}^{-1}(\mathbf{x}). \tag{66}$$

The function $\mathbf{f}$ describes the kinematic model of the robot and the geometric relations between coordinates in the work space and internal coordinates. For a planar two-link robot this function is bijective.

7.1 Determination of the reference trajectories

An efficient approach to solve the inverse problem is the differential method [7] where by differentiation of (65) with respect to time we obtain

$$\dot{\mathbf{x}} = \frac{\partial \mathbf{f}(\boldsymbol{\theta})}{\partial \boldsymbol{\theta}} \frac{\mathrm{d}\boldsymbol{\theta}}{\mathrm{d}t} = \frac{\partial \mathbf{f}(\boldsymbol{\theta})}{\partial \boldsymbol{\theta}} \dot{\boldsymbol{\theta}}. \tag{67}$$

Respectively, we have

$$\dot{\mathbf{x}} = \mathbf{J}(\boldsymbol{\theta}) \dot{\boldsymbol{\theta}}, \tag{68}$$

or

$$\mathrm{d}\mathbf{x} = \mathbf{J}(\boldsymbol{\theta}) \mathrm{d}\boldsymbol{\theta}. \tag{69}$$

$\frac{\mathrm{d}\mathbf{x}}{\mathrm{d}\boldsymbol{\theta}} = \mathbf{J}(\boldsymbol{\theta})$ is a $m \times n$ Jacobi-matrix where m is the number of work space coordinates and n the number of internal link coordinates. A change in position is usually given incrementally by $\Delta \mathbf{x}$ as approximation of $\mathrm{d}\mathbf{x}$. That is

$$\Delta \mathbf{x} = \mathbf{J}(\boldsymbol{\theta}) \Delta \boldsymbol{\theta}. \tag{70}$$

The incremental angular change $\Delta \boldsymbol{\theta}$ then follows as

$$\Delta \boldsymbol{\theta} = \mathbf{J}^{-1}(\boldsymbol{\theta}) \Delta \mathbf{x}. \tag{71}$$

Because of discretization effects, an angular correction has to be introduced during the incremental approach in order to guarantee exact following of the robot trajectory according to (65). If a smooth differential velocity profile is given in terms of work space coordinates, then it is transformed through the matrix $\mathbf{J}$ into a smooth, respectively differentiable, angular velocity profile as long as the matrix $\mathbf{J}$ is not singular.

As reference trajectories, a square and a circle have been chosen. In the case when a trapezoidal velocity profile is given, then this results in an non-smooth, non-continuous acceleration profile. As the required acceleration cannot be supplied immediately by the motor, control oscillations will occur for the velocity profile. In order to avoid such osciallations, a continuously differentiable velocity profile is chosen. The acceleration then is chosen as

$$\ddot{\theta}_{\mathrm{d}} = \frac{\pi^2}{\tau^2} \sin\left(\frac{t}{\tau}\pi\right). \tag{72}$$

The corresponding velocity profile is therewith

$$\dot{\theta}_{\mathrm{d}} = \frac{\pi}{\tau}\left(1 - \cos\left(\frac{t}{\tau}\pi\right)\right), \tag{73}$$

if the motor starts from zero initial condition. After reaching the maximal velocity $\dot{\theta}_{\mathrm{d,max}}$, the arm is not longer accelerated and the angle inreases proportionally with time. The deceleration is analogous to the acceleration procedure.

7.2 Control of the robot arm

In the following the quality of the controller design, consisting of a FSMC and fuzzy sliding-mode adaptation block is examined for each link of a simulated two-link robot with the angles θ_1 and θ_2. The results obtained by two adaptive fuzzy sliding-mode controllers are given. The sampling interval is chosen as 0.008s. Using planetary drives, the respective arms can be regarded as decoupled since they have a high inertia due to the drives. The two-link robot can then be regarded as being composed of two SISO-systems. Thus, every arm can be modelled effectively as a double integrator. Nonlinear effects in the form of position-dependent friction and remaining coupling effects are neglected and will be rejected during control.

In Fig. 10, the reference trajectory in the form of a square is given (−−). The task of the controller is to make the system (−) follow the desired system trajectory from an initial state which does not lie on the reference trajectory. In Fig. 11, the output values of the two angles of the robot arm θ_1 (−−) and θ_2 (−) and their respective velocities are given. At the same time, the manipulated variables u_1 (−−) and u_2 (−) are shown. In Fig. 12, the respective velocity profiles for the desired anlges are presented. In Fig. 13, the adaptation rates for the four parameters of the first fuzzy sliding-mode controller are shown, whereas in Fig. 14 the adaptation parameters themselves are given. In Fig. 13, one can see that the initial adaptation rate is limited, leading nevertheless to quickly convergent parameter values in Fig. 14. By using the fuzzy sliding-mode controller in the adaptation block, a robust adaptation behaviour can be introduced while the adaptation rate is limited. At the same time, the parameter convergence is fast enough. In Figs. 15 and 16 the respective values for the adaptation rate and parameters of the second FSMC are given.

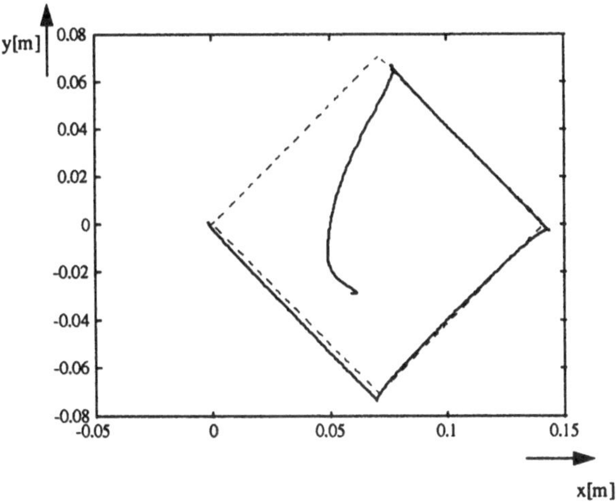

Figure 10: Reference (– –) and actual (-) square trajectory

8 Conclusions

By choosing the moments of the membership functions as adaptation parameters of the fuzzy sliding-mode controller, a linear adaptation problem is obtained, where the static transfer characteristic of the controller can be adapted locally. The adaptation strategies are defined linguistically and reveal to be of a fuzzy sliding-mode type. The robust behaviour of the FSMC then subsequently is incorporated in the adaptation block by limiting the adaptation rate.

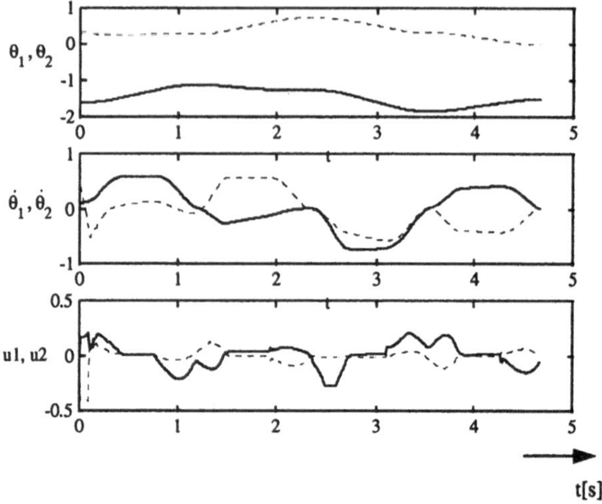

Figure 11: State space and manipulated variable for the arm

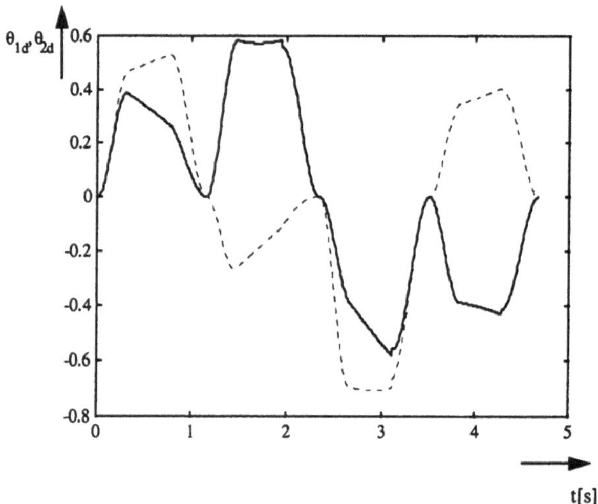

Figure 12: Reference velocity profile

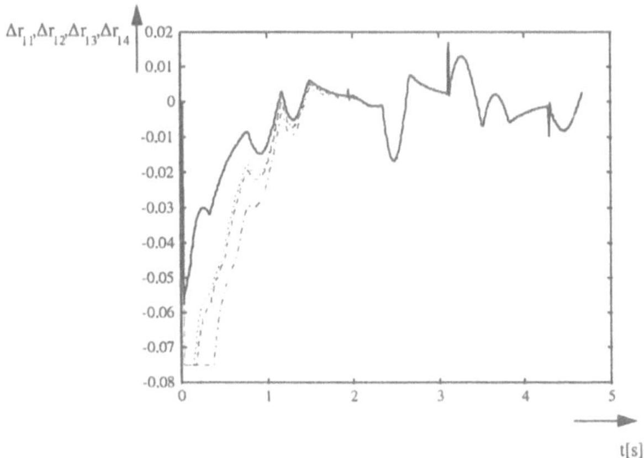

Figure 13: Parameter adaptation rate for the first controller

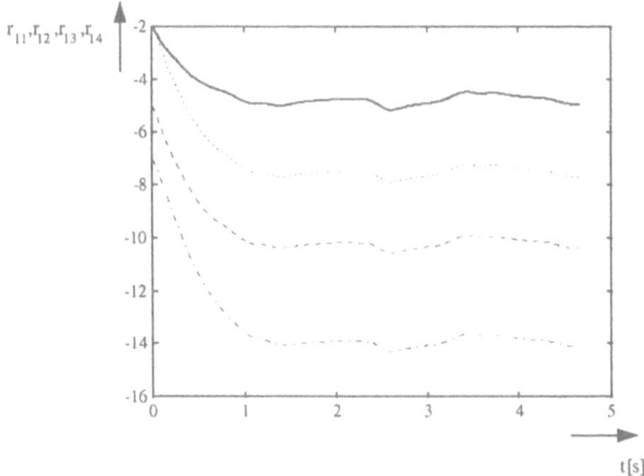

Figure 14: Parameter adaptation for the first controller

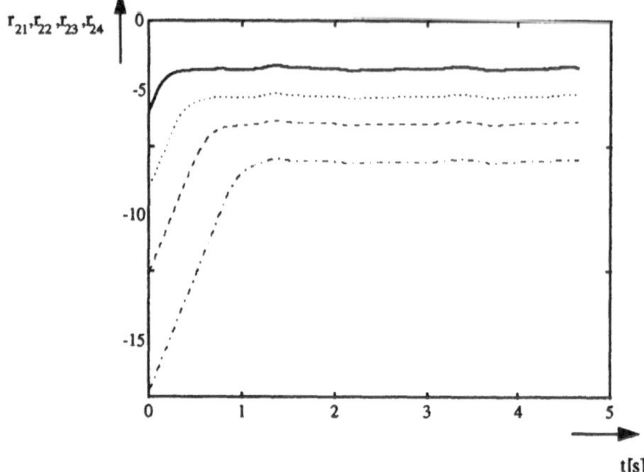

Figure 15: Parameter adaptation rate for the second controller

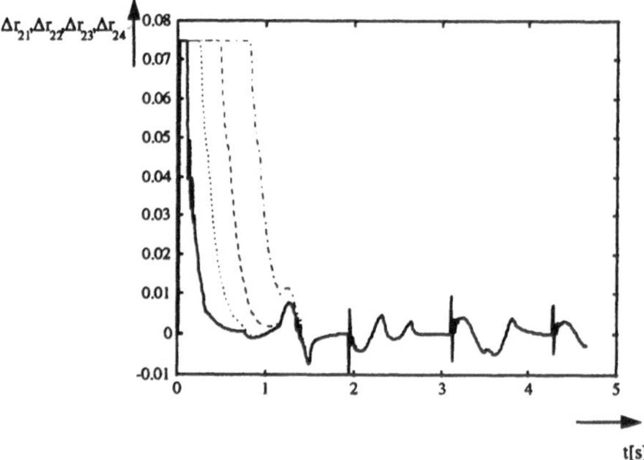

Figure 16: Parameter adaptation for the second controller

References

[1] K.J. Åström and B. Wittenmark. A survey of adaptive control applications. In *Proceedings of the* 34th *IEEE Conference on Decision and Control, New Orleans*, pages 649–654, 1995.

[2] P. Eykhoff. "Every good regulator of a system must be a model of that system". *Modeling, Identification and Control*, 15(3):135–139, 1994.

[3] G.-C. Hwang and S.C. Lin. A stability approach to fuzzy control design for nonlinear systems. *Fuzzy Sets and Systems, Nord-Holland*, 48:279–287, 1992.

[4] M. Jansen. *Globale Modellbildung und garantiert stabile Regelung von Robotern mit strukturierten neuronalen Netzen.* Fortschrittberichte, VDI-Verlag, 1995.

[5] H. Kiendl. Fuzzy control. *Automatisierungstechnik*, 41(1):A1–A4, 1993.

[6] J. P. Lasalle and S. Lefschetz, editors. *Die Stabilitätstheorie von Ljapunow - Die direkte Methode mit Anwendungen.* Bibliographisches Institut, Mannheim, 1967.

[7] R. Palm. Control of a redundant manipulator using fuzzy rules. *Fuzzy Sets and Systems, Nord-Holland*, 45:279–298, 1992.

[8] J.-J. Slotine and W. Li. *Applied Nonlinear Control.* Prentice Hall, Englewood Cliffs, 1991.

[9] Z.-M. Yeh. A performance approach to fuzzy control design for nonlinear systems. *Fuzzy Sets and Systems, Nord-Holland*, 64:339–352, 1994.

[10] T. Yoshizawa. Stability and boundedness of solutions. *Archives for rationale Mechanics and Analysis*, 6:409–421, 1960.

[11] T. Yoshizawa. Stability of sets and perturbed systems. *Funkcialaj Ekvacioj*, 5:31–69, 1962.

Discrete-Time Adaptive Fuzzy Logic Control of Feedback Linearizable Systems

S. Jagannathan
Automated Analysis Corporation
423 S. W. Washington Street
Peoria, IL 61602, USA

1 Introduction

In this article we utilize the use of fuzzy systems as function approximators for the design of an adaptive fuzzy controller (FC). The approximation properties of fuzzy systems per se have been extensively studied in [Buckley, 92; Kosko, 92; Kosko, 94; Langari and Tomizuka, 91; Wang and Mendel, 92; Wang, 94, Ying, 93; Ying, 94; Zeng and Singh, 95]. The results reported in [Wang and Mendel, 92; Zeng and Singh, 95] show that fuzzy associate memory functions (FAM) are universal approximators for certain classes of functions.

In [Vandegrift et al., 95], the adaptive control of a class of nonlinear systems in discrete-time was presented using the approximation properties of fuzzy systems. Although Lyapunov stability and passivity properties were analysed, this analysis was limited to a class of nonlinear systems of the form $x(k+1) = f(x(k)) + v_c(k)$, where there are no uncertainties in the coefficient of the control input $v_c(k)$. However, if a system in continuous-time of the form $\dot{x} = f_1(x) + v_{c1}(k)$ is discretized, the resulting discrete-time system will be of the form $x(k+1) = f(x(k)) + g(x(k))v_c(k)$ for some functions $f(.)$ and $g(.)$. In [Jagannathan, 96c] an adaptive FC was proposed for this wider class of nonlinear discrete-time systems.

For the purpose of control, even if the open-loop system is stable, it must be shown that inputs, outputs, and states remain bounded when a feedback loop is designed. In addition, to achieve a *well-defined* controller, the control input, $v_c(k) = \frac{N(x)}{D(x)}$, must be properly selected to ensure that $D(x)$ will be non zero for all time. For feedback linearization, a controller with these properties is usually needed. Furthermore, if any adaptation scheme is implemented in order to provide an estimate for $\hat{D}(x)$ of $D(x)$, then extra precaution is required to guarantee that $\hat{D}(x) \neq 0$ for all time.

In [Jagannathan et al. 1996c], a FC is designed to perform feedback linearization of continuous-time Brunovsky form systems. It takes into account all of the above mentioned issues. Note, that at this point there is no simple and straightforward way to naturally extend results from continuous to discrete-time systems. The motivation for the work reported in this article is to provide similar results for discrete-time systems. The traditional problems with discrete-time adaptive control are overcome by using a *single Lyapunov function containing parameter identification errors and control errors.* This guarantees at once both stable identification and stable tracking. However, it leads to complex proofs and avoids the need for the certainty equivalence assumption. Along the way, various other standard assumptions in discrete-time adaptive control are also overcome, including persistence of excitation, linearity-in-the-parameters, and the need for tedious computation of a regression matrix. The problem of $\hat{g}(x(k)) \neq 0$ is confronted by appropriately selecting particular fuzzy rules' parameters, as well as the control input.

We treat the design of FC [Jagannathan, 96b] where the weights (or the values of the FC's output) enter linearly. We discuss the controller structure, various weights update algorithms, and persistence of excitation definitions. Note that linearity in the FC's weights is far milder than the usual adaptive control restriction of linearity in the *unknown system parameters.* This is so since the universal approximation property of a fuzzy system means that any smooth nonlinear function can be reconstructed. Next, modified weights tuning methods are proposed for feedback linearization, accompanyied by rigorous stability analysis.

2 Background

For the general class of nonlinear systems defined below, a FC is designed so that the output follows a prescribed trajectory with a bounded error. Some mathematical and system theory notions are given in this section.

Let $\Re$ denote the real numbers, $\Re^n$ denote the real $n \times 1$-vectors, $\Re^{mxn}$ the real mxn matrices. Let S be a compact simply-connected set in $\Re^n$. With maps f into $\Re^k$, define $C^k(S)$ as the space such that $f(\cdot)$ is continuous in it. We denote by $\|\cdot\|$ any suitable norm. Given a matrix $A = [a_{ij}]$, $A \in \Re^{nxm}$ the Frobenius norm is defined by

$$\| A \|_F^2 = trA^T A = \sum_i \sum_j a_{ij}^2, \tag{1}$$

with $tr(\cdot)$ being the trace operation. The associated inner product is $<$

$A, B >_F = trA^TB$. The Frobenius norm $\| A \|_F$, which is denoted by $\| \cdot \|$ throughout this article unless stated otherwise, is nothing but the 2-norm over the space defined by stacking the matrix columns into a vector, so that it is compatible with the vector 2-norm. That is, $\| Ax \| \leq \| A \| \| x \|$.

2.1 Fuzzy systems

In this section we discuss particular aspects of fuzzy systems related to their use in closed-loop control applications. An analytic expression for the fuzzy system's output in terms of its inputs is provided. The membership functions (MF) are selected to obtain suitable approximation of functions (see Kosko, 92 and Buckley, 92 for further details).

2.2 Fuzzy system structure and reasoning surface

A standard fuzzy system [Lewis and Liu, 95] takes a crisp input x, fuzzifies it using prescribed MF, uses a fuzzy rule-base to determine associated output values, and finally defuzzifies the output values into a single crisp value y. The output y of the fuzzy system, plotted as a function of the input x is called the reasoning surface.

Given an integer N, define the set $\overline{N} \equiv 1, 2, \cdots, N$. For each component x_j of the input vector $x = [x_1\, x_2\, \cdots x_n]^T$ one selects MF, e.g., triangular membership functions. Let N_j represent the number of MFs defined for x_j. Then the MF for x_j may be denoted as $X_j^1, X_j^2, \cdots, X_j^{N_j}$, or generally as $X_j^{i_j}; i_j \in \overline{N}_j$. Normally, the MF, $X_j^{i_j}(x_j)$, is distinguished from its value $\mu_j^{i_j}(x_j)$ evaluated for a given value of x_j. Given a prescribed value of x_j, the value $\mu_j^{i_j}(x_j)$ is called the degree of membership of x_j in $X_j^{i_j}$.

The fuzzy rule-base consists of R rules, the ρ-th of which often has the form

$$\text{IF } x_1\, is\, X_1^{i_{1\rho}})\, and\, x_2\, is\, X_2^{i_{2\rho}}\, and\, \cdots\, and\, x_1\, is\, X_n^{i_{n\rho}}\;\; THEN\;\; y_i\, is\, u_{i\rho}.$$

This sort of rule (Mamdani-type) covers many applications and is sufficiently general to give us the key function approximation capability that is so essential for control purposes. In this rule, the observed or measured values of the inputs at a specific sampling time are x_j. Furthermore, $x_j\, is\, X_j^{i_{j\rho}}$, means the extent to which x_j belongs to the MF $X_j^{i_{j\rho}}$, which is given by computing the degree of membership $\mu_j^{i_{j\rho}}(x_j)$. The values $u_{i\rho}$ associated with y_i are often defined using MFs prescribed for each y_i. The values $u_{i\rho}$ will also be called here the weights for the output component y_i and rule ρ.

The degrees of membership $\mu_j^{i_{j\rho}}(x_j), j \in \overline{n}$, that are associated with rule ρ must be combined into a single degree of membership, to be used for computing the value of y_i. This is done here by forming the product of these membership degrees, that is

$$\mu_{i_{1\rho},i_{2\rho},\cdots,i_{n\rho}}(x) = \mu_1^{i_{1\rho}}(x_1)\mu_2^{i_{2\rho}}(x_2)\cdots\mu_n^{i_{n\rho}}(x_1). \tag{2}$$

Though not normally done, this product of membership degrees can be interpreted as an n-dimensional MF, $X_{i_{1\rho},i_{2\rho},\cdots,i_{n\rho}}(x) = X_1^{i_{1\rho}}(x_1)\cdots X_n^{i_{n\rho}}(x_n)$. Given a specific value of x, several rules may produce values for y_i, perhaps all different. The issue of combining these different values in order to determine a single value for y_i is known as defuzzification. A standard technique, centroid-defuzzification, computes this single value as

$$y_i = \frac{\sum_{\rho=1}^{R} u_{i\rho}\mu_{i_{1\rho},i_{1\rho},\cdots,i_{n\rho}}(x)}{\sum_{\rho=1}^{R} \mu_{i_{1\rho},i_{1\rho},\cdots,i_{n\rho}}(x)}, \quad i = 1,2,\cdots,m. \tag{3}$$

This equation computes the outputs y_i of the fuzzy system in terms of the input $x \in \Re^n$. It is also known as the Fuzzy Associate Memory (FAM) function or the reasoning surface.

Defining

$$\varphi_{i\rho}(x) = \frac{\mu_{i_{1\rho},i_{1\rho},\cdots,i_{n\rho}}(x)}{\sum_{\rho=1}^{R} \mu_{i_{1\rho},i_{1\rho},\cdots,i_{n\rho}}(x)}, \tag{4}$$

for each rule ρ and output component i, one may write the FAM function as

$$y_i = \sum_{\rho=1}^{R} u_{i\rho}\varphi_{i\rho}(x), \quad i = 1,2,\cdots,m. \tag{5}$$

It has been shown by several authors [Ying, 93; Wang and Mendel, 92; Wang, 94; Kosko, 94] that MFs provide a basis for continuous functions if the MF and the fuzzy rules are properly chosen. This justifies their name as fuzzy basis functions (FBF), and implies a universal approximation result for fuzzy systems. One may order the weights $u_{i\rho}$ into a vector U_i and the values of the FBF, $\varphi_{i\rho}(x)$ into a vector $\varphi_i(x)$. This can be expressed as

$$y = U^T\varphi(x). \tag{6}$$

The ordering of the weights and the FBF into matrices is not unique, but a convenient ordering is provided by left-hand odometer order.

The function approximation result can now be stated as follows. Let $f(x(k)) : \Re^n \to \Re^m$ be a smooth function. Then, given a compact set $S \in \Re^n$ and a positive number ϵ_N, there exists a fuzzy system such that

$$f(x(k)) = U^T \varphi(x) + \epsilon, \tag{7}$$

with $\| \epsilon \| < \epsilon_N$ for all $x \in S$. The value ϵ (generally a function of x) is the approximation error, and it decreases as the number N_j of MFs in each dimension increases. The approximation result has been shown for triangular, trapezoidal, Gaussian MF, and other types of MFs. Note that this result only says that "there exists a fuzzy system that approximates f(x(k))", but it does not show how to determine the required weights. For given MFs, the issue of finding the weights such that a fuzzy system does indeed approximate a given function $f(x(k))$ closely enough is not an easy one. This is also the case when the fuzzy system is in a closed-loop control configuration. In this article, methods are provided to tune the weights so that they are adapted on-line in order to obtain suitable approximation.

2.3 Stability

To formulate the discrete-time FC, the following stability notions are needed. Consider the nonlinear system given by

$$\begin{aligned} x(k+1) &= f(x(k), v_c(k)), \\ y(k) &= h(x(k)), \end{aligned} \tag{8}$$

where $x(k)$ is a state vector, $v_c(k)$ is the input vector and $y(k)$ is the output vector. The solution is said to be *uniformly ultimately bounded (UUB)* if for all $x(k_0) = x_0$, there exists an $\epsilon \geq 0$ and a number $N(\epsilon, x_0)$ such that $||x(k)|| \leq \epsilon$ for all $k \geq k_0 + N$.

Consider now the linear discrete time-varying system given by

$$x(k+1) = A(k)x(k) + B(k)v_c(k), y(k) = C(k)x(k). \tag{9}$$

Lemma 2.1 *Define $\psi(k_1, k_0)$ as the state-transition matrix corresponding to $A(k)$ for the system (9), i.e., $\psi(k_1, k_0) = \prod_{k=k_0}^{k_1-1} A(k)$. Then if $\| \psi(k_1, k_0) \| \leq 1, \forall k_1, k_0 \geq 0$, the system (9) is exponentially stable.*

Proof: See [Sadegh, 93]. ■

3 System dynamics and the tracking problem

In this section we describe the class of systems to be dealt with in this article and study the error dynamics using a specific feedback linearizing controller.

3.1 Tracking error dynamics for a class of feedback linearizable systems

Consider an mnth-order multi-input/multi-output discrete-time state feedback linearizable minimum phase nonlinear system, given as

$$\begin{aligned} x_1(k+1) &= x_2(k), \\ &\vdots \\ x_{n-1}(k+1) &= x_n(k), \\ x_n(k+1) &= f(x(k)) + g(x(k))v_c(k) + d(k), \end{aligned} \tag{10}$$

with state $x(k) = [x_1^T(k), \cdots, x_n^T(k)]^T$, $x_i(k) \in \Re^m, i = 1, \cdots, n$, and control $v_c(k) \in \Re^m$. The nonlinear functions $f(\cdot)$ and $g(\cdot)$ are assumed unknown. The disturbance vector acting on the system at the instant k is $d(k) \in \Re^m$. We assume that this disturbance vector is unknown but bounded so that $\| d(k) \| \leq d_M$ is a known constant. Further, the unknown smooth function satisfies the mild assumption

$$\| g(x(k)) \| \geq \mathrm{g} > 0, \tag{11}$$

with g a known lower bound. The assumption given above on the smooth function $g(x(k))$ implies that $g(x(k))$ is strictly either positive or negative for all x. From now on, without loss of generality, we will assume that $g(x(k))$ is strictly positive. Note that at this point there is no general approach to analyze this class of unknown nonlinear systems. Adaptive control, for instance needs an additional linear-in-the-parameters assumption [Åström and Wittenmark, 89; Goodwin and Sin, 84].

Feedback linearization will be used for the purpose of output tracking. That is, given a desired trajectory in terms of output, $x_{nd}(k)$, and its delayed values, find a control input $v_c(k)$, so that the system tracks the desired trajectory with an acceptable bounded error in the presence of disturbances while all the states and controls remain bounded. In order to continue, the following assumptions are required.

Assumption 3.1 (Bounds for system and desired trajectory) :

1. *The sign of $g(x(k))$ is known.*

2. The desired trajectory vector with its delayed values is assumed to be available for measurement and bounded by an upper bound q_B.

■

Given a desired trajectory $x_{nd}(k)$ and its delayed values, we define the tracking error as

$$e_n(k) = x_n(k) - x_{nd}(k). \tag{12}$$

The filtered tracking error, $r(k) \in \Re^m$, [Slotine and Li, 89] is

$$r(k) = e_n(k) + \lambda_1 e_{n-1}(k) + \cdots + \lambda_{n-1} e_1(k), \tag{13}$$

where $e_{n-1}(k), \cdots, e_1(k)$ are the delayed values of the error $e_n(k)$ and $\lambda_1, \cdots, \lambda_{n-1}$ are constant matrices selected so that

$$| \, z^{n-1} + \lambda_1 z^{n-2} + \cdots + \lambda_{n-1} \, |,$$

is stable. Equation (13) can be expressed as

$$r(k+1) = e_n(k+1) + \lambda_1 e_{n-1}(k+1) + \cdots + \lambda_{n-1} e_1(k+1). \tag{14}$$

Using (10) in (14), the dynamics of the filtered tracking error system (14) can be written in terms of the tracking error as

$$r(k+1) = f(x(k)) - x_{nd}(k+1) + \lambda_1 e_n(k) + \cdots + \lambda_{n-1} e_2(k) + g(x(k)) v_c(k) + d(k). \tag{15}$$

Equation (15) can be expressed as

$$r(k+1) = f(x(k)) + g(x(k)) v_c(k) + d(k) + Y_d, \tag{16}$$

where

$$Y_d = -x_{nd}(k+1) + \sum_{i=0}^{n-2} \lambda_{i+1} e_{n-i}. \tag{17}$$

If we know the functions $f(x(k))$ and $g(x(k))$, and if no disturbances are present, the control input $v_c(k)$ can be selected as the feedback linearization controller

$$v_c(k) = \frac{1}{g(x(k))}(-f(x(k)) + v(k)), \tag{18}$$

where $v(k)$ is taken as an auxiliary input given by

$$v(k) = k_v r(k) - Y_d. \tag{19}$$

Then the filtered tracking error $r(k)$ goes to zero exponentially by properly selecting the gain matrix k_v. Since the system functions are not known a priori, the control input $v_c(k)$ can be selected as

$$v_c(k) = \frac{1}{\hat{g}(x(k))}(-\hat{f}(x(k)) + v(k)), \tag{20}$$

with $\hat{f}(x(k))$ and $\hat{g}(x(k))$ being the estimates of $f(x(k))$ and $g(x(k))$ respectively. Note that even in adaptive control of linear systems, guaranteeing the boundedness of $\hat{g}(x(k))$ away from zero, becomes an important issue for this type of controller.

Equation (16) can be rewritten as

$$r(k+1) = v(k) - v(k) + f(x(k)) + g(x(k))v_c(k) + d(k) + Y_d. \tag{21}$$

Subsituting (19), (20) for $v(k)$ in (21), (21) can be rewritten as

$$r(k+1) = k_v r(k) + \tilde{f}(x(k)) + \tilde{g}(x(k))v_c(k) + d(k), \tag{22}$$

where the function estimation errors are given by

$$\tilde{f}(x(k)) = f(x(k)) - \hat{f}(x(k)), \tag{23}$$

and

$$\tilde{g}(x(k)) = g(x(k)) - \hat{g}(x(k)). \tag{24}$$

This is an error system wherein the filtered tracking error is driven by the function estimation errors and unknown disturbances.

In this article, discrete-time fuzzy systems are used to provide the estimate $\hat{f}(\cdot)$ and $\hat{g}(\cdot)$. The error system (22) is used to focus on selecting discrete-time fuzzy weight tuning algorithms that guarantee the stability of the filtered tracking error $r(k)$. Then (13), with the input being $r(k)$ and the output given as $e(k)$, describes a stable system. In this case, using the notion of operator gain [Jagannathan and Lewis, 96a] one can guarantee that $e(k)$ exhibits stable behavior.

4 FC design for feedback linearization

In this section we derive the error system dynamics and present the FC's structure.

4.1 Approximation of unknown functions

It will be necessary to review the notation for a fuzzy system. Assume that there exist some constant ideal weights U_f and U_g so that the nonlinear functions in (10) can be written as

$$f(x(k)) = U_f^T \varphi_f(k) + \epsilon_f(k), \tag{25}$$

and

$$g(x(k)) = U_g^T \bar{\varphi}_g(k) + \epsilon_g(k). \tag{26}$$

The membership functions $\varphi_f(k)$ and $\bar{\varphi}_g(k)$ provide a suitable basis set for $f(\cdot)$ and $g(\cdot)$ respectively. In addition, we have that $\| \epsilon_f(k) \| < \epsilon_{Nf}$ and $\| \epsilon_g(k) \| < \epsilon_{Ng}$ with the bounding constants ϵ_{Nf} and ϵ_{Ng} being known.

Let $x \in U$ be a compact subset of R^n. Assume that $h(x(k)) \in C^\infty[U]$, i.e. a smooth function $U \longrightarrow R$, so that the Taylor series expansion of $h(x(k))$ exists. One can derive that $\| x(k) \| \leq d_{01} + d_{11} \| r(k) \|$. Then using the bound on $x(k)$ and expressing $h(x(k))$ as (27), yields an upper bound on $h(x(k))$. This upper bound is given as

$$\| h(x(k)) \| = \| U_h^T \varphi_h(k) + \epsilon_h(k) \| \leq C_{01} + C_{11} \| r(k) \|, \tag{27}$$

with C_0 and C_{11} being computable constants. In addition, the MFs are bounded by a known upper bound

$$\begin{aligned} \| \varphi_f(k) \| &\leq \varphi_{fmax}, \\ \| \varphi_g(k) \| &\leq \varphi_{gmax}. \end{aligned} \tag{28}$$

Lemma 4.1 *For each time k, $x(k)$ is bounded by*

$$\| x(k) \| \leq d_{01} + d_{11} \| r(k) \| \leq q_B + l_1 \| r(0) \| + d_{11} \| r(k) \| . \tag{29}$$

Proof: Use the solution of (13). ■

4.2 Error system dynamics

The function estimates, employed to select the control input presented in (20), are defined here as

$$\hat{f}(x(k)) = \hat{U}_f^T(k) \varphi_f(k), \tag{30}$$

and

$$\hat{g}(x(k)) = \hat{U}_g^T(k) \varphi_g(k), \tag{31}$$

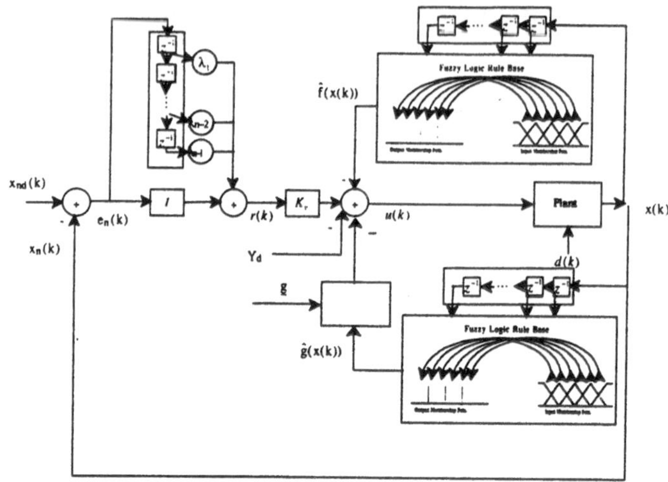

Figure 1: Discrete-time FCr structure for feedback linearization.

with $\hat{U}_f(k)$ and $\hat{U}_g(k)$ being the current values of the weights. This yields the controller structure shown in Fig. 1. The output of the plant is processed through a series of delays to obtain the past values of the output, and these are fed as inputs to the system so that the nonlinear function in (10) can be suitably approximated. Thus, the FC derived by using filtered error notions, naturally provides a dynamical fuzzy system structure. Note that neither the input $v_c(k)$ nor its past values are needed by the fuzzy system. The next step is to determine the weight updates so that the tracking performance of the closed-loop filtered error dynamics is guaranteed.

Let U_f and U_g be the unknown ideal weights required for the approximation to hold in (30) and (31). Assume they are bounded by known values so that

$$\| U_f(k) \| \leq U_{fmax}. \tag{32}$$

and

$$\| U_g(k) \| \leq U_{gmax}. \tag{33}$$

Then the error in the weights during estimation is given by

$$\tilde{U}_f(k) = U_f - \hat{U}_f(k), \tag{34}$$

and

$$\tilde{U}_g(k) = U_g - \hat{U}_g(k). \tag{35}$$

The closed-loop filtered dynamics (22) becomes

$$r(k+1) = k_v r(k) + \tilde{U}_f^T(k)\varphi_f(k) + \tilde{U}_g^T(k)\overline{\varphi}_g(k)v_c(k) + \epsilon_f(k) + \epsilon_g(k)v_c(k) + d(k). \tag{36}$$

4.3 Well-defined control problem

In general, the boundedness of $x(k), \hat{U}_f(k)$ and $\hat{U}_g(k)$ does not indicate the stability of the closed-loop system, because control law (20) is not well defined when $\hat{g}(\hat{U}_g, x) = 0$. Therefore, one has to guarantee the boundedness of the controller as well. To overcome the problem, several techniques exist in the literature that assure local or global stability with the help of an additional knowledge. First, if the bounds on the function $g(x(k))$ are known, then $\hat{g}(\hat{U}_g, x)$ may be set to a constant and a robust adaptive controller bypasses this problem. This is not an accurate solution because the bounds on the function are not known a priori.

If $g(x(k))$ is reconstructed by an adaptive scheme, then a local solution can be generated by assuming that the initial estimates are close to the actual values. Furthermore, it is also assumed that these values do not leave a feasible invariant set in which $\hat{g}(\hat{U}_g, x)$ is not equal to zero, or is inside a region of attraction of a stable equilibrium point which forms a feasible set [Kanellakopoulos et al., 91]. Unfortunately, even with a very good knowledge of the system, it is not easy to choose initial weight values such that the fuzzy system approximates it.

The way to avoid the problem is to project $\hat{U}_g(k)$ onto an estimated feasible region by properly selecting the weight values [Polycarpou and Ioannou, 91]. A shortcoming of this solution is that the actual $\hat{U}_g(k)$ does not necessarily belong to this set which results in a sub-optimal solution.

4.4 Proposed controller

In order to guarantee the boundedness of $\hat{g}(x(k))$ away from zero for all well-defined values of $x(k), \hat{U}_f(k)$, and $\hat{U}_g(k)$, the control input in (20) is selected in terms of another control input $u_c(k)$ and a robust control term $u_r(k)$ as

$$\begin{aligned} v_c(k) &= u_c(k) + \frac{u_r(k) - u_c(k)}{2} e^{\gamma(\|u_c(k)\| - s)}, \quad I = 0, \\ &= u_r(k) - \frac{u_r(k) - u_c(k)}{2} e^{-\gamma(\|u_c(k)\| - s)}, \quad I = 1, \end{aligned} \tag{37}$$

where

$$u_c(k) = \frac{1}{\hat{g}(x(k))}(-\hat{f}(x(k)) + v(k)), \tag{38}$$

and

$$u_r(k) = -\mu \frac{\| u_c(k) \|}{\text{g}} sgn(r(k)). \tag{39}$$

The indicator I in (37) is

$$\begin{aligned} I &= 1, \; if \;\; \| \hat{g}(x(k)) \| \geq \underline{g} \; and \;\; \| u_c(k) \| \leq s, \\ &= 0, \; otherwise. \end{aligned} \tag{40}$$

where $\gamma < \ln \frac{2}{s}, \mu > 0$, and $s > 0$ are design parameters. These modifications in the control input are necessary in order to ensure that the function estimate $\hat{g}(x(k))$ is bounded away from zero.

The intuition behind this controller is as follows. When $\| \hat{g}(x(k)) \| \geq \underline{g}$ and $\| u_c(k) \| \leq s$, then the total control input is set to $u_c(k)$, otherwise the control is smoothly switched to the auxiliary input $u_r(k)$ due to the additional term in (37). This results in a well-defined control everywhere and the uniform ultimate boundedness of the closed-loop system can be shown by appropriately selecting the weight tuning algorithms.

5 FC for feedback linearization

In this section, a fuzzy system, without membership function tuning, is considered as a first step to bridging the gap between discrete-time adaptive control and FC. In this case, the tunable weights of the fuzzy system enter in a linear fashion. The assumptions made are mild enough and also, rigorous Lyapunov stability analysis is presented. Furthermore, a family of weight tuning paradigms that guarantee the stability of the closed-loop system (36) is presented in this section. One has to demonstrate here that the tracking error $r(k)$ is small enough and that the weights $\hat{U}_f(k)$, and $\hat{U}_g(k)$ remain bounded.

Stability analysis by Lyapunov's direct method is performed using a novel weight tuning algorithm for a FC. Persistency of excitation (PE) , which cannot be tested or guaranteed for a fuzzy system, is generally needed for suitable performance. Therefore, modified tuning paradigms are proposed in subsequent subsections to make the FC robust. To guarantee stability, it is shown that the tuning rate for the weights must decrease when there is an increase in the number of MFs. By employing a projection algorithm it is shown that the tuning rate can be made independent of the number of MF. To proceed further, we review now the required formal machinery.

Lemma 5.1 *If $A(k) = I - \alpha\varphi(x(k))\varphi^T(x(k))$ in (9), where $0 < \alpha < 2$ and $\varphi(x(k))$ is a vector of basis functions, then $\| \psi(k_1, k_0) \| < 1$ is guaranteed if there is an $L > 0$ such that $\sum_{k=k_0}^{k_1+L-1} \varphi(x(k))\varphi^T(x(k)) > 0$ for all k. Then, Lemma 2.1 guarantees the exponential stability of the system (9).*

Proof: *Follow the steps in [Sadegh, 93].* ∎

Definition 5.1 *An input sequence x(k) is said to be* persistency exciting (PE) *if there are* $\lambda > 0$ *and an integer* $k_1 \geq 1$ *such that*

$$\lambda_{min}[\sum_{k=k_0}^{k_1} \varphi(x(k))\varphi^T(x(k))] > \lambda, \forall k_0 \geq 0, \tag{41}$$

where $\lambda_{min}(P)$ *represents the smallest eigenvalue of* P. ∎

Note that PE is exactly the stability condition needed in Lemma 5.1.

Assumption 5.1 *(Initial Condition Requirement). Suppose the desired trajectory is bounded from above and define the constants* l_1 *and* d_{11} *by Lemma 4.1. Let the approximation property (7) hold for the functions* $f(x(k))$ *and* $g(x(k))$ *with an accuracy of* ϵ_{Nf} *and* ϵ_{Ng} *for all* x *inside the ball of radius* $b_x > q_B$. *Let the initial tracking error satisfy* $\| r(0) \| < \frac{b_x - q_B}{l_1 + d_{11}}$.

The set S_r *specifies the set of allowed initial tracking error* $r(0)$. *Note that the approximation accuracy determines the allowed magnitude of the initial tracking error* $r(0)$. *By suitably selecting the MFs the accuracy is smaller for larger* b_x. ∎

5.1 Weight updates with PE

Table 1 presents a discrete-time weight tuning algorithm based on the filtered tracking error. The algorithm guarantees (see following theorem) that the tracking error and the error in the weight estimates are bounded if a PE condition holds. (This PE requirement is later on relaxed in Theorem 5.2).

Theorem 5.1 (Discrete-time FC requiring PE) *Let the desired trajectory* $x_{nd}(k)$ *be bounded, the function reconstruction error bounds,* ϵ_{Nf} *and* ϵ_{Ng}, *and the disturbance bound* d_M *be known constants. Take the control input for (10) as (37) with weight tuning for* $f(x(k))$ *provided by (42) and for* $g(x(k))$ *as in (43).*

Assume that the initial error in weight estimates is bounded and let the MF vector, $\varphi_f(k)$ *and* $\overline{\varphi}_g(k)u_c(k)$ *be subject to PE. Then the filtered tracking error* $r(k)$ *and the error in weight estimates,* $\tilde{U}_f(k)$, *and* $\tilde{U}_g(k)$ *are UUB,*

Table 1: Discrete-Time FC with PE

The control input is

$$\begin{aligned} v_c(k) &= u_c(k) + \frac{u_r(k) - u_c(k)}{2} e^{\gamma(\|u_c(k)\| - s)}, \; I = 1, \\ &= u_r(k) - \frac{u_r(k) - u_c(k)}{2} e^{-\gamma(\|u_c(k)\| - s)}, \; I = 0. \end{aligned}$$

The weight tuning for f(x(k)) is given by

$$\hat{U}_f(k+1) = \hat{U}_f(k) + \alpha \varphi_f(k) r^T(k+1), \tag{42}$$

and for g(x(k)) as

$$\begin{aligned} \hat{U}_g(k+1) &= \hat{U}_g(k) + \beta \varphi_g(k) r^T(k+1), \; I = 1, \\ &= \hat{U}_g(k), \; I = 0, \end{aligned} \tag{43}$$

with $\alpha > 0$ and $\beta > 0$ denoting constant learning rate parameters or adaptation gains.

---

with bounds given by (76) or (96) with (58) or (98) and (59) provided that the following conditions hold:

$$(1) \quad \beta \| \overline{\varphi}_g(k) u_c(k) \|^2 = \beta \| \varphi_g(k) \|^2 < 1, \tag{44}$$

$$(2) \quad \alpha \| \varphi_f(k) \|^2 < 1, \tag{45}$$

$$(3) \quad \eta < 1, \tag{46}$$

$$(4) \quad max(a_4, b_0) < 1, \tag{47}$$

where η is given as

$$\eta = \alpha \| \varphi_f(k) \|^2 + \beta \| \overline{\varphi}_g(k) u_c(k) \|^2, \tag{48}$$

and for $I = 1$, and for $I = 0$, it is defined as

$$\eta = \alpha \| \varphi_f(k) \|^2, \tag{49}$$

where a_4, b_0 are design parameters chosen using the gain matrix k_{vmax}. Observe here that the parameters α, β, and η are dependent on the trajectory.

Proof: Let the initial condiitons are bounded by q_B. Suppose that the approximation property (7) holds for the functions $f(x(k))$ and $g(x(k))$ with a given accuracy of ϵ_{Nf} and ϵ_{Ng} in the compact set $S_x \equiv (x(k) \mid \| x(k) \| < b_x)$ with $b_x > q_B$. Define $S_r \equiv (r(k) \mid \| r(k) \| < \frac{b_x - q_B}{l_1 + d_{11}})$. Let $r(0) \in S_r$. Then the approximation property holds.

Define the Lyapunov function candidate

$$J = r^T(k)r(k) + \frac{1}{\alpha}tr(\tilde{U}_f^T(k)\tilde{U}_f(k)) + \frac{1}{\beta}tr(\tilde{U_g}^T(k)\tilde{U}_g(k)). \tag{50}$$

The first difference is given by

$$\begin{aligned}\triangle J &= r^T(k+1)r(k+1) - r^T(k)r(k) + \frac{1}{\alpha}tr(\tilde{U}_f^T(k+1)\tilde{U}_f(k+1) \\ &\quad -\tilde{U}_f^T(k)\tilde{U}_f(k))\frac{1}{\beta}tr(\tilde{U}_g^T(k+1)\tilde{U}_g(k+1) - \tilde{U}_g^T(k)\tilde{U}_g(k)).\end{aligned} \tag{51}$$

<u>*Region I:*</u> $\| \hat{g}(x(k)) \| \geq \underline{g}$ *and* $\| u_c(k) \| \leq s$.

The filtered error dynamics (36) can be rewritten as

$$\begin{aligned}r(k+1) &= k_v r(k) + (f(x(k)) - \hat{f}(x(k))) + (g(x(k)) - \hat{g}(x(k)))u_c(k) \\ &\quad + d(k) + g(x(k))u_d(k),\end{aligned} \tag{52}$$

where $u_d(k) = v_c(k) - u_c(k)$. *Substituting (42), (43) in (52), one obtains*

$$\begin{aligned}r(k+1) &= k_v r(k) + \tilde{U}_f^T(k)\varphi_f(k) + \tilde{U}_g^T(k)\overline{\varphi}_g(k)u_c(k) \\ &\quad + \epsilon(k) + d(k) + g(x(k))u_d(k),\end{aligned} \tag{53}$$

where

$$\epsilon(k) = \epsilon_f(k) + \epsilon_g(k)u_c(k). \tag{54}$$

Equation (53) can be rewritten

$$r(k+1) = k_v r(k) + \overline{e}_f^T(k) + \overline{e}_g^T(k) + \epsilon(k) + d(k) + g(x(k))u_d(k), \tag{55}$$

where

$$\overline{e}_f(k) = \tilde{U}_f^T(k)\varphi_f(k), \tag{56}$$

$$\overline{e}_g(k) = \tilde{U}_g^T(k)\overline{\varphi}_g(k)u_c(k). \tag{57}$$

The error in dynamics for the weight update laws are given for this region as

$$\begin{aligned}\tilde{U}_f(k+1) &= (I - \alpha\varphi_f(k)\varphi_f^T(k))\tilde{U}_f(k) - \alpha\varphi_f(k)(k_v r(k) + \overline{e}_g(k) \\ &\quad + g(x(k))u_d(k) + \epsilon(k) + d(k))^T,\end{aligned} \tag{58}$$

and

$$\begin{aligned}\tilde{U}_g(k+1) &= (I - \beta\varphi_g(k)\varphi_g^T(k))\tilde{U}_g(k) - \beta\varphi_g(k)(k_v r(k) + \overline{e}_f(k) \\ &\quad + g(x(k))u_d(k) + \epsilon(k) + d(k))^T.\end{aligned} \tag{59}$$

Substituting (55) (58) and (59) in (51), and simplifying, we obtain

$$\begin{aligned}
\Delta J &= -r^T(k)[I - k_v^T k_v]r(k) + 2\eta(k_v r(k))^T(g(x(k))u_d(k) + \epsilon(k) + d(k)) \\
&\quad (1+\eta)(g(x(k))u_d(k) + \epsilon(k) + d(k))^T(g(x(k))u_d(k) + \epsilon(k) + d(k)) \\
&\quad -(1-\eta) \parallel (\bar{e}_f(k) + \bar{e}_g(k)) \\
&\quad -\frac{\eta}{(1-\eta)}(k_v r(k) + g(x(k))u_d(k) + \epsilon(k) + d(k)) \parallel^2 \\
&\quad +\frac{\eta}{(1-\eta)}(k_v r(k) + g(x(k))u_d(k) + \epsilon(k) + d(k))^T \\
&\quad (k_v r(k) + g(x(k))u_d(k) + \epsilon(k) + d(k)),
\end{aligned} \tag{60}$$

where

$$\eta = \alpha \parallel \varphi_f(k) \parallel^2 + \beta \parallel \varphi_g(k) \parallel^2 . \tag{61}$$

Equation (60) can be rewritten as

$$\begin{aligned}
\Delta J &= -(1 - a_1 k_{vmax}^2) \parallel r(k) \parallel^2 + 2a_2 k_{vmax} \parallel r(k) \parallel \\
&\quad (g(x(k))u_d(k) + \epsilon(k) + d(k)) + a_3(g(x(k))u_d(k) + \epsilon(k) + d(k))^T \\
&\quad (g(x(k))u_d(k) + \epsilon(k) + d(k)) \\
&\quad -(1-\eta) \parallel (\bar{e}_f(k) + \bar{e}_g(k)) - \frac{\eta}{(1-\eta)} \\
&\quad (k_v r(k) + g(x(k))u_d(k) + \epsilon(k) + d(k)) \parallel^2,
\end{aligned} \tag{62}$$

where

$$a_1 = 1 + \eta + \frac{\eta}{(1-\eta)}, \tag{63}$$

$$a_2 = \eta + \frac{\eta}{(1-\eta)}, \tag{64}$$

$$a_3 = 1 + \eta + \frac{\eta}{(1-\eta)}. \tag{65}$$

Now having in mind that the function $g(x(k))$ is defined on a compact set, one can conclude that

$$\parallel g(x(k)) \parallel \leq C_{01} + C_{12} \parallel r(k) \parallel, \tag{66}$$

with C_{01}, C_{12} being computable constants.

In this region, the bound on $u_d(k)$ can be obtained as

$$\begin{aligned}
\parallel u_d(k) \parallel &\leq \parallel v_c(k) - u_c(k) \parallel \\
&\leq \parallel \frac{(u_r(k) - u_c(k))}{2} e^{\gamma(\parallel u_c(k) \parallel - s)} \parallel .
\end{aligned} \tag{67}$$

Since in this region, $\| u_c(k) \| \leq s$, *and the other input* $u_r(k)$ *is given by (39), the bound in (67) can be obtained as a constant since all the terms on the right side are bounded and this bound is denoted by*

$$\| u_d(k) \| \leq C_2. \tag{68}$$

Now the bound for $g(x(k))u_d(k)$ *is obtained as*

$$\begin{aligned} \| g(x(k))u_d(k) \| &\leq C_2(C_{01} + C_{12} \| r(k) \|) \\ &\leq C_0 + C_1 \| r(k) \| . \end{aligned} \tag{69}$$

Using the bound presented in (69) for $g(x(k))u_d(k)$, *the first difference of the Lyapunov function (62) is rewritten as*

$$\begin{aligned} \Delta J = \; & -(1 - a_1 k_{vmax}^2) \| r(k) \|^2 + 2a_2 k_{vmax} \| r(k) \| \\ & (C_0 + C_1 \| r(k) \| + \epsilon_N + d_M) + \\ & a_3(C_0 + C_1 \| r(k) \| + \epsilon_N + d_M)^T (C_0 + C_1 \| r(k) \| + \epsilon_N + d_M) - \\ & (1-\eta) \| (\overline{e}_f(k) + \overline{e}_g(k)) - \\ & \frac{\eta}{(1-\eta)} (k_v r(k) + g(x(k))u_d(k) + \epsilon(k) + d(k)) \|^2, \end{aligned} \tag{70}$$

where the bound for $\epsilon(k)$ *can be obtained as*

$$\begin{aligned} \| \epsilon(k) \| & \leq \| \epsilon_f \| + \| \epsilon_g u_c(k) \| \\ & \leq (\epsilon_{Nf} + s\epsilon_{Ng}) \\ & \leq \epsilon_N. \end{aligned} \tag{71}$$

Simplifying (70), one obtains

$$\begin{aligned} \Delta J = \; & -(1 - a_4) \| r(k) \|^2 + 2a_5 \| r(k) \| + a_6 \\ & -(1-\eta) \| (\overline{e}_f(k) + \overline{e}_g(k)) - \\ & \frac{\eta}{(1-\eta)} (k_v r(k) + g(x(k))u_d(k) + \epsilon(k) + d(k)) \|^2, \end{aligned} \tag{72}$$

where

$$a_4 = a_1 k_{vmax}^2 + 2a_2 C_1 k_{vmax} + a_3 C_1, \tag{73}$$

$$a_5 = a_2 k_{vmax}(\epsilon_N + d_M + C_0) + a_3 C_1(\epsilon_N + d_M) + a_3 C_0 C_1, \tag{74}$$

and

$$a_6 = a_3C_0^2 + 2a_3C_0(\epsilon_N + d_M) + (\epsilon_N + d_M)^2. \tag{75}$$

The second term in (72) is always negative as long as the conditions (44) through (47) hold. Since a_4, a_5 and a_6 are positive constants, $\Delta J \leq 0$ as long as (44) through (47) hold and

$$\| r(k) \| > \delta_{r1}, \tag{76}$$

where

$$\delta_{r1} > \frac{1}{(1-a_4)}[a_3 + \sqrt{a_5^2 + a_6(1-a_4)}]. \tag{77}$$

$\| \sum_{k=k_0}^{\infty} \Delta J(k) \| = \| J(\infty) - J(0) \| < \infty$ *since* $\Delta J \leq 0$ *as long as (44) through (47) hold. The definition of J and inequality (76) imply that every initial condition in the set χ will evolve entirely within χ. In other words, whenever the tracking error $\| r(k) \|$ is outside the region defined by (76), $J(r(k), \tilde{U}_f(k), \tilde{U}_g(k))$ will decrease. This further implies that the tracking error $r(k)$ is UUB for all $k \geq 0$. It remains to be shown that the weight estimation errors, $\tilde{U}_f(k)$ and $\tilde{U}_g(k)$ or equivalently $\hat{U}_f(k)$ and $\hat{U}_g(k)$ are bounded.*

In general, to show the boundedness of the weight estimation errors, one uses the error in weight updates (58) and (59), the tracking error bound (76), the PE condition and Lemma 5.1. From (58) and (59), it can be seen that the output of each fuzzy system is driving the other. Therefore, if the initial weight estimation errors for both fuzzy systems are bounded, then applying the bound for the tracking error (76), the PE condition and Lemma 5.1, one can show that the weight estimation errors $\tilde{U}_f(k)$ and $\tilde{U}_g(k)$ or equivalently $\hat{U}_f(k)$ and $\hat{U}_g(k)$ are bounded. This concludes the boundedness of both tracking error and weight estimates for both fuzzy systems in this region. On the other hand, a similar and elegant way to show the boundedness of the tracking error and weight estimates is to apply passivity theory. The proof using passivity theory is shown in [Jagannathan, 96b; Jagannathan, 96c].

Region II: $\| \hat{g}(x(k)) \| \leq \underline{g}$ *and* $\| u_c(k) \| > s$.

Since the input $u_c(k)$ may not be defined in this region, for the purpose of notational simplicity we will use it in the form of either $\hat{g}(x(k))u_c(k)$ or $u_c(k)e^{-\gamma(\|u_c(k)\|-s)}$. Therefore, in this region, the tracking error system given in (53) is rewritten as

$$r(k+1) = k_v r(k) + \bar{e}_f^T(k) + \overline{g(x(k))u_d(k)} + \epsilon_f(k) + d(k), \tag{78}$$

where

$$\overline{g(x(k))u_d(k)} = g(x(k))v_c(k) - \hat{g}(x(k))u_c(k). \tag{79}$$

Note that the extremum of the function $ye^{-\gamma y}$ *for* $\forall y > 0$ *can be found as a solution to the following equation*

$$\frac{\partial(y^{-\gamma y})}{\partial y} = (1 - \gamma y)e^{-\gamma y} = 0, \tag{80}$$

which is $y = \frac{1}{\gamma}$, *and it is a maximum solution. Evaluating the function* $u_c(k)e^{-\gamma u_c(k)}$ *yields an upper bound for* $u_c(k) = \frac{1}{\gamma e}$ *and this bound is used in the forthcoming equations.*

Let us compute the bound for $g(x(k))u_c(k)$ *and* $\hat{g}(x(k))u_c(k)$. *Consider the following cases in this region when* $\| u_c(k) \| \leq s$ *and* $\| u_c(k) \| > s$. *The bound on* $v_c(k)$ *from (36) can be written as*

$$\| v_c(k) \| \leq \frac{u_r(k) - u_c(k)}{2} e^{-\gamma(\|u_c(k)\| - s)}. \tag{81}$$

Using (39) for $u_r(k)$, *equation (81) can be rewritten as*

$$\| v_c(k) \| \leq \frac{1}{2}(\frac{\mu}{g} \| u_c(k) \| + \| u_c(k) \|)e^{-\gamma(\|u_c(k)\| - s)}. \tag{82}$$

If $\| u_c(k) \| \leq s$, *then* $e^{\gamma s} \leq 2$ *and equation (82) can be written as*

$$\| v_c(k) \| \leq d_1, \tag{83}$$

where

$$d_1 = (\frac{\mu}{\underline{g}} s + d_0 s), \tag{84}$$

is bounded from above by some positive constant. On the other hand, if $\| u_c(k) \| > s$, *equation (82) can be expressed as*

$$\| v_c(k) \| \leq d_1, \tag{85}$$

where

$$d_1 = \frac{1}{2}(\frac{\mu}{\underline{g}} \frac{1}{\gamma e} + d_0 \frac{1}{\gamma e}), \tag{86}$$

Note here that for simplicity, the upper bound for $\| v_c(k) \|$ *is denoted as* d_1 *for both cases. Now the bound for* $g(x(k))v_c(k)$ *can be obtained as*

$$\| g(x(k))v_c(k) \| \leq C_0 + C_1 \| r(k) \|, \tag{87}$$

where $C_0 = d_1 C_{01}$ *and* $C_1 = d_1 C_{12}$. *Similarly, the bound for* $\hat{g}(x(k))u_c(k)$ *can be deduced as*

$$\begin{aligned} \| \hat{g}(x(k))u_c(k) \| &\leq gs, \quad if \ \| u_c(k) \| \leq s, \\ &\leq \frac{g}{\gamma e}, \quad if \ \| u_c(k) \| > s, \end{aligned} \tag{88}$$

and is denoted as C_2.

Using the individual upper bounds of $g(x(k))v_c(k)$ *and* $\hat{g}(x(k))u_c(k)$, *the upper bound for* $\| \overline{g(x(k))u_d(k)} \|$ *can be obtained as*

$$\| \overline{g(x(k))u_d(k)} \| = \| g(x(k))v_c(k) - \hat{g}u_c(k) \| \leq C_3 + C_4 \| r(k) \|, \tag{89}$$

where $C_3 = C_0 + C_2$ *and* $C_4 = C_1$.

Now using the Lyapunov function (50), the first difference (51) can be obtained as

$$\begin{aligned} \Delta J &= -r(k)^T(I - k_v^T k_v)r(k) \\ &+ 2(k_v r(k) + \overline{g(x(k))u_d(k)} + \epsilon_f(k) + d(k))^T \\ &(\overline{g(x(k))u_d(k)} + \epsilon_f(k) + d(k)) \\ &+ \frac{1}{(1 - \alpha\varphi_f^T(k)\varphi_f(k))}(k_v r(k) + \overline{g(x(k))u_d(k)} + \epsilon_f(k) + d(k))^T \\ &(\overline{g(x(k))u_d(k)} + \epsilon_f(k) + d(k)) - (1-\eta) \\ &\| \bar{e}_f(k) - \frac{\eta}{(1-\eta)}(k_v r(k) + \overline{g(x(k))u_d(k)} + \epsilon_f(k) + d(k)) \|^2, \end{aligned} \tag{90}$$

where η *is given in (49). Subsituting for* $\overline{g(x(k))u_d(k)}$ *from (89) in (90) and rearranging the terms in (90) we obtain*

$$\begin{aligned} \Delta J &= -(1 - b_0) \| r(k) \|^2 + 2b_1 \| r(k) \| + b_2 - (1-\eta) \\ &\| \bar{e}_f(k) - \frac{\eta}{(1-\eta)}(k_v r(k) + \overline{g(x(k))u_d(k)} + \epsilon_f(k) + d(k)) \|^2 \end{aligned} \tag{91}$$

where

$$b_0 = k_{vmax}^2 + 2C_4(C_4 + k_{vmax}) + \frac{(C_4 + k_{vmax})^2}{(1 - \alpha \| \varphi_f(k) \|)^2}, \tag{92}$$

$$\begin{aligned} b_1 &= C_3(C_4 + k_{vmax}) + C_3C_4 + (C_4 + k_{vmax})(\epsilon_{Nf} + d_M) + \\ &\frac{C_3(C_4 + k_{vmax})}{(1 - \alpha \| \varphi_f(k) \|^2)} + \frac{(C_4 + k_{vmax})(\epsilon_{Nf} + d_M)}{(1 - \alpha \| \varphi_f(k) \|^2)}, \end{aligned} \tag{93}$$

$$\begin{aligned} b_2 &= 2C_3^2 + 2C_3(\epsilon_{Nf} + d_M) + (\epsilon_{Nf} + d_M)^2 \\ &\frac{C_3^2 + 2C_3(\epsilon_{Nf} + d_M) + (\epsilon_{Nf} + d_M)^2}{(1 - \alpha \| \varphi_f(k) \|^2)}, \end{aligned} \tag{94}$$

and

$$\| \epsilon_f(k) \| \leq \epsilon_{Nf}. \tag{95}$$

The second term in (91) is always negative as long as conditions (44) through (47) hold. Since b_0, b_1 and b_2 are positive constants, $\Delta J \leq 0$ as long as

$$\| r(k) \| > \delta_{r2}, \tag{96}$$

where

$$\delta_{r2} = \frac{1}{(1-b_0)}[b_1 + \sqrt{b_1^2 + b_2(1-b_0)}]. \tag{97}$$

Then one has $|\sum_{k=k_0}^{\infty} \Delta J(k)| = | J(\infty) - J(0) | < \infty$ since $\Delta J \leq 0$ as long as (44) through (47) hold. The definition of J and inequality (96) imply that every initial condition in the set χ will evolve entirely within χ. In other words, whenever the tracking error $\| r(k) \|$ is outside the region defined by (96), $J(r(k), \tilde{U}_f(k), \tilde{U}_g(k))$ will decrease. This further implies that the tracking error $r(k)$ is UUB for all $k \geq 0$ and it remains to be shown that the weight estimation errors, $\tilde{U}_f(k)$ or equivalently $\hat{U}_f(k)$ are bounded.

To show the boundedness of the weight estimation errors, one uses the error in weight updates (58) for $f(\cdot)$, the tracking error bound (96), the PE condition and Lemma 5.1. Since the weight estimates for $\hat{g}(x(k))$ are not updated in this region, the bounds for the weight estimates need not to be shown. However, to show the boundedness of the weight estimates for $\hat{f}(x(k))$, the dynamics relative to the error in weight estimates using (58) are given by

$$\begin{aligned} \tilde{U}_f(k+1) &= (I - \alpha\varphi_f(k)\varphi_f^T(k))\tilde{U}_f(k) - \alpha\varphi_f(k) \\ & \quad (k_v r(k) + C_3 + C_4 \| r(k) \| + \epsilon_f(k) + d(k))^T, \end{aligned} \tag{98}$$

where the tracking error $r(k)$ is shown to be bounded. Applying the PE condition and Lemma 5.1, the boundedness of the weight estimation errors $\tilde{U}_f(k)$ or equivalently $\hat{U}_f(k)$ are guaranteed. Let us denote the bound by δ_{f2}. This concludes the boundedness of both tracking error and weight estimates for both fuzzy systems in this region.

Reprise:

Combining the results from region I and II, one can readily set $\delta_r = \max(\delta_{r1}, \delta_{r2}), \delta_f = \max(\delta_{f1}, \delta_{f2})$, and δ_g. Thus, for both regions, if $\| r(k) \| > \delta_r$, then $\Delta J \leq 0$ and $v_c(k)$ is bounded. Let us express $(\| r(k) \|, \| \tilde{U}_f(k) \|, \| \tilde{U}_g(k) \|)$ by the new variables (ξ_1, ξ_2, ξ_3). Define the region

$$\Xi : \xi | \xi_1 < \delta_r, \xi_2 < \delta_f, \xi_3 < \delta_g.$$

Then there exists an open set

$$\Omega : \xi | \xi_1 < \overline{\delta}_r, \xi_2 < \overline{\delta}_f, \xi_3 < \overline{\delta}_g,$$

where $\overline{\delta}_i > \delta_i$ *implies that* $\Xi \subset \Omega$. *In other words, we have proved that whenever* $\xi_i > \delta_i$, *then* $J(\xi)$ *will not increase and will remain in the region* Ω *which is an invariant set. Therefore all the signals in the closed-loop system remain bounded. This concludes the proof.* ■

In applications, the right-hand sides of (76) or (96), (58) or (98) and (59) may be taken as *practical bounds* on the norms of the error $r(k)$, the weight estimation errors $\tilde{U}_f(k)$ and $\tilde{U}_g(k)$. Since the target weight values are bounded, it follows that the weights, $\hat{U}_f(k)$ and $\hat{U}_g(k)$ provided by the tuning algorithms are bounded, hence the control input is bounded.

Observe from (76) or (96) that the tracking error increases with the reconstruction error bound ϵ_N and the disturbance bound d_M. Yet small tracking errors (but not arbitrary small) may be achieved by selecting small gains k_v. In other words, placing the closed-loop poles near the origin inside the unit circle forces smaller tracking errors. Again, selecting $k_{vmax} = 0$ results in a deadbeat controller, but it should be avoided as it is not robust.

It is important to note that the problem of initializing the weights does not arise, since when $\hat{U}_f(0)$ are taken as zero the term $k_v r(k)$ stabilizes the plant on an interim basis for a restricted class of nonlinear systems such as robotic systems.

5.2 Projection algorithm

The adaptation gains $\alpha > 0$ and $\beta > 0$, are constant parameters in the update laws presented in (58) and (59). These update laws have a major drawback. In fact, using (44), the upper bound on the adaptation gain for g(x(k)) can be obtained as

$$\beta < \frac{1}{\| \varphi_g(k) \|^2}, \tag{99}$$

Since $\varphi_g(k) \in \Re^{N_2}$ (N_2 being the number of MFs), it is evident that the upper bound on the adaptation gain β depends upon the number of MFs. Specifically, if there are N_2 MFs and the maximum value of the each MF is taken as unity (as for the sigmoid-shaped MF), then the bounds on the adaptation gain which assures the stability of the closed-loop system are given by

$$0 < \beta < \frac{1}{N_2}. \tag{100}$$

In other words, the upper bound on the adaptation gain decreases with an increase in the number of MFs, so that learning must slow down for guaranteed performance. This major drawback can be easily overcome by modifying the update rule to obtain a projection algorithm [Goodwin and Sin, 84]. To wit, replace the constant adaptation gain at each layer by

$$\beta = \frac{\xi}{\zeta + \| \varphi_g(k) \|^2}, \tag{101}$$

where

$$\zeta > 0, \tag{102}$$

and

$$0 < \xi < 1, \tag{103}$$

are constants. Note that ξ is now the new adaptation gain and it is always true that

$$\frac{\xi}{\zeta + \| \varphi_g(k) \|^2} \| \varphi_g(k) \|^2 \quad < \quad 1, \tag{104}$$

hence guaranteeing (44) for every N_2. Similarly, using (44) and (46), it can be shown that the adaptation gain α should satisfy $\alpha < \frac{1}{\|\varphi_f(k)\|^2}$.

For guaranteed closed-loop stability, it is necessary that the MF outputs $\varphi_f(k)$ and $\varphi_g(k)$ be PE. However, it is difficult to verify the PE of the MFs $\varphi_f(k)$ and $\varphi_g(k)$. Further, if the MFs are tuned to improve approximation accuracy, guaranteeing PE becomes even more difficult. In the next section, improved weight tuning paradigms are presented so that PE is relaxed.

5.3 Weight tuning modification relaxing the PE condition

Methods like σ-modification [Polycarpou and Ioannou, 91] or ϵ-modification [Narendra and Annaswamy, 87], are available in robust adaptive control of continuous systems wherein the persistency of excitation condition is not needed. On the other hand, modification to the standard weight tuning mechanisms in discrete-time to avoid the necessity of PE is also investigated in [Jagannathan and Lewis, 96a] and [Jagannathan, 96b].

In [Jagannathan, 96b] an approach similar to ϵ-modification was derived for discrete-time fuzzy systems for feedback linearization. The following theorem from that paper shows tuning algorithms which do not require persistence of excitation. The controller derived therein is given in Table 2.

Table 2: Discrete-Time Controller Using FL system: PE not Required

$$
\begin{aligned}
v_c(k) &= u_c(k) + \frac{u_r(k) - u_c(k)}{2} e^{\gamma(\|u_c(k)\| - s)}, \quad I = 1, \\
&= u_r(k) - \frac{u_r(k) - u_c(k)}{2} e^{-\gamma(\|u_c(k)\| - s)}, \quad I = 0.
\end{aligned}
$$

The weight tuning for f(x(k)) is given by

$$
\hat{U}_f(k+1) = \hat{U}_f(k) + \alpha\varphi_f(k) r^T(k+1) - \delta \parallel I - \alpha\varphi_f(k)\varphi_f^T(k) \parallel \hat{U}_f(k), \tag{105}
$$

and the weight tuning for g(x(k)) is provided by

$$
\begin{aligned}
\hat{U}_g(k+1) &= \hat{U}_g(k) + \beta\varphi_g(k) r^T(k+1) - \rho \parallel I - \beta\varphi_g(k)\varphi_g^T(k) \parallel, \quad I = 1, \\
&= \hat{U}_g(k), \quad I = 0,)
\end{aligned}
\tag{106}
$$

with $\alpha = \frac{\xi_f}{\zeta_f + \|\varphi_f(k)\|^2}$ and $\beta = \frac{\xi_g}{\zeta_g + \|\varphi_g(k)\|^2}$ where $\alpha > 0$, $\beta > 0$, $\delta > 0$, $\rho > 0$, $\zeta_f > 0, \zeta_g > 0$ and $0 < \xi_f < 1, 0 < \xi_g < 1$ denote learning rate parameters or adaptation gains.

Theorem 5.2 (FC feedback linearization without PE) *Assume the hypotheses in Theorem 5.1 and consider the modified weight tuning algorithms provided for f(x(k)) by (105) and for g(x(k)) by (106). Then the filtered tracking error r(k) and the weight estimates $\hat{U}_f(k)$ and $U_g(k)$ are UUB, with the bounds given by equations (128) or (149), (132) or (153) and (136) provided the following conditions hold:*

$$
\begin{aligned}
&(1) \quad \beta \parallel \bar{\varphi}_g(k) u_c(k) \parallel = \beta \parallel \varphi_g(k) \parallel^2 < 1, && (107) \\
&(2) \quad \alpha \parallel \varphi_f(k) \parallel^2 < 1, && (108) \\
&(3) \quad \eta + max(P_1, P_3, P_4) < 1 && (109) \\
&(4) \quad 0 < \delta < 1, && (110) \\
&(5) \quad 0 < \rho < 1, && (111) \\
&(6) \quad max(a_2, b_0) < 1, && (112)
\end{aligned}
$$

with $P_1, P_3,$ and P_4 being constants that depend upon η, δ and ρ where

$$
\begin{aligned}
\eta &= \alpha \parallel \varphi_f(k) \parallel^2 + \beta \parallel \bar{\varphi}_g(k) u_c(k) \parallel^2 \\
&= \alpha \parallel \varphi_f(k) \parallel^2 + \beta \parallel \varphi_g(k) \parallel^2, \quad I = 1, \\
&= \alpha \parallel \varphi_f(k) \parallel^2, \quad I = 0,
\end{aligned}
\tag{113}
$$

and a_2, b_0 are design parameters chosen using the gain matrix k_v. Observe *here that he parameters α, β and η depend upon the trajectory.*

Proof: Let the initial condiitons are bounded by q_B. Assume that the approximation property (7) holds for the functions $f(x(k))$ and $g(x(k))$ with a given accuracy of ϵ_{Nf} and ϵ_{Ng} in the compact set $S_x \equiv (x(k) \mid \| x(k) \| < b_x)$ with $b_x > q_B$. Define $S_r \equiv (r(k) \mid \| r(k) \| < \frac{b_x - q_B}{l_1 + d_{11}})$. Let $r(0) \in S_r$. Then the approximation property holds.

Region I: $\| \hat{g}(x(k)) \| \geq \underline{g}$ *and* $\| u_c(k) \| \leq s$.

Select the Lyapunov function candidate (50) whose first difference is given by (51). The error in dynamics for the weight update laws are given for this region as

$$\begin{aligned} \tilde{U}_f(k+1) &= (I - \alpha\varphi_f(k)\varphi_f^T(k))\tilde{U}_f(k) \\ &\quad -\alpha\varphi_f(k)(k_v r(k) + \bar{e}_g(k) + g(x(k))u_d(k) + \epsilon(k) + d(k))^T \\ &\quad +\delta \| I - \alpha\varphi_f(k)\varphi_f^T(k) \| \hat{U}_f(k), \end{aligned} \tag{114}$$

and

$$\begin{aligned} \tilde{U}_g(k+1) &= (I - \beta\varphi_g(k)\varphi_g^T(k))\tilde{U}_g(k) \\ &\quad -\beta\varphi_g(k)(k_v r(k) + \bar{e}_f(k) + g(x(k))u_d(k) + \epsilon(k) + d(k))^T \\ &\quad +\rho \| I - \beta\varphi_g(k)\varphi_g^T(k) \| \hat{U}_g(k). \end{aligned} \tag{115}$$

Substituting (114), (115), in (51) and after some rewriting we obtain

$$\begin{aligned} \Delta J &= -r(k)^T[I - (2+\eta)k_v^T k_v]r(k) + 2(2+\eta)(k_v r(k))^T \\ &\quad (g(x(k))u_d(k) + \epsilon(k) + d(k)) \\ &\quad + (2+\eta)(g(x(k))u_d(k) + \epsilon(k) + d(k))^T \\ &\quad (g(x(k))u_d(k) + \epsilon(k) + d(k)) \\ &\quad + 2P_2 \| k_v r(k) + g(x(k))u_d(k) + \epsilon(k) + d(k) \| \\ &\quad + 2\eta(g(x(k))u_d(k))^T(\epsilon(k) + d(k)) + 2\eta\epsilon(k)d(k) \\ &\quad - (1 - \eta - P_3) \| \bar{e}_f(k) \|^2 - (1 - \eta - P_4) \| \bar{e}_g(k) \|^2 \\ &\quad - 2(1 - \eta - P_1) \| \bar{e}_f(k) \| \| \bar{e}_g(k) \| \\ &\quad - \frac{1}{\eta} \| I - \alpha\varphi_f(k)\varphi_f^T(k) \|^2 [\delta(2-\delta) \| \tilde{U}_f(k) \|^2 \\ &\quad - 2\delta(1-\delta) \| \tilde{U}_f(k) \| U_{fmax} - \delta^2 U_{fmax}^2] \\ &\quad - \frac{1}{\beta} \| I - \beta\varphi_g(k)\varphi_g^T(k) \|^2 [\rho(2-\rho) \| \tilde{U}_g(k) \|^2 \end{aligned}$$

$$- 2\rho(1-\rho) \parallel \tilde{U}_g(k) \parallel U_{gmax} - \rho^2 U_{gmax}^2], \tag{116}$$

where η is given in (109) and

$$P_1 = 2(\delta \parallel I - \alpha\varphi_f(k)\varphi_f(k)^T \parallel + \rho \parallel I - \beta\varphi_g(k)\varphi_g(k)^T \parallel), \tag{117}$$

$$\begin{aligned} P_2 \;&=\; 2(\delta \parallel I - \alpha\varphi_f(k)\varphi_f(k)^T \parallel U_{fmax}\varphi_{fmax} + \rho \parallel I \\ &\quad -\beta\varphi_g(k)\varphi_g(k)^T \parallel U_{gmax}\varphi_{gmax}), \end{aligned} \tag{118}$$

$$P_3 = (\eta + \delta \parallel I - \alpha\varphi_f(k)\varphi_f(k)^T \parallel)^2, \tag{119}$$

and

$$P_4 = (\eta + \rho \parallel I - \beta\varphi_g(k)\varphi_g(k)^T \parallel)^2. \tag{120}$$

Now in this region, the bound on $u_d(k)$ can be obtained as

$$\begin{aligned} \parallel u_d(k) \parallel \;&\le\; \parallel v_c(k) - u_c(k) \parallel \\ &\le\; \parallel \frac{u_r(k) - u_c(k)}{2} e^{\gamma(\parallel u_c(k) \parallel - s)} \parallel . \end{aligned} \tag{121}$$

In this region, since $\parallel u_c(k) \parallel \le s$ and the auxiliary input $u_r(k)$ is given by (39), the bound in (121) is taken as a constant since the right side terms are bounded. This bound is denoted as

$$\parallel u_d(k) \parallel \le C_2. \tag{122}$$

Then the bound for $g(x(k))u_d(k)$ is written as (69). Using the bound for $g(x(k))u_d(k)$, substituting it in (116),and completing the squares for $\parallel \tilde{U}_f(k) \parallel$ and $\parallel \tilde{U}_g(k) \parallel$ we obtain

$$\begin{aligned} \Delta J \;\le\;& -(1-a_2) \parallel r(k) \parallel^2 + 2a_3 \parallel r(k) \parallel + a_4 \\ & -(1-\eta-P_3) \parallel \bar{e}_f(k) \parallel^2 - (1-\eta-P_4) \parallel \bar{e}_g(k) \parallel^2 \\ & -2(1-\eta-P_1) \parallel \bar{e}_f(k) \parallel\parallel \bar{e}_g(k) \parallel \\ & - \parallel (\sqrt{P_3}\bar{e}_f(k) + \sqrt{P_4}\bar{e}_g(k)) \\ & -(k_v r(k) + g(x(k))u_d(k) + \epsilon(k) + d(k)) \parallel^2 \\ & -\frac{1}{\alpha} \parallel I - \alpha\varphi_f(k)\varphi_f^T(k) \parallel^2 \delta(2-\delta) \\ & [\parallel \tilde{U}_f(k) \parallel - \frac{(1-\delta)}{(2-\delta)} U_{fmax}^2]^2 \\ & -\frac{1}{\beta} \parallel I - \beta\varphi_g(k)\varphi_g^T(k) \parallel^2 \rho(2-\rho) \\ & [\parallel \tilde{U}_g(k) \parallel - \frac{(1-\rho)}{(2-\rho)} U_{gmax}]^2, \end{aligned} \tag{123}$$

where

$$a_2 = (2+\eta)k_{vmax}^2 + 2(1+\eta)C_1 k_{vmax} + (2+\eta)C_1^2 + 2k_{vmax}C_1, \tag{124}$$

$$\begin{aligned} a_3 &= (1+\eta)k_{vmax}(\epsilon_N + d_M + C_0) \\ &+ P_2 k_{vmax} + P_2 C_1 + \eta C_1(\epsilon_N + d_M) \\ &+ \frac{1}{2}(2+\eta)C_1(\epsilon_N + d_M + C_0) + 2k_{vmax}(\epsilon_N + d_M + C_0), \end{aligned} \tag{125}$$

$$\begin{aligned} a_{44} &= 2P_2(\epsilon_N + d_M + C_0) + 2\eta C_0(\epsilon_N + d_M) \\ &+ (2+\eta)(\epsilon_N + d_M + C_0)^2 + 2\eta\epsilon_N d_M, \end{aligned} \tag{126}$$

and

$$\begin{aligned} a_4 &= a_{44} + \frac{1}{\alpha} \parallel I - \alpha\varphi_f(k)\varphi_f^T(k) \parallel^2 \frac{\delta^2}{(2-\delta)} U_{fmax}^2 \\ &+ \frac{1}{\beta} \parallel I - \beta\varphi_g(k)\varphi_g^T(k) \parallel^2 \frac{\rho^2}{(2-\rho)} U_{gmax}^2. \end{aligned} \tag{127}$$

All the terms in (123) are always negative except for the first term as long as the conditions (107) through (112) hold. Since a_2, a_3 and a_4 are positive constants, $\Delta J \leq 0$ as long as (107) through (112) hold with

$$\parallel r(k) \parallel > \delta_{r1}, \tag{128}$$

where

$$\delta_{r1} = \frac{1}{(1-a_2)}[a_3 + \sqrt{a_3^2 + a_4(1-a_2)}]. \tag{129}$$

Similarly, completing the squares for $\parallel r(k) \parallel, \parallel \tilde{U}_g(k) \parallel$ and using (116) yields

$$\begin{aligned} \Delta J &= -(1-a_2)[\parallel r(k) \parallel - \frac{a_3}{(1-a_2)}]^2 \\ &- (1-\eta-P_3) \parallel \bar{e}_f(k) \parallel^2 - (1-\eta-P_4) \parallel \bar{e}_g(k) \parallel^2 \\ &- 2(1-\eta-P_1) \parallel \bar{e}_f(k) \parallel \parallel \bar{e}_g(k) \parallel \\ &- \parallel (\sqrt{P_3}\bar{e}_f(k) + \sqrt{P_4}\bar{e}_g(k)) \\ &- (k_v r(k) + g(x(k))u_d(k) + \epsilon(k) + d(k)) \parallel^2 \\ &- \frac{1}{\alpha} \parallel I - \alpha\varphi_f(k)\varphi_f^T(k) \parallel^2 \delta(2-\delta) \\ &[\parallel \tilde{U}_f(k) \parallel - \frac{(1-\delta)}{(2-\delta)} U_{fmax} - a_4]^2 \end{aligned}$$

$$
\begin{aligned}
& -\frac{1}{\beta} \parallel I - \beta\varphi_g(k)\varphi_g^T(k) \parallel \rho(2-\rho) \\
& [\parallel \tilde{U}_g(k) \parallel - \frac{(1-\rho)}{(2-\rho)} U_{gmax}]^2, \qquad (130)
\end{aligned}
$$

where

$$
\begin{aligned}
a_4 &= \delta^2 U_{fmax}^2 + \frac{\alpha}{\parallel I - \alpha\varphi_f(k)\varphi_f^T(k) \parallel^2}[a_{44} + \frac{a_3^2}{(1-a_2)} \\
&+ \frac{1}{\beta} \parallel I - \beta\varphi_g(k)\varphi_g^T(k) \parallel^2 \frac{\rho^2}{(2-\rho)} U_{gmax}^2]. \qquad (131)
\end{aligned}
$$

Then $\Delta J \leq 0$ as long as (107) through (112) hold and the quadratic term for $\tilde{U}_f(k)$ in (130) is positive, which is guaranteed when

$$\parallel \tilde{U}_f(k) \parallel > \delta_{f1}, \qquad (132)$$

where

$$\delta_{f1} = \frac{1}{(2-\delta)}[(1-\delta) + \sqrt{(1-\delta)^2 + a_4(2-\delta)}]. \qquad (133)$$

Similarly, completing the squares for $\parallel r(k) \parallel, \parallel \tilde{U}_f(k) \parallel$ and using (116) yields

$$
\begin{aligned}
\Delta J &= -(1-a_2)[\parallel r(k) \parallel - \frac{a_3}{(1-a_2)}]^2 \\
& -(1-\eta-P_3) \parallel \overline{e}_f(k) \parallel^2 -(1-\eta-P_4) \parallel \overline{e}_g(k) \parallel^2 \\
& -2(1-\eta-P_1) \parallel \overline{e}_f(k) \parallel\parallel \overline{e}_g(k) \parallel \\
& - \parallel (\sqrt{P_3}\overline{e}_f(k) + \sqrt{P_4}\overline{e}_g(k)) \\
& -(k_v r(k) + g(x(k))u_d(k) + \epsilon(k) + d(k)) \parallel^2 \\
& -\frac{1}{\alpha} \parallel I - \alpha\varphi_f(k)\varphi_f^T(k) \parallel^2 \delta(2-\delta) \\
& [\parallel \tilde{U}_f(k) \parallel - \frac{(1-\delta)}{(2-\delta)} U_{fmax}]^2 \\
& -\frac{1}{\beta} \parallel I - \beta\varphi_g(k)\varphi_g^T(k) \parallel^2 \rho(2-\rho) \\
& [\parallel \tilde{U}_g(k) \parallel - \frac{(1-\rho)}{(2-\rho)} U_{gmax} - a_4], \qquad (134)
\end{aligned}
$$

where

$$
a_4 = \rho^2 U_{gmax}^2 + \frac{\beta}{\parallel I - \beta\varphi_g(k)\varphi_g^T(k) \parallel^2}[a_{44} + \frac{a_3^2}{(1-a_2)}
$$

$$+\frac{1}{\alpha} \| I - \alpha\varphi_f(k)\varphi_f^T(k) \|^2$$
$$\frac{\delta^2}{(2-\delta)} U_{fmax}^2]. \tag{135}$$

Then $\Delta J \leq 0$ *as long as (107) through (112) hold and the quadratic term for* $\tilde{U}_g(k)$ *in (134) is positive, which is guaranteed when*

$$\| \tilde{U}_g(k) \| > \delta_g, \tag{136}$$

where

$$\delta_g = \frac{1}{(2-\rho)}[(1-\rho) + \sqrt{(1-\rho)^2 + a_4(2-\rho)}]. \tag{137}$$

This shows the upper bounds for the tracking and weight estimation errors for this region for all $\| u_c(k) \| \leq s$.

Region II: $\| \hat{g}(x(k)) \| < \underline{g}$ *and* $\| u_c(k) \| > s$.

Select the Lyapunov function candidate (50) whose first difference is given by (51). The tracking error system in (36) can be rewritten as

$$r(k+1) = k_v r(k) + \bar{e}_f^T(k) + \overline{g(x(k))u_d(k)} + \epsilon_f(k) + d(k), \tag{138}$$

where

$$\overline{g(x(k))u_d(k)} = g(x(k))v_c(k) - \hat{g}(x(k))u_c(k). \tag{139}$$

For the case of modified weight tuning (114) - (115) in this region, let us denote the bound given in (69) as d_1. *The bound for* $\hat{g}u_c(k)$ *can be obtained as*

$$\begin{aligned} \| \hat{g}(x(k))u_c(k) \| &\leq \underline{g}s, \quad \| u_c(k) \| \leq s, \\ &\leq \frac{\underline{g}}{\gamma e}, \quad \| u_c(k) \| > s, \end{aligned} \tag{140}$$

whose upper bound in either case is denoted by C_2. *Using the individual upper bounds, the upper bound for* $\overline{g(x(k))u_d(k)}$ *can be obtained as (89).*

Consider the first difference of the Lyapunov function and substitute the bounds for $\overline{g(x(k))u_d(k)}$. *Then complete the squares and rearrange terms to obtain*

$$\Delta J = -(1-b_0) \| r(k) \|^2 + 2b_1 \| r(k) \| + b_2$$

$$-(1-\eta)\parallel \overline{e}_f(k)-\frac{(\eta+\delta\parallel I-\alpha\varphi_f(k)\varphi_f^T(k)\parallel)}{(1-\eta)}$$
$$(k_v r(k)+\overline{g(x(k))u_d(k)}+\epsilon_f(k)+d(k))\parallel^2$$
$$-\frac{1}{\alpha}\parallel I-\alpha\varphi_f(k)\varphi_f^T(k)\parallel^2[\delta(2-\delta)\parallel \tilde{U}_f(k)\parallel^2$$
$$-2\delta(1-\delta)\parallel \tilde{U}_f(k)\parallel U_{fmax}-\delta^2U_{fmax}^2], \quad (141)$$

where

$$b_0=a_0k_{vmax}^2+2a_0C_4K_{vmax}+a_0C_4^2, \quad (142)$$

$$\begin{aligned} b_1 &= a_0k_{vmax}(C_3+\epsilon_{Nf}+d_M) \\ &\quad +a_0(C_3+\epsilon_{Nf}+d_M)C_4(C_3+\epsilon_{Nf}+d_M) \end{aligned} \quad (143)$$

$$\begin{aligned} b_2 &= a_0(C_3+\epsilon_{Nf}+d_M)^2 \\ &\quad +2\delta\parallel I-\alpha\varphi_f(k)\varphi_f^T(k)\parallel(C_3+\epsilon_{Nf}+d_M)U_{fmax}\varphi_{fmax}, \end{aligned} \quad (144)$$

$$a_0=1+\alpha\varphi_f^T(k)\varphi_f(k)+\frac{(\alpha\varphi_f(k)\varphi_f^T(k)+2\delta\parallel I-\alpha\varphi_f(k)\varphi_f^T(k)\parallel)^2}{(1-\alpha\varphi_f^T(k)\varphi_f(k))}, \quad (145)$$

and

$$\parallel \epsilon_f(k)\parallel\leq\epsilon_{Nf}. \quad (146)$$

The second term in (141) is always negative as long as the conditions (107) through (112) hold. Completing the squares for $\parallel \tilde{W}_f(k)\parallel$ results in

$$\begin{aligned} \Delta J &= -(1-b_0)\parallel r(k)\parallel^2+2b_1\parallel r(k)\parallel+b_3 \\ &\quad -(1-\eta)\parallel \overline{e}_f(k)-\frac{(\eta+\delta\parallel I-\alpha\varphi_f(k)\varphi_f^T(k)\parallel)}{(1-\eta)} \\ &\quad (k_v r(k)+\overline{g(x(k))u_d(k)}+\epsilon_f(k)+d(k))\parallel^2 \\ &\quad -\frac{1}{\alpha}\parallel I-\alpha\varphi_f(k)\varphi_f^T(k)\parallel^2\delta(2-\delta) \\ &\quad [\parallel \tilde{U}_f(k)\parallel-\frac{(1-\delta)}{(2-\delta)}U_{fmax}]^2, \end{aligned} \quad (147)$$

with

$$b_3=b_2+\frac{1}{\alpha}\parallel I-\alpha\varphi_f(k)\varphi_f^T(k)\parallel^2\frac{\delta}{(2-\delta)}U_{fmax}^2. \quad (148)$$

Since b_0, b_1 and b_3 are positive constants in (147) and the second and third terms are always negative, $\Delta J \leq 0$ as long as

$$\| r(k) \| > \delta_{r2}, \tag{149}$$

where

$$\delta_{r2} = \frac{1}{(1-b_0)}[b_1 + \sqrt{b_1^2 + b_3(1-b_0)}]. \tag{150}$$

Similarly, completing the squares for $\| r(k) \|$ using (141) yields

$$\begin{aligned} \Delta J &= -(1-b_0)[\| r(k) \| - \frac{b_1}{(1-b_0)}]^2 \\ &\quad -(1-\eta) \| \bar{e}_f(k) - \frac{(\eta + \delta \| I - \alpha\varphi_f(k)\varphi_f^T(k) \|)}{(1-\eta)} \\ &\quad (k_v r(k) + \overline{g(x(k))u_d(k)} + \epsilon_f(k) + d(k)) \|^2 \\ &\quad -\frac{1}{\alpha} \| I - \alpha\varphi_f(k)\varphi_f^T(k) \|^2 \delta(2-\delta)[\| \tilde{U}_f(k) \| \\ &\quad -2\frac{(1-\delta)}{(2-\delta)} \| \tilde{U}_f(k) \| U_{fmax} - b_4], \end{aligned} \tag{151}$$

with

$$b_4 = \frac{1}{(2-\delta)}[\frac{\alpha}{\| I - \alpha\varphi_f(k)\varphi_f^T(k) \|^2} \frac{b_1^2}{(1-b_0)} + \delta^2 U_{fmax}]. \tag{152}$$

Since b_0, b_1 and b_4 are positive constants in (151) and the second and third terms are always negative, $\Delta J \leq 0$ as long as

$$\| \tilde{U}_f(k) \| > \delta_{f2}, \tag{153}$$

where

$$\delta_{f2} = \frac{1}{(2-\delta)}[(1-\delta) + \sqrt{(1-\delta)^2 + b_4(2-\delta)}]. \tag{154}$$

$| \sum_{k=k_0}^{\infty} \Delta J(k) | = | J(\infty) - J(0) | < \infty$ *since $\Delta J \leq 0$ as long as (107) through (112) hold. The definition of J and inequalities (149) and (153) imply that every intial conditon in the set χ will evolve entirely within χ. Therefore, using standard Lyapunov extension, the tracking error $r(k)$ and the error in weight updates are UUB.*

Reprise:

Combining the results from region I and II, one can readily set $\delta_r = max(\delta_{r1}, \delta_{r2}), \delta_f = max(\delta_{f1}, \delta_{f2})$, *and* δ_g.

Thus for both regions, if $\| r(k) \| > \delta_r$, *then* $\Delta J \leq 0$ *and* $v_c(k)$ *is bounded. Let us denote* $(\| r(k) \|, \| \tilde{U}_f(k) \|, \| \tilde{U}_g(k) \|)$ *by the new variables* (ξ_1, ξ_2, ξ_3). *Define the region*

$$\Xi : \xi | \xi_1 < \delta_r, \xi_2 < \delta_f, \xi_3 < \delta_g.$$

Then there exists an open set

$$\Omega : \xi | \xi_1 < \overline{\delta}_r, \xi < \overline{\delta}_f, \xi_3 < \overline{\delta}_g,$$

where $\overline{\delta}_i > \delta_i$ *implies that* $\Xi \subset \Omega$. *In other words, it was shown that whenever* $\xi_i > \delta_i$ *then* $V(\xi)$ *will not increase and will remain in the region* Ω *which is an invariant set. Therefore all the signals in the closed-loop system remain uniformly ultimately bounded. This concludes the proof.* ■

Remarks:

1. For practical purposes, equations (128) or (149), (132) or (153) and (136) can be considered as bounds for $r(k), \tilde{U}_f(k)$ and $\tilde{U}_g(k)$ in both regions.

2. The reconstruction errors and the bounded disturbances are all embodied in the constants given by δ_r, δ_f and δ_g. Note that the bound for the tracking error can be kept small if the closed-loop poles are placed near origin.

3. If the switching parameter s is small, it will limit the control input and this will result in a large tracking error leading into undesirable closed-loop performance. Large value of s saturates the control input $v_c(k)$.

4. Uniform ultimate boundedness of the closed-loop system is shown without persistency of excitation on the input signals. The certainty equivalence principle is not used. The fuzzy system can be easily initialized as $\hat{U}_f(k) = 0$, and $\hat{U}_g(k) > g^{-1}$(g). No assumptions such as the existence of an invariant set, region of attraction, or a feasible region is needed.

Note that the reconstruction error bound ϵ_N and the bounded disturbances d_M increase the bounds for $\| r(k) \|$ and $\| \tilde{U}_f(k) \|$ and $\| U_g(k) \|$ in a very interesting way. Note that small tracking error bounds, but not arbitrarily small, may be achieved by placing the closed-loop poles inside the unit circle and near the origin through the selection of the largest eigenvalue, k_{vmax}. On the other hand, the weight error estimates are fundamentally bounded by U_{fmax}, and U_{gmax} the known bound on ideal weights U_f and

U_g. The parameters δ and ρ offer a design tradeoff between the relative eventual magnitudes of $\| r(k) \|$ and $\| \tilde{U}_f(k) \|$ and $\| \tilde{U}_g(k) \|$; a smaller δ yields a smaller $\| r(k) \|$ and a larger $\| \tilde{U}_f(k) \|$, and vice versa. For the tuning weights $\| \tilde{U}_g(k) \|$, similar effects are observed.

The effect of the adaptation gains α and β on the weight estimation errors, $\tilde{U}_f(k)$ and $\tilde{U}_g(k)$, and the tracking error, $r(k)$, can be easily observed by using the bounds presented in (132) or (153) and (136). Large values of α and β force smaller tracking error, but larger weight estimation errors. In contrast, a small value of α and β force larger tracking and smaller weight estimation errors.

6 Simulation results

Consider a planar two link robot arm whose state equations in continuous-time are written as [Slotine and Li, 89]

$$\begin{aligned} \dot{X}_1 &= X_2, \\ \dot{X}_2 &= F(X_1, X_2) + G_1(X_1, X_2)V_c, \end{aligned} \tag{155}$$

where $X_1 = [q_1, q_2]^T$ are joint angles, $X_2 = [x_3, x_4]^T$ are joint velocities, and $V_c = [v_1, v_2]^T$ is the input vector. The nonlinear functions in (155) are described by $F(X_1, X_2) = [M(X_1)]^{-1}G(X_1, X_2)$, where $M(X_1)$ is equal to

$$\begin{bmatrix} (b_1+b_2)a_1^2 + b_2 a_2^2 + 2b_2 a_1 a_2 cos(q_2) & b_2 a_2^2 + b_2 a_1 a_2 cos(q_2) \\ b_2 a_2^2 + b_2 a_1 a_2 cos(q_2) & b_2 a_2^2 \end{bmatrix}$$

and $G(X_1, X_2)$ is equal to

$$\begin{bmatrix} -b_2 a_1 a_2 (2q_3 q_4 + q_4^2) sin(q_2) + 9.8(b_1+b_2) a_1 cos(q_1) + 9.8 b_2 a_2 cos(q_1+q_2) \\ b_2 a_1 a_2 q_1^2 sin(q_2) + 9.8 b_2 a_2 cos(q_1+q_2) \end{bmatrix}$$

and

$$G_1(X_1, X_2) = M^{-1}(X_1, X_2). \tag{156}$$

The above system when discretized using Euler's approximation can be expressed as

$$\begin{aligned} x_1(k+1) &= x_2(k), \\ x_2(k+1) &= f(x_1(k), x_2(k)) + g(x_1(k), x_2(k))v_c(k), \end{aligned} \tag{157}$$

where

$$f(\cdot) = -x_1(k) + 2x_2(k) + T^2 F(x_1(k), x_2(k)),$$

and

$$g(\cdot) = T^2 G(x_1(k), x_2(k)).$$

The matrix used to transform the system (155) to (156) is given by

$$\begin{bmatrix} 1 & 0 \\ 1 & T \end{bmatrix}. \tag{158}$$

The parameters for the nonlinear system were selected as $a_1 = a_2 = 1, b_1 = b_2 = 1$. Desired sinusoidal, $\sin(\frac{2\pi t}{25})$, and cosine inputs, $\cos(\frac{2\pi t}{25})$, were preselected for the axis 1 and 2 respectively. The gains of the PD controller were chosen as $k_v = \mathrm{diag}(0.1,\ 0.1)$ and the design parameters were set to $s = 10, \Gamma = 0.05$. Sampling interval of $10ms$ was considered. Seventeen traingular MFs were selected for each axis. The initial conditions for X_1 were chosen to be $[0.5, 0.1]^T$, and the weights for $F(\cdot)$ were initialized to zero whereas the weights for $G(\cdot)$ were initialized to the identity matrix. Figure 2 presents the tracking response of the FC using (155) through (156) with $\alpha_3 = 0.1, \alpha_i = 1.0; \forall i = 1,2$ and $\beta_3 = 0.1, \beta_i = 1.0; \forall i = 1,2$. From the figure, it can be seen that the FC performs impressively. ■

7 Conclusions

A novel adaptive FC is proposed that guarantees prescribed performance for the general class of feedback linearizable, but unknown, discrete-time nonlinear systems with mild assumptions. This FC has a multi-loop structure with inner fuzzy logic approximation loops and a outer tracking loop. The structure of the FC and adaptation law are selected such that a Lyapunov-based proof of uniform ultimate boundedness is obtained.

The benefits of this novel adaptive FC can be summarized as follows. First, it requires no certainty equivalence assumption. Second, it is model-free and requires no regression matrix computation because the MFs form a set of universal basis functions. Third, the PE condition is not required to show the boundedness of the fuzzy parameters. Furthermore, this method for controller synthesis is systematic and the proposed adaptive FC performs remarkably well. Finally, this approach unlike most FC methods, is easily applied to high dimensional MIMO nonlinear systems. Future work includes extending the FC to a wider class of nonlinear systems.

References

[1] Åström, K. J. and Wittenmark, B. (1989), *Adaptive control*, Addison-Wesley Company, Reading, Massachusetts.

[2] Buckley, J. J. (1992), "Universal fuzzy controllers," *Automatica* , vol. 28, no.6, pp.1245-1248.

[3] Goodwin, G. C. and Sin, K. S., (1984) *Adaptive Filtering, Prediction, and Control*, Prentice-Hall, Englewood Cliffs, NJ.

[4] Jagannathan, S. and Lewis, F. L. (1996a), "Robust implicit self tuning regulator," *Automatica*, vol.12, no. 12, pp.1629-1644.

[5] Jagannathan, S. (1996b), "Adaptive fuzzy logic control of feedback Linearizable nonlinear systems,", *Proc. of the IEEE Conf. on Robotics and Automation*,vol.1, pp. 258-263.

[6] Jagannathan, S., Lewis, F. L., Vandegrift, M., and Commuri, S. (1996c) "Adaptive fuzzy logic control of feedback linearizable nonlinear systems," *Proceedings of the ISAI/IFIS*, Cancun, Mexico, pp.385-392.

[7] Kanellakopoulos, I., Kokotovic, P.V., and Morse, A. S. (1991) "Systematic design of adaptive controllers for feedback linearizable systems", *IEEE Trans. on Automatic Control*, vol.36, pp.1241-1253.

[8] Kosko, B. (1994), " Fuzzy systems as universal approximators ," *IEEE Trans. on Computers* , vol. 43, no. 10.

[9] Kosko, B. (1992), Neural Networks and Fuzzy Systems, New Jersey: Prentice-Hall.

[10] Langari, R. and Tomizuka, M. (1991), "Analysis and stability of a class of fuzzy linguistic controllers with internal dynamics," *Proceedings ASME Winter Annual Meeting*, paper 91-WA-DSC-13.

[11] Lewis, F. L. and Liu, K. (1995), "Towards a paradigm for fuzzy logic control," *Automatica.*

[12] Narendra, K. S. and Annaswamy, A. M. (1987), "A new adaptive law for robust adaptation without persistent excitation," *IEEE Trans. on Automatic Control*, vol.AC-32, no.2, pp. 134-145.

[13] Polycarpou, M. M. and Ioannou, P. A. (1991), "Identification and control using neural network models: design and stability analysis," *Dept. of Elec. Engg.*, Tech Report.91-09-01.

[14] Sadegh, N. (1993), "A perceptron network for functional identification and control of nonlinear systems," *IEEE Trans. on Neural Networks*, vol.4, no.6, pp. 982-988.

[15] Slotine, J-J.E. and Li, W. (1991), *Applied Nonlinear Control*, Englewood Cliffs, NJ, Prentice-Hall.

[16] Vandegrift, M., Lewis, F. L., Jagannathan, S., and Liu, K. (1995), "Fuzzy logic control of a class of discrete-time nonlinear systems ", *Proc. of the IEEE Symposium on Intelligent Control*, pp.395-401.

[17] Wang, L-X. and Mendel, M. (1992), "Fuzzy basis functions, universal approximators, and orthogonal least-squares learning", *IEEE Trans. on Neural Networks*, pp. 807-814, vol. 3, no. 5.

[18] Wang, L-X. (1994), *Adaptive Fuzzy Systems and Control -Design and Stability Analysis*, New Jersey:Prentice-Hall.

[19] Ying, H. (1993), "General analytical structure of typical fuzzy controllers and their limiting structure theorems ", *Automatica*, vol. 29, no. 4 , pp. 1139-1143.

[20] Ying, H. (1994), "Sufficient conditions on general fuzzy systems as function ", *Automatica*, vol. 30, no. 3, pp. 521-525.

[21] Zeng, X-J. and Singh, M. G. (1995) "Approximation theory of fuzzy systems-MIMO case ", *IEEE Trans. on Fuzzy Systems*, vol. 3, no. 2, pp. 219-235.

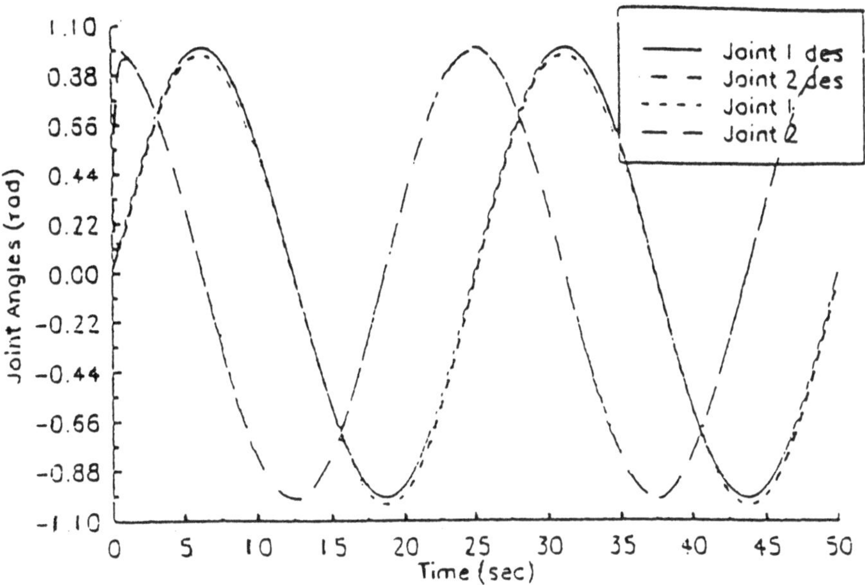

Figure 2: Response of FL controller.

Fuzzy Model Reference Learning Control

J.R. Layne
WL/AACF
2185 Avionics Circle
Wright Patterson AFB, OH 45433-7301, USA

K. M. Passino
Dept. of Electrical Engineering
The Ohio State University
2015 Neil Avenue, Columbus, OH 43102, USA

1 Introduction

Over recent years, fuzzy control has emerged as a practical alternative to classical control schemes when one is interested in controlling certain time-varying, non-linear, and ill-defined processes. There have in fact been several successful commercial and industrial applications of fuzzy control [1] - [5]. Despite this success, there exist several significant drawbacks of this approach:

1. The design of fuzzy controllers is usually performed in an *ad hoc* manner; hence, it is often not clear exactly how to justify the choices for many parameters in the fuzzy controller (e.g., the membership functions, defuzzification strategy, and fuzzy inference strategy).
2. The fuzzy controller constructed for the nominal plant may later perform inadequately if significant and unpredictable plant parameter variations, structural changes, or environmental disturbances occur.

In this article a "learning" control algorithm is presented which helps to resolve some of these fuzzy controller design issues. This algorithm employs a reference model (a model of how you would like the plant to behave) to provide closed-loop performance feedback for synthesizing and tuning a

[†]J. Layne gratefully acknowledges the support of the U.S. Air Force Palace Knight Program.

fuzzy controller's knowledge-base. Consequently, this algorithm is referred to as a "fuzzy model reference learning controller" (FMRLC). The FMRLC grew from research on how to improve Procyk and Mamdani's linguistic self-organizing controller (SOC) [6, 7, 8, 9, 10, 11, 12, 13] by utilizing certain general ideas in conventional adaptive control [14, 15]. The first advantage that the FMRLC has over the SOC is that it does not rely on the specification of an explicit inverse model of the process (which can be difficult/impossible to determine for many applications). In addition, the performance criteria for the linguistic SOC can only characterize what is essentially a compromise between rise-time and overshoot (and not the relative importance of each) and hence it provides little flexibility in specifying what performance is to be achieved/maintained (this is the case even if the "optimized" fuzzy performance evaluator introduced in [7, 8, 16] is used). Via the use of a reference model, in the FMRLC framework we incorporate a capability for accurately quantifying virtually any form of desired performance. Next, note that the knowledge-base modification algorithm of Procyk and Mamdani [6] relies on modification of a fuzzy relation table which describes the relationship between the fuzzy controller inputs and outputs. Often, this automatically implies that all input and output universes of discourse must be quantized into discrete levels to implement the fuzzy relation in a computer. Unfortunately, this will generally result in large memory requirements and computational demands since a fuzzy relation table often contains many entries for real world applications (some progress has been made at addressing the computational complexity of knowledge-base modification for the SOC in [7]). In this article, we use a knowledge-base modification algorithm (similar to the one in [16]) which reduces computation time and memory requirements by utilizing a rule base array table rather than a fuzzy relation table. The knowledge-base modification approach is flexible enough to be used in both the conventional SOC approach and the FMRLC (this is shown in [17, 18]).

Using conventional adaptive control terminology, the FMRLC and SOC are "direct" adaptive control schemes since they directly update the parameters of the controller without explicit identification of the plant parameters. Other relevant literature that focuses on direct adaptive fuzzy control includes the work in [19] where an adaptive fuzzy system is developed for a continuous casting plant, and the approach in [20] where a fuzzy system adapts itself to driver characteristics for an automotive speed control device. The use of fuzzy systems for estimation/identification [21, 22, 23, 24, 25, 26, 27] is relevant, especially if "indirect adaptive" [14, 15] fuzzy control techniques (i.e., ones where plant parameters are identified and used to tune the parameters of the controller) such as those in [28, 29, 30] are used. Also, it is interesting to note that in [25, 26, 29] there are inherent uses of inverse dynamics of the plant; however, our use of the fuzzy inverse model is significantly different. Finally, the authors note that since the initial results in FMRLC were introduced in [18, 17] some other relevant new fuzzy/neural adaptive/learning techniques have been developed. Several of these are also described in this book.

In Sect. 2, the detailed description of the FMRLC algorithm is presented. Then in Sect. 3 the FMRLC will be used as a learning controller for a two degree-of-freedom robot manipulator to illustrate the application of FMRLC for a multi-input, multi-output (MIMO) process. We briefly review other applications of the FMRLC in Sect. 4. Finally, in the concluding remarks in Sect. 5 we will discuss the advantages and disadvantages of FMRLC and highlight some important future research directions.

2 Fuzzy model reference learning control

The FMRLC, which is shown in Fig. 1, utilizes a learning mechanism that (i) observes data from a fuzzy control system (i.e., $r(kT)$ and $y(kT)$), (ii) characterizes its current performance, and (iii) automatically synthesizes and/or adjusts the fuzzy controller so that some pre-specified performance objectives are met. These performance objectives are characterized via the *reference model* shown in Fig. 1. In an analogous manner to conventional MRAC where conventional controllers are adjusted, the learning mechanism seeks to adjust the fuzzy controller so that the closed-loop system (the map from $r(kT)$ to $y(kT)$) acts like a pre-specified reference model (the map from $r(kT)$ to $y_m(kT)$). Next we describe each component of the FMRLC in more detail.

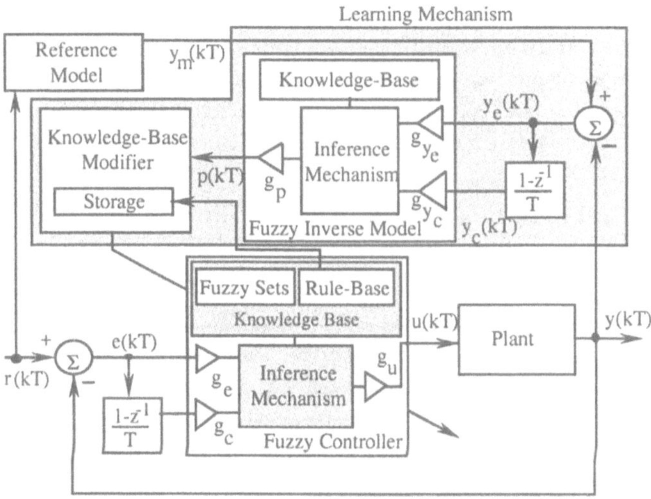

Figure 1: Architecture for the FMRLC.

2.1 The fuzzy controller

The process in Fig. 1 is assumed to have q inputs denoted by the q - dimensional vector $u(kT) = [u_1(kT) \ldots u_q(kT)]^t$ (T is the sample period) and s outputs denoted by the s - dimensional vector $y(kT) = [y_1(kT) \ldots y_s(kT)]^t$. Most often the inputs to the fuzzy controller are generated via some function of the plant output $y(kT)$ and reference input $r(kT)$. Fig. 1 shows a special case of such a map that was found useful in many applications. The inputs to the fuzzy controller are the error $e(kT) = [e_1(kT) \ldots e_s(kT)]^t$ and change in error $c(kT) = [c_1(kT) \ldots c_s(kT)]^t$ defined as $e(kT) = r(kT) - y(kT)$ and $c(kT) = \frac{e(kT)-e(kT-T)}{T}$ respectively, where $r(kT) = [r_1(kT) \ldots r_s(kT)]^t$ denotes the desired process output.

In fuzzy control theory, the range of values for a given controller input or output is often called the "universe of discourse". Often, for greater flexibility in fuzzy controller implementation, the universes of discourse for each process input are "normalized" to the interval $[-1, +1]$ by means of constant scaling factors. For our fuzzy controller design, the gains $\underline{g}_e$, $\underline{g}_c$, and $\underline{g}_u$ were employed to normalize the universe of discourse for the error $e(kT)$, change in error $c(kT)$, and controller output $u(kT)$, respectively (e.g., $\underline{g}_e = [g_{e_1}, \ldots, g_{e_s}]^t$ so that $g_{e_i} e_i(kT)$ is an input to the fuzzy controller). The gains $\underline{g}_e$ are chosen so that the range of values of $g_{e_i} e_i(kT)$ lie on $[-1, 1]$ and $\underline{g}_u$ is chosen by using the allowed range of inputs to the plant in a similar way. The gains $\underline{g}_c$ are determined by experimenting with various inputs to the system to determine the normal range of values that $c(kT)$ will take on; then $\underline{g}_c$ is chosen so that this range of values is scaled to $[-1, 1]$.

We utilize q multiple-input single-output (MISO) fuzzy controllers, one for each process input u_n (equivalent to using one MIMO controller). The knowledge base for the fuzzy controller associated with the nth process input is generated from **if-then** control rules of the form:

If $\tilde{e}_1$ is $\tilde{E}_1^j$ **and** ... **and** $\tilde{e}_s$ is $\tilde{E}_s^k$ **and** $\tilde{c}_1$ is $\tilde{C}_1^l$ **and** ... **and** $\tilde{c}_s$ is $\tilde{C}_s^m$ **Then** $\tilde{u}_n$ is $\tilde{U}_n^{j,\ldots,k,l,\ldots,m}$,

where $\tilde{e}_a$ and $\tilde{c}_a$ denote the *linguistic variables* associated with controller inputs e_a and c_a, respectively, $\tilde{u}_n$ denotes the linguistic variable associated with the controller output u_n, $\tilde{E}_a^b$ and $\tilde{C}_a^b$ denote the bth *linguistic value* associate associated with $\tilde{e}_a$ and $\tilde{c}_a$, respectively, and $\tilde{U}_n^{j,\ldots,k,l,\ldots,m}$ denotes the *consequent linguistic value* associated with $\tilde{u}_n$. Hence, as an example, one fuzzy control rule could be

If *error* is *positive-large* **and** *change-in-error* is *negative-small* **Then** *plant-input is positive-big*

(in this case $\tilde{e}_1$ = *"error"*, $\tilde{E}_1^4$ = *"positive-large"*, etc.). A set of such rules forms the rule-base which characterizes how to control a dynamical system.

The above control rule may be quantified by utilizing fuzzy set theory to obtain a fuzzy implication of the form:

$$\textbf{If } E_1^j \textbf{ and } \ldots \textbf{ and } E_s^k \textbf{ and } C_1^l \textbf{ and } \ldots \textbf{ and } C_s^m \textbf{ Then } U_n^{j,\ldots,k,l,\ldots,m},$$

where E_a^b, C_a^b, and $U_n^{j,\ldots,k,l,\ldots,m}$ denote the fuzzy sets that quantify the *linguistic statements* "$\tilde{e}_a$ is $\tilde{E}_a^b$", "$\tilde{c}_s$ is $\tilde{C}_s^m$", and "$\tilde{u}_n$ is $\tilde{U}_n^{j,\ldots,k,l,\ldots,m}$", respectively. For the example above we may use fuzzy sets on the $e_i(t)$ normalized universes of discourse as shown in Fig. 2. Assume that we use the same fuzzy

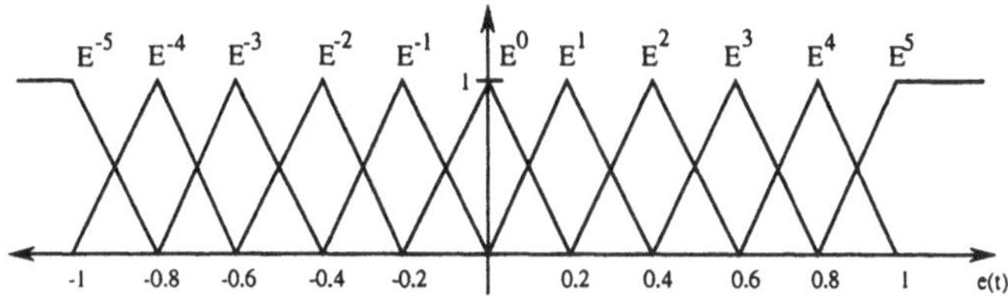

Figure 2: Fuzzy Sets on a Universe of Discourse.

sets on the $c_i(t)$ normalized universes of discourse. The membership functions on the output universe of discourse are assumed to be unknown; they are what the FMRLC will automatically synthesize. In fact, we will initialize the fuzzy controller knowledge base with 121 rules (for our examples we utilize the fuzzy sets shown in Fig. 2 and use all possible combinations of rules) where all the right-hand-side membership functions are triangular with base widths of 0.4 and centers at zero (to model that the fuzzy controller initially knows nothing about how to control the plant; of course one can often make a reasonable best guess at how to specify a fuzzy controller that is "more knowledgable"). For example, if $s = 1$ then all rules in our controller will take on the form

$$\textbf{If } E_1^j \textbf{ and } C_1^l \textbf{ Then } U_n^{j,l},$$

where the membership functions for E_1^j and C_1^l are shown in Fig. 2 and $U_n^{j,l}$ is a fuzzy set with triangular membership function with base width 0.4 centered at zero. In conventional direct fuzzy controller development the designer specifies a set of such control rules where $U_n^{j,l}$ are also specified *a priori*; for the FMRLC, the system will automatically specify and/or modify the fuzzy sets $U_n^{j,l}$ to improve/maintain performance. Finally, we note that we use min to represent the premise and implication and the standard center-of-gravity (COG) defuzzification technique [31].

2.2 The reference model

The reference model provides a capability for quantifying the desired performance. In general, the reference model may be any type of dynamical system (linear or non-linear, time-invariant or time-varying, discrete or continuous time, etc.). The performance of the overall system is computed with respect to the reference model by generating an error signal $y_e(kT) = [y_{e_1} \dots y_{e_s}]^t$ where $y_e(kT) = y_m(kT) - y(kT)$. Given that the reference model characterizes design criteria such as rise time and overshoot and the input to the reference model is the reference input $r(kT)$, the desired performance of the controlled process is met if the learning mechanism forces $y_e(kT)$ to remain very small for all time; hence, the error $y_e(kT)$ provides a characterization of the extent to which the desired performance is met at time $t = kT$. If the performance is met ($y_e(kT) \approx 0$) then the learning mechanism will not make significant modifications to the fuzzy controller. On the other hand if the components of $y_e(kT)$ are big, the desired performance is not achieved and the learning mechanism must adjust the fuzzy controller. Next we describe the operation of the learning mechanism.

2.3 The learning mechanism

As previously mentioned, the learning mechanism performs the function of modifying the knowledge base of a direct fuzzy controller so that the closed loop system behaves like the reference model. These knowledge base modifications are made by observing data from the controlled process, the reference model, and the fuzzy controller. The learning mechanism consists of two parts: a fuzzy inverse model and a knowledge base modifier. The fuzzy inverse model performs the function of mapping $y_e(kT)$ (representing the deviation from the desired behavior), to changes in the process inputs $p = [p_1, \dots p_r]^t$ that are necessary to force $y_e(kT)$ to zero. The knowledge base modifier performs the function of modifying the fuzzy controller's knowledge base to affect the needed changes in the process inputs. More details of this process are discussed next.

Using the fact that most often a control engineer will know how to roughly characterize the inverse model of the plant, the authors in [17] introduce the idea of using a fuzzy system to map $y_e(kT)$ and possibly functions of $y_e(kT)$ (or process operating conditions), to the necessary changes in the process inputs $p(kT)$. This map is called the *fuzzy inverse model* since information about the plant inverse dynamics is used in its specification. Note that similar to the fuzzy controller, the fuzzy inverse model shown in Fig. 1 contains normalizing scaling factors, namely $\underline{g}_{y_e}$, $\underline{g}_{y_c}$, and $\underline{g}_p$, for each universe of discourse. Given that $g_{y_{e_i}} y_{e_i}$ and $g_{y_{c_i}} y_{c_i}$ are inputs to the fuzzy inverse model, the knowledge base for the fuzzy inverse model associated with the nth process input is generated from fuzzy implications of the form:

$$\textbf{If } Y_{e_1}^j \text{ and } ... \text{ and } Y_{e_s}^k \text{ and } Y_{c_1}^l \text{ and } ... \text{ and } Y_{c_s}^m \textbf{ Then } P_n^{j,...,k,l,...,m},$$

where $Y_{e_a}^b$ and $Y_{c_a}^b$ denote the bth fuzzy set for the error y_{e_a} and change in error y_{c_a}, respectively, associated with the ath process output and $P_n^{j,...,k,l,...,m}$ denotes the consequent fuzzy set for this rule describing the necessary change in the nth process input. As with the fuzzy controller, we utilize membership functions for the normalized input universes of discourse as shown in Fig. 2, triangular membership functions for the output universes of discourse, min to represent the premise and implication, and COG defuzzification.

Given the information about the necessary changes in the input as expressed by the vector $p(kT)$, the *knowledge base modifier* changes the knowledge base of the fuzzy controller so that the previously applied control action will be modified by the amount $p(kT)$. Therefore, consider the previously computed control action $u(kT-T)$, which contributed to the present good/bad system performance. Note that $e(kT-T)$ and $c(kT-T)$ would have been the process error and change in error, respectively, at that time. By modifying the fuzzy controller's knowledge base we may force the fuzzy controller to produce a desired output $u(kT-T)+p(kT)$. Assume that only symmetric membership functions are defined for the fuzzy controller's output so that $c_n^{j,...,k,l,...,m}$ denotes the center value of the membership function associated with the fuzzy set $U_n^{j,...,k,l,...,m}$ (initially, $c_n^{j,...,k,l,...,m}(0)=0$). Knowledge base modification is performed by shifting centers of the membership functions of the fuzzy sets $U_n^{j,...,k,l,...,m}$ which are associated with the fuzzy implications that contributed to the previous control action $u(kT-T)$ (initially possibly shifting them away from having centers at zero). This modification involves shifting these membership functions by an amount specified by $p(kT)=[p_1(kT), \; ... \, p_r(kT)]^t$ so that

$$c_n^{j,...,k,l,...,m}(kT) = c_n^{j,...,k,l,...,m}(kT-T) + p_n(kT). \tag{1}$$

The degree of contribution for a particular fuzzy implication whose fuzzy relation is denoted $R_n^{j,...,k,l,...,m}$ is determined by its "activation level", defined

$$\begin{aligned} \delta_n^{j,...,k,l,...,m}(t) = \min\{&\mu_{E_1^j}(e_1(t)), ..., \mu_{E_s^k}(e_s(t)), \\ &\mu_{C_1^l}(c_1(t)), ..., \mu_{C_s^m}(c_s(t))\}, \end{aligned} \tag{2}$$

where μ_A denotes the membership function of the fuzzy set A. Only those rules whose activation level $\delta_n^{j,...,k,l,...,m}(kT-T)>0$ are modified; all others remain unchanged. It is important to note that our rule-base modification procedure implements a form of *local* learning and hence utilizes memory. In other words, different parts of the rule-base are "filled in" based on different operating conditions for the system, and when one area of the rule-base is updated, other rules are not affected. Hence, the controller adapts to new situations and also remembers how it has adapted to past situations. This justifies the use of the term "learning" rather than "adaptive" (for more details on this point see [32, 17, 33]). Note that while the above statement of the knowledge base update procedure is for one step in the past, it is easy to generalize for d steps in the past. You need to change it to d steps if it takes d steps for a plant input to affect the plant output to make sure that you are updating rules that actually contributed directly to the error $y_e(kT)$.

As an example, assume that all the normalizing gains for both the direct fuzzy controller and the fuzzy inverse model are unity and that the fuzzy inverse model produces an output $p_n(kT) = 0.5$ indicating that the value of the output to the plant at time $kT - T$ should have been $u(kT - T) + 0.5$ to improve performance (i.e., to force $y_{e_1} \approx 0$). Next, suppose that $e_1(kT-T) = 0.75$ and $c_1(kT - T) = -0.2$. Then rules

$$\textbf{If } E_1^3 \textbf{ and } C_1^{-1} \textbf{ Then } U_n^{3,-1}, \text{ and}$$

$$\textbf{If } E_1^4 \textbf{ and } C_1^{-1} \textbf{ Then } U_n^{4,-1},$$

are the only rules with activation levels greater than zero ($\delta_n^{3,-1} = 0.25$ and $\delta_n^{4,-1} = 0.75$) so these rules will be the only ones that have their consequent fuzzy sets ($U_n^{3,-1}$, $U_n^{4,-1}$) modified (See Fig. 2). To modify these fuzzy sets we simply shift their centers according to (1).

2.4 Design procedure and guidelines

Selection of the scaling gains can impact the overall performance so we provide a gain selection procedure in the following. Note that although it is often not highlighted, most learning/adaptive control approaches assume that you are given an initial controller structure and parameters (e.g., initial controller gains must be chosen in adaptive control approaches). In what follows we provide a procedure to pick such initial parameters for the FMRLC.

1. Select the controller gains $\underline{g}_{y_e}$ associated with the desired output change $y_e(kT)$ such that each universe of discourse is mapped to the interval $[-1, 1]$.

2. Choose the gain $\underline{g}_p$ to be the same as for the fuzzy controller output gain $\underline{g}_u$. This will allow the elements of $p(kT)$ to take on values as large as the largest possible inputs.

3. Assign the numerical value 0 to the scaling factors associated with the changes in the desired output changes (i.e., all elements of $\underline{g}_{y_c}$ are set equal to 0).

4. Apply a step input to the process which is of a magnitude that may be typical for the process during normal operation. Observe the process response and the reference model response.

5. Three cases:

   (a) If there exist unacceptable oscillations in a given process output response about the reference model response, then increase the associated element of $\underline{g}_{y_c}$. Go to step 4.

   (b) If a given process output response is unable to "keep up" with the reference model response, then decrease the associated element of $\underline{g}_{y_c}$. Go to step 4.

   (c) If the process response is acceptable with respect to the reference model response, then the controller design is completed.

For the robot application presented in this paper, the above gain selection procedure has proven very successful. However, given that the procedure is a result of practical experience with the FMRLC rather than strict mathematical analysis, it is possible that it will not work for all processes. For some applications (although none of the ones studied in [18, 17, 33, 34, 35]), the procedure may result in an unstable process. In such situations, it may be necessary to modify other controller parameters such as the controller sampling period T or the number of fuzzy controller rules.

While the selection of the scaling gains for the fuzzy controller and fuzzy inverse model is important, there are many other issues to pay attention to. The initialization of the direct fuzzy controller can drastically effect performance for some applications (especially the initial transient response). The fuzzy inverse model design can be complex and critical to the performance you can obtain for some applications. Moreover the choice of the inputs to the fuzzy controller and fuzzy inverse model can significantly impact its ability to learn and remember control actions. You want to choose the inputs to the fuzzy inverse model so that if it needs to change, for example, the learning rate for different plant conditions it can do this. For the fuzzy controller inputs, if you need different direct fuzzy controllers for different operating conditions, the fuzzy controller inputs must be chosen so that they characterize the changes in the operating conditions. When the operating condition changes, the FMRLC will try to learn a new controller and will not completely forget the old controller for the well learned operating condition. Since the inputs to the fuzzy controller will change and hence move the knowledge updating procedure to a new region of the rule-base. Without appropriate inputs the knowledge-base modifier will sometimes tend to update well-learned parts of the rule-base so that while the FMRLC may be able to continually adapt and maintain high performance it will not be able to learn and remember its actions.

3 Two-degree of freedom robot manipulator

Figure 3 illustrates the physical model of a two degree of freedom manipulator. It consists of two links where link #1 is mounted on a rigid base by means frictionless hinge and link #2 is mounted at the end of link #1 by means of a frictionless ball bearing. This control problem is provided

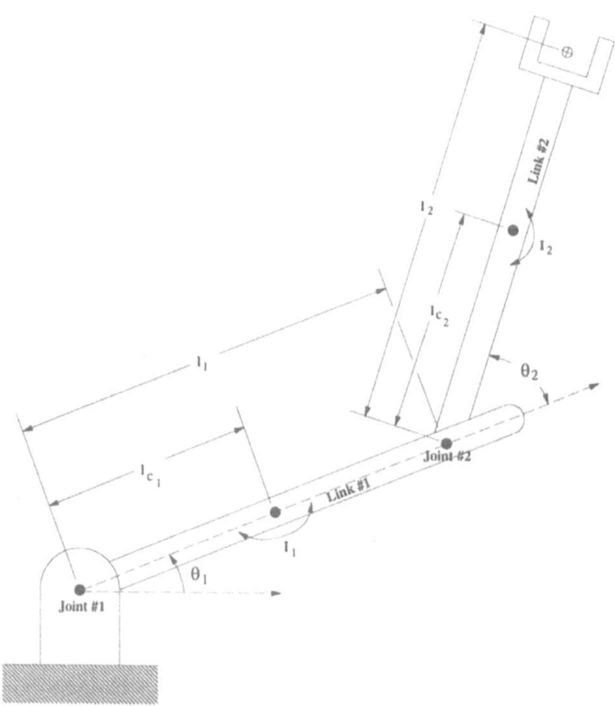

Figure 3: Graphical representation of a 2-link robot.

to illustrate the application of the FMRLC to a nonlinear MIMO system. The inputs to the process are the torques τ_1 and τ_2 which are applied to the links at joints #1 and #2. The outputs are the joint positions θ_1 and θ_2. The model for the robotic system was developed using the well-known Lagrangian equations in classical dynamics and is expressed by the matrix differential equation [36, 15]:

$$\begin{bmatrix} H_{11} & H_{12} \\ H_{21} & H_{22} \end{bmatrix} \begin{bmatrix} \ddot{\theta}_1 \\ \ddot{\theta}_2 \end{bmatrix} + \begin{bmatrix} -h\dot{\theta}_2 & -h\dot{\theta}_1 - h\dot{\theta}_2 \\ h\dot{\theta}_1 & 0 \end{bmatrix} \begin{bmatrix} \dot{\theta}_1 \\ \dot{\theta}_2 \end{bmatrix} + \begin{bmatrix} g_1 \\ g_2 \end{bmatrix} = \begin{bmatrix} \tau_1 \\ \tau_2 \end{bmatrix}, \quad (3)$$

where

$$H_{11} = m_1 l_{c_1}^2 + I_1 + m_2[l_1^2 + l_{c_2}^2 + 2l_1 l_{c_2} \cos(\theta_2)] + I_2, \quad (4)$$
$$H_{22} = m_2 l_{c_2}^2 + I_2, \quad (5)$$
$$H_{12} = H_{21} = m_2 l_1 l_{c_2} \cos(\theta_2) + m_2 l_{c_2}^2 + I_2, \quad (6)$$
$$h = m_2 l_1 l_{c_2} \sin(\theta_2), \quad (7)$$
$$g_1 = m_1 l_{c_1} g \cos(\theta_1) + m_2 g[l_{c_2} \cos(\theta_1 + \theta_2) + l_1 \cos(\theta_1)], \quad (8)$$
$$g_2 = m_2 l_{c_2} g \cos(\theta_1 + \theta_2), \quad (9)$$

and where $\theta = [\theta_1,\ \theta_2]^t$ are the two joint angles, $\tau = [\tau_1,\ \tau_2]^t$ are the input joint torques. For purposes of simulation the robot parameters are given by: (i) $m_1 = 1.0\ kg$ - mass of link #1, (ii) $m_2 = 1.0\ kg$ - mass of link #2, (iii) $l_1 = 1.0\ meters$ - length of link #1, (iv) $l_2 = 1.0\ meters$ - length of link #2, (v) $l_{c_1} = 0.5\ meters$ - distance from joint #1 to the center of gravity of link #1, (vi) $l_{c_2} = 0.5\ meters$ - distance from joint #2 to the center of gravity of link #2, (vi) $I_1 = 0.2\ kg - m^2$ - lengthwise centroidal inertia of link #1, and (vii) $I_2 = 0.2\ kg - m^2$ - lengthwise centroidal inertia of link #2.

3.1 FMRLC design

For this application, the process contains two inputs, namely τ_1 and τ_2. Consequently, two MISO fuzzy controllers are needed for this process (one for each process input). The inputs to the fuzzy controller are the robot joint position error $e = [e_1,\ e_2]^T$ and change in error $c = [c_1,\ c_2]^T$. The fuzzy controllers have outputs τ_1 for the first controller and τ_2 for the second controller. For both fuzzy controller designs, 11 fuzzy sets are defined for each controller input such that the membership functions are triangular shaped (with base widths of 0.4) and evenly distributed on appropriate universes of discourse (the outer-most membership functions saturate in the usual manner). Also, the controller gains for the error, change error, and the controller output are chosen to be $\underline{g}_e = [\frac{1}{2\pi}, \frac{1}{2\pi}]^t$, $\underline{g}_c = [\frac{1}{20}, \frac{1}{20}]^t$, and $\underline{g}_u = [100, 25]^t$, respectively. The knowledge-base array for both fuzzy controllers was initially chosen with all zero entries. The fuzzy controller sampling period was chosen to be $T = 5$ milliseconds.

The reference model for this FMRLC design is given by the following differential equation

$$\begin{bmatrix} \dot{y}_{m_1}(t) \\ \dot{y}_{m_2}(t) \end{bmatrix} = \begin{bmatrix} -0.75 & 0.0 \\ 0.0 & -1.5 \end{bmatrix} \begin{bmatrix} y_{m_1}(t) \\ y_{m_2}(t) \end{bmatrix} + \begin{bmatrix} +0.75 & 0.0 \\ 0.0 & +1.5 \end{bmatrix} \begin{bmatrix} r_1(t) \\ r_2(t) \end{bmatrix}, \quad (10)$$

where y_{m_1} and y_{m_2} specify the system performance for θ_1 and θ_2, respectively. For FMRLC implementation, the inputs to the reference model r_1 and r_2 are equal to the desired position of joints #1 and #2, respectively.

For this FMRLC design, two fuzzy inverse models are needed, one for each fuzzy controller. In general, both process inputs will affect both process outputs. However, for this fuzzy inverse model design we will assume that the cross-coupling between the inputs is negligible (i.e., τ_1 affects only θ_1 and τ_2 affects only θ_2). As a result, the input to a given fuzzy inverse model includes the error and change in error between the associated reference model output and robot position. Therefore, for the ith fuzzy inverse model, these inputs may be expressed as $y_{e_i}(kT) = y_{m_i}(kT) - \theta_i(kT)$ and $y_{c_i}(kT) = \frac{y_{e_i}(kT) - y_{e_i}(kT-T)}{T}$ respectively. For these inputs, 11 fuzzy sets are defined with triangular shaped membership functions which are evenly distributed on the appropriate universe of discourse. The normalizing fuzzy system gains associated with $y_e(kT)$, $y_c(kT)$, and $p(kT)$ are chosen to be $\underline{g}_{y_e} = [\frac{1}{2\pi}, \frac{1}{2\pi}]^t$, $\underline{g}_{y_c} = [1, \frac{1}{2}]^t$, and $\underline{g}_p = [100, 25]^t$, respectively. For the robot process for an increase in the torque τ_1 we would generally expect an increase in the process output θ_1. Likewise, for an increase in the torque τ_2 we would generally expect an increase in the process output θ_2. Consequently, the knowledge-base array shown in Table 1 was employed for both fuzzy inverse models.

Table 1: Rule base array table for the fuzzy inverse models for the robot.

$P_i^{j,k}$		Y_c^k										
		−5	−4	−3	−2	−1	+0	+1	+2	+3	+4	+5
Y_e^j	−5	−1.0	−1.0	−1.0	−1.0	−1.0	−1.0	−0.8	−0.6	−0.4	−0.2	0.0
	−4	−1.0	−1.0	−1.0	−1.0	−1.0	−0.8	−0.6	−0.4	−0.2	0.0	+0.2
	−3	−1.0	−1.0	−1.0	−1.0	−0.8	−0.6	−0.4	−0.2	0.0	+0.2	+0.4
	−2	−1.0	−1.0	−1.0	−0.8	−0.6	−0.4	−0.2	0.0	+0.2	+0.4	+0.6
	−1	−1.0	−1.0	−0.8	−0.6	−0.4	−0.2	0.0	+0.2	+0.4	+0.6	+0.8
	0	−1.0	−0.8	−0.6	−0.4	−0.2	0.0	+0.2	+0.4	+0.6	+0.8	+1.0
	+1	−0.8	−0.6	−0.4	−0.2	0.0	+0.2	+0.4	+0.6	+0.8	+1.0	+1.0
	+2	−0.6	−0.4	−0.2	0.0	+0.2	+0.4	+0.6	+0.8	+1.0	+1.0	+1.0
	+3	−0.4	−0.2	0.0	+0.2	+0.4	+0.6	+0.8	+1.0	+1.0	+1.0	+1.0
	+4	−0.2	0.0	+0.2	+0.4	+0.6	+0.8	+1.0	+1.0	+1.0	+1.0	+1.0
	+5	0.0	+0.2	+0.4	+0.6	+0.8	+1.0	+1.0	+1.0	+1.0	+1.0	+1.0

3.2 Simulation results

The simulation results for the FMRLC of the two degree-of-freedom robot manipulator are shown below in Fig. 4 for joint #1 and Fig. 5 for joint #2. The FMRLC provides good system tracking with respect to the reference model. As a result, the system exhibits good steady state and transient response. In fact, the response for joint #1 in Fig. 4 was so close to the response of the reference model that the two almost perfectly overlap. Note that this performance is achieved even though we initialize the direct fuzzy controller with a rule-base table of all zeros. Clearly the FMRLC quickly learns how to control the robot.

4 FMRLC applications

The FMRLC has been used for:

1. Control of a cart-pendulum system where certain improvements over SOC were illustrated [17] and for a rotational inverted pendulum.
2. Anti-skid brake system control to enhance performance when there are significant variations in the road conditions [35, 37].
3. Cargo ship steering where in [34] it is shown to have certain advantages over conventional model reference adaptive control.
4. Vibration damping in a two-link flexible robot where in [38, 39] the authors develop a fuzzy controller and show how its performance can be enhanced if it is tuned with the FMRLC (experimental results are also provided for both the direct fuzzy controller and the FMRLC to illustrate its ability to compensate for the effects of a payload variation).
5. For aircraft control law reconfiguration in case of failures [40, 41].
6. Slip power recovery control of a variable speed drive [42].
7. Speed control of indirect field oriented induction machine drives [43, 44].
8. Control of a liquid tank process experiment [45].
9. Velocity control of a non-linear, time-varying rocket system [33],
10. Torque regulation for a base braking control problem to reduce the effect of process variations [46].

Furthermore, in [47], the performance of the FMRLC was enhanced by dynamically focusing on the current operating condition.

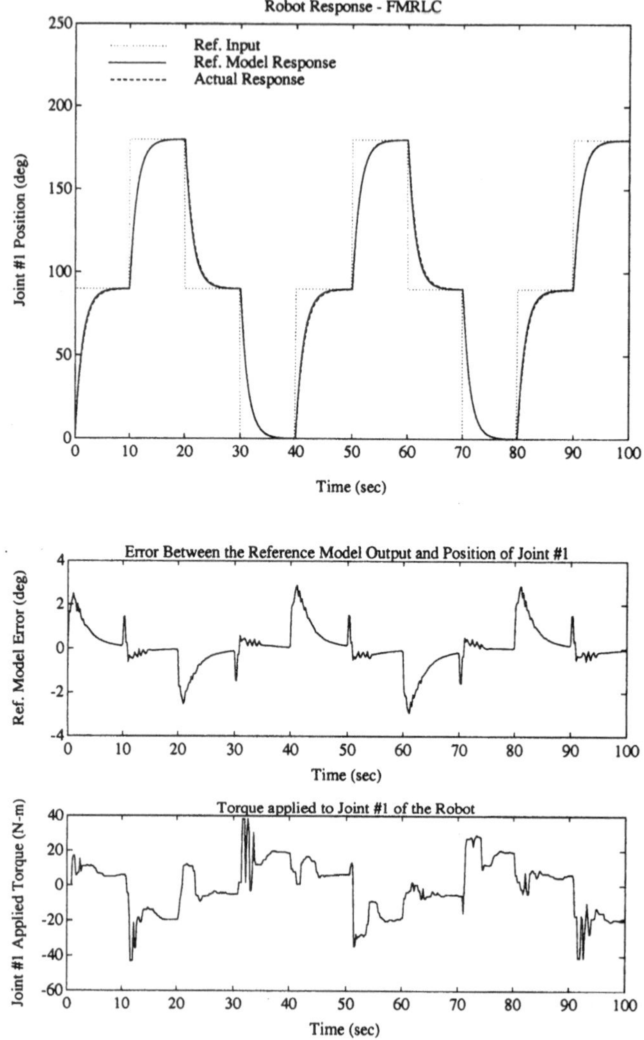

Figure 4: Simulation results for joint #1 of FMRLC controlled robot system.

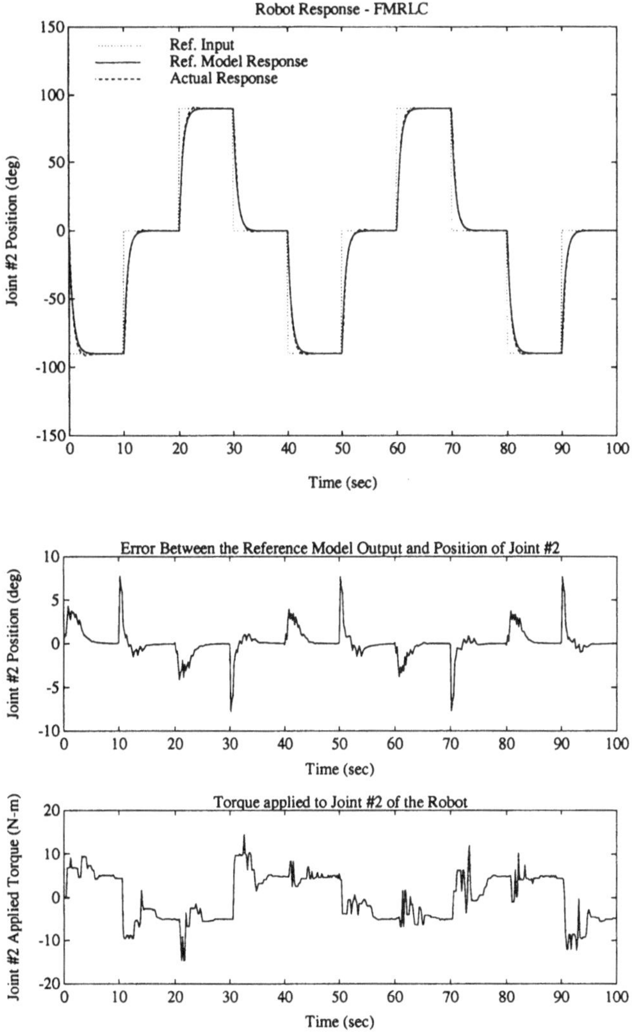

Figure 5: Simulation results for joint #2 of FMRLC controlled robot system.

5 Concluding remarks

The principal objectives of this article were to: (i) introduce the FMRLC, (ii) provide a *design methodology* for the FMRLC, and (iii) develop a MIMO FMRLC for a nonlinear two degree-of-freedom robot manipulator. The key advantages that the FMRLC seems to offer as a learning controller may be summarized as follows:

- A detailed analytical model of the process is not needed to develop the FMRLC although it may be needed for simulation-based training.
- The FMRLC provides an automatic method to synthesize a portion of the knowledge-base (specifically, the right-hand-sides of the rules) for the direct fuzzy controller while at the same time it ensures that the system will behave in a desirable fashion (in particular, there is no need to "learn from drastic failures" as is often the case for other learning control techniques - e.g., as is often done for the inverted pendulum).
- The learning/adaptation mechanism in the FMRLC dynamically and continually updates the rule-base in the direct fuzzy controller in response to process parameter variations and/or disturbances, In this way if unpredictable changes occur within the plant, the FMRLC can make on-line adjustments to a direct fuzzy controller to maintain adequate performance levels.

Basically, by combining learning/adaptive control concepts with fuzzy system theory, we have developed a control scheme which often has a fast rate of convergence and often provides an appropriate nonlinear mapping between controller inputs and outputs (i.e., it automatically performs "function approximation" [32] to achieve learning control).

Despite these apparent advantages of the FMRLC algorithm, several drawbacks do exist: (1) the design procedure (e.g., selection of the fuzzy inverse model, initial fuzzy controller, and the scaling gains) tends to be somewhat *ad hoc*, (2) there have been no investigations for the FMRLC to theoretically show that the fuzzy controller can in fact be tuned so that the performance specified in the reference model can be achieved (this problem is very well studied in conventional adaptive control where linear controllers are tuned so that performance specified in linear reference models is achieved), (3) conditions for stability and convergence of the FMRLC algorithm are yet to be found (although conditions for similar algorithms have been found), (4) *persistent excitation* [14, 15] issues for the FMRLC need to be mathematically investigated (since the reference input affects the ability of the fuzzy controller parameters to converge to values that result in the reference model behavior being achieved), and (5) although it provides certain improvements over SOC (as shown in [17]), the FMRLC algorithm is still computationally intensive for some applications. These disadvantages provide several future research directions. For example, future research involving the FMRLC algorithm should include a mathematical analysis of the controller to better quantify the effect of controller design parameters and work needs to be directed towards developing faster algorithms for FMRLC computations to ensure that the FMRLC can be employed in even more complex real world applications then have already been investigated.

Finally we want to emphasize the importance of laying foundations for comparative analyses of conventional and "intelligent" control techniques such as fuzzy control. Many concepts and results from conventional control (e.g., stability and stability analysis) can provide for a more careful engineering evaluation of intelligent control techniques and provide for productive research directions. At the same time, the intelligent control techniques have much to offer conventional control by infusing new concepts, approaches to control, and new design methodologies. In this article we make a small move in the direction of bridging the gap between fuzzy learning control and conventional adaptive control; it is hoped that this will be beneficial to both fields.

References

[1] W. Kickert and H. V. N. Lemke, "Application of a fuzzy controller in a warm water plant," *Automatica*, vol. 12, no. 4, pp. 301–308, 1976.

[2] E. Mamdani and S. Assilian, "An experiment in linguistic synthesis with a fuzzy logic controller," *Intl. Journal of Man-Machine Studies*, vol. 7, no. 1, pp. 1–13, 1975.

[3] J. Bernard, "Use of a rule-based system for process control," *IEEE Control Systems Magazine*, vol. 8, pp. 3–13, October 1988.

[4] Y. Li and C. Lau, "Development of fuzzy algorithms for servo systems," *IEEE Control Systems Magazine*, vol. 9, pp. 65–72, April 1989.

[5] K. Self, "Designing with fuzzy logic," *IEEE Spectrum*, pp. 42–105 and 105, November 1990.

[6] T. Procyk and E. Mamdani, "A linguistic self-organizing process controller," *Automatica*, vol. 15, no. 1, pp. 15–30, 1979.

[7] E. Scharf and N. Mandic, "The application of a fuzzy controller to the control of a multi-degree-of-freedom robot arm," in *Industrial Applications of Fuzzy Control*, pp. 41–62, Amsterdam, the Netherlands: M. Sugeno (ed.), 1985.

[8] R. Tanscheit and E. Scharf, "Experiments with the use of a rule-based self-organising controller for robotics applications," *Fuzzy Sets and Systems*, vol. 26, pp. 195–214, 1988.

[9] S. Shao, "Fuzzy self-organizing controller and its application for dynamic processes," *Fuzzy Sets and Systems*, vol. 26, pp. 151–164, 1988.

[10] S. Isaka, A. Sebald, A. Karimi, N. Smith, and M. Quinn, "On the design and performance evaluation of adaptive fuzzy controllers," *Proceedings, 1988 IEEE Conference on Decision and Control*, pp. 1068–1069, Austin, Texas, December 1988.

[11] S.Daley and K. F. Gill, "Comparison of a fuzzy logic controller with a P+D control law," *Journal of Dynamical System, Measurement, and Control*, vol. 111, pp. 128–137, June 1989.

[12] S.Daley and K. F. Gill, "Altitude control of a spacecraft using an extended self-organizing fuzzy logic controller," *Proc. I. Mech. E.*, vol. 201, no. 2, pp. 97–106, 1987.

[13] S.Daley and K. F. Gill, "A design study of a self-organizing fuzzy logic controller," *Proc. I. Mech. E.*, vol. 200, pp. 59–69, 1986.

[14] K. Åström and B. Wittenmark, *Adaptive Control.* Reading, Massachusetts: Addison-Wesley Publishing Company, 1989.

[15] K. Narendra and A. Annaswamy, *Stable Adaptive Systems.* Englewood Cliffs, New Jersey: Prentice Hall, 1989.

[16] T. Yamazaki, *An improved algorithm for a self-organizing controller and its experimental analysis.* PhD thesis, London University, 1982.

[17] J. Layne and K. Passino, "Fuzzy model reference learning control," *Proceedings of the 1st IEEE Conference on Control Applications*, pp. 686–691, Dayton, Ohio, September 1992.

[18] J. Layne, "Fuzzy model reference learning control," Master's thesis, Department of Electrical Engineering, The Ohio State University, March 3 1992.

[19] G. Bartolini, G. Casalino, F. Davoli, R. M. M. Mastretta, and E. Morten, "Development of performance adaptive fuzzy controllers with applications to continuous casting plants," *Industrial Application of Fuzzy Control*, pp. 73–86, 1985.

[20] H. Takahashi, "Automatic speed control device using self-tuning fuzzy logic," *1988 IEEE Workshop on Automotive Applications of Electronics*, pp. 65–71, Dearborn, Michigan, October 1988.

[21] T. Takagi and M. Sugeno, "Fuzzy identification of systems and its applications to modeling and control," *IEEE Transactions on systems, Man, and Cybernetics*, vol. 15, no. 1, pp. 116–132, 1985.

[22] L. Wang and J. Mendel, "Generating fuzzy rules by learning from examples," *Proceedings, 1991 IEEE International Symposium on Intelligent Control*, pp. 263–268, Arlington, Virginia, August 1991.

[23] R. M. Tong, "Some properties of fuzzy feedback systems," *IEEE Transactions on Systems, Man, and Cybernetics*, vol. 10, pp. 327–330, June 1980.

[24] A. Cumani, "On a possibilistic approach to the analysis of fuzzy feedback systems," *IEEE Transactions on Systems, Man, and Cybernetics*, vol. 12, pp. 417–422, May/June 1982.

[25] E. Czogała and W. Pedrycz, "On identification in fuzzy systems and its applications in control problems," *Fuzzy Sets and Systems*, vol. 6, pp. 73–83, 1981.

[26] E. Czogała and W. Pedrycz, "Control problems in fuzzy systems," *Fuzzy Sets and Systems*, vol. 7, pp. 257–273, 1982.

[27] C. Batur, A. Srinivasan, and C. Chan, "Automatic rule based model generation for uncertain complex dynamical systems," *Proceedings, 1991 IEEE International Symposium on Intelligent Control*, pp. 275–279, 1991.

[28] F. V. D. Rhee, H. V. N. Lemke, and J. Dijkman, "Knowledge based fuzzy control of systems," *IEEE Transactions on Automatic Control*, vol. 35, pp. 148–155, February 1990.

[29] P. Graham and R. Newell, "Fuzzy adaptive control of a first-order process," *Fuzzy Sets and Systems*, vol. 31, pp. 47–65, 1989.

[30] P. Graham and R. Newell, "Fuzzy identification and control of a liquid level rig," *Fuzzy Sets and Systems*, vol. 8, pp. 255–273, 1988.

[31] C. Lee, "Fuzzy logic in control systems: Fuzzy logic controller-part I," *IEEE Trans. on Systems, Man. and Cybernetics*, vol. 20, pp. 404–418, March/April 1990.

[32] J. Farrell and W. Baker, "Learning control systems," in *An Introduction to Intelligent and Autonomous Control Systems* (P. Antsaklis and K. Passino, eds.), Kluwer Academic Publishers; Norwell MA, 1993.

[33] J. Layne and K. Passino, "Fuzzy model reference learning control," *Journal of Intelligent and Fuzzy Systems*, vol. 4, pp. 33–47, 1996.

[34] J. Layne and K. Passino, "Fuzzy model reference learning control for cargo ship steering," *IEEE Control Systems*, vol. 13, no. 6, pp. 23–34, 1993.

[35] J. Layne, K. Passino, and S. Yurkovich, "Fuzzy learning control for anti-skid braking systems," *IEEE Trans. on Control System Technology*, vol. 1, pp. 122–129, June 1993.

[36] J. Slotine and W. Li, *Applied Nonlinear Control.* Englewood Cliffs, New Jersey: Prentice Hall, 1991.

[37] J. Layne, K. Passino, and S. Yurkovich, "Fuzzy learning control for anti-skid braking systems," *Proc. IEEE Conf. on Decision and Control*, pp. 2523–2528, Tucson, AZ, December 1992.

[38] V. Moudgal, W. Kwong, and K. Passino, "Learning control for a two-link flexible mechanism," *Proc. of the American Control Conference*, pp. Baltimore, MD, June 1994.

[39] V. Moudgal, W. Kwong, K. Passino, and S. Yurkovich, "Fuzzy learning control for a flexible-link robot," *IEEE Transactions on Fuzzy Systems*, vol. 3, no. 2, pp. 199–210, 1995.

[40] W. Kwong and K. Passino, "Fuzzy learning systems for aircraft control law reconfiguration," *Proceedings of the IEEE Int. Symp. on Intelligent Control*, pp. Columbus, Ohio, Aug. 16-18 1994.

[41] W. Kwong, K. Passino, E. Laukonen, and S. Yourkovich, "Expert supervision of fuzzy learning systems for fault tolerant aircraft control," *Proceedings of the IEEE*, vol. 83, no. 3, pp. 466–483, 1995.

[42] Y. Tang and L. Xu, "Fuzzy logic application for intelligent control of a variable speed drive," *IEEE Transactions on Energy Conversion*, pp. 679–685, 1994.

[43] L. Zhen and L. Xu, "A comparison study of three fuzzy schemes for indirect vector control of induction machines," *Preceedings, IEEE Industrial Application Society Annual Meeting*, pp. 1725–1732, 1996.

[44] L. Zhen and L. Xu, "Fuzzy learning enhanced speed control of an indirect field oriented induction machine drive," *Preceedings, IEEE International Symposium on Intelligent Control*, pp. 109–114, 1996.

[45] J. Zumberge and K. Passino, "A case study in intelligent vs. conventional control for a process control experiment," *Proceedings of the 1996 IEEE International Symposium on Intelligent Control*, pp. 37–42, 1996.

[46] W. Lennon and K. Passino, "Intelligent control for brake systems," *Proceedings of the 1995 IEEE International Symposium on Intelligent Control*, pp. 499–504, 1995.

[47] W. Kwong and K. Passino, "Dynamically focused fuzzy learning control," *IEEE Transactions on Systems, Man, and Cybernetics - Part B*, vol. 26, no. 1, pp. 53–74, 1996.

Development of a Fuzzy Relational-Based Predictive Controller

Mary M. Bourke and D. Grant Fisher
Department of Chemical and Materials Engineering
University of Alberta
Edmonton, Canada T6G 2G6

1 Introduction

The most widely used feedback controller for continuous industrial processes is still the simple Proportional, Integral plus Derivative (PID) controller. However, the widespread availability of digital process control computers and the need to meet international competition in product quality and production efficiency has resulted in a significant increase in the use of more advanced discrete control algorithms. Included in these advanced techniques is fuzzy logic control, and in particular control through the use of fuzzy model-based predictive controllers [1], [9], [10], [19], [24].

This article presents a fuzzy predictive controller that has a structure similar to conventional model-based predictive controllers. It uses relational-based models which are superior to rule-based models in that they are developed more quickly and easily than rule-based systems and permit identification directly from input/output data. Another advantage is that predictive control, either one-step or multi-step ahead, can be implemented in a manner similar to classical predictive controllers. This article demonstrates the parallels between *classical* predictive control and *fuzzy* predictive control.

As will be shown later, the proposed fuzzy predictive model-based controller structure is similar to conventional predictive control. The design has been generalized so that the controller is not application specific. It contains enough flexibility so that it can be easily modified for different classes of applications and can incorporate self learning and adaptive capabilities. The predictive controller employs the *max-product* compositional operator which gives *better* results, using a minimum distance criterion, than the more widely used *max-min* compositional operator.

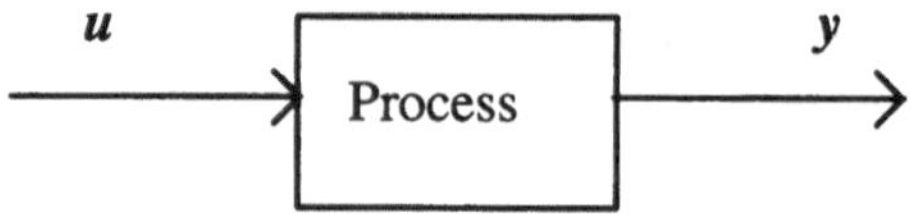

Figure 1: Standard process model.

2 Fuzzy process structures

Starting with the basics, a process of interest can be represented by a standard input/output discrete model as illustrated in Fig. 1. That is, a crisp (non-fuzzy) input, u, is fed to a deterministic process and a crisp (non-fuzzy) output, y, is produced. It should be noted here that the notation throughout this article shows fuzzy variables (i.e., $\tilde{u}$, $\tilde{y}$) with tildas and crisp variables (i.e., u, y) without.

Fuzzy processes can be illustrated by a schema similar to Fig. 1 where the input/output is inherently fuzzy or is appropriately fuzzified/defuzzified. The same figure can also be used to describe the three basic problem scenarios, discussed below, for which fuzzy logic is ideally suited, i.e., (1) Fuzzy Process Dynamics (*fuzzy* model); (2) Fuzzy Process Input (*fuzzy* input); (3) Fuzzy Process Measurement (*fuzzy* output).

2.1 Fuzzy process dynamics

The *fuzzy* dynamic system model is encountered widely in the literature. This case typically arises when the input and output data are discrete but the process is considered too complicated to be modeled by conventional techniques. The interface techniques of *fuzzification* and *defuzzification* are introduced in order to model the process from a *fuzzy* perspective. The system model for this scenario is illustrated in Fig. 2, with the hatched box marking the *fuzzy* boundaries, i.e., the input and output of the hatched box are fuzzy).

In many instances in the literature, clearly deterministic processes are artificially fuzzified to demonstrate the capabilities of various *fuzzy* controllers. This artificial application of fuzzy logic leads to some confusion and questions

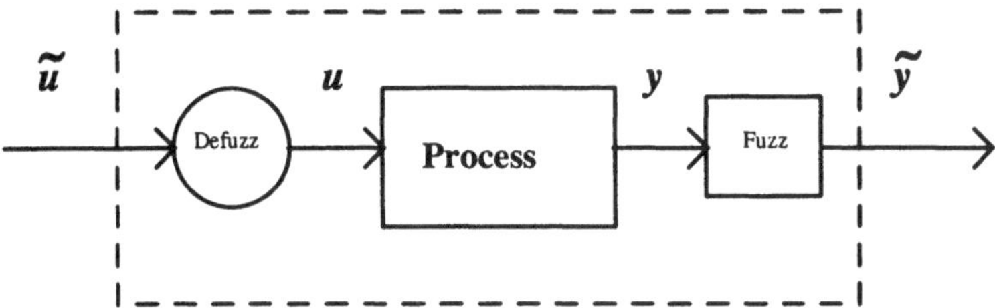

Figure 2: Fuzzy process dynamics model.

about the need for fuzzy technology. However, there are several application areas where the process dynamics are inherently fuzzy in the sense that the process input, output and/or the model are inherently uncertain (e.g., tar sands extraction, bio-treatment and mineral processing). These example processes are all being controlled today using conventional technologies with crisp data. However, fuzzy logic can compliment these areas by using vague or indexed data which conventional control systems can not include. When crisp and fuzzy data are to be considered together, it is customary to fuzzify the crisp data so that all information is evaluated on the same basis.

2.2 Fuzzy process input

The *fuzzy* process input scenario is encountered when the process is inherently fuzzy due to the control law. In these situations the process output measurement is crisp and accurate, and it is possible to set the crisp process input to any desired value. However, the problem is that the control law or relationship between the input and output of the controller is only approximate (e.g., whether the error is 1.234 or 1.888 the control law may still call for only a *small* change in process input). Although the error is known precisely there is not enough information available to determine a precise value for the manipulated variable, u. An industrial example would be mineral processing using flotation. The current error (or setpoint change) may be known exactly but it is not possible to calculate precise values for the flotation chemicals required to drive the error to zero. Another example is where a manipulated flow variable can be accurately set to any crisp value but some key component (e.g., composition) of the manipulated variable changes with time and is currently unknown. Thus the manipulated variable can be set accurately to a crisp value but the effect on the process is *approximate* or *fuzzy*. This scenario could be represented by Fig. 2, without defuzzification of $\tilde{u}$.

2.3 Fuzzy process measurement

The final *fuzzy* process scenario involves *fuzzy* process measurements. In this situation the process measurement is inherently fuzzy (i.e., qualitative or indexed). The manipulated input into the process is crisp, but the controller is fuzzy because of the *fuzziness* inherent in the process measurement. An industrial example would be in tar sands or mineral processing where the composition of the process output (controlled variable) can only be estimated approximately on-line. This process model could be represented by Fig. 2, without a fuzzification block for y.

2.4 Summary of fuzzy structures

The three most common *fuzzy* process situations (i.e., fuzzy process, input or output) can be combined to produce other process situations, such as a totally *fuzzy* process where the dynamics, inputs and measurements are all fuzzy. The modeling and control would be executed the same as in the other problem scenarios, but without the need to artificially fuzzify or defuzzify.

Although the *fuzzification* and *defuzzification* may seem artificial in some instances, the power or benefit of the technique is that it permits the handling of both fuzzy and non-fuzzy data in the same model.

Since it is clear that the fuzzy model can handle both crisp and fuzzy data, via fuzzification and defuzzification interfaces, these interfaces will be assumed to be an implicit part of the modeling and control structure and will not be shown explicitly in the block diagrams during the controller synthesis. This will simplify the controller diagrams and facilitate a one-to-one comparison between the conventional predictive control structure and the fuzzy logic predictive control structure developed in this article. The fuzzification and defuzzification interfaces are included, however, in the final controller design schema (Fig. 10).

3 Fuzzy predictive control vs. classical predictive control

This section shows that the development of *fuzzy* relational model-based control parallels the development of conventional model-based control.

3.1 Model-based control

Model-based control design is a modified formulation of traditional feedback control, and can be transformed into the standard servo-regulatory form, as will be shown below. The principles inherent in the following controller development are valid for both crisp and *fuzzy* systems as will be confirmed by the *fuzzy predictive controller* design presented later.

A convenient starting point to explain the development of model-based control (e.g., Internal Model Control (IMC) [22]) is the standard feedback control system, shown in Fig. 3. It is translated into a model-based control strategy by adding and subtracting a process model, G_m, between u and the feedback signal, as shown in Fig. 4. One process model, G_m, is then combined with the controller, G_c, to form a new controller, G_c^*. This new control scenario is illustrated in Fig. 5. Note that G_c^* is simply the closed-loop transfer function for a standard feedback loop containing G_c and G_m.

If $G_m = G_p$ then $\hat{L} = L$ and Fig. 5 reduces to the simplified, but equivalent system in Fig. 6. Fig. 6 can be interpreted as an open-loop servo-regulatory control scheme from which it follows that for perfect control $G_c^* = G_p^{-1}$. Thus the basic design procedure for model-based feedback control [22] is to : (1) derive the process model, G_m; (2) set $G_c^* = G_m^{-1}$; and, if desired, convert to the traditional (Fig. 3) form using $G_c = G_c^*/(1 - G_c^* G_m)$.

The observation that the controller for "perfect" model-based control is equal to the inverse of the process emphasizes the importance of selecting and identifying a fuzzy model that has a "well-behaved" and accurate *inverse*. As shown later, this is often difficult.

The next step in this development is to show that in addition to more straight forward *design* the model-based control system of Fig. 5 gives *better* control *performance* than conventional feedback control for many processes (e.g., those containing time delays). This will be done in the next subsection by analyzing the Smith predictor control scheme from the perspective of model-based control. Extending the model-based control system of Fig. 5, to include *predictive control systems* will be done following the discussion of Smith predictors.

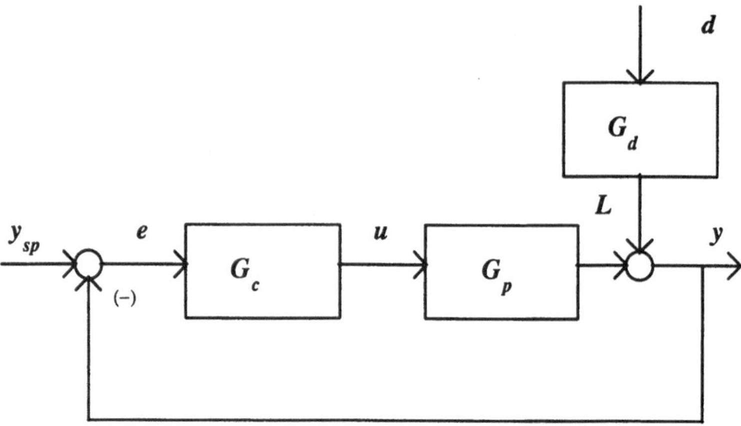

Figure 3: Standard feedback control.

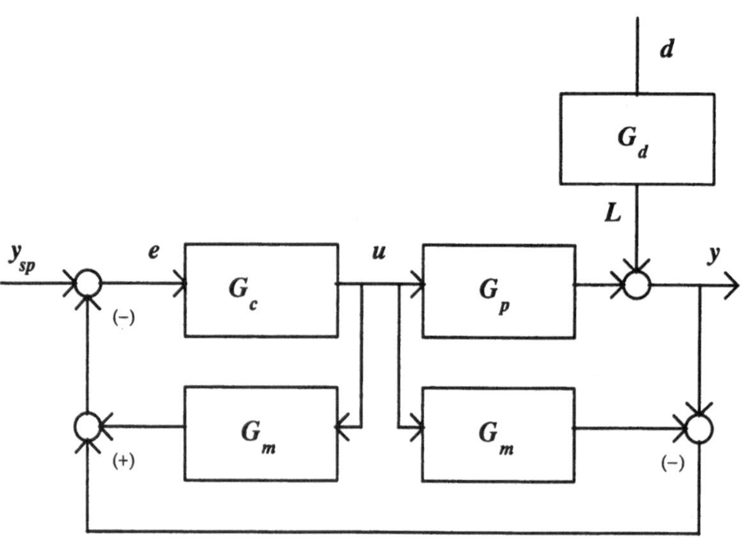

Figure 4: Feedback control with process model added/subtracted.

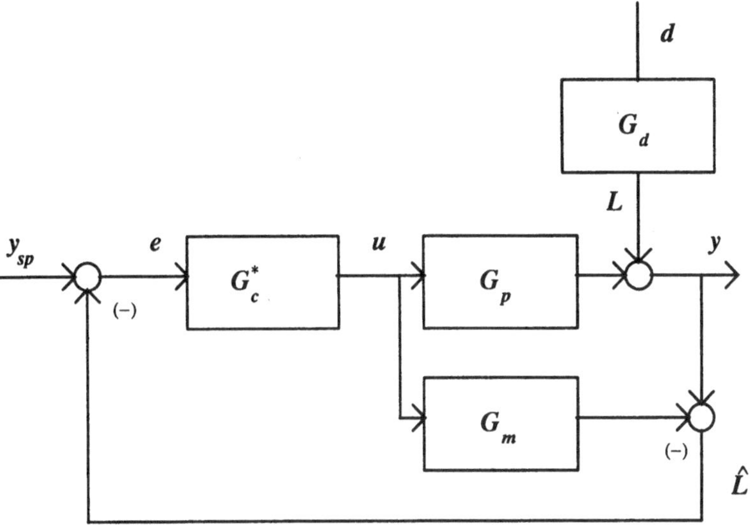

Figure 5: Model-based feedback control.

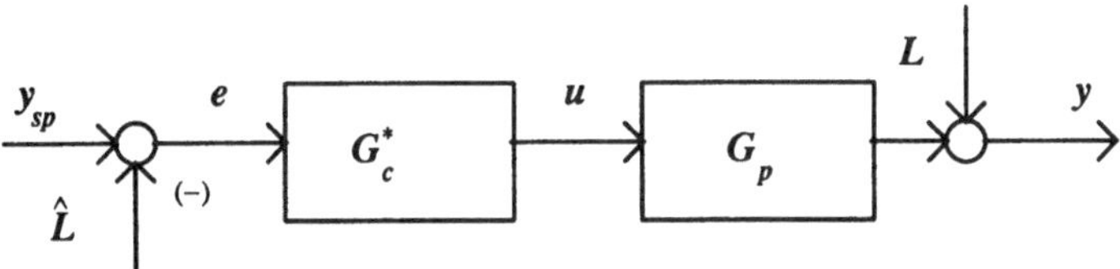

Figure 6: Simplification of servo-regulator and model-based control (perfect model).

3.2 Time delay compensation using a Smith predictor

Now consider again the model-based feedback control illustrated in Fig. 5. Perfect control for setpoint and load disturbances requires *model-inverse control* (i.e., $G_c^* = G_p^{-1}$). This is not physically possible if the process contains a time delay since a time delay is not invertible.

The Smith predictor technique, shown in Fig. 7, handles time delay processes by using two process models, one *with* dead time, G_m, and the other *without*, G_m^*. The controller output is sent to the actual process, G_p, as well as to the models. If $G_m = G_p$ then $r = L$ and the *design* problem becomes one of designing a standard feedback controller (G_c in Fig. 7) for the *delay-free process* G_m^*. This removes the time delay from the *design* problem. Note however that the actual process output is a delayed version of the output of G_m^* or conversely the output of G_m^* is a *prediction* $\hat{y}(k+\tau+1)$ of the actual plant output (assuming $L =$ constant over the period of prediction). Therefore control of the actual process output, y, is a delayed version of the control of the corresponding delay-free system, (i.e., time delays are invariant under feedback). Note that Fig. 7 is equal to Fig. 4 except that $G_m^* \neq G_m$. In many of the controllers proposed in the literature the time delay is approximated and the two approaches become approximately equivalent. Note that the inversion of the process model is not always handled analytically or explicitly. For example, it is easy to show that the closed-loop transfer function of the feedback loop in Fig. 4 containing G_c and G_m approaches G_m^{-1} as the gain of the controller ($G_c = K_c$) increases.

3.3 Predictive control

Conventional feedback controllers are designed to reduce the *current* control error, $e(k) = y_{sp}(k) - \hat{y}(k)$. This is adequate for simple applications but problems arise with more difficult applications, (e.g., those with time-delays, non-minimum-phase behaviour, etc.). Consider Fig. 8 where a step $L(k)$ is introduced. The controller sees the disturbance (i.e., $e(k) = -L(k)$ if $y_{sp} = 0$ and $\hat{y}_2 = 0$) and under *ideal* (model-based) conditions generates control action $u(k)$ that would exactly compensate for $L(k)$. However, the time delay, τ, in G_p delays the effect of u on y. Therefore, for the next τ control intervals, the controller sees $e = -L$ and generates additional changes in $\{u(k+i), i = 1, 2, \ldots, \tau - 1\}$ even though none are required. However, if a *model* is used to *predict* the future process output such that $\hat{y}_2(k+1) = y(k+\tau+1)$ then the problem is eliminated (i.e., $e(k) = -L(k)$ but if $u(k)$ exactly compensates for $L(k)$ then the error will be zero at the next control interval and no further control action will be taken). This concept of predictive model-based control can be extended to formulate controllers

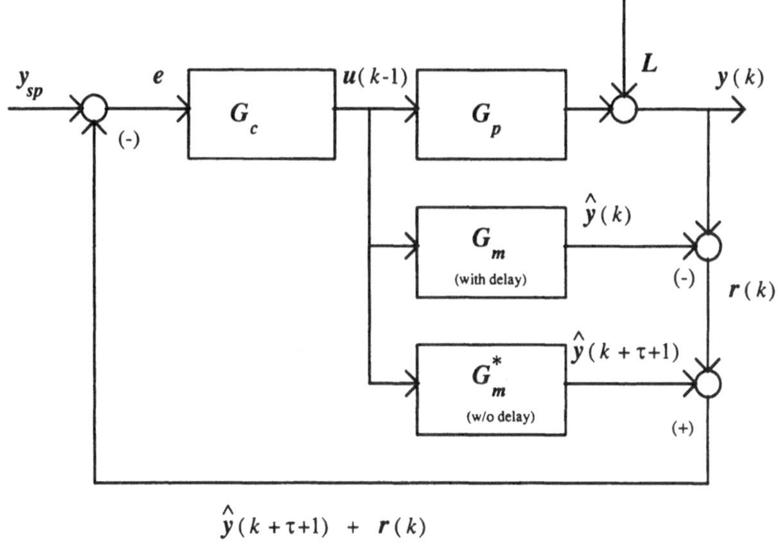

Figure 7: Smith predictor.

based on *trajectories* (e.g., $\{e(k+i), i = \tau, \ldots, N\}$) rather than point values.

The basic concepts required for derivation of a fuzzy controller can be deduced from the simplified representation shown in Fig. 8. The basic structure of Fig. 8 is similar to that of Fig. 7 (Smith Predictor) *except that a single block with two outputs is used to represent the model rather than two separate blocks.*

The key difference in predictive control is that the control error $e(k+\tau+1)$ used by the controller is based on a future *predicted* value of the process output, $\hat{y}(k + \tau + 1)$, rather that the *current* value $\hat{y}(k)$ used in the figures preceding Fig. 7.

The advantage of *model-based* control is that it makes design easier and more transparent (e.g., it makes the conditions required for "perfect control" obvious). Consider the *block diagram* shown in Fig. 8 with a step disturbance $L(k)$. The "*Filter*" in the feedback path is required for *practical* applications to accommodate noise and modeling errors but, for simplicity, is omitted in the following analysis. "Perfect" *servo* control, such that $y = y_{sp}\,\forall t$ is obtained if $G_c = G_p^{-1}$ and no feedback is required (i.e., $r(k) = 0$). With perfect modeling (i.e., $G_m = G_p$), $\hat{y}_1 = y_p$ and $r = L$ so Fig. 8, with $\hat{y}_2 = 0$, can be interpreted as *feedforward* control with L obtained by estimation rather that direct measurement. (Conventional feedforward design techniques

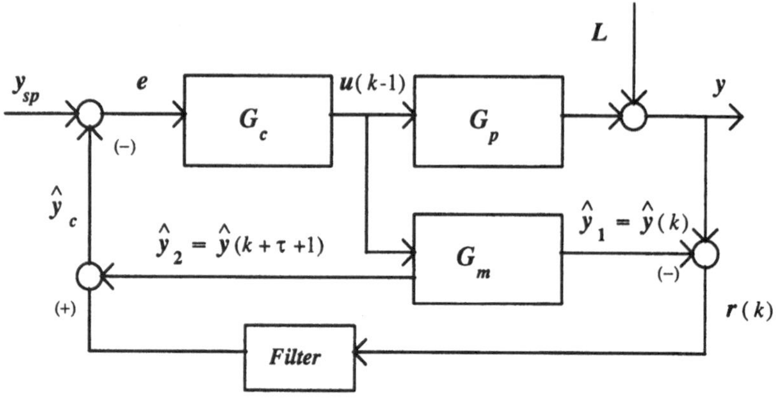

Figure 8: Predictive control at control interval k.

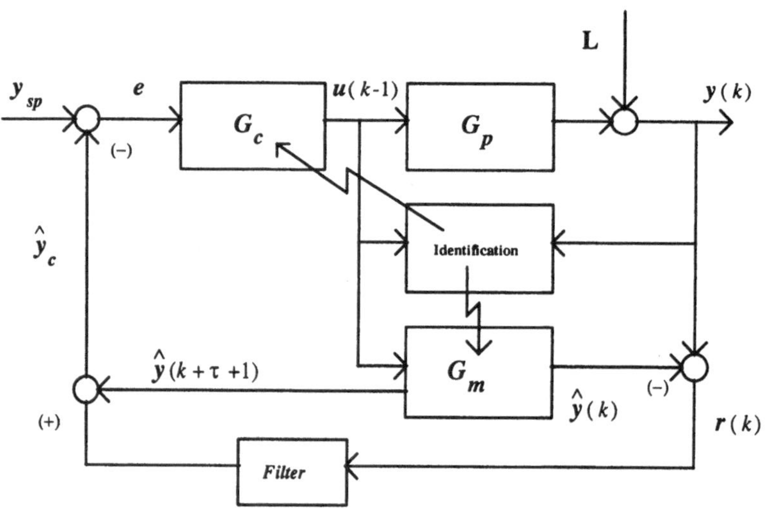

Figure 9: Self-learning predictive control.

also lead to $G_c = G_p^{-1}$ for perfect control). This interpretation leads to the obvious result that "perfect" *regulatory* control, such that $y = y_{sp} = 0 \, \forall t$ and $\forall L$ is achieved if $G_c = G_p^{-1}$. Note that with the addition of fuzzification and defuzzification Fig. 8 can also be interpreted as a fuzzy control system.

From the preceding development it is obvious that model-based control techniques require a process model for use in the output prediction and/or control calculation steps. For simple, *time-invariant* processes it may be sufficient to derive a suitable model *a priori* and include it, for example in the G_m block of Fig. 8. However, in many practical applications it is necessary to update the model *on-line* due to changes in operating conditions, raw materials, product specifications, etc. In principle, this is easily accomplished by adding an identification block to Fig. 8. As shown in Fig. 9, the *identification* block supplies an updated process model to the prediction and control blocks (and, in some practical applications, to the *Filter*). This can be done at some regular multiple of the control interval or on an *as required* basis.

The (pseudo) block diagram of the proposed *Self-Learning Predictive Fuzzy Relational Controller* is shown in Fig. 10 and parallels that of classical and modern control techniques as outlined above. Figure 10 follows from Fig. 9 by the addition of fuzzification and defuzzification blocks. The specific design of the *fuzzy logic controller* developed below assumes that the process variables, both input and output, are crisp and control optimization is performed through the minimization of the crisp scalar distance, $|y_{sp} - \hat{y}_c|$.

The self-learning portion of the control scheme can be achieved through the use of any identification algorithm, such as the averaging technique in [7]. The identification algorithm is a separate module from the controller and can be omitted. If used, it should be chosen for model accuracy for the specific system being controlled and, to be consistent with the following controller development, should generate a *fuzzy first order plus delay model.*

3.4 Optimal model-based control calculations

The availability of a process model makes the control calculation much more direct and in the ideal case *eliminates the need for controller tuning.* Consider a predictive model, G_m, that gives a one step ahead prediction of the process output based on past inputs (e.g., $\hat{y}(k+1) = G_m \cdot u(k)$). At control interval, k, $\hat{y}(k+1)$ in the model is replaced by the desired setpoint value, $y_{sp}(k+1)$, and the control action is calculated *analytically* (in the simplest ideal case) as $u(k) = G_p^{-1} \cdot y_{sp}(k+1)$. An alternative implementation of the control calculation is to use an *optimization* or *search* procedure to find the $u(k)$ that minimizes a user defined objective function, J (e.g., $[y_{sp}(k+1) - \hat{y}(k+1)]^2$). Note that the optimization procedure can in principle be extended to include

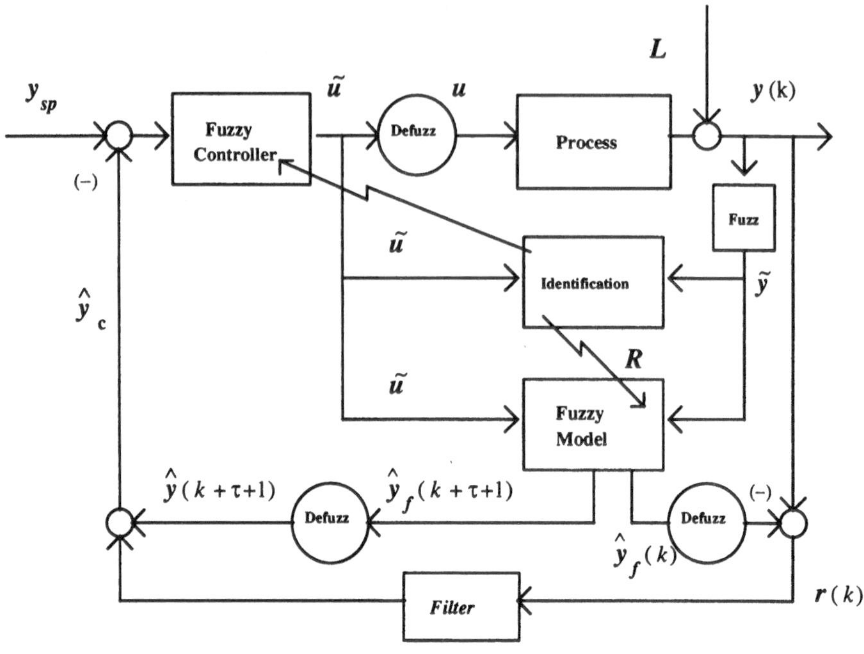

Figure 10: Self-learning predictive fuzzy logic control (f subscript = *fuzzy* variable).

constraints, trajectories, a wide range of models, etc. and in the ideal "perfect control" case gives the same results as using G_m^{-1}.

A generalized outline of model-based, predictive control based on the above is:

1. Develop a feedback predictor capable of estimating the current and future process output, $\{\hat{y}_c(k+i), i = 0, 1, 2, \ldots\}$. This should include an estimate of the future effect of current disturbances or residuals (i.e., $\hat{y}_c = \hat{y}_2 + r(k)$ in Fig. 10 if *Filter*=1).
2. Define an objective function (e.g., minimize the error between the desired output (trajectory), y_{sp}, and the predicted output, $\hat{y}_c$, at some point(s) in the future).
3. Calculate the control action $u(k)$ that minimizes the objective function using analytical and/or (constrained) optimization procedures and implement $u(k)$. Repeat the calculation of the current control action at each control interval.

Techniques to develop optimal, model-based, predictive control are well developed in the discrete control literature and the basic steps of modeling, identification, optimization, etc. are well established in the *fuzzy* systems literature. It would therefore appear straightforward to develop a fuzzy-model-based predictive controller. However, as discussed in the next section, it was found [3] that *fuzzy* systems have unique properties that necessitate a different approach than used with discrete-time systems.

3.5 A direct fuzzy analogy of discrete control

The original intention of the work was to develop a fuzzy controller simply by replacing the steps of modeling, identification and control used in discrete model-based controllers by their fuzzy analogs. However, experience [3] showed that this did *not* give satisfactory results. In the interest of spurring further developments in fuzzy methodology and controller design, a brief summary of our observations about the differences between crisp and fuzzy systems is given below.

1. In fuzzy applications the effect of the disturbances (i.e., effect of d on y in Fig. 3) is usually not modeled separately from the process input/output (*i.e.*, the effect of u on y). In self-learning control applications (e.g., Fig. 10) this means, for example, that an unmeasured

disturbance will produce changes in the fuzzy model which may later result in poorer servo performance. Also since the disturbances are not modeled explicitly it is difficult to do *robustness* or *stability* analysis (see Sect. 5.6 (h)).

2. Long-range fuzzy prediction is usually less accurate than with crisp systems due to the nonlinear effect of the membership functions, the composition used, the relational model not capturing the process dynamics very well, etc.

3. Formal, direct inversion of a (relational) fuzzy model does not usually give satisfactory control results, (.g., $G_c^* = G_p^{-1}$, where G is a relational model, does not give the "perfect control" results referred to in the preceding discussion).

4. Fuzzy optimization algorithms when applied to the calculation of the *optimal* control action (see Sect. 3.4) frequently find a very "flat optimum" so the calculated control, $u(k)$, tends to be "sluggish" and to "wander".

4 Fuzzy predictive controller design

The following fuzzy predictive controller design has the same structure as shown in Fig. 9 and 10, but the details of the formulation are different than the discrete controller design.

4.1 Process and prediction models

For the controller design, define $u \in U$ and $y \in Y$ as the crisp input and output respectively. Note again that the fuzzy variables possess a tilda. Also, let $\tilde{u} = \{\tilde{u}_i | i = \{1, 2, ..., m\}\} \in U$ and $\tilde{y} = \{\tilde{y}_l | l = \{1, 2, ..., n\} \in Y$ be the *fuzzy* spaces of the input and output respectively, all defined on the finite *fuzzy* universes of discourses indicated.

The *Process Model* is assumed to be the first order fuzzy state model with time delay, τ

$$\tilde{y}(k+1) = \tilde{R}^{\circ}\tilde{y}(k)^{\circ}\tilde{u}(k-\tau), \tag{1}$$

where $\tilde{R}$ is the relational process model and $^{\circ} \in O$ stands for the family of compositions (i.e., *max-min*, *max-product*, etc.).

The *Prediction Model* is also a first order plus delay fuzzy state model which, since there is no explicit disturbance modeling, is obtained by simply replacing k by $(k + \tau)$ in the *Process Model* of (1). Note that for *practical* applications, performance can be improved by incorporating some method of estimating the future effect of current disturbances and handling modeling errors. In Fig. 10, $r(k)$ represents the combined effect of disturbances (e.g., $L(k)$), noise and errors between the model-based prediction versus the measured process output. The simplest method of prediction sets $\{r(k + i) = r(k), i = 1, 2, \ldots\}$. In some implementations the *Filter* block of Fig. 10 is expanded to include a model-based prediction or a time series forecasting algorithm.

4.2 Calculation of the controller output

The control calculation for the proposed single-point, τ-step ahead, predictive controller assumes that $\{u(k+i) = u(k), i = 1, 2, \ldots, \tau - 1\}$ so that only one control value, $\Delta u(k)$, needs to be calculated at time, k. As shown in [3], a *fuzzy* prediction model can be derived from (1) by successive back substitution and put in the form

$$\tilde{y}(k + \tau + 1) = \tilde{Q}^{\circ}\tilde{u}(k), \tag{2}$$

where $\tilde{Q} = \tilde{K}_{k-1}{}^{\circ}\tilde{K}_{k-2}{}^{\circ}\tilde{K}_{k-3}{}^{\circ}\ldots{}^{\circ}\tilde{K}_{k-\tau}{}^{\circ}\tilde{y}(k)$. In ideal model-based control, as discussed in Sect. 3.1, if the predicted future output is replaced by the desired value, y_{sp}, then the control action can be calculated from $u(k) = Q^{-1} \cdot y_{sp}(k+\tau+1)$. However, in *practical* control algorithms it is necessary to include a tuning parameter to accommodate individual user preferences and practical factors such as model-process-mismatch, noise, robustness, etc. The approach adopted for the proposed *fuzzy predictive controller* is as follows:

1. Calculate the *mean-level* or *steady state* control action, $u_{gain}(k)$, which will result in $y(k + i) \to y_{sp}(k + i)$ for large values of i, (i.e., at steady state). This $u_{gain}(k)$ is the smallest, single-step control action that will achieve the desired setpoint.

2. Calculate the *one-step-ahead* or *deadbeat* control action, $u_{dync}(k)$, that (allowing for the time delay, τ) will drive the output from its current state to the desired state in one control interval (i.e., $y(k + \tau + 1) = y_{sp}(k + \tau + 1)$). This is the strongest single-step control action that needs to be taken at time k.

3. Define the *actual* controller output at time k as a linear combination of $u_{gain}(k)$ and $u_{dync}(k)$

$$u(k) = \alpha \cdot u_{gain}(k) + (1 - \alpha) \cdot u_{dync}(k), \tag{3}$$

so that the user can specify conservative, mean-level control ($\alpha = 1$), aggressive deadbeat control ($\alpha = 0$) or any combination ($0 < \alpha < 1$).

The procedure for calculating u_{gain}, u_{dync}, and the actual controller output, $u(k)$, is outlined next and detailed in [3] . The procedure for calculating the controller output is illustrated by Fig. 11.

4.2.1 Calculation of the mean-level control action, u_{gain}

The mean-level control action, $\tilde{u}_{gain}(k)$, is calculated using a relational *gain* matrix, $\tilde{G}(\tilde{y}(k+\tau+1), \tilde{u}(k))$

$$\tilde{u}_{gain}(k) = \tilde{G}^{-1} \circ \tilde{y}_{sp}(k+\tau+1). \tag{4}$$

Thus the discrete control action, $u_{gain}(k) = defuzz(\tilde{u}_{gain}(k))$, would result (with no feedback) in an output trajectory that is equal to the natural *open-loop step response* of the *stable process* with a final value of $y_{sp}(k+\tau+1)$.

4.2.2 Calculation of the one-step-ahead or deadbeat control action, u_{dync}

The calculation of $u_{dync}(k)$ uses a predictive fuzzy algorithm based on the identified relational matrix, $\tilde{R}$, so these calculations take into account the process dynamics. The strategy for the development of this dynamic input calculation, $u_{dync}(k)$, is that the controller output (calculated using a first order predictive model) should move the process from the current state to the desired setpoint state within one control interval.

Information available at time k includes the process model and the predictive model [3] which by iteration can be put in the form of (2) where $\tilde{Q} = \tilde{K}_{k-1} \circ \tilde{K}_{k-2} \circ \ldots \circ \tilde{K}_{k-\tau} \circ \tilde{y}(k) = known$ and $\tilde{K}_{k-i} = \tilde{R} \circ \tilde{y}(k+\tau-i) \circ \tilde{u}(k-i)$. It has been shown [7] that minimizing the fuzzy control error does not minimize the crisp control error (i.e., $min|\tilde{y}_{sp}(k) - \hat{\tilde{y}}(k)| \nRightarrow min|y_{sp}(k) - \hat{y}(k)|$). Because of this, the calculated value of the controller output is adjusted iteratively by minimizing the crisp difference, $|y_{sp}(k+\tau+1) - \hat{y}(k+\tau+1) - r(k)|$ at each step. The residual, $r(k)$, includes the structured disturbance, $L(k)$, measurement noise, and model-process-mismatch. In the proposed fuzzy controller, the *filtered* value of $r(k)$ is calculated from

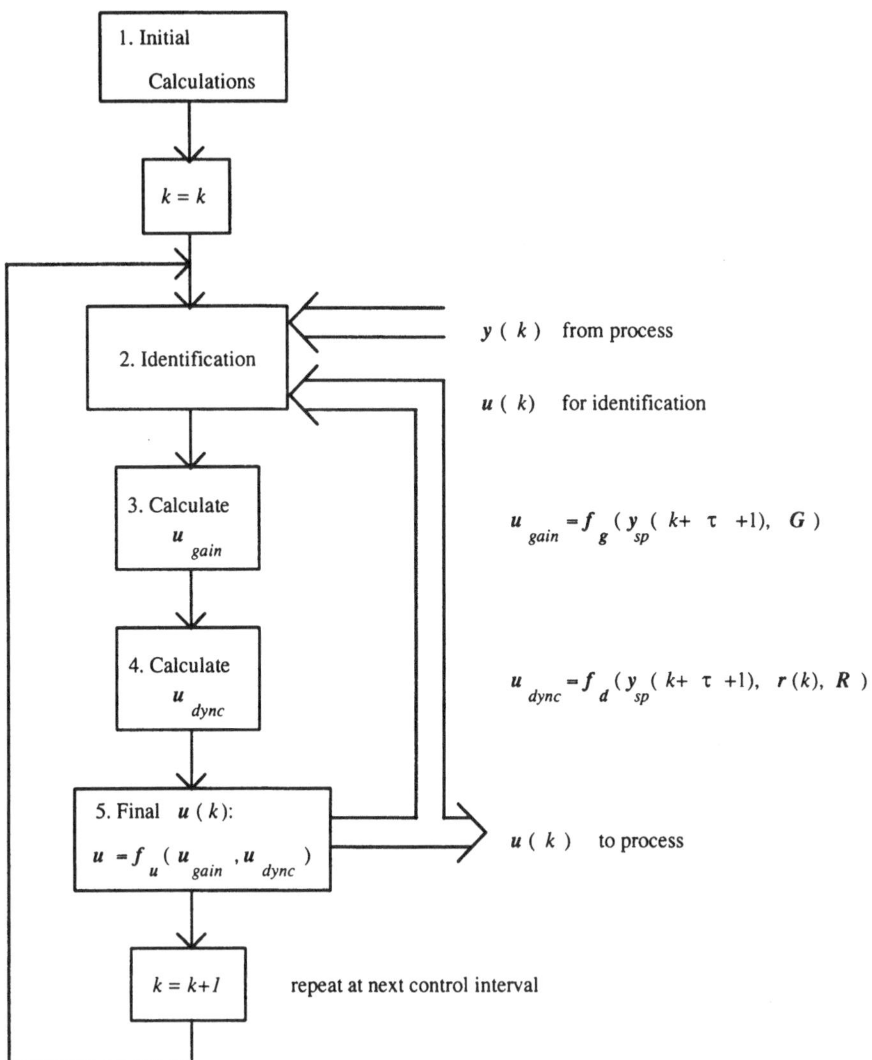

Figure 11: Flow diagram of the predictive controller.

$$r(k) = \eta \cdot \{y(k) - \hat{y}(k)\} + (1-\eta) \cdot \{y_{sp}(k) - \hat{y}(k)\} - \{y_{sp}(k+\tau+1) - \hat{y}(k+\tau+1)\}. \tag{5}$$

In (5), the first difference $y(k) - \hat{y}(k)$, would be equal to the disturbance, $L(k)$, if the modeling were exact and the second difference $y_{sp}(k) - \hat{y}(k)$, is a measure of the *prediction or modeling* error. The parameter $0 < \eta < 1$ is chosen by the control system designer to determine the weighting of the disturbance versus the modeling error in the calculation of $r(k)$. The third difference is added to improve the prediction capacity of the fuzzy logic controller. The value $\hat{y}(k+\tau+1)$ in (5) and Fig. 10 is the defuzzified value of the current prediction estimate

$$\tilde{y}(k+\tau+1) = \tilde{R}^{\circ}\tilde{y}(k)^{\circ}\tilde{u}(k-1). \tag{6}$$

The fuzzy input $u_{dync}(k)$ is calculated by iteratively searching for the process input that minimizes the discrete error, $e(k) = |y_{sp}(k+\tau+1) - \hat{y}(k+\tau+1) - r(k)|$. The search is made by adjusting the values of the individual components of the fuzzy input vector as a function of the discrete error. For illustration purposes, assume that the control application has a fuzzy input vector of dimension 5 with an initial value of [0, a, b, c, 0]. The initial value could be equal to $\tilde{u}_{gain}(k)$ or the previous $\tilde{u}_{dync}(k-1)$ or $\tilde{u}(k)$. Let the crisp error be, $e = y_{sp}(k+\tau+1) - \hat{y}(k+\tau+1) - r(k)$. Then

$$u_{dync}(k) = [0,\, a + f(e),\, b,\, c - f(e),\, 0] \;\text{ if }\; e > 0, \tag{7}$$

$$u_{dync}(k) = [0,\, a - f(e),\, b,\, c + f(e),\, 0] \;\text{ if }\; e < 0, \tag{8}$$

where

$$f(e) = s \cdot \gamma \cdot |e|, \tag{9}$$

and

$$s = \left\{ \begin{array}{ccccc} +1 & if & \tilde{u}\uparrow & \Longrightarrow & \tilde{y}\uparrow \\ -1 & if & \tilde{u}\uparrow & \Longrightarrow & \tilde{y}\downarrow \end{array} \right\}. \tag{10}$$

$\gamma \geq 1$ is a tuning parameter for the convergence rate.

The adjustment of the fuzzy input vector, $u_{dync}(k)$, by (7) to (10) is carried out iteratively until the error, $e(k)$, is sufficiently small but with a maximum on the iterations per control interval (set to 10 for examples in this article).

4.2.3 Calculation of the actual controller output, $u(k)$

The process input, $u(k)$, calculated from the weighted average of the *gain* (mean-level) input, $u_{gain}(k)$, and the *dynamic* (deadbeat) input, $u_{dync}(k)$ where $0 < \alpha < 1$, given by (3), still may not be aggressive enough to force the process to reach the setpoint within the required performance specifications. This is mainly due to the *poor* predictive capabilities of the relational matrix for large τ. To compensate for any shortfall that may result from the calculated process input, $u(k)$, an input tuning factor, $a(k)$, is introduced based on the final error, $e(k)$, of the iterative procedure for $u_{dync}(k)$

$$a(k) = 1 + \beta \cdot |e|, \tag{11}$$

where $0 \leq \alpha \leq 1$ determines the aggressiveness of the input tuning factor. The final calculated process input value, $u_{final}(k) = a(k) \cdot u(k)$, is obviously a function of the accuracy of the predictive model and will not normally give "perfect control". If the final predictive error, e, is very small, the tuning factor $a(k) \approx 1$, and there is no adjustment to the process input, $u(k)$.

5 A comparison of fuzzy relational versus PI control

The proposed fuzzy-relational controller and a conventional Proportional plus Integral (PI) controller were both applied to a non-linear, simulated process defined such that the large process gain variations made feedback control very difficult.

5.1 Simulated process model

The process to be controlled is [13]

$$y(k) = \sigma(a_1 y(k-1) + b_1 sinh(0.8 \cdot u(k-1))^4), \tag{12}$$

where

$$a_1 = 0.0, b_1 = 10.0,$$

and

$$\sigma(x) = \frac{2}{0.995 \cdot (1 + \exp(-x))}.$$

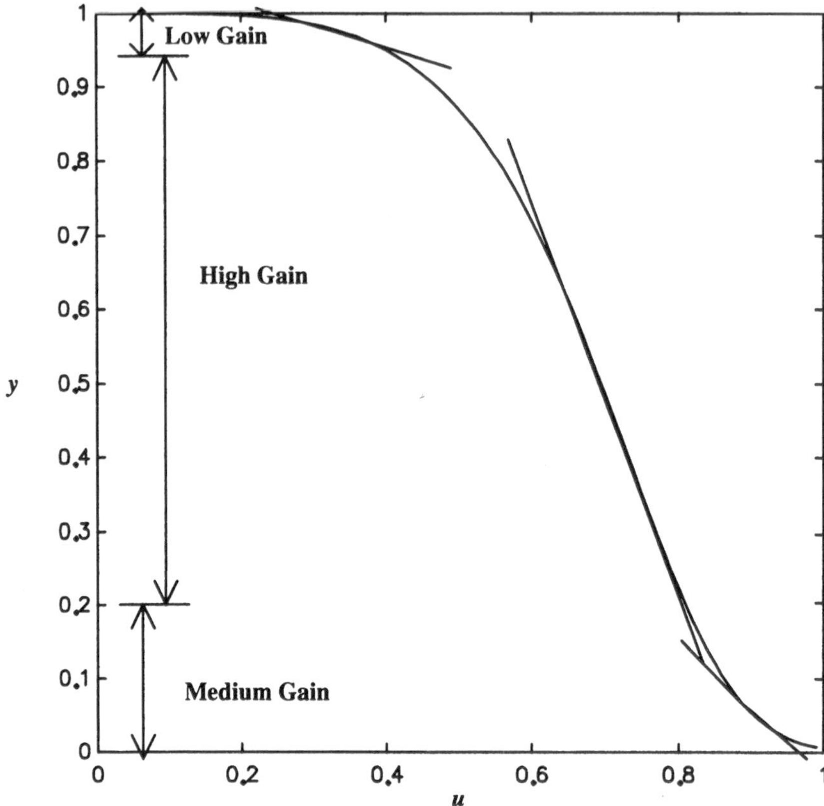

Figure 12: Nonlinear processes showing regions with low, medium and high gain.

As shown in Fig. 12, the steady state gain of this process varies significantly. In later discussions reference is made to *low*, *high* and *medium* gain regions as defined in Fig. 12.

5.2 Conventional PI control

The incremental PI controller used for these studies is

$$\Delta u(k) = K_c[\{e(k) - e(k-1)\} + \frac{\Delta t}{\tau_i}e(k)], \tag{13}$$

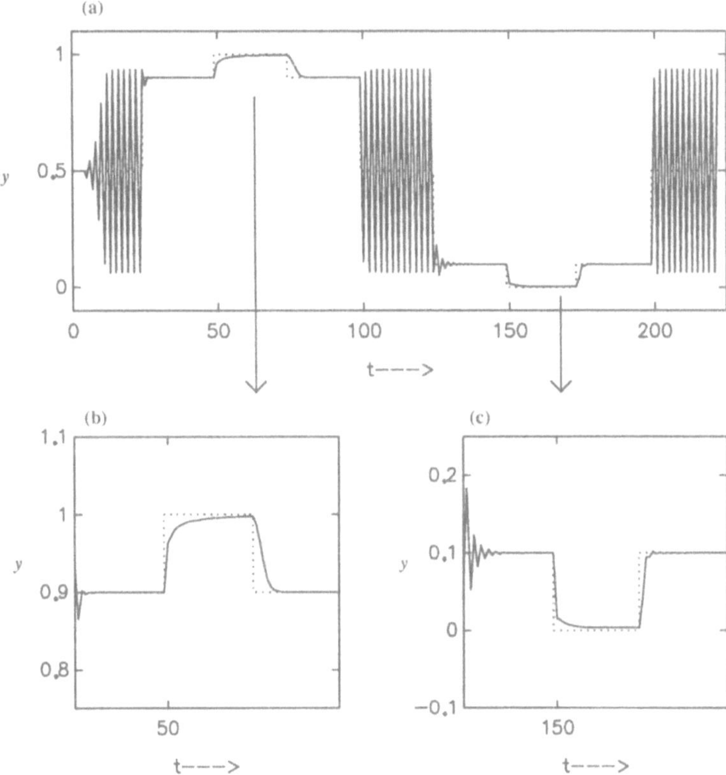

Figure 13: PI control tuned for Medium Region (a) Output profile; (b) Magnification of low gain region; (c) Magnification of medium gain region; ($K_c = -0.0875; \tau_i = 0.1$) ($J = 0.1332$) (Actual–; Setpoint $\cdots$).

where $\Delta t \equiv$ one sampling interval and K_c and τ_i are the proportional and integral tuning constants, respectively. Closed-loop control performance is a function of the *loop-gain* (i.e., a function of the product of the controller gain K_c and the process gain). Therefore, as expected, a single value of the controller gain K_c does not give good control over the full range of process operation due to the large changes in process gain shown in Fig. 12. If the PI controller is tuned for good performance in the *medium gain* region of the process ($y \approx 0.05$), then performance in the *high gain* region ($0.2 \leq y \leq 0.8$) is highly oscillatory and obviously unsatisfactory, as shown by the oscillations in Fig. 13.

If the PI controller is *detuned* (i.e., lower K_c) to provide satisfactory control in the *high gain* region of the process, then performance in the *low gain* region ($0.95 \leq y \leq 1.0$) is sluggish and less precise. It should be noted that if the process gain variations (Fig. 12) were known and non-time varying then the controller gain, K_c, could be varied as a function of the process conditions, such that the product of the process and controller gains remains approximately constant. Control performance would then be more uniform over the full operating range.

5.3 The fuzzy model

A first order plus delay fuzzy state space model, (1), is used in the identification and controller calculations regardless of the order of the actual process. Such an approximation is common for the control of over-damped process systems. Since the $\{u, y\}$ data obtained from the process simulations is crisp, it must be *fuzzified* and the resulting controller output *defuzzified.* Partitioning of the data is subjective, but $\rho = 5$ referential fuzzy sets (or clusters) appeared adequate for both the input and the output for this application. The shape of the referential fuzzy sets was chosen to be isosceles triangles, which satisfy the requirements of being normal and convex. These fuzzy membership functions were placed over the input-output space so that the space was *completely covered* and the referential sets had a 50% overlap. The *defuzzification* formula was chosen to be the weighted average of the center of area method which uses all the information in the *fuzzy* output vector. The minimization criterion used to judge the performance of the controller (for a time series containing differences between the *defuzzified* (crisp) output and the setpoint as defined by

$$J = \frac{1}{N - \tau - 1} \sum_{k=\tau+2}^{N} |y_{sp}(k) - \hat{y}(k)|. \tag{14}$$

5.4 Tuning and implementation of the fuzzy predictive controller

There are six tuning parameters associated with the proposed fuzzy predictive controller:

- α varies the weighting between u_{gain} and u_{dync} in the final value of u [in (3), $\alpha = 0.45$].
- β varies the intensity of the tuning factor a for u_{dync} [in (11), $\beta = 0.25$].

- η is a filter parameter that varies the weighting between current actual error and predicted error [in (5), $\eta = 0.95$].
- ε is the difference tolerance (i.e., $|y_{sp} - \hat{y}| \leq \varepsilon$) for testing the convergence of the u_{dync} search algorithm [Sect. 4.2.2, $\varepsilon = 0.01$].
- γ is the tuning factor for the iterative input search algorithm [Sect. 4.2.2, $\gamma = 3.0$].
- ω is a filter parameter that varies the length of the window over which the error, e, is averaged [$\omega = 1$].

The first three parameters regulate the smoothness and accuracy of the control. These parameters required tuning in order to obtain acceptable control. The last three parameters were arbitrarily set and maintained constant. That is not to say that these latter parameters can not be tuned, just that the values determined for these parameters during preliminary testing were found to be adequate for the simulations presented. The tuning objective for these simulations was to minimize the J values of (14) for each composition, while maintaining a *smooth* (i.e. non-oscillatory) control action.

Two relational matrix models are required for this predictive controller, the dynamic matrix model, $\tilde{R}$, and the gain matrix model, $\tilde{G}$. Since the gain model is a steady state model which relates steady state input to steady state output, the dimensions of this relational matrix can be larger than the dynamic matrix model. For example an 11×11 two dimensional *gain* model requires less computation than a 5×5×5 three dimensional *dynamic* model. With the increased dimension of the gain model there is an increase in the accuracy of the controller, particularly when the process is close to the setpoint.

The averaging identification technique [3] for the *gain* matrix is given by (15) below. It is noted that when the gain model, $\tilde{G}$, is identified by an averaging technique then the same model is valid for prediction of either input or output (i.e., $\tilde{G}(\tilde{u}, \tilde{y}) = \tilde{G}^{-1}(\tilde{y}, \tilde{u})$).

$$G(\tilde{u}_{k-\tau-1}, \tilde{y}_k) = \frac{\sum_{k=\tau+2}^{N} \prod(\tilde{u}_i^{(k-\tau-1)}, \tilde{y}_l^{(k)})}{\sum_{k=\tau+2}^{N} \kappa(\tilde{u}_i^{(k-\tau-1)}, \tilde{y}_l^{(k)})}, \tag{15}$$

where $1 \leq l \leq p$, $1 \leq i \leq m$ and $\kappa(u, y) = 1$ if $u \cdot y > 0$, otherwise $\kappa(u, y) = 0$.

Finally, for tuning processes with a fast response the tuning factor $a(k)$ for the dynamic input calculation, $u_{dync}(k)$, can be adjusted so that it is not as aggressive as required for highly overdamped processes. Thus the parameter a in (11) can be redefined for underdamped processes by replacing the 1 by

0 such that $a(k) = \beta \cdot |e|$. This adjustment reduces the aggressiveness of the calculated input and therefore reduces overshoot.

5.5 Control performance

The proposed fuzzy logic controller was applied to the non-linear process problem described in Sect. 5.1 with the objective of *good* overall control, minimum overshoot and non-oscillatory control action.

For this study, the PI controller was initially tuned for the *medium gain* region of the given process and the controller was then required to provide servo control over the entire process range. Fig. 13 shows the results of this application. Clearly from Fig. 13(c), *good* control is evident in the *medium gain* region. Figure 13(b) illustrates the slower yet stable control in the *low gain* region. However, control in the *high gain* region is oscillatory, as shown in Fig. 13(a). The output y remains bounded because the fuzzy control algorithm inherently imposes finite limits on the control action, u. Thus PI controllers produce poorer and/or "unstable" results in regions outside the tuned range.

The obvious direction to take, based on the results of Fig. 13 is to reduce the controller gain. Thus the PI controller was retuned for the *high gain* region and servo control was again tested over the entire process range. Fig. 14 shows the results of the retuning. The PI controller produces *good* results in the *high gain* region, however, setpoint tracking is slower in both the *medium* and *low* gain regions.

Three-level gain scheduling of the PI controller was then applied in an attempt to obtain *good* control response in all regions. As shown in Fig. 15, the response has improved in the *low gain* region (i.e., is faster than in Fig. 14). However, response is *poorer* in the *medium gain* region due to the detuning required to improve control in the *high gain* region (i.e., to eliminate the oscillatory response at the 200th sampling instance). Overall control results using PI gain scheduling improved to $J = 0.015$, compared to with the results of tuning for only the high gain region ($J = 0.017$) or the medium gain region ($J = 0.133$).

The proposed fuzzy logic controller was then applied to the same non-linear process problem. The tuning objective for fuzzy logic control was *good* overall control, with minimum overshoot. Fig. 16 illustrates the results obtained using the fuzzy logic controller. *Clearly good overall control is obtained over the entire process range* and the manipulated variable, u, is relatively smooth and does not show any sudden jumps with setpoint changes. Comparing the J results of the fuzzy logic controller, $J = 0.013$, with the best PI

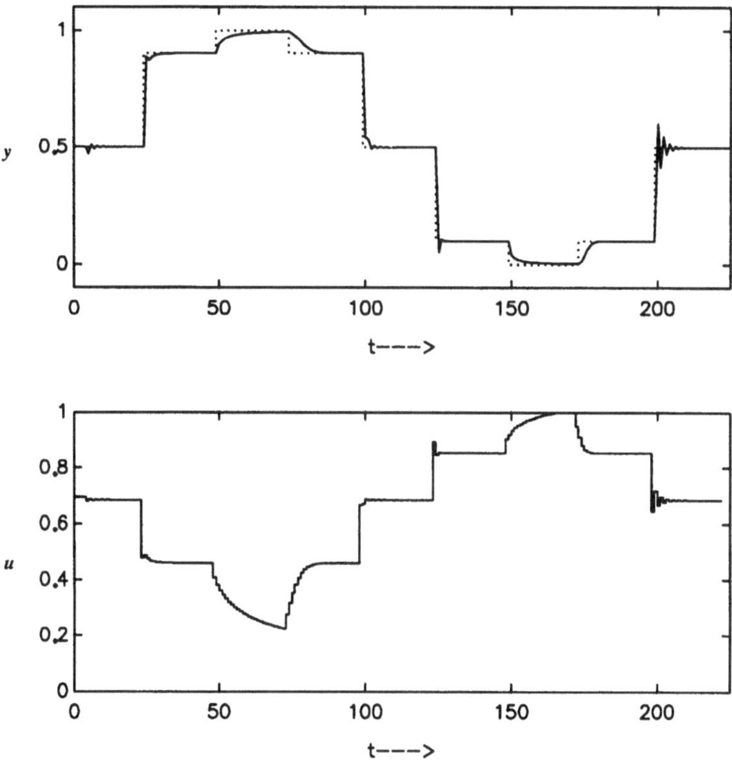

Figure 14: PI control tuned for high gain region ($K_c = -0.0472$; $\tau_i = 0.1$) ($J = 0.0170$) ($Actual -$; $Setpoint \cdots$).

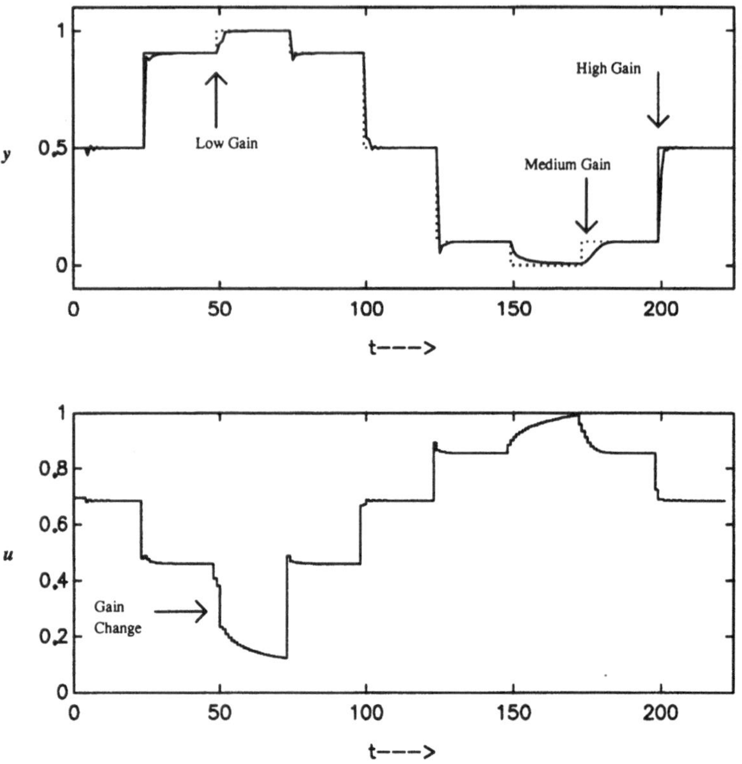

Figure 15: PI control with gain scheduling ($K_c(Low) = -0.3416$; $K_c(High) = -0.0098$; $K_c(Medium) = -0.0294$; $\tau_i = 0.1$) ($J = 0.0150$) (*Actual* –; *Setpoint* $\cdots$).

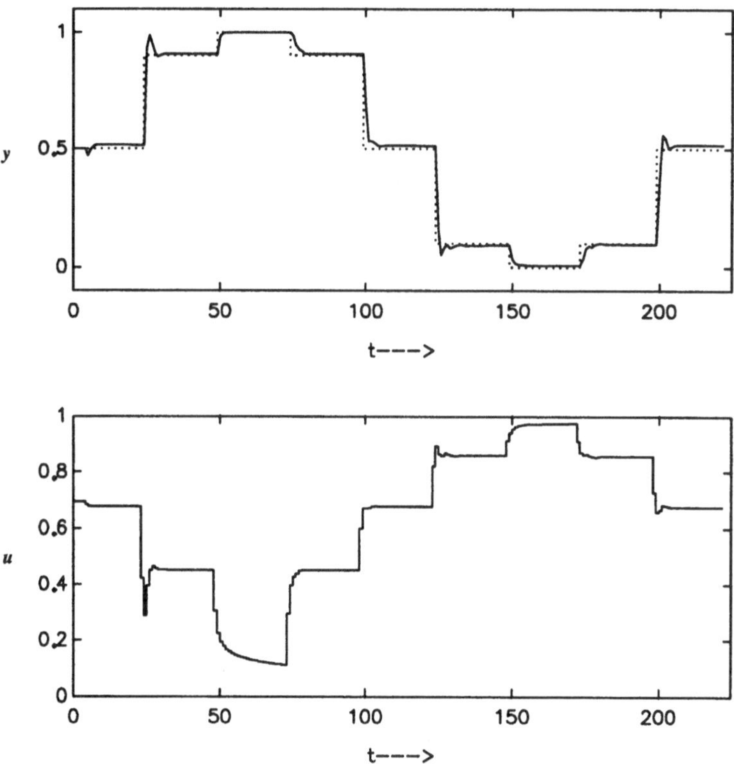

Figure 16: Fuzzy control with *max-product* composition ($\alpha = 0.45; \beta = 0.25; \eta = 0.95; \varepsilon = 0.01; \gamma = 3.0; \omega = 1$) ($J = 0.013$) (*Actual* –; *Setpoint* $\cdots$).

controller, with gain scheduling, $J = 0.015$, shows that the fuzzy logic controller operates *better* over the entire process range, based on this minimum distance criterion. *Both the gain scheduled PI controller and the fuzzy predictive controller require a priori knowledge of the process gain variations. However, the fuzzy predictive controller gains are inherent in the process model and the activation of self-learning permits the controller to handle the same full range of process gain variations for slowly changing processes.*

For all the simulations performed with the fuzzy controller the absolute control error, $e = y_{sp} - \hat{y}_c$, was less than 5% of the process output. Thus the fuzzy predictive control algorithm has demonstrated that it is capable of successfully controlling a variety of process situations. As well, additional simulation results presented in [3] demonstrated that the algorithm can handle larger ($\tau \approx 6$) process time delays.

Subsequently, Wong [26], [27] tested the fuzzy predictive controller developed here against the fuzzy predictive controller proposed by Postlethwaite [19]. Both controllers utilize a numerical search to obtain the control action that minimizes the predicted future error. The two controllers were laboratory tested for both servo and regulatory performance on a pH neutralization process [12] and a higher order process [15]. Based on the mean square of the error, both controllers were found to give good performance, however the controller presented here was slightly better.

5.6 Controller convergence and stability

Two important objectives for controller design in any domain (continuous, discrete, fuzzy, etc.) are *convergence* and *stability*. Unfortunately, it is not possible to get general results or to rigorously test for these characteristics when a *fuzzy* controller is used in *practical* applications. However, the model-based approach outlined in this article does give some important insights based on simple, idealized cases that provide guidelines for practical applications. For example (in over-simplified, incomplete terms):

(a) Perfect open-loop servo-control (Fig. 8 with $G_m = Filter = 0$), such that $y = y_{sp}$, is obtained if $G_c = G_p^{-1}$ and sufficient conditions for stability are that G_c and G_p be stable.

(b) Similar conclusions apply to feedforward control (see Sect. 3.3).

(c) The principles of *internal model control* [22] show that Fig. 8 (with $\hat{y}_2 = 0$, $Filter = 1$, $G_m = G_p$ and no prediction) can be simplified to Fig. 6 and is essentially an open-loop system (similar to (a) and (b) above). Sufficient conditions for stability are again that G_c and G_p be stable. Furthermore, it can be shown that if $G_m \neq G_p$, then there always exists a filter such that the (continuous/discrete) system is stable [22]. (The trivial filter for servo control of stable processes is that $Filter = 0$).

(d) If the model-based controller, implied by Figs. 8 and 10, calculates the *increment* in control action $\Delta u(k) = u(k) - u(k-1)$ required to eliminate the (predicted) error, then the equivalent of integral action is introduced into the feedback system and guarantees zero offset (i.e., convergence of $y(k)$ to $y_{sp}(k)$ as time $k \to \infty$) for sustained (step) changes in y_{sp} and/or L. Even without incremental control action, offset to step disturbances can be eliminated by "search" or "adjustment" mechanisms built into the fuzzy controller to ensure that the discrete error is eliminated. Also the self-learning implementation will normally modify the process model until the offset is (approximately) eliminated.

(e) For discrete single-point, τ-step ahead, predictive control ($\hat{y}_c(k+\tau+1)$ = accurate prediction in Fig. 10) it is reasonable to conclude for perfect modeling ($G_m = G_p$) that the same conditions for stability and convergence apply as discussed above.

(f) When an *optimization* algorithm is used to calculate the control action at time k, rather than a direct, analytical calculation (e.g., $u(k) = G_p^{-1} \cdot y_{sp}(k+1)$ as per Sect. 3.4) then stability/convergence analysis of the control system expands to include all the stability, convergence, reproducibility etc. problems associated with weighted, constrained optimization and is, in general, intractable.

(g) For multi-point predictive discrete controllers (formulated in terms of $\{\hat{y}(k+i), i = 1, 2, \ldots, N\}$) it is possible to *prove* stability as $N \to \infty$ [21] but when N is finite there is no general proof.

(h) *Stability* and *convergence* analysis for the general case of feedback control in the presence of model-process-mismatch ($G_m \neq G_p$) leads to the large and difficult body of control literature on *robustness* (e.g., [2], [14], [16]]). In general, guarantees of stability and/or convergence are not possible unless there are known *bounds* or *characteristics* of the model-process-mismatch expressed in terms of model parameters, regions in the frequency domain, etc. This information is not normally available for fuzzy systems.

Summary: Since fuzzy models are never perfect (and the model-process-mismatch in fuzzy formulations is not usually characterized or within known bounds), the direct, quantitative analysis of the stability and convergence of fuzzy feedback systems requires extensions to the complex area of control theory referred to in (f) through (h). However, points (a) through (e) provide useful guidelines for the design and tuning of fuzzy, model-based controllers of the type developed in this work.

6 Discussion

The *fuzzy* relational model-based, self-learning, and predictive controller presented in this article gave better servo performance than a conventional gain-scheduled PIcontroller when applied to a very non-linear process. It was also found as part of this work that the *max-product* composition gave consistently better results than the traditional *max-min* composition [7].

The *fuzzy* controller was developed based on the same fundamental concepts used to develop predictive controllers based on discrete mathematical models but several important differences were noted [3] between fuzzy and discrete systems as discussed in Sect. 3.5. For example, unlike *deterministic* models that may start with an initially erroneous but complete model, *fuzzy* models require a large amount of process data that covers the entire operating range in order to be considered complete. Also *deterministic* models that include explicit modeling of the noise/disturbance terms (e.g., the widely used Box-Jenkins type models) require formulation of a predictor to account for the future effect, on the predicted process output, of current and past disturbances. This is not required in fuzzy controllers of the type presented here because the noise/disturbance effects are "lumped" into the relational process model $\tilde{R}$.

A *deterministic*, optimal, model-based controller will always outperform the corresponding fuzzy controller if the model is accurate. The types of applications where *fuzzy* controllers may prove advantageous are those containing modeling and/or implementation uncertainties (e.g., high-order or nonlinear processes, inherently-fuzzy measured or manipulated variables, poorly defined and variable noise and/or disturbance characteristics, control strategies that include both mathematical and fuzzy decision logic, etc.). Since the structure and concepts of the fuzzy and deterministic (discrete) predictive controllers are so similar, a self-learning fuzzy controller implemented at plant startup can be improved iteratively and evolve into the corresponding deterministic/crisp controller as more experience and more accurate models are obtained.

6.1 Future work

The final result of this work is a *practical fuzzy logic controller* suitable for industrial applications. However, future work in the area of fuzzy model-based control could include the following.

1. Modification of the algorithms presented in this article to handle a larger dynamic matrix, $\tilde{R}$. To save computation time, the calculation of the "dynamic" control action, $u_{dync}(k)$, could be done only when the error was large. *Near* the setpoint, *mean level control* based on the 2-dimensional gain matrix, $\tilde{G}$, could be employed.

2. The control algorithm presented in this article does not include the defuzzification algorithm in the overall optimization. Thus the defuzzification procedure chosen may result in some auxiliary error which in turn would weaken the fuzzy model at the numeric level. Modification of the control algorithm to include joint optimization of the defuzzification and the relational model might give better performance.

3. There are several features documented in the *deterministic* predictive control literature that could be transformed into the *fuzzy* domain. For example, the formulation could be extended to include output trajectories, $\{\hat{y}(k+i), i = 1, 2, \ldots, N\}$ and/or control trajectories $\{u(k+j-1), j = 1, 2, \ldots, N_u \leq N\}$.

4. Most fuzzy control formulations implicitly impose bounds on the control action, $u(\cdot)$, (e.g., via the defuzzification step). The physical process always has physical limits on the input and output variables. Output constraints can be mapped into input constraints using the process model. It therefore appears straight forward to include the physical input/output constraints in the fuzzy control algorithm.

References

[1] Batur, C., Srinivasan, A., Chan, C.-C. (1995), "Fuzzy Model Based Fuzzy Predictive Controllers", *Journal of Intelligent and Fuzzy Systems*, Vol. 3, pp. 117-130.

[2] Bitmead, R.R., Gevers, M., Wertz, H. (1990), *Adaptive Optimal Control - The Thinking Mans GPC*, Prentice Hall, USA.

[3] Bourke, M.M. (1995), *Self-Learning Predictive Control using Relational-Based Fuzzy Logic*, Ph.D.Thesis, University of Alberta. (via anonymous ftpsite prancer.eche.ualberta.ca in /pub/reports/BOURKE/THESIS.)

[4] Bourke, M.M., Fisher, D.G. (1994), "The Complete Resolution of Cartesian Products of Fuzzy Sets", *Fuzzy Sets and Systems*, Vol. 63, pp. 111-115.

[5] Bourke, M.M., Fisher, D.G. (1995a), "Calculation and Application of a Minimum Fuzzy Relational Matrix", *Fuzzy Sets and Systems*, Vol. 74, pp. 225-236.

[6] Bourke, M.M., Fisher, D.G. (1995b), "Convergence, Eigen Fuzzy Sets and Stability Analysis of Relational Matrices", *Fuzzy Sets and Systems*, Vol. 81, pp. 227-234.

[7] Bourke, M.M., Fisher, D.G. (1997), "A Comparative Analysis of Identification Algorithms for Fuzzy Relational Matrices", *Fuzzy Sets and Systems*, under review.

[8] Bremner, H., Postlethwaite, B. (1994), "The Development of a Relational Fuzzy Model Based Controller for an Industrial Process", *IEEE International Conference on Fuzzy Systems*, Orlando, Fl, Vol. 1, pp. 539-543.

[9] De Oliveira, J.V., Lemos, J.M. (1994), "Fuzzy Model Based Long-Range Predictive Control", *IEEE International Conference on Fuzzy Systems*, Vol. 1, pp. 378-381.

[10] De Oliveira, J.V., Lemos, J.M. (1995), "Long-range Predictive Adaptive Fuzzy Relational Control", *Fuzzy Sets and Systems*, Vol. 70, pp. 337-357.

[11] Graham, B.P., Newell, R.B. (1989), Fuzzy Adaptive Control of a First-Order Process, *Fuzzy Sets and Systems*, Vol. 31, pp. 47-65.

[12] Hall, R.C., Seborg, D.E. (1989), Modeling and Self-Tuning Control of a Multivariable pH Process Part I: Modeling and Multiloop Control. In: *Proceedings of the American Control Conference*, Pittsburgh, pp. 1822-1827.

[13] Hernández, E., Arkun, Y. (1993), "Control of Nonlinear Systems Using Polynomial ARMA Models", *AIChE Journal*, Vol. 39, pp. 446-460.

[14] Kwakernaak, H., Sivan, R. (1972), *Linear Optimal Control Systems*, Wiley Interscience, USA.

[15] McIntosh, A.R. (1987), *Performance and Tuning of Adaptive GPC*, Masters Thesis, University of Alberta.

[16] Morari, M., Zafiriou, E. (1989), *Robust Process Control*, Prentice Hall, USA.

[17] Pedrycz, W. (1993), *Fuzzy Control and Fuzzy Systems*, Second Extended Edition, John Wiley & Sons, Inc., New York, N.Y.

[18] Pedrycz, W. (1994), "Why Triangular Membership Functions?", *Fuzzy Sets and Systems*, Vol. 64, pp. 21-30.

[19] Postlethwaite, B. (1994), "A Model-Based Fuzzy Controller", *Trans IChemE.*, Vol. 72, pp. 38-46.

[20] Postlethwaite, B. (1996), "Building a Model-Based Fuzzy Controller", *Fuzzy Sets and Systems*, Vol. 79, pp. 3-13.

[21] Rawlings, J.B., Muske, K.R. (1993), "The Stability of Constrained Receding Horizon Control", *IEEE Transactions of Auto Control*, Vol. 38, pp. 1512-1516.

[22] Rivera, D.E., Morari, M. (1986), "Internal Model Control: PID Controller Design", *Ind. Eng. Chem. Process Des. Dev.*, Vol. 25, pp. 252-265.

[23] Shaw, I.S., Krüger, J.J. (1992), "New Fuzzy Learning Model with Recursive Estimation for Dynamic Systems",*Fuzzy Sets and Systems*, Vol. 48, pp. 217-229.

[24] Skrjanc, I., Kavsek-Biasizzo, K., Matko, D. (1996), "Fuzzy Predictive Control Based on Fuzzy Model", EUFIT'96, pp. 1864-1869.

[25] Song, J.J., Park, W. (1993), "A Fuzzy Dynamic learning Controller for Chemical Process Control", *Fuzzy Sets and Systems*, Vol. 54, pp. 121-133.

[26] Wong, C.H., Shah, S.L., Fisher, D.G. (1995), *Fuzzy Process Identification and Control*, M. Sc. Thesis, University of Alberta, 1995.

[27] Wong, C.H., Shah, S.L., Fisher, D.G. (1996), "Adaptive Fuzzy Relational Predictive Control", *Fuzzy Sets and Systems*, under review.

A Simplified Fuzzy Relational Structure for Adaptive Predictive Control

J. Valente de Oliveira
UBI – Dept. Mathematics and Computer Science
Rua Marquês d'Ávila e Bolama, 6200 Covilhã, Portugal

J. M. Lemos
INESC – Research Group on Control of Dynamic Systems
Rua Alves Redol 9, Apartado 13069, 1000 Lisboa, Portugal

1 Introduction

Model based predictive controllers have a number of appealing features such as:

- The ability to take into account the impact of the current control action on the future process state. This is a useful when dealing with non-minimum phase behaviors (e.g., to stablize plants whose open-loop response to a positive input step results first in a decrement of the output and only afterwards, in an increment), unknown or partially unknown dynamics.
- The ability to accomodate knowledge about future requirements on the plant state represented in terms of a pre-defined tracking reference signal.
- Effectiveness of control even when the predictor is a coarse approximator of the plant dynamics
- The ability to deal with multiple objectives and constraints, e.g., on the manipulated variable.

The basic idea behind predictive control is conceptually simple (see Fig. 1). A multi-step (or long-range) cost functional is defined. It reflects the tracking error between the reference signal and the plant's output starting from the current time instant up to a choosen future time horizon. The cost functional is minimized at each time step relative to either the control signal, or some adjustable parameters of the controller. Based on measured input-output data, a suitable plant model (or predictor) is built and used for prediction.

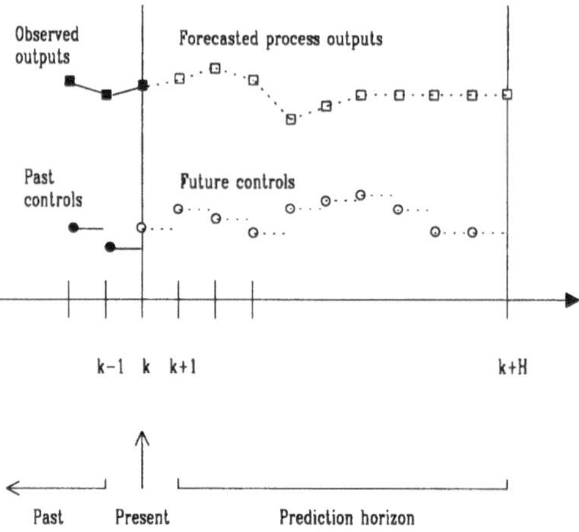

Figure 1: Minimizing a cost function over a long-range time horizon.

Classical examples of predictive control schemes are the Multistep Multivariable Adaptive Regulator (MUSMAR) [4] and the Generalized Predictive Controller (GPC) [3]. Both are based on a linear, time-invariant plant model and are closely related to Linear Quadratic Gaussian (LQG) control. Within this framework, robustness with respect to unmodelled plant dynamics has been extensively studied [8]. Adaptive predictive controllers incorporating linear predictors are designed on the assumption that the dynamic behavior of a plant can be considered linear around a given operating point. When wider operating ranges are to be considered, these control systems attempt to automatically adjust their parameters with sometimes unaffordable control costs. For extending the considered operating region, a nonlinear model may be required.

Recently, nonlinear plant models have been introduced for this class of control schemes. For instance, Nonlinear Auto Regressive with eXogenous inputs (NARX) models [10], artificial neural networks [9], and fuzzy models [15] have been exploited. Fuzzy models, besides being able to provide the required nonlinear plant model, can also provide, under suitable constraints, a linguistic interpretation for the relevant dynamic behavior of the plant. This has been considered a definite advantage over other nonlinear representations such as NARX, or artificial neural networks.

In adaptive control, a suitable recursive parameter estimation algorithm for the identification of the model is required. To address this issue we describe in this article a simplified fuzzy framework for adaptive predictive control. Namely, a re-parametrization of a fuzzy relational structure, surving as a plant model, is proposed. This re-parametrization allows the use of the Recursive Least Squares (RLS) algorithm for the purpose of speeding up identification. The RLS algorithm has a quadratic convergence rate. Also, it provides centered estimates in the presence of white noise under the assumption that full compatibility exists between the actual plant and its model [7]. The re-parametrization, under the above assumption, gives rise to a simplified and computationally more efficient fuzzy relational structure. It is shown that such a re-parametrized fuzzy relational structure is a universal approximator with clear semantics: it can be interpreted as a set of simplified fuzzy if-then rules.

The article is organized as follows. In Sect. 2, a tracking control problem is formulated. Different assumptions are considered giving rise to different control laws. In Sect. 3, fuzzy relational models are developed as plant descriptors. The focus is on the possibilities of speeding up identification using RLS algorithm. Section 4 presents experimental results in the context of the control of an electric furnace. Section 5 concludes this article.

2 Predictive control: formulation and control strategies

In this section the predictive control problem is introduced and the underlying basic assumptions are discussed. Several possible predictive control strategies are described.

2.1 Plant description

For the sake of convenience, consider a single-input/single-output plant whose dynamics can be represented by a discrete time input-output model of the form

$$\begin{aligned} y[k+1] &= \Upsilon(y[k], \ldots, y[k-p+1], u[k], \ldots, u[k-q+1]) \\ &= \Upsilon_{k+1}, \end{aligned} \tag{1}$$

where $y[k]$ and $u[k]$ represent discrete-time inputs and outputs, respectively; p and q are positive integer constants referring to the order of the plant and such that $0 < q \leq p$. Furthermore, the plant is assumed to be time-invariant, causal and stable with known structure, but with unknown parameters. The input $u[k]$ is a uniformly bounded function of the time k. Notice that Υ is a continuous nonlinear mapping.

2.2 The predictive control problem

The basic predictive control problem can be stated as follows. Consider a plant described by (1) and an *a priori* known reference trajectory $\vec{r}$. Find the optimal control sequence $\vec{u}^* = (u^*[k], u^*[k+1], \cdots, u^*[k+H-1])'$, where H refers to the chosen time horizon and the prime $'$ denotes the transpose of a vector, such that the following multi-step cost functional is minimized

$$
\begin{aligned}
Q &\triangleq \frac{1}{2}\sum_{i=1}^{H}[(r[k+i]-y[k+i])^2+\rho_i(\Delta u[k+i-1])^2] \\
&\triangleq \frac{1}{2}(\vec{e}'\vec{e}+\Delta\vec{u}'\vec{\rho}\Delta\vec{u}),
\end{aligned}
\tag{2}
$$

where

$$
\begin{aligned}
\vec{y} &\triangleq (y[k+1], y[k+2], \ldots, y[k+H])', \\
\vec{r} &\triangleq (r[k+1], r[k+2], \ldots, r[k+H])', \\
\vec{u} &\triangleq (u[k], u[k+1], \ldots, u[k+H-1])', \\
\vec{e} &\triangleq \vec{r}-\vec{y} \\
\vec{\rho} &\triangleq \mathrm{diag}(\rho_1, \rho_2, \ldots, \rho_H)', \\
\Delta\vec{u} &\triangleq (\Delta u[k], \Delta u[k+1], \ldots, \Delta u[k+H-1])'.
\end{aligned}
$$

Moreover, define $\Delta u[k+i] \triangleq u[k+i]-u[k+i-1]$, $r[k+i]$ as the reference signal to be tracked at $(k+i)$ -th sampling time and $y[k+i]$ as the plant output at the same time instant. The parameters $\rho_i \geq 0$ are penalty factors associated with each control action for the purpose of preventing eventual excessive oscillations.

Since no assumption on the linearity of the plant is made, no closed form solution should be expected for the general case. Thus, the explicit minimization of the cost functional (2) has to be carried out at each sampling time. Here the gradient descent method, one of the simplest minimization techniques, is used to search for solutions of the above minimization problem.

2.3 Control strategies

The minimization of Q over the time horizon of H steps may be realized by different control strategies.

One possible strategy is to assume that all of the $u[k+j]$, $(j=0,\ldots,H-1)$ are free. This is in the spirit of the GPC strategy, cf. [3]. Therefore, the minimization of (2) relative to the control sequence $\vec{u}$, is carried out at each sampling time according to

$$\vec{u}_{o+1} = \vec{u}_o + \alpha\Delta\vec{u}_o, \tag{3}$$

where the index o refers to the o-th iteration of the optimization process performed at each sampling time k; α is an *a priori* fixed step size gain such that $0 < \alpha \leq 1$. The increments $\Delta\vec{u}_o$ are computed according to what is believed to be the descent direction of the multi-step criterion, i.e., the negative gradient direction

$$\begin{aligned}\Delta\vec{u}_o &= -\nabla_u Q_o \\ &= \Phi\vec{e}_o - \vec{\rho}\Delta\vec{u}_o \\ &= (I+\Gamma)^{-1}\Phi\vec{e}_o,\end{aligned}$$

where I is the $H \times H$ identity matrix, Γ is given by

$$\Gamma = \begin{bmatrix} \rho_1 & -\rho_2 & 0 & 0 & \cdots & 0 \\ 0 & \rho_2 & -\rho_3 & 0 & \cdots & 0 \\ \vdots & \vdots & \vdots & \vdots & \ddots & \vdots \\ 0 & 0 & 0 & 0 & \cdots & \rho_H \end{bmatrix}, \tag{4}$$

and Φ is the Jacobian matrix

$$\begin{aligned}\Phi &= \begin{bmatrix} \frac{\partial\Upsilon_{k+1}}{\partial u[k]} & \frac{\partial\Upsilon_{k+2}}{\partial u[k]} & \cdots & \frac{\partial\Upsilon_{k+H}}{\partial u[k]} \\ \frac{\partial\Upsilon_{k+1}}{\partial u[k+1]} & \frac{\partial\Upsilon_{k+2}}{\partial u[k+1]} & \cdots & \frac{\partial\Upsilon_{k+H}}{\partial u[k+1]} \\ \vdots & \vdots & \ddots & \vdots \\ \frac{\partial\Upsilon_{k+1}}{\partial u[k+H-1]} & \frac{\partial\Upsilon_{k+2}}{\partial u[k+H-1]} & \cdots & \frac{\partial\Upsilon_{k+H}}{\partial u[k+H-1]} \end{bmatrix} \\ &= \begin{bmatrix} \frac{\partial\Upsilon_{k+1}}{\partial u[k]} & \frac{\partial\Upsilon_{k+2}}{\partial u[k]} & \cdots & \frac{\partial\Upsilon_{k+H}}{\partial u[k]} \\ 0 & \frac{\partial\Upsilon_{k+2}}{\partial u[k+1]} & \cdots & \frac{\partial\Upsilon_{k+H}}{\partial u[k+1]} \\ \vdots & \vdots & \ddots & \vdots \\ 0 & \cdots & 0 & \frac{\partial\Upsilon_{k+H}}{\partial u[k+H-1]} \end{bmatrix}. \end{aligned} \tag{5}$$

Notice that Φ is triangular, i.e., for $j \geq i$, we have that $\frac{\partial\Upsilon_{k+i}}{\partial u[k+j]} = 0$ due to the assumption of causality of the plant.

After having met a given stop criterion, the optimization process (3) yields the control sequence $\vec{u}$, (hopefully) such that $\vec{u} \approx \vec{u}^*$. However, the control action $u[k]$ actually applied to the plant is

$$u[k] = (\underbrace{1, 0, \cdots, 0}_{H \text{ elements}})\vec{u} + \zeta. \tag{6}$$

This means that only the first control action is actually used - all other control actions are discarded. Notice that the discarded control actions $u[k+j]; j = 1, \ldots, H-1$ refer to control actions in the future. By ignoring them, at the current sampling time, open-loop control actions are prevented. This is known as the receding time horizon strategy. The whole optimization process is then repeated at time instant $k+1$.

In (6), $\{\zeta\}$ is a white noise dither sequence with low power injected to persistently excite the plant and thus make identification possible. This is needed because closed-loop parameter estimation is governed by a trade-off between control performance and the performance of the identification procedure: the better the control the poorer is the model excitation.

It is worth noting here that by making (for all $i = 1, \ldots, H$, and $j = 0, \ldots, H-1$),

$$\frac{\partial \Upsilon_{k+i}}{\partial u[k+j]} = \frac{\partial \Upsilon_{k+i}}{\partial u[k]},$$

the control strategy presented in [15] (the first control value $u[k]$ left is free while all the others are let constant and equal to the first) emerges as a particular case of the above strategy.

The remaining issues connected with predictive control pertain to the information about the plant dynamics needed to compute the required derivatives, specially those defining the matrix Φ in (5). A fuzzy model acting as a predictor is used, and the derivatives are computed according to the *certainty equivalence principle*, i.e., by taking the current estimates of the model parameters as if they were the true values of the parameters.

3 Fuzzy relational models and simplified fuzzy structures

In this section fuzzy relational models are developed as plant descriptors. Both the notation and terminology are introduced. The identification of this type of models is then described. The focus is on the possibilities for speeding up identification using the RLS algorithm.

3.1 Fuzzy relational models

A fuzzy relational model aiming at capturing the dynamics of (1) is characterized by a numeric-to-linguistic (N/L) interface, a fuzzy processing stage (FPS), and a linguistic-to-numeric (L/N) interface.

The N/L interface has n primary linguistic terms (reference fuzzy sets). The i-th of these fuzzy sets is represented by the membership function $X_i(x)$, defined in $\mathbf{X}$, a closed interval of $\Re$. The membership function $X_i(x)$ is parametrized in θ_i^X so that $X_i(x) = X_i(x;\theta_i^X)$. Usually θ_i^X is a two-element vector, $\theta_i^X = (\theta_{i1}^X, \theta_{i2}^X)'$, with θ_{i1}^X being the center, and θ_{i2}^X being the width of the i-th input membership function.

Whenever x is a real-valued variable, the conversion, denoted by $\mathcal{L}_x$ and performed by the N/L interface, is given as

$$\begin{aligned}\mathcal{L}_x(x) &= [X_1(x), \cdots, X_i(x), \cdots, X_n(x)] \\ &= [X_1, \ldots, X_i, \ldots, X_n] \\ &= X.\end{aligned}$$

The L/N interface converts the fuzzy set Y produced by the fuzzy processing stage into a numeric sample. Using for output membership functions the same notation, the j-th output membership function is denoted by $Y_j(y) = Y_j(y;\theta_j^Y)$, θ_j^Y being a vector of parameters defined in the model output space $\mathbf{Y}$. The output fuzzy set Y is thus represented by an array of membership degrees such that $Y = (Y_1, \ldots, Y_j, \ldots, Y_m)$, m being the number of output membership functions.

The output mapping is denoted by $\mathcal{N}_y : [0,1]^m \to \mathbf{Y}$, and is implemented by an injective differentiable defuzzification method undefined for the empty set $Y = \emptyset = [0, \cdots, 0]$. For instance, using the center-of-gravity defuzzification, the numeric model output $\hat{y}$ is given by

$$\hat{y} = \mathcal{N}_y(Y) = \frac{\sum_{j=1}^{m} Y_j \theta_{j1}^Y}{\sum_{j=1}^{m} Y_j}, \tag{7}$$

where θ_{j1}^Y denotes the center of the j-th output membership function.

The fuzzy processing stage of the model is given by

$$\begin{aligned}Y[k+1] &= \overline{Y}[k] \circ \overline{Y}[k-1] \circ \cdots \circ \overline{Y}[k-p+1] \circ \\ & \quad U[k] \circ U[k-1] \circ \cdots \circ U[k-q+1] \circ R,\end{aligned} \tag{8}$$

where $\circ$ is the fuzzy composition operator and represents a general $\oplus - \odot$ composition, $U[k] = \mathcal{L}_u(u[k])$ stands for the linguistic representation of the model input at the discrete time instant k , and $\overline{Y}[k] = \mathcal{L}_y(\mathcal{N}_y(Y[k]))$. The fuzzy relation R is a mapping such that

$$R : \underbrace{\mathbf{U} \times \cdots \times \mathbf{U}}_{q \text{ times}} \times \underbrace{\mathbf{Y} \times \cdots \times \mathbf{Y}}_{p \text{ times}} \to [0,1].$$

in a point wise manner (8) reads as [12]

$$\begin{aligned}Y_j &= \oplus_{i_1=1}^{n_1} \oplus_{i_2=1}^{n_2} \cdots \oplus_{i_p=1}^{n_p} \oplus_{i_{p+1}=1}^{n_{p+1}} \cdots \oplus_{i_{p+q}=1}^{n_{p+q}} \bar{Y}_{i_1}[k] \odot \bar{Y}_{i_2}[k-1] \odot \ldots \\ & \quad \odot \bar{Y}_{i_p}[k-p+1] \odot U_{i_{p+1}}[k] \odot \ldots \odot U_{i_{p+q}}[k-q+1] \odot R_{i_1 i_2 \ldots i_{p+q} j},\end{aligned}$$

where $\oplus$ denotes a triangular-conorm (s-norm) such as maximum, and $\odot$ denotes a triangular-norm (t-norm) such as minimum or product [12].

Notice that (8) represents a topology where the fuzzy output of the model at the time instant k, $Y[k]$, is converted by the output interface to its numeric format, $\hat{y}[k]$, delayed, and then sent back to its input interface. This scheme is much more general than using only the fuzzy output of the model as feedback [18].

3.2 Simplified fuzzy relational structures

Identification schemes based on or inspired by neural nets learning schemes, especially those based on the backpropagation algorithm, exhibit slow parameter convergence rates, cf. [14]. This is an inherent drawback of first order optimization schemes such as the gradient descent method. In adaptive contro, fast convergence rates of the identifier are usefull while drastic forgetting should be avoided. From this depends both the amplitude and duration of the initial transient, and the effective handling of time-varying systems. In an attempt to provide a suitable recursive parameter estimation algorithm for the type of model used here, a new approach to the identification of fuzzy relational models based on the RLS algorithm is summarized [17].

The RLS algorithm [7] is used for estimating the elements of the fuzzy relation. RLS has a quadratic convergence rate and it can be made numerically robust, or able to deal with time-varying processes. It provides centered estimates in the presence of white noise and full structural compatibility between the model and the target process. The RLS algorithm requires a model *linear in the parameters.* However, this does not imply that the model should be linear. A linear in the parameters model is written as

$$\hat{y}[k] = \phi[k]'\Omega, \tag{9}$$

where $\hat{y}[k]$ is the output of the model, $\phi[k]$ is a vector containing observed data, and Ω is the vector of parameters subject to estimation.

One of the basic versions of the RLS algorithm is as follows:

$$\begin{aligned}
\Omega[k] &= \Omega[k-1] + K[k](y[k] - \Omega'[k-1]\phi[k-1]), \\
K[k] &= P[k]\phi[k-1], \\
P[k] &= (P[k-1] - \frac{P[k-1]\phi[k-1]\phi'[k-1]P[k-1]}{\lambda + \phi'[k-1]P[k-1]\phi[k-1]})/\lambda,
\end{aligned}$$

where P is the error covariance squared matrix, K is the Kalman gain, and $0 < \lambda \leq 1$ is a constant forgetting factor.

One simple way to apply RLS to a fuzzy relational model is to assume the following transformation at the model output

$$\hat{y}[k] = \sum_{j=1}^{m} a_j Y_j[k],$$

where $a_j (j = 1, \ldots, m)$ are simply *unconstrained* real coefficients, instead of the parameters of the output membership functions. This means that the model predictions are merely numerical, in the same way as in Takagi-Sugeno fuzzy rules where the consequents are purely numerical. Depending on the composition operator used, this allows the grouping of all elements in R into the vector of parameters Ω.

The selection of a pair of triangular norms and conorms in the definition of the composition operator is crucial. While some of these pairs simply do not allow the definition of a linear in the parameters model, other pairs generate memory expensive models, e.g., this is case of $x \oplus y = x + y - xy$, and $x \odot y = xy$. Some other composition operators give rise to slow convergence models. To overcome both the memory requirements, the slow convergence rates, and still provide a linear in the parameters model, the following averaging-product operator is used [5]

$$z = \sqrt[w]{\frac{1}{N} \sum_{i=1}^{N} x_i^w}, \quad w \neq 0.$$

The averaging-product composition operator allows a re-parametrization of a fuzzy relational structures giving rise to one with less parameters and thus computationally more efficient. Consider the simplest static fuzzy relational model under this composition ($w = 1$)

$$\begin{aligned} \hat{y} &= a_1 Y_1 + \cdots + a_j Y_j + \cdots + a_m Y_m, \\ Y_j &= \frac{1}{n} \sum_{i=1}^{n} X_i R_{ij} \quad j = 1, \ldots, m. \end{aligned} \tag{10}$$

Replacing Y_j with their equals we obtain

$$\begin{aligned} \hat{y} &= \frac{a_1}{n}(X_1 R_{11} + X_2 R_{21} + \cdots + X_n R_{n1}) + \\ &\quad \frac{a_2}{n}(X_1 R_{12} + X_2 R_{22} + \cdots + X_n R_{n2}) + \cdots + \\ &\quad \frac{a_m}{n}(X_1 R_{1m} + X_2 R_{2m} + \cdots + X_n R_{nm}) \\ &= \underbrace{(\frac{1}{n} \sum_{l=1}^{m} a_l R_{1l})}_{\Omega_1} X_1 + \underbrace{(\frac{1}{n} \sum_{l=1}^{m} a_l R_{2l})}_{\Omega_2} X_2 + \cdots + \underbrace{(\frac{1}{n} \sum_{l=1}^{m} a_l R_{nl})}_{\Omega_n} X_n. \end{aligned}$$

This is a new fuzzy relational structure where the number of parameters ($N = n$) is m times lower than the number of parameters in the original fuzzy relational structure. Bearing in mind that RLS requires the updating of the error covariance matrix involving the square of the total number of parameters, the new structure besides being simpler is also computationally more efficient. This new fuzzy relational structure will be denoted as $\hat{y} = X\Omega$.

Consider now a fuzzy relational model given as in (8) under the averaging-product composition, that is

$$\begin{aligned} \hat{y}[k+1] &= a_1 Y_1 + \cdots + a_j Y_j + \cdots + a_m Y_m, \\ Y_j &= \frac{1}{n_1 n_2 \ldots n_{p+q}} \sum_{i_1=1}^{n_1} \sum_{i_2=1}^{n_2} \cdots \sum_{i_p=1}^{n_p} \sum_{i_{p+1}=1}^{n_{p+1}} \cdots \sum_{i_{p+q}=1}^{n_{p+q}} \bar{Y}_{i_1}[k] \end{aligned}$$

$$\bar{Y}_{i_2}[k-1] \ldots \bar{Y}_{i_p}[k-p+1] U_{i_{p+1}}[k] \ldots U_{i_{p+q}}[k-q+1] R_{i_1 i_2 \ldots i_{p+q} j}.$$

This type of a fuzzy relational model will be denoted as

$$\hat{y}[k+1] = Y[k]Y[k-1] \cdots Y[k-p+1]U[k] \cdots U[k-q+1]\Omega, \tag{11}$$

and it represents the relationship (9) where ϕ and Ω are defined as follows

$$\begin{aligned} \phi &= \begin{bmatrix} \phi_1 \\ \phi_2 \\ \vdots \\ \phi_N \end{bmatrix} \\ &= \begin{bmatrix} Y_1[k]Y_1[k-1] \cdots Y_1[k-p+1]U_1[k] \cdots U_1[k-q+1] \\ Y_1[k]Y_1[k-1] \cdots Y_1[k-p+1]U_1[k] \cdots U_2[k-q+1] \\ \vdots \\ Y_{n_1}[k]Y_{n_2}[k-1] \cdots Y_{n_p}[k-p+1]U_{n_{p+1}}[k] \cdots U_{n_{p+q}}[k-q+1] \end{bmatrix}, \end{aligned}$$

$$\Omega = \begin{bmatrix} \Omega_1 \\ \Omega_2 \\ \vdots \\ \Omega_N \end{bmatrix} = \begin{bmatrix} \frac{1}{N} \sum_{l=1}^{m} a_l R_{11\ldots 1l} \\ \frac{1}{N} \sum_{l=1}^{m} a_l R_{11\ldots 2l} \\ \vdots \\ \frac{1}{N} \sum_{l=1}^{m} a_l R_{n_1 n_2 \cdots n_{p+q} l} \end{bmatrix},$$

where $N = \prod_{i=1}^{p+q} n_i$ is the length of both vectors.

Notice that the number of parameters in the simplified fuzzy relational structure is still m times lower that the number of parameters in the original structure. The price to be paid for this significant reduction in the number of parameters is that the knowledge of the algebraic relationships between the elements of R is lost.

To illustrate the performance of these two fuzzy relational structures in terms of convergence let us consider the following example. Consider the fuzzy relational system

$$\begin{aligned} \hat{y}[k] &= a_1 Y_1[k] + a_2 Y_2[k] + a_3 Y_3[k], \\ Y_j[k+1] &= \bigvee_{i_1=1}^{3} \bigvee_{i_2=1}^{3} U_{i_1}[k] \bar{Y}_{i_2}[k] R_{i_1 i_2 j} \quad j = 1, 2, 3, \end{aligned} \tag{12}$$

where $\bigvee$ denotes maximum. The parameters of this system are: $a_1 = 10$, $a_2 = 100$, and $a_3 = 1000$. All the elements of R are zero except for $R_{111} = 1.0$, $R_{121} = 0.7$, $R_{131} = 0.5$, $R_{222} = 0.1$, $R_{232} = 0.9$, $R_{323} = 0.8$, $R_{333} = 0.6$.

This fuzzy relational system is to be identified using the following simplified fuzzy relational structure

$$\begin{aligned} \hat{y} &= U[k] Y[k] \Omega \\ &= \sum_{i_1=1}^{3} \sum_{i_2=1}^{3} U_{i_1}[k] Y_{i_2}[k] \Omega_{3(i_1-1)+i_2} \\ &= \sum_{j=1}^{N=9} \phi_j \Omega_j, \end{aligned} \tag{13}$$

where

$$\phi = \begin{bmatrix} \phi_1 \\ \phi_2 \\ \phi_3 \\ \phi_4 \\ \vdots \\ \phi_9 \end{bmatrix} = \begin{bmatrix} U_1[k]Y_1[k] \\ U_1[k]Y_2[k] \\ U_1[k]Y_3[k] \\ U_2[k]Y_1[k] \\ \vdots \\ U_3[k]Y_3[k] \end{bmatrix}, \tag{14}$$

$$\Omega = \begin{bmatrix} \Omega_1 \\ \Omega_2 \\ \Omega_3 \\ \Omega_4 \\ \vdots \\ \Omega_9 \end{bmatrix} = \begin{bmatrix} \frac{1}{9}\sum_{l=1}^{m=3} a_l R_{11l} \\ \frac{1}{9}\sum_{l=1}^{m=3} a_l R_{12l} \\ \frac{1}{9}\sum_{l=1}^{m=3} a_l R_{13l} \\ \frac{1}{9}\sum_{l=1}^{m=3} a_l R_{21l} \\ \vdots \\ \frac{1}{9}\sum_{l=1}^{m=3} a_l R_{33l} \end{bmatrix}. \tag{15}$$

Figure 2 shows a) data from system (12) *versus* model (13) output, and b) the parameter estimates convergence. The estimates converge in about 20 iterations. It can be seen that the estimates have converged to their true value, i.e., to the value found when Ω is computed via (15) for the system (12).

In the context of simplified fuzzy relational structures the following questions can be stated: i) How can a simplified structure be interpreted? and ii) Are these simplified structures general enough, i.e., are they universal approximators?

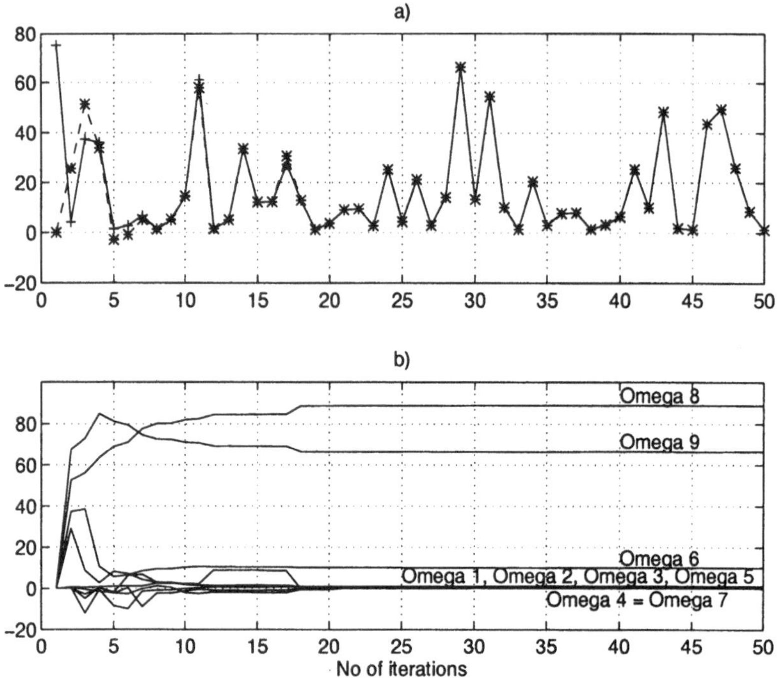

Figure 2: Performance of a simplified fuzzy structure: a) data (solide line) versus model output (dashed line), and b) the evolution of parameter estimates b).

3.3 On interpretation

With the above described re-parametrization one moves from relational systems to functional systems. A simplified fuzzy relational structure can be understood as: i) A set of fuzzy if-then rules with linear defuzzification, and where the j-th of $N = n_1 n_2 \cdots n_{p+q}$ rules has the form

Rule$_j$: IF $y[k]$ is Y_{i_1} AND $\cdots$ AND $u[k-q+1]$ is $U_{i_{p+q}}$ THEN $y_j[k+1] = \Omega_j$.

The above fuzzy if-then rule can be interpreted as (i) a simplified Takagi-Sugeno rule where only the first term of the consequent is considered and ii) as a radial basis function network, whenever Gaussian membership functions are used.

3.4 On approximation

There is a number of fuzzy systems that have the property of being universal approximators, i.e., that are able to approximate any real-valued continuous function defined on a compact domain to any degree of accuracy. In [19] these fuzzy systems bellong to a class characterized by Mamdani-type of fuzzy rules with fuzzy antecedents and fuzzy consequents, Gaussian membership functions, product inference, and center-of-gravity defuzzification. At the same time an additive fuzzy system is also an universal approximator [6]. Buckley [1, 2] has shown that a modification of a Takagi-Sugeno fuzzy system iresults in yet another universal approximator. The modifications introduced are as follows: the consequent part of each fuzzy rule is a polynomial function and a linear defuzzification is used. In [20] the approximation properties of SISO fuzzy systems is studied in detail. The focus is on the properties of fuzzy basis functions and how these relate to approximation . The center-of-gravity defuzzification method is assumed.

The simplified fuzzy relational structures introduced here do not bellong to any of the fuzzy system studied so far.

In [11] it was shown that the family of radial basis function networks with the *same* smoothing factor in each kernel is dense in the set of continuous real functions, no matter the type of metric considered. Therefore, a simplified fuzzy relational structure equipped with Gaussian membership functions all sharing the *same* width is an universal approximator.

In the appendix, it is shown that the family of simplified fuzzy relational structures with Gaussian membership functions with *independent* widths is dense in the set of continuous real-valued functions (the *sup-metric* is used).

4 Control of an electric furnace

In this section, the application of an adaptive predictive control algorithm based on simplified fuzzy relational structures is presented. The process is an electric furnace used in ceramic manufacturing, see figure 3. Its main features are: maximum temperature 1300°C, maximum power 3KW, the power control is via a signal with values between 0 and 100% using Pulse Width Modulation.

The furnace belongs to the class of plants where the desired set-point to be followed (temperature reference) is known in advance - a convenient feature in predictive control schemes. In this case, the reference to be followed has nearly 10 hours of duration. The sampling period is $T_s = 6.7s$.

The linearized dynamics varies with the operating point: heating dynamics are significantly faster than the colling dynamics since no active cooling is available. Moreover, within the same operating point, the dynamics varies with time: the cooling rate depends on how long the furnace has been at a given temperature. That is, the process is both nonlinear and time-variant.

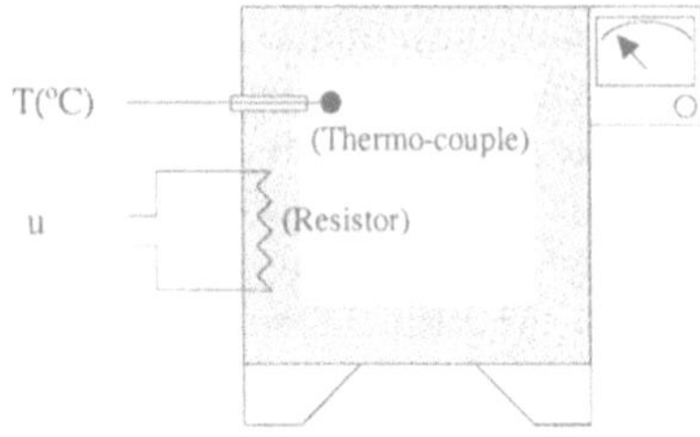

Figure 3: Schematic view of the electric furnace.

The following simplified fuzzy relational structure is taken as plant descriptor

$$\hat{y}[k+i] = Y[k+i-1]Y[k+i-2]U[k+i-1-\tau]\Omega. \tag{16}$$

The required derivatives for this fuzzy relational structure are

$$\frac{\partial \bar{Y}_l[k+i]}{\partial u[k+j]} = 0 \quad \text{if } i < j+\tau+1$$

$$\frac{\partial \bar{Y}_l[k+i]}{\partial u[k+j]} = \sum_{i_1=1}^{m}\sum_{i_2=1}^{m}\sum_{i_3=1}^{n} Y_{i_1}Y_{i_2}\Omega_f \frac{\partial U_{i_3}[k+i-1-\tau]}{\partial u[k+j]} \quad \text{if } i = j+\tau+1$$

$$\frac{\partial \bar{Y}_l[k+i]}{\partial u[k+j]} = \sum_{i_1=1}^{m}\sum_{i_2=1}^{m}\sum_{i_3=1}^{n} Y_{i_2}U_{i_3}\Omega_f \frac{\partial Y_{i_1}[k+i-1]}{\partial u[k+j]} \quad \text{if } i = j+\tau+2$$

and finally, if $i > j+\tau+2$,

$$\frac{\partial \bar{Y}_l[k+i]}{\partial u[k+j]} = \sum_{i_1=1}^{m}\sum_{i_2=1}^{m}\sum_{i_3=1}^{n} U_{i_3}\Omega_f (Y_{i_2}\frac{\partial Y_{i_1}[k+i-1]}{\partial u[k+j]} + Y_{i_1}\frac{\partial Y_{i_2}[k+i-2]}{\partial u[k+j]}.$$

In the above, $\Omega_f = \Omega_{f(i_1,i_2,i_3)} = \Omega_{m^2(i_1-1)+m(i_2-1)+i_3)}$.

Figure 4 shows the performance during temperature tracking for ten hours long experiment with $\tau = 2$, and $H = 10$. For the same experiment, Fig. 5 shows the temperature reference output versus the observed output temperature and the control signal for the first 2 hours. The Ω vector of (16) is initialized to zero in the beginning of the experiment. One can observe both fast convergence rates and good tracking.

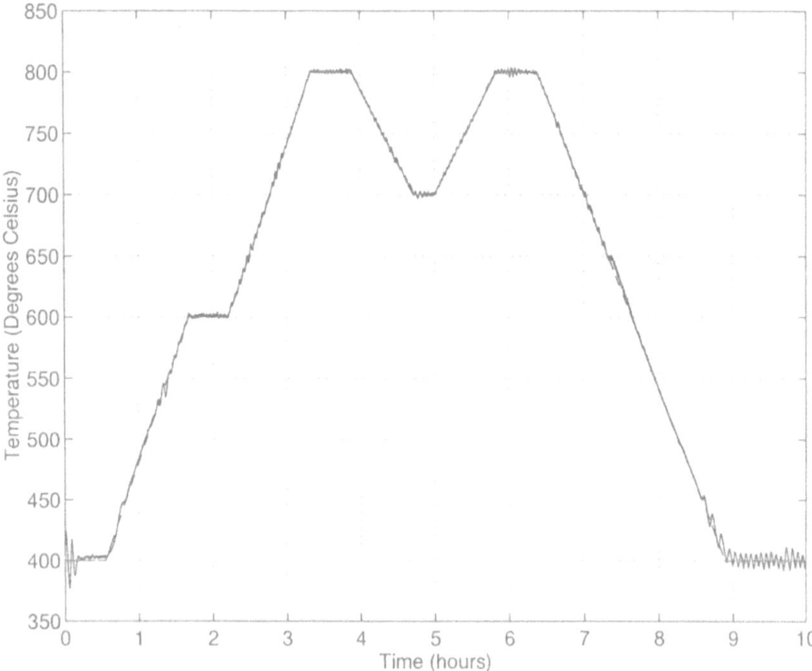

Figure 4: Controlling the furnace: Reference - dashed line, and observed output temperature - solid line.

5 Conclusions

In an attempt to extend the operating range of linear-based long-range predictive controllers, and to contribute to the current state of model based fuzzy control design methods, a simplified fuzzy framework for predictive control was developed, discussed, and illustrated.

When an adaptive predictive control scheme is concerned fast convergence rates of the identifier are useful while drastic forgetting should be avoided. To address this issue a re-parametrization of a fuzzy relational structure serving as a plant description is proposed. For particular composition operator, this simplified fuzzy relational structure requires less parameters, is computationally efficient, has the property of being a universal approximator, and has clear semantics in terms of a particular type of fuzzy rules. Furthermore, a simplified fuzzy relational structure allows for the use of the RLS algorithm for the purpose of speeding identification using on-line closed-loop observations. For the control law the free control actions assumptions is considered for on-line minimization of the cost functional and a receding time horizon strategy is used.

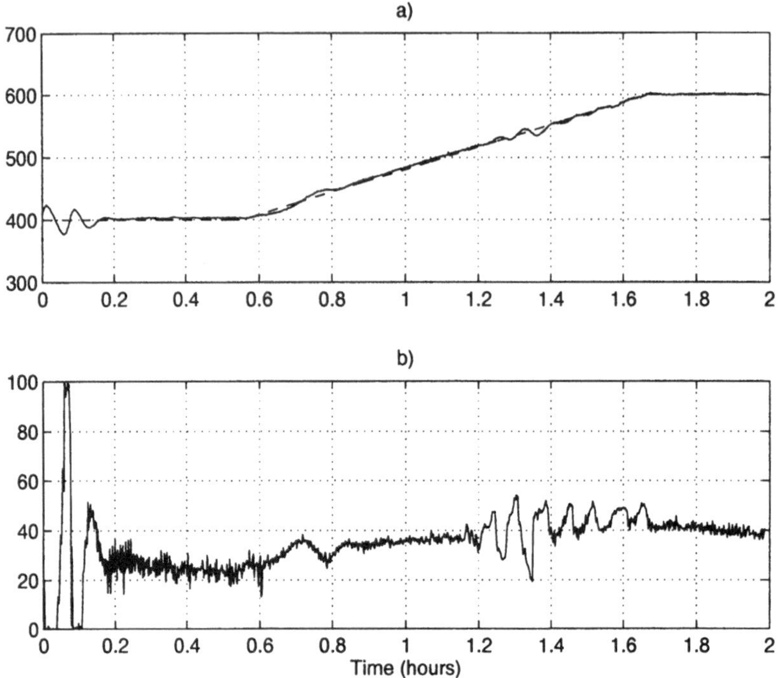

Figure 5: Control results for the two first hours: a) Reference and furnace output temperature; b) Control signal.

References

[1] J. J. Buckley, "Universal fuzzy controller", *Automatica*, vol. 28, pp.1245-1248, 1992.

[2] J. J. Buckley, "Sugeno type controllers are universal controllers", *Fuzzy Sets and Systems*, **53**, pp. 299-304, 1993.

[3] D. W. Clarke, C. Mohtadi, and P.S. Tuffs, "Generalized Predictive Control – Part I & Part II", *Automatica*, **23**, 1987, 137-160.

[4] C. Greco, G. Menga, E. Mosca, and G. Zappa, "Performance improvements of self-tuning controllers by multi-step horizons: The MUSMAR approach", *Automatica*, **20**, (1984), pp. 681-699.

[5] G. F. Klir, and T. A. Folger, *Fuzzy Sets, Uncertainty, and Information*, Englewood Cliffs, NJ: Prentice-Hall, 1988.

[6] B. Kosko, "Fuzzy systems as universal approximators", *Proc. of IEEE Int. conf. on Fuzzy Systems*, San Diego, USA, 1992, pp. 1153-1162.

[7] L. Ljung, *System Identification: theory for the user*, (Prentice-Hall, Englewood Cliffs, N.J. 1987)

[8] E. Mosca, G. Zappa, and J. M. Lemos, "Robustness of multipredictor adaptive regulators: MUSMAR", *Automatica*, **25**, 1989, 521-529.

[9] D. W. Nguyen and B. Widrow, "Neural networks for self-learning control systems", *IEEE Control Systems Magazine*, Apr. 1990, 18-23.

[10] A. Patwardhan, J. Rawlings, and T. Edgar, "Non-linear model predictive control", *Chemical Engineering Communications*, **87** (1990), pp. 123-141.

[11] J. Park, and I. W. Sandberg, "Universal approximation using radial-basis-functions networks" *Neural Computation*, **3**, pp. 246-257, 1991.

[12] W. Pedrycz, *Fuzzy Control and Fuzzy Systems* (Research Studies Press/Wiley, Chichester 1993).

[13] W. Rudin, *Principles of Mathematical Analysis*, McGraw-Hill, Inc. 1976.

[14] J. Valente de Oliveira, "Neuron inspired learning rules for fuzzy relational structures", *Fuzzy Sets and Systems* 57 (1993) 41-53.

[15] J. Valente de Oliveira, J.M. Lemos, "Long-range predictive adaptive fuzzy relational control", *Fuzzy Sets and Systems* **70** (1995) 337-357.

[16] J. Valente de Oliveira, "A design methodology for fuzzy system interfaces", *IEEE Trans. on Fuzzy Sys*, vol. 3, no. 4, 1995, pp. 404-414.

[17] J. Valente de Oliveira and J. M. Lemos, "Speeding up fuzzy relational identification: The RLS approach", *Proc. of the Sixt International Fuzzy Systems Association World Congress*, São Paulo, Brasil, 1995, pp. 121-124.

[18] J. Valente de Oliveira, "Prediction using relational systems", in *Fuzzy Modelling: Paradigms and Practice* W. Pedrycz (ed), Kluwer Academic Publishers, 1996, pp. 91-113.

[19] L.X. Wang, "Fuzzy systems are universal approximators", *Proc. of IEEE Inter. Conf. on Fuzzy Systems.* San Diego, USA, 1992, pp. 1163-1170.

[20] X.-J. Zeng and M. Singh, "Approximation theory of fuzzy systems - SISO case", *IEEE Trans. on Fuzzy Systems*, vol. 2, no. 2, pp. 162-176.

Appendix A

In this apendix the Stone-Weierstrass theorem is used for showing that a simplified fuzzy relational structure with Gaussian membership functions with independent widths is able of approximating any real continuous function on a compact set to arbitrary accuracy. Essentially, [19] is followed.

Definition 1 *A set of MISO simplified fuzzy relational structures, denoted by C in the sequel, consists of all functions of the form*

$$\begin{aligned} f(\vec{x}) &= \sum_{j=1}^{N} \Phi_j \Omega_j \\ &= \sum_{j=1}^{N} \Omega_j \prod_{i=1}^{n} \mu_{A_i^j}(x_i), \end{aligned} \tag{17}$$

where $f : U \subset \Re^N \to \Re$, $\vec{x} = [x_1 \;\; x_2 \;\; \cdots \;\; x_n]'$and the prime denotes the transpose of a vector. Furthermore, $\mu_{A_i^j}(x)$ is the membership function of the i-th fuzzy set of the j-th rule, A_i^j, and is given as

$$\mu_{A_i^j}(x) = \exp\{-\frac{(x-\theta_{i1}^j)^2}{(\theta_{i2}^j)^2}\}, \tag{18}$$

with $\theta_{i2}^j \neq 0$, $(i = 1,\ldots,n)$, $(j = 1,\ldots,N)$. N is the total number of fuzzy rules, each one of the form

$Rule_j$: If x_1 is A_1^j and $\cdots$ and x_n is A_n^j then y_j is Ω_j.

It is assumed that $N \geq 1$ and U is compact (closed and bounded).

The design parameters of the simplified structures in C are:

1. The number of fuzzy sets defined in the i-th input space, n_i $(i = 1, 2, \ldots, n)$.
2. The total number of rules N.
3. The parameters of the Gaussian membership functions, $\theta_i^j = [\theta_{i1}^j \;\; \theta_{i2}^j]'$, $(i = 1,\ldots,n)$, $(j = 1,\ldots,N)$, θ_{i1}^j, being the center, and θ_{i2}^j the width.
4. The coeficients of the consequents, Ω_j, $j = 1,\ldots,N$.

Definition 2 *The sup-metric d_∞ is given by*

$$d_\infty = \sup_{\vec{x}\in U}(|f_1(\vec{x}) - f_2(\vec{x})|), \tag{19}$$

where $f1, f2 : U \subset \Re^N \to \Re$.

Lemma 1 *(C, d_∞) is a metric space.*

Proof:

First, $\mathcal{C}$ is not empty since $N \geq 1$ by definition. Furthermore, from (17) it is obvious that $(\mathcal{C}, d_\infty)$ is well-defined, i.e. $\forall_{\vec{x}\in U}\forall_{f\in\mathcal{C}}\exists_{z\in\Re} z = f(\vec{x})$. □

Stone-Weierstrass theorem [13]: Let Z be a set of real continuous functions on a compact set U. If i) Z is an algebra, ii) Z separates points on U, and iii) Z vanishes at no point of U, then the uniform closure of Z consists of all real continuous functions on U, i.e., Z is dense in $C[U]$, $C[U]$ being the set of all real continuous functions defined on U.

Lemma 2 *$\mathcal{C}$ is an algebra, i.e., the set $\mathcal{C}$ is closed under addition, multiplication, and scalar multiplication.*

Proof: Let $f_1, f_2 \in \mathcal{C}$ such that

$$\begin{aligned} f_1(\vec{x}) &= \sum_{j1=1}^{N1} \Omega 1_{j1} \prod_{i=1}^{n} \mu_{A1_i^{j1}}(x_i), \\ f_2(\vec{x}) &= \sum_{j2=1}^{N2} \Omega 2_{j2} \prod_{i=1}^{n} \mu_{A2_i^{j1}}(x_i), \end{aligned} \tag{20}$$

$$f_1(\vec{x}) + f_1(\vec{x}) = \sum_{j=1}^{N1+N2} \Omega\xi(j)_j \prod_{i=1}^{n} \mu_{A\xi(j)_i^{\zeta(j)}}(x_i), \tag{21}$$

where $\xi(j) = 1 + \text{integer}(j/N2)$ and $\zeta(j) = \text{integer}(j/N2) + \text{modulus}(j/N2)$. Since (21) is the same form as (17), $f_1 + f_2 \in \mathcal{C}$.

Similarly

$$f_1(\vec{x}) f_2(\vec{x}) = \sum_{j1=1}^{N1} \sum_{j2=1}^{N2} \Omega 1_{j1} \Omega 2_{j2} \prod_{i=1}^{n} \mu_{A1_i^{j1}}(x_i) \mu_{A2_i^{j2}}(x_i).$$

Since the product of Gaussians is a Gaussian, the above expression is also in the form of (17), thus $f_1 f_2 \in \mathcal{C}$.

Finally, for any $a \in \Re$ we have that

$$a f_1(\vec{x}) = \sum_{j1=1}^{N1} a \Omega 1_{j1} \prod_{i=1}^{n} \mu_{A1_i^{j1}}(x_i), \tag{22}$$

which is again in the form of (17). □

Lemma 3 *$\mathcal{C}$ separates point in U, i.e.*

$$\forall_{\vec{x}', \vec{x}'' \in U} \ \exists_{f\in\mathcal{C}} \ \vec{x}' \neq \vec{x}'' \Rightarrow f(\vec{x}') \neq f(\vec{x}'').$$

Proof: The proof is completed by constructing an f with the required property. Let $\vec{x^0} = [x_1^0 \cdots x_n^0]'$ and $\vec{y^0} = [y_1^0 \cdots y_n^0]'$, and define two fuzzy sets in the i-th input space of U with the membership functions:

$$\mu_{A_i^1}(x_i) = \exp\{-\frac{(x_i - x_i^0)^2}{2}\},$$
$$\mu_{A_i^2}(x_i) = \exp\{-\frac{(x_i - y_i^0)^2}{2}\}.$$

Moreover consider two fuzzy rules, i.e., $N = 2$. All parameters have been specified except Ω_j $(j = 1, 2)$. Under these conditions $f(\vec{x^0})$ and $f(\vec{y^0})$ read as

$$f(\vec{x^0}) = \Omega_1 + \Omega_2 \prod_{i=1}^{n} \exp\{-\frac{(x_i^0 - y_i^0)^2}{2}\}, \quad (23)$$

and

$$f(\vec{y^0}) = \Omega_2 + \Omega_1 \prod_{i=1}^{n} \exp\{-\frac{(y_i^0 - x_i^0)^2}{2}\}. \quad (24)$$

Set $\Omega_1 = 1$ and $\Omega_2 = 0$ then

$$f(\vec{x^0}) = 1,$$
$$f(\vec{y^0}) = \prod_{i=1}^{n} \exp\{-\frac{(x_i^0 - y_i^0)^2}{2}\}.$$

Since $\vec{x^0} \neq \vec{y^0}$, there must exists some i such that $x_i^0 \neq y_i^0$ hence

$$\exp\{-\frac{(x_i^0 - y_i^0)^2}{2}\} \neq 1$$

and

$$f(\vec{x^0}) = 1 \neq f(\vec{y^0}). \quad \square$$

Lemma 4 *C vanishes at no point of U, i.e.*

$$\forall_{\vec{x} \in U} \exists_{f \in C} f(\vec{x}) \neq 0.$$

Proof: Any $f \in C$ with all its $\Omega_j > 0$ satisfy the required f property. $\square$

Theorem 1 *For any given real continuous function g in the compact set set $U \subset \Re^n$ and arbitrary $\epsilon > 0$, there exists $f \in C$, where C is given by Definition 1, such that*

$$\sup_{\vec{x} \in U}(|g(\vec{x}) - f(\vec{x})|) < \epsilon \quad (25)$$

Proof: By Definition 1, C is a set of real continuous functions on U. The proof is thus a direct consequence of the Stone-Weierstrass theorem, and Lemmas 1-4. $\square$

Predictive Control Based on a Fuzzy Model

Igor Škrjanc, Katarina Kavšek-Biasizzo, Drago Matko
Faculty of Electrical and Computer Engineering
University of Ljubljana
Tržaška 25, 61000 Ljubljana, Slovenia

1 Introduction

In predictive control the output signal y is predicted at each sampling time. This prediction is made implicitly or explicitly according to the model of the controlled process. Next, a control action is selected that is intended to bring the predicted process output back to a given reference signal so that the difference between the reference signal and the output is minimized. Control methods essentially based on the principle of predictive control are Richalet's method (*Model Algorithmic Control*), Cutler's method (*Dynamic Matrix Control*), De Keyser's method (*Extended Prediction Self-Adaptive Control*), and Ydstie's method (*Extended Horizon Adaptive Control*).

With respect to the process model, two main approaches have been developed in the area of predictive control. The first one is based on a parametric model of the controlled process. The parametric model can be in the form of a transfer function or in a state space form. An important disadvantage of the parametric model is that it represents a linearized model of the process in which case the control of strongly nonlinear processes could be unsatisfactory. The second approach proposed in the literature is based on nonparametric models. The advantage of this approach is that the model coefficients can be obtained directly from samples of the input and output responses without assuming a particular model structure. Predictive control based on a fuzzy model of the controlled process is a combination of the nonparametric and parametric approaches.

In Sect. 2 of this article we deal with the concept of fuzzy identification. In Sect. 3 we describe the concept of conventional predictive control and predictive control based on fuzzy model. In Sect. 4 we present a simulation based example which illustrates the application of fuzzy model based predictive control. Finally, in Sect. 5, the implementation of the proposed predictive control technique on a real temperature plant is discussed.

2 The identification of a fuzzy model

A fuzzy model represents a static nonlinear mapping between input and output variables. Dynamic systems are usually modelled by feeding back delayed input and output signals. The common nonlinear model structure is NARX (Nonlinear AutoRegressive with eXogenous input) model, which gives the mapping between past input-output data and the predicted output

$$\hat{y}(k+1) = \mathcal{F}(y(k), y(k-1), \ldots, y(k-n+1), u(k), \ldots, u(k-m+1)), \quad (1)$$

where $(y(k), y(k-1), \ldots, y(k-n+1)$ and $u(k), u(k-1) \ldots, u(k-m+1)$ denote the delayed model output and input signals, respectively. A fuzzy model therefore approximates the function $\mathcal{F}$. Structure identification, in the context of a fuzzy model, means specifying the operators for the logical connectives, fuzzification, inference and defuzzification computational procedures. Once the structure is determinated the fuzzy model parameters can be estimated using least squares methods [1], [2], [3], [4], [5].

The fuzzy identification algorithm used in this article is based on the Takagi-Sugeno (TS) fuzzy model of the first order.

2.1 The identification algorithm

The TS fuzzy model consist of fuzzy rules of the type

$$\boldsymbol{R}_i : \textbf{IF } x_1 \textit{ is } \boldsymbol{A}_i \textit{ and } x_2 \textit{ is } \boldsymbol{B}_i \textbf{ THEN } y = f_i(x_1, x_2), \qquad i = 1, \ldots N, \quad (2)$$

where x_1 and x_2 are the input variables, y is the output variable, and A_i, B_i are fuzzy sets characterized by their membership functions. The if-parts (antecedents) of the rules describe fuzzy regions in the space of input variables. The then-parts (consequents) are functions of the inputs, usually defined as

$$f_i(x_1, x_2) = a_i x_i + b_i x_2 + r_i, \quad (3)$$

where a_i, b_i are the parameters of the consequent from the i-th fuzzy rule from above. For $a_i = b_i = 0$ the model becomes a TS fuzzy model of the zeroeth order. This fuzzy model can be regarded as a collection of several linear models defined locally in the fuzzy regions determined by the rule antecedents. Smooth transition from one fuzzy region to another is guaranteed by the overlapping of the fuzzy regions which in turn is provided by the overlapping membership functions from the antecedents.

Fuzzy identification of the TS fuzzy model of the zeroeth order requires rules of the form

$$\boldsymbol{R}_i : \textbf{IF } x_1 \textit{ is } \boldsymbol{A}_i \textit{ and } x_2 \textit{ is } \boldsymbol{B}_i \textbf{ THEN } y = r_i, \qquad i = 1, \ldots N, \quad (4)$$

where r_i are crisp (pointwise) values of the outputs.

The antecedents are fuzzy relations defined on the cartesian product $X = X_1 \times X_2$. The fuzzification of a crisp input x_i $(i = 1, 2)$ produces an input column vector (a fuzzy input)

$$\boldsymbol{\mu}(x_i) = [\mu_{A_1}(x_i), \mu_{A_2}(x_i), \ldots \mu_{A_m}(x_1)]^T . \tag{5}$$

The degrees of fulfillment for all possible fuzzy relations representing the antecedents are calculated and recorded into the matrix $\boldsymbol{S}$. If algebraic product is used to represent the "and" connective from the antecedents, this matrix can be directly obtained by multiplication. That is,

$$\boldsymbol{S} = \boldsymbol{\mu}_1 \otimes \boldsymbol{\mu}_2^T = \boldsymbol{\mu}_1 \cdot \boldsymbol{\mu}_2^T , \tag{6}$$

where $\boldsymbol{\mu}_1$ and $\boldsymbol{\mu}_2$ are the fuzzy inputs.

A crisp output value y is computed as a weighted mean value of sigletons (center-of-singeltons), that is

$$y = \frac{\sum_{i=1}^{n} \sum_{j=1}^{m} s_{ij} r_{ij}}{\sum_{i=1}^{n} \sum_{j=1}^{m} s_{ij}} . \tag{7}$$

The dimension of the matrix $\boldsymbol{S}(m \times n)$, which represents the structure of the model, depends on the dimensions of input fuzzy sets $\boldsymbol{\mu}_1(m \times 1)$ and $\boldsymbol{\mu}_2(n \times 1)$. The fuzzy relational matrix $\boldsymbol{R}$ consists of elements r_{ij}.

In order to apply a standard least-squares method to estimate the parameters r_{ij}, the vectors $\boldsymbol{s}$ and $\boldsymbol{r}$ are constructed from $\boldsymbol{S}$ and $\boldsymbol{R}$, respectively

$$\begin{aligned} \boldsymbol{s} &= (s_{11}\; s_{12} \ldots s_{1n} \ldots s_{m1}\; s_{m2} \ldots s_{mn})^T , \\ \boldsymbol{r} &= (r_{11}\; r_{12} \ldots r_{1n} \ldots r_{m1}\; r_{m2} \ldots r_{mn})^T . \end{aligned} \tag{8}$$

Using these vectors, equation 7 can be rewritten as

$$y = \frac{\boldsymbol{s}^T \cdot \boldsymbol{r}}{\boldsymbol{s}^T \cdot \boldsymbol{I}} = \frac{\boldsymbol{s}^T (x_1, x_2) \cdot \boldsymbol{r}}{\boldsymbol{s}^T (x_1, x_2) \cdot \boldsymbol{I}} , \tag{9}$$

where $\boldsymbol{I}$ defines a vector of ones of the same dimension $(n \cdot m \times 1)$ as $\boldsymbol{s}$ and $\boldsymbol{r}$. The elements r_{ij} are estimated on the basis of observations made at equidistant time intervals by measuring the process input and output. A system of linear equations is constructed from the above equations for the time intervals $t = t_1, t = t_2, \ldots, t = t_N$

$$\begin{bmatrix} \boldsymbol{s}^T (t_1) \\ \boldsymbol{s}^T (t_2) \\ \vdots \\ \boldsymbol{s}^T (t_N) \end{bmatrix} \cdot \boldsymbol{r} = \begin{bmatrix} \boldsymbol{s}^T (t_1) \cdot \boldsymbol{I} y (t_1) \\ \boldsymbol{s}^T (t_2) \cdot \boldsymbol{I} y (t_2) \\ \vdots \\ \boldsymbol{s}^T (t_N) \cdot \boldsymbol{I} y (t_N) \end{bmatrix} . \tag{10}$$

This system is of the form

$$\boldsymbol{\Psi} \cdot \boldsymbol{r} = \boldsymbol{\Omega}, \tag{11}$$

with a known nonsquare matrix $\boldsymbol{\Psi}$ and a known vector $\boldsymbol{\Omega}$. The solution of this overdetermined system is obtained by taking the pseudo-inverse as an optimal solution for the vector $\boldsymbol{r}$ in a least-squares sense. That is

$$\boldsymbol{r} = (\boldsymbol{\Psi}^T \boldsymbol{\Psi})^{-1} \boldsymbol{\Psi}^T \boldsymbol{\Omega}, \tag{12}$$

where $\boldsymbol{\Psi}$ stands for a fuzzified data matrix with dimension $N \times (n \cdot m)$ and $\boldsymbol{\Omega}$ is of dimension $N \times 1$.

In the case of more than two input variables (multi-input/single-output fuzzy system), $\boldsymbol{S}$ and $\boldsymbol{R}$ are no longer matrices, but both become a tensor defined in the total product space of the inputs.

3 Model based predictive control

Model based Predictive Control (MBPC) is a control strategy based on the explicit use of a dynamic model of the process. This model is used to predict the future behaviour of the process output signal over a certain (finite) time horizon and to evaluate control actions by minimizing a certain cost function. MBPC stands for a collection of several different techniques all based on the same principles. Originally, the algorithms have been developed for linear systems, but the basic idea of prediction has been extended to nonlinear systems as well. The approach based on optimization of a cost function is accompanied by a major drawback: due to the nonlinear nature of the models (fuzzy models, neural network based models, or any other nonlinear model) a nonlinear and nonconvex optimization problem has to be solved at each sampling time. In real-time applications, it can not be guaranted that the global optimum is found within one sampling period. In order to overcome this disadvantage, some attempts at using approximate local linear models for each operating range and a corresponding linear MBPC algorithm have been reported in the literature [12], [13].

3.1 Dynamic matrix control algorithm

In this section the basics of predictive control based on the so-called convolution theorem are introduced. The convolution model is described using the equation

$$y(k) = \sum_{i=1}^{\infty} g_i \Delta u_{k-i} + n(k), \tag{13}$$

where $y(k)$ represents the output signal, $\Delta u(k)$ is the input signal, g_i are coeficients of the process step response, and $n(k)$ stands for unmodelled dynamics.

The control signal in the case of model-reference predictive control is obtained by optimizing a cost function with respect to $\Delta u(k+j)$. That is,

$$J = \sum_{j=N_1}^{N_2} (\hat{y}(k+j) - y_m(k+j))^2 + \lambda \sum_{j=0}^{N_u-1} (\Delta u(k+j))^2, \tag{14}$$

where it is assumed that $\Delta u(k+j) = 0$ for $j \geq N_u$. Here $y_m(k+j)$ is the j steps ahead prediction of the reference signal, $\hat{y}(k+j)$ is the predicted output signal obtained using the process model, and $\Delta u(k+j)$ stands for the predicted control signal. Furthermore, N_1 and N_2 are the lower and upper prediction horizons and N_u denotes the control prediction horizon. The parameter λ gives the weight of the control signal.

The output signal prediction can be said to depend on the free response of the process $y_p(k+j)$ and the forced response $y_v(k+j)$. That is,

$$\hat{y}(k+j) = y_p(k+j) + y_v(k+j) + n(k). \tag{15}$$

Free response of the process means behavior where $\Delta u(k+j) = 0$ for $j = 1, \ldots, N_u$ is assumed. Forced response means behavior in the case of an input signal $\Delta u(k+j)$ for $j > 0$. The output signal prediction is then described in the form

$$\hat{y}(k+j) = \sum_{i=1}^{j} g_i \Delta u(k+j-i) + \sum_{i=j+1}^{\infty} g_i \Delta u(k+j-i) + n(k) \tag{16}$$

$$= y(k) + \sum_{i=1}^{j} g_i \Delta u(k+j-i) + \sum_{i=j+1}^{\infty} g_i \Delta u(k+j-i) - \sum_{i=1}^{\infty} g_i \Delta u(k-i). \tag{17}$$

The above equation can be rewritten in the compact form

$$\hat{y}(k+j) = \boldsymbol{g}_j \Delta u(k+j) + p_j, \tag{18}$$

where

$$\boldsymbol{g}_j = [g_1 g_2 \cdots g_j], \tag{19}$$

denotes the vector of step response coefficients and p_j represents the free response of the system given by

$$p_j = y(k) + \sum_{i=j+1}^{\infty} g_i \Delta u(k+j-i) - \sum_{i=1}^{\infty} g_i \Delta u(k-i), \tag{20}$$

$$p_j = y(k) + \sum_{i=1}^{\infty} (g_{j+i} - g_i)\Delta u(k-i). \tag{21}$$

In the case of an asymptoticaly stable process, the second equation from above becomes

$$p_j = y(k) + \sum_{i=1}^{N} (g_{j+i} - g_i)\Delta u(k-i), \tag{22}$$

where the maximum prediction horizon N is chosen to fulfill the equation

$$g_{j+i} - g_i \cong 0, \tag{23}$$

for $i > N$ and $j = N1, \ldots N_2$. This equation can be fulfilled only for asymptoticaly stable processes.

(14) can be rewritten in the compact matrix form

$$J = (\boldsymbol{y} - \boldsymbol{y}_m)(\boldsymbol{y} - \boldsymbol{y}_m)^T + \lambda \Delta \boldsymbol{u} \Delta \boldsymbol{u}^T, \tag{24}$$

where

$$\boldsymbol{y} = [\hat{y}(k+N_1), \ldots, \hat{y}(k+N_2)]^T, \tag{25}$$

denotes the predicted output vector and

$$\boldsymbol{y}_m = [y_m(k+N_1), \ldots, y_m(k+N_2)]^T, \tag{26}$$

is the predicted reference signal vector where the prediction is confined between the lower and upper prediction horizons. The vector

$$\Delta \boldsymbol{u} = [\Delta u(k) \ldots \Delta u(k+N_u-1)]^T, \tag{27}$$

represents the sequence of control signals.

The prediction of the output signal in a compact matrix form is given as

$$\boldsymbol{y} = \boldsymbol{G}\Delta\boldsymbol{u} + \boldsymbol{p}, \tag{28}$$

where

$$\boldsymbol{G} = \begin{bmatrix} g_{N_1} & \cdots & g_1 & 0 & \cdots & 0 \\ g_{N_1+1} & \cdots & g_2 & g_1 & 0 & \vdots \\ \vdots & \vdots & \vdots & \vdots & \vdots & \vdots \\ g_{N_2} & g_{N_2-1} & \cdots & \cdots & \cdots & g_{N_2-N_u+1} \end{bmatrix}, \tag{29}$$

and the vector $\boldsymbol{p}$ represents a sequence of process free response

$$\boldsymbol{p} = [p_{N_1} \ldots p_{N_2}]. \tag{30}$$

Considering all said above, the cost function can be represented as

$$J = (\boldsymbol{G}\Delta\boldsymbol{u} + \boldsymbol{p} - \boldsymbol{y}_m)(\boldsymbol{G}\Delta\boldsymbol{u} + \boldsymbol{p} - \boldsymbol{y}_m)^T + \lambda\Delta\boldsymbol{u}\Delta\boldsymbol{u}^T. \tag{31}$$

The optimal solution for the above cost function gives the control law of the DMC algorithm in the following form

$$\Delta\boldsymbol{u} = (\boldsymbol{G}^T\boldsymbol{G} + \lambda\boldsymbol{I})^{-1}\boldsymbol{G}^T(\boldsymbol{y}_m - \boldsymbol{p}). \tag{32}$$

The solution is given in a vector form and provides the means for computing an input signal for values of N_u in advance where only the first value of the input signal is applied to the process. In the next sampling period the solution is computed again and another set of N_u values is obtained in accordance to a receding horizon strategy.

3.2 Calculation of a dynamic matrix based on a fuzzy model

The main idea of fuzzy predictive control using a dynamic matrix is to combine the advantages of fuzzy modeling and predictive control. The idea is based on the *on-line* computation of the dynamic matrix $\boldsymbol{G}$. The described method offers some advantages in the case of nonlinear processes where the dynamics depends on the operating point and can be represented as $\boldsymbol{G}(u, y)$. The dynamic matrix is computed on the basis of the fuzzy model, $\boldsymbol{r}$, and the fuzzy inverse model, $\boldsymbol{r}_{inv}$, of the process whenever the operating point is changed. The vector $\boldsymbol{g}_j$ consists of j normalized step response coefficients. These coefficient are calculated using the algorithm which is given below.

First, N-steps ahead prediction of the control signal $u_{pred}(k+N)$ is done according to

$$u_{pred}(k+N) = \frac{\boldsymbol{s}^T(y_m(k+N), y_m(k+N)) \cdot \boldsymbol{r}_{inv}}{\boldsymbol{s}^T(y_m(k+N), y_m(k+N)) \cdot \boldsymbol{I}}. \tag{33}$$

The predicted future control signal is than given as the difference beetween the current control signal $u(k)$ and the predicted control signal $u_{pred}(k+N)$. That is

$$\Delta u_{pred} = u_{pred}(k+N) - u(k). \tag{34}$$

The predicted control signal $u_{pred}(k+N)$ is than used to calculate the future behaviour of the system in the sense of a predicted step response, namely

$$g_1 = y(k), \tag{35}$$

$$g_j = \frac{\boldsymbol{s}^T(g_{j-1}, u_{pred}(k+N)) \cdot \boldsymbol{r}}{\boldsymbol{s}^T(g_{j-1}, u_{pred}(k+N)) \cdot \boldsymbol{I}} \quad j = 1, \cdots N_2. \tag{36}$$

The vector $\boldsymbol{g}_j$ is then normalized, i.e.,

$$\boldsymbol{g}_j = \frac{|\boldsymbol{g}_j - g_1|}{|\Delta u_{pred}|}. \tag{37}$$

Consequently, the dynamic matrix $\boldsymbol{G}$ is formed from the coefficient vectors $\boldsymbol{g}_j$ for $j = N_1, \ldots, N_2$.

A new dynamic matrix $\boldsymbol{G}$ is computed when the reference signal $w(k)$ is changed or the difference between the process output $y(k)$ and the model-reference signal $y_m(k)$ becomes significant.

4 Application to a simulated liquid level control

This section illustrates the proposed predictive control algorithm when applied to a simulated nonlinear process concerning liquid level control. The simulation is intended to give an insight into how predictive control based on a fuzzy model actually works. It can be said that nonlinear controllers that are designed on the basis of a nonlinear process model describing the actual nonlinear process behavior accurately enough, show an acceptable and efficient control performance. The predictive control based on a fuzzy model is also robust with respect to the process-model mismatch.

4.1 Modelling of the liquid level process

The highly nonlinear liquid level process consists of a closed spherical tank, valve and motor-driven valve.

The liquid level in the closed spherical tank is manipulated by the incoming flow rate $\Phi_{in}(t)$. The flow rate of the stream is controled by the motor-driven valve. The system is disturbed by the outlet flow rate $\Phi_{out}(t)$.

The nonlinear dynamics of the system is mainly due to the spherical shape of the tank, the inner pressure p_i, and the valve characteristics.

The process can be theoreticaly described on the basis of the equilibrium equation, which describes the known volume balance as follows

$$\Phi_{in}(t) - \Phi_{out}(t) = S(h(t))\frac{dh(t)}{dt}. \tag{38}$$

In the above equation, $h(t)$ is the liquid level, and $S(h(t))$ is the transverse section of the tank which changes as a function of the liquid level. This change is described by

$$S(h(t)) = \pi(2Rh(t) - h^2(t)), \tag{39}$$

where R stands for the radius of the spherical tank.

The manipulated, or input variable, in this case is the incoming flow rate $\Phi_{in}(t)$. The controled, or output variable, is the liquid level in the tank $h(t)$. The disturbance by the outlet flow rate $\Phi_{out}(t)$ can be described as

$$\Phi_{out}(t) = c_p\sqrt{2gh(t)}, \tag{40}$$

where g represents the gravitational acceleration and c_p is the constant of the outlet valve which denotes the transverse section of the outlet valve.

The nonlinear model of the liquid level process can then be described by

$$\frac{dh(t)}{dt} = \frac{\Phi_{in}(t)}{\pi(2Rh(t) - h^2(t))} - \frac{c_p\sqrt{2gh(t)}}{\pi(2Rh(t) - h^2(t))}, \tag{41}$$

$$\frac{dh(t)}{dt} = f(\Phi_{in}(t), h(t)), \tag{42}$$

where the process parameters have the following values: $R = 1m, c_p = 0.05m^2$, and $g = 9.81ms^{-2}$.

The linearized process dynamics is of the first order and can be described with the following differential equation

$$\frac{dh(t)}{dt} = A(\Phi_{in_0}, h_0)h(t) + B(\Phi_{in_0}, h_0)\Phi_{in}(t). \tag{43}$$

Linearization is performed arround the equilibrium, or operating point, given by

$$f(\Phi_{in_0}(t), h_0(t)) = 0, \tag{44}$$

$$\frac{dh(t)}{dt} = \frac{\Phi_{in}0(t)}{\pi(2Rh_0(t) - h_0^2(t))} - \frac{c_p\sqrt{2gh_0(t)}}{\pi(2Rh_0(t) - h_0^2(t))} = 0. \tag{45}$$

In (45), the operating point is determined by $h_0(t)$ and Φ_{in_0}. These are related to each other in the following manner

$$\Phi_{in_0}(t) = c_p\sqrt{2gh(t)}. \tag{46}$$

The constants $A(\Phi_{in_0}, h_0)$ and $B(\Phi_{in_0}, h_0)$ of the linearized process model are the partial derivatives

$$A(\Phi_{in_0}, h_0) = \left.\frac{\partial f(\Phi_{in}(t)}{\partial h(t)}\right|_{h_0(t),\Phi_{in}0(t)}, \tag{47}$$

$$B(\Phi_{in_0}, h_0) = \left.\frac{\partial f(\Phi_{in}(t)}{\partial \Phi_{in_0}(t)}\right|_{h_0(t),\Phi_{in}0(t)}, \tag{48}$$

and their values are determined as

$$A(\Phi_{in_0}, h_0) = \frac{-c_p g(2Rh_0(t) - h_0^2(t))}{\pi\sqrt{2gh_0(t)}(2Rh_0(t) - h_0^2(t))^2}, \tag{49}$$

$$B(\Phi_{in_0}, h_0) = \frac{1}{\pi(2Rh_0(t) - h_0^2(t))}. \tag{50}$$

To provide for the classical representation of the process, the time constant T and the process gain K are given too. They are both strongly dependent on the current operating point $h_0(t)$ and $\Phi_{in}0(t)$. The time constant of the process is given as a function of the operating point

$$T(h_0(t)) = \frac{\pi\sqrt{2gh_0(t)}(2Rh_0(t) - h_0^2(t))}{c_p g}, \tag{51}$$

and the process gain is

$$K(h_0(t)) = \frac{\sqrt{2gh_0(t)}}{c_p g}. \tag{52}$$

Figures 1 and 2 represent the behavior in time of the process gain and process time constant, both depending on the operating point.

Figures 1 and 2 show the strong nonlinear behavior of the liquid level process. The nonlinear analitical model is used to simulate the actual process and a sampling period $T_s = 1s$ is used to obtain the simulated data.

4.2 Fuzzy modelling of the process

The Takagi-Sugeno fuzzy model of first order is constructed from input-output simulated data. Only parameter identification is performed because the structure of the model is known and thus only the parameters of the model have to be estimated. To obtain the input-output data a step input

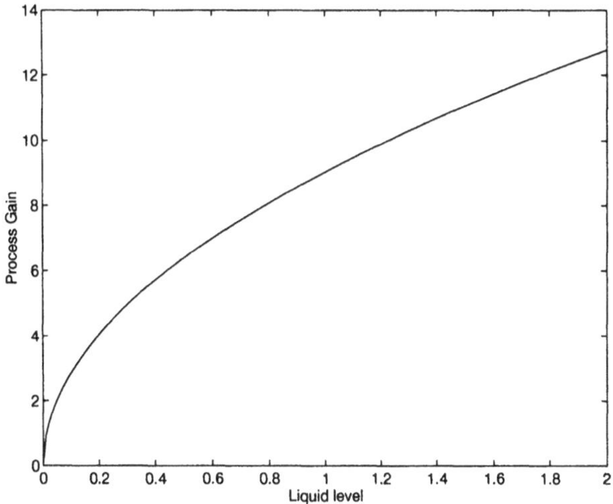

Figure 1: The process gain as a function of the liquid level.

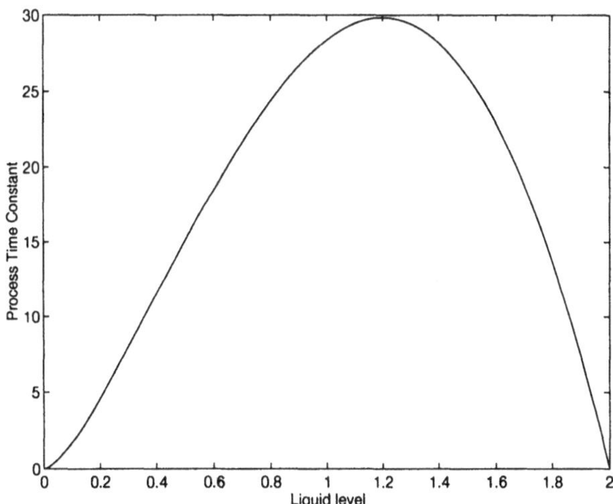

Figure 2: The process time constant as a function of the liquid level.

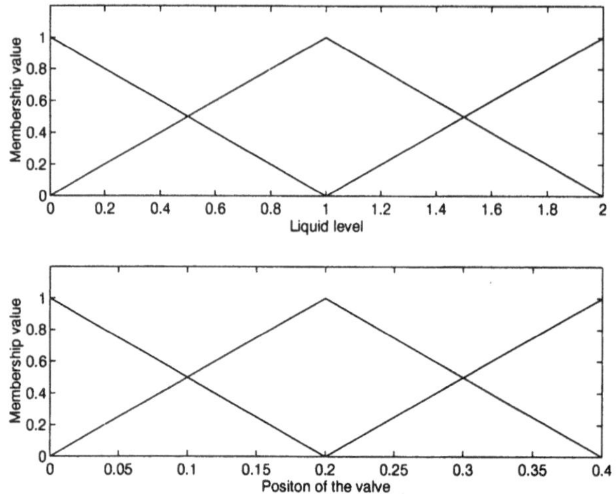

Figure 3: Membership functions for Φ_{in} and h.

excitations are used. The dynamics of the process can be represented as a first order NARX output model

$$h(k+1) = \mathcal{F}(\Phi_{in}(k), h(k)), \tag{53}$$

where $\mathcal{F}$ is an unknown nonlinear function approximated by the fuzzy model. The fuzzy model is given in the form of the fuzzy matrix $\boldsymbol{R}$. To apply the predictive control algorithm based on this fuzzy model the inverse process model has to be obtained too. The inverse dynamics of the process can be described using a first order NARX input model

$$\Phi_i n(k) = \mathcal{G}(h(k-1), h(k)), \tag{54}$$

where $\mathcal{G}$ stands for the unknown nonlinear function approximated by the fuzzy model. This fuzzy model is given in the form of the fuzzy matrix $\boldsymbol{R}_{inv}$.

The domains of the variables of the fuzzy model and the fuzzy inverse model represent the partition of the operating domain into sub-domains. Each variable takes three fuzzy values described by corresponding membership functions. The upper part of Fig. 3 represents the fuzzy values of the signal Φ_{in}. The lower part of Fig. 3 represents the fuzzy values of the signal h.

In general, the number, position and the shape of the membership functions is chosen according to the system's behavior and can be done using several methods, e.g., clustering techniques, neural networks, genetic algorithms. The consequent parameters of the fuzzy model and the fuzzy inverse model have been determinated by the ordinary least-squares algorithm.

The fuzzy model of liquid level process given by the fuzzy matrix

$$\boldsymbol{R} = \begin{bmatrix} -0.0015 & 0.8978 & 0.0000 \\ 0.0296 & 0.9942 & 1.9448 \\ -0.0116 & 1.0575 & 2.0460 \end{bmatrix}, \tag{55}$$

and can be re-expressed, column by column, into the vector form $\boldsymbol{r}$

$$\boldsymbol{r} = (-0.0015\ 0.0296\ -0.0116\ 0.8978\ 0.9942\ 1.0575\ 0\ 1.9448\ 2.046)^T. \tag{56}$$

The inverse fuzzy model of the liquid level process given in a fuzzy matrix form has the following parameters

$$\boldsymbol{R}_{inv} = \begin{bmatrix} -0.0150 & -0.8468 & 0.0000 \\ 1.4088 & 0.1774 & -1.1082 \\ 0.0013 & 1.7615 & 0.2903 \end{bmatrix}, \tag{57}$$

and can also be re-expressed, column by column, into the vector form $\boldsymbol{r}_{inv}$

$$\boldsymbol{r}_{inv} = (-0.015\ 1.4088\ 0.0013 - 0.8468\ 0.1774\ 1.7615\ 0 - 1.1082\ 0.2903)^T. \tag{58}$$

The validation of the fuzzy model which is presented in Fig. 4 has been done using a different data set, namely using input-output data for a different excitation signal. Better approximation of the process output can be obtained using more membership functions for both input variables. Additionally, the steady-state characteristic of the fuzzy model is compared with the steady-state characteristic of the simulated process in Fig. 5.

4.3 The dynamic matrix control based on a fuzzy model

The identified fuzzy model is integrated into a DMC predictive control scheme in the manner described earlier in this article. The value of the lower prediction horizon is $N_1 = 1$ and this of the upper prediction horizon is $N_2 = 6$. The control horizon is $N_u = 3$ and the maximum prediction horizon is $N = 100$. Also ,the penalty for the control effort is $\lambda = 0.25$. The reference model transfer function is chosen as

$$G_m(z) = \frac{0.0952z^{-1}}{1 - 0.9048z^{-1}}, \tag{59}$$

for the sampling time $T_s = 1s$ which is the discrete equivalent of the continuous transfer function

$$G_m(s) = \frac{1}{10s + 1}. \tag{60}$$

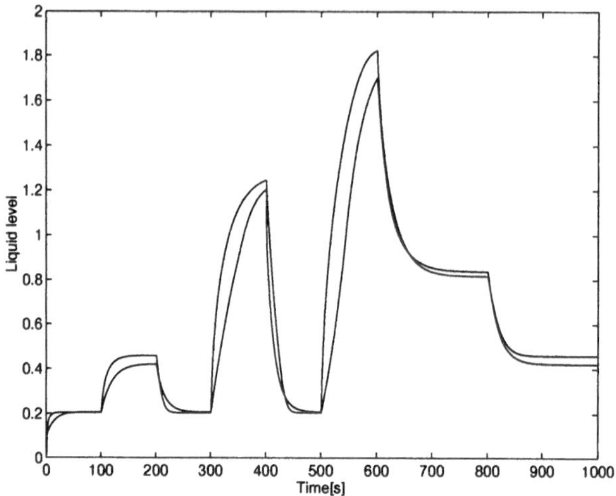

Figure 4: Validation of the fuzzy model $\boldsymbol{R}$.

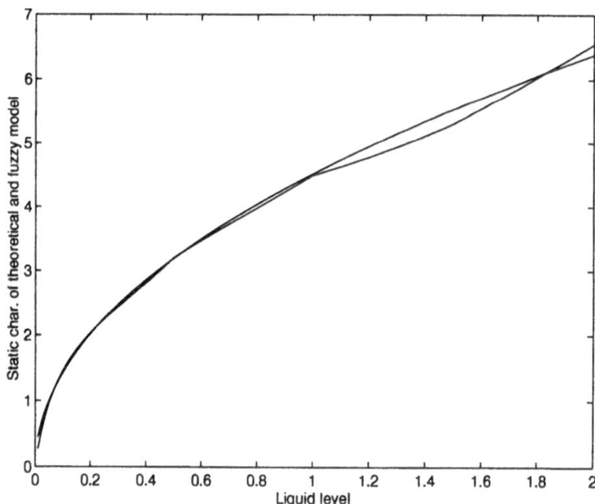

Figure 5: Comparison of the steady-state characteristics of the fuzzy model and the simulated process.

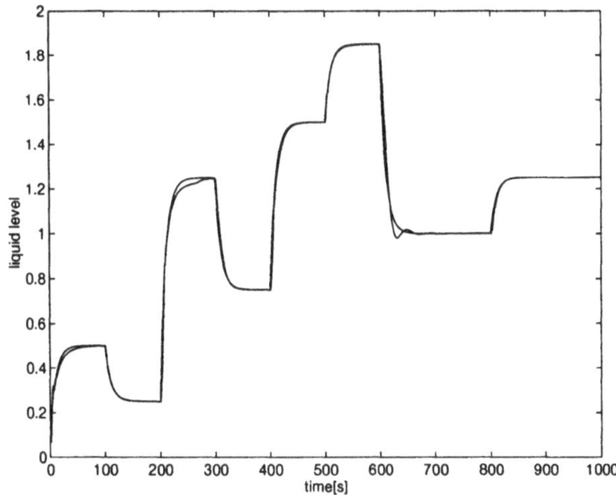

Figure 6: Simulation results for fuzzy predictive control of the spherical tank.

Figures 6 and 7 show the simulation results for the spherical tank output, model-reference signal, and the control signal, achieved by the fuzzy predictive control scheme. The effectiveness of this control scheme is illustrated by the perfect reference model tracking in the whole control domain despite of the strong nonlinearity of the simulated process. This is shown in Fig. 6 where the output signals $y(k)$ and $y_m(k)$ are given. The control signal $u(k)$ is shown in Fig. 7. The proposed novel fuzzy model-based dynamic matrix control algorithm exhibits very good tracking properties and also good disturbance rejection properties.

5 Fuzzy model based predictive control of an industrial temperature plant

The heart of the temperature plant is a tubular heat exchanger, through which steam from an electrically heated steam generator continuously circulates in a counter-current flow to a water circuit. A schematic diagram is shown in Fig. 8. The temperature of the steam is kept constant by local pressure control in the steam generator and the flow of the steam is controlled by the position of the steam valve. After being heated in the exchanger, the water passes through a pneumatic valve into the air cooler and then re-enters the exchanger.

The output of the whole process is the temperature of the water leaving the exchanger (T31). The output is controlled by the position of the steam

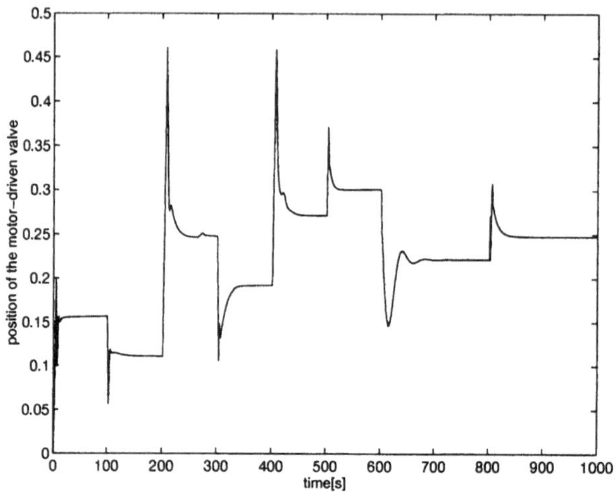

Figure 7: The control signal for the fuzzy predictive control of the spherical tank.

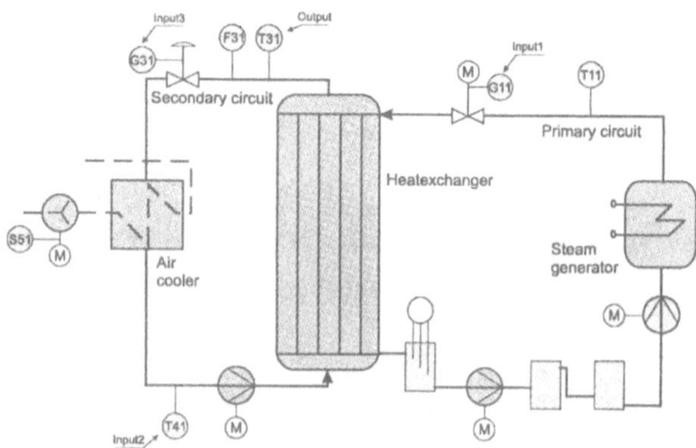

Figure 8: Temperature plant.

valve (G11) in the primary circuit which position represents an input signal.

The behaviour of the process strongly depends on the operating conditions, which are defined by other signals applied to the process: temperature at the outlet of the air cooler (T41) and pneumatic valve position (G31). With approximately constant temperature T41, the setting of the pneumatic valve position causes significant changes of the process gain.

5.1 Fuzzy modeling of the temperature process

Similar to static feedforward neural network, fuzzy models actually represent a static mapping between model input and output fuzzy sets. Thus, dynamic systems are modelled as a nonlinear static mapping between the fuzzy sets defined in the space of the lagged model inputs and outputs. Note that in the same way the system dynamics are captured in other types of models like linear regression models, or neural networks.

For the temperature process, a MISO fuzzy model is identified. The fuzzy rules constituting the model approximate a first-order nonlinear regression model where the new temperature T31 is a function of the current temperature T31, the steam valve position G11, the current temperature T41 and the current position of the pneumatic valve G31. That is,

$$y(k+1) = f(y(k), u_1(k), u_2(k), u_3(k), u_4(k)). \tag{61}$$

In each fuzzy input space only three equally spaced and triangulary shaped membership functions are chosen. Then the fuzzy model of the process is given by a four-dimensional hyperspace structure $\boldsymbol{R}$.

For the validation of the identified model, the verification signals, shown in Fig. 9 were used. To test the model, we performed simulation recursively. In contrast to the one-step-ahead prediction, the calculated model output is fed back. If the model is to be used for the prediction of the process outputs over a long-range horizon, it is necessary to validate it in the recursive simulation mode instead of the one-step ahead prediction.

5.2 Real time experiment

Although the process is very complex, it could be presented by a model with dynamics of the first order, small time delay, significantly time varying parameters, and nonlinearities due to the operating point. The results in the

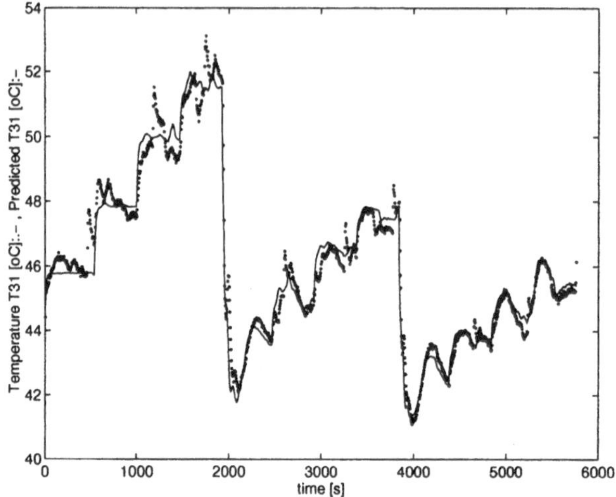

Figure 9: The verification of the identified model.

case of fuzzy predictive control are shown in Fig. 2, where the output $y(t)$, the reference-model output $y_m(t)$ and the control signal $u(t)$ are shown for two different operating conditions determinated by the position of the pneumatic valve G31 (0% and 12.5% opened).

6 Conclusion

In this chapter a predictive control algorithm based on a fuzzy model was presented. In the context of the real-time experiments on an industrial scale temperature it can be seen that the algorithm has a good performance. The main advantage in comparison with other conventional techniques is the use of a fuzzy model which enables the implementation of a predictive control scheme to nonlinear processes.

References

[1] Takagi, T. and M. Sugeno (1985). *Fuzzy Identification of Systems and its Application to Modelling and Control.* IEEE Trans. on Systems, Man and Cybernetics, Vol. 15, No. 1, pp.116-132.

[2] Sugeno, M. and K. Tanaka (1991). *Successive Identification of a Fuzzy Model and its Application to Prediction of a Complex System.* Fuzzy Sets and Systems, Vol. 42, pp. 315-334.

[3] Czogala, E. and Pedrycz, W. (1981). *On Identification in Fuzzy Systems and its Applications in Control Problems.* Fuzzy Sets and Systems, Vol. 6, pp. 73-83.

[4] Pedrycz, W. (1984). *An Identification Algorithm in Fuzzy Relational Systems.* Fuzzy Sets and Systems, Vol. 15, pp. 153-167.

[5] Tong, R.M. (1980). *The Evaluation of Fuzzy Models Derived from Experimental Data.* Fuzzy Sets and Systems, Vol.4, pp 1-12.

[6] D.Clarke, Advances in Model-Based Predictive Control, Oxford Science Publication, 1994

[7] J.L. Marchetti, D.A.Mellicamp, D.E.Seborg, Predictive Control Based on Discrete Convolution Models, Ind. Eng. Chem. Process Des. Dev., 1983, Vol. 22, pp.488-495

[8] R.m.C De Keyser, P.G.A. Van de Valde, F.A.G. Dumortier A Comparative Study of Self-adaptive Long-range Predictive Control Methods, Automatica, Vol. 24, No.2, pp. 149-163, 1988

[9] E.Czogala, W.Pedrycz, On identification in fuzzy systems and its applications in control problems, Fuzzy Sets and Systems, North Holland Publishing Company, Vol.6, No.1, Page 73-83, 1981

[10] B.M.Pfeiffer Identifikation von Fuzzy-Regeln aus Lerndaten, Beitrag zum 3. Workshop Fuzzy-Control des VDE/GMA-Unterausschusses, Dortmund, 1993

[11] W.Pedrycz, An identification algorithm in fuzzy relational systems, Fuzzy sets and systems, North Holland Publishing Company, Vol.15, Page 153-167, 1984

[12] M.Ayala-Botto, T.J.J.van den Boom, A.Krijgsman, J.S.da Costa, Constrained Nonlinear Predictive Control Based on Input-Output Linearization Using a Neural Network. IFAC World Congress'96, San Francisco. USA. pp.175-180, 1996

[13] H.M.Ritt, P.Krauss, H.Rake, Predictive Control of pH-plant Using Gain Scheduling. IMACS CESA'96 Multiconference. Lille, France. pp.473-478, 1996

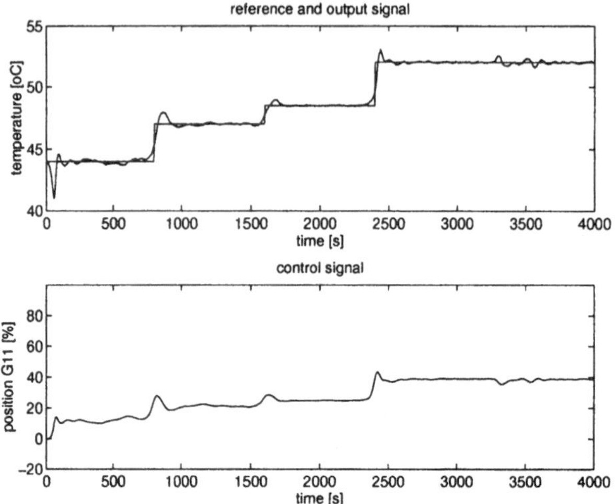

Figure 10: The process output and the reference-model signal in the case of fuzzy predictive control.

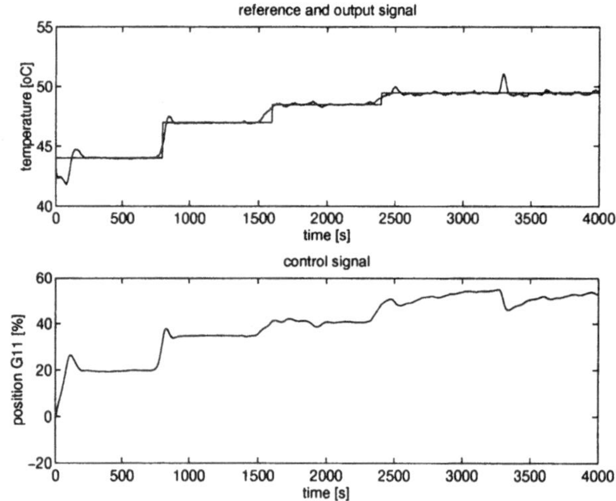

Figure 11: The process output and the reference-model signal in the case of fuzzy predictive control.

Transient Performance, Robustness and Off-Equilibrium Linearisation in Fuzzy Gain Scheduled Control

Tor A. Johansen
SINTEF Electronics and Cybernetics, Automatic Control
N-7034 Trondheim, Norway

Kenneth J. Hunt
Daimler-Benz R & D, Intelligent Systems Group
Alt-Moabit 96 A, D-10559 Berlin, Germany

Peter J. Gawthrop
Centre for Systems and Control, University of Glasgow
Glasgow G12 8QQ, Scotland

1 Introduction

The ability to perform well during transient operation will be an unavoidable requirement in future control systems where one pursues higher control performance and operational flexibility closer to the fundamental limitations of the physical system.

In this article we address design of Takagi-Sugeno (TS) fuzzy gain scheduled control systems with particular emphasis on transient performance. Notice that we do not distinguish between gain scheduled control and Takagi-Sugeno fuzzy control using multiple local linear controllers [18, 19, 21, 14, 3, 23, 22]. The Takagi-Sugeno fuzzy inference mechanism is simply viewed as a generic and efficient means of interpolating local linear controllers, see also [8, 7]. Takagi-Sugeno fuzzy control and gain scheduled control designs typically fall into one of the following categories:

- **Design based on linearizations around a set of equilibrium points:** The controllers are typically scheduled on the output, state or auxiliary variable. This approach is described in, e.g., [16, 15, 1, 12]. Application of classical linearization theory requires that the linearizations correspond to equilibria of the nonlinear system. This is a serious restriction that may lead to poor transient performance of the fuzzy gain scheduled control system. The system may even become unstable when it is operating far away from the equilibrium manifold.

- **Design based on dynamic linearization along a nominal trajectory:** Typically, the controllers are scheduled in open loop on the reference trajectory or on auxiliary inputs which are used to define the nominal trajectory, e.g., [16]. This covers transient operation, but suffers from the drawback that the performance may be poor when the system is operating far away from the reference trajectory, due to the open loop scheduling. Furthermore, control design for the resulting linear time-varying (LTV) systems is considerably more difficult than control design for the linear time-invariant (LTI) systems that result from linearization around equilibrium points. Multiple LTI design based on off-equilibrium linearizations at fixed points in time along the reference trajectory (scheduled in open loop) is suggested in [14, 3]. This approach relies on the assumption that the reference trajectory is slowly time-varying.

- **Design based on multiple linearizations around a single equilibrium point:** This approach is studied in, e.g., [19, 21, 2, 23, 22]. It may lead to controllers that performs well during transient operation because the framework allows some of the local linear controllers to be associated with transient operating regimes, although this is not explicitly discussed in the referenced work. However, since only a single equilibrium point is considered, this approach aims at regulation and stabilization, and is not well suited for tracking and control problems with multiple equilibria.

In [6] it was suggested to schedule on a variable composed of the output, state, control input and auxilliary inputs. Gain scheduled control is then designed by linearizing the nonlinear system at states that may not be equilibria. The benefit from this approach is that the transient (off-equilibrium) dynamics of the control system may be significantly improved using local LTI design and closed loop scheduling. The purpose of the present article is to extend and formulate the idea of off-equilibrium linearization.

The importance of the transient behavior of gain scheduled control systems is also addressed in [13], where it is described how to implement linear controllers based on classical linearizations (at equilibria only) such that the "linearization property" [10] also holds during transient operation.

2 Classical and off-equilibrium linearization

Consider the time-invariant nonlinear system

$$\dot{x} = f(x,u), \qquad y = g(x), \tag{1}$$

where $x \in R^n$, $u \in R^r$, $y \in R^m$ and f and g are smooth. The equilibrium manifold for this system is given by

$$\mathcal{E} \;=\; \{(x,u) \in R^{n+r} \mid f(x,u) = 0\}.$$

Now, for any equilibrium point $(x_0, u_0) \in \mathcal{E}$ the classical linearization is defined by the LTI system

$$\begin{aligned}\dot{x} &= A_0(x - x_0) + B_0(u - u_0),\\ A_0 &= \frac{\partial f}{\partial x}(x_0, u_0),\\ B_0 &= \frac{\partial f}{\partial u}(x_0, u_0).\end{aligned}$$

One interpretation of the linearized dynamics is that it approximates the dynamics of the nonlinear system for initial values and inputs that deviate slightly from x_0 and u_0 respectively. A major reason for the popularity of the classical linearization approach is: (i) the existence of a complete and well-developed theory for analysis of the local stability properties of equilibria of nonlinear systems in terms of their linearized versions, e.g., [11], and (ii) a well established theory for design of LTI control systems based on linearization.

Suppose instead that the dynamics of the nonlinear system are approximated near an arbitrary point $(x_0, u_0) \in R^{n+r}$ (which need not be an equilibrium point). That is,

$$\dot{x} \;=\; A_0(x - x_0) + B_0(u - u_0) + f(x_0, u_0). \tag{2}$$

This equation can also be written in the form

$$\dot{x} \;=\; A_0 x + B_0 u + d_0, \tag{3}$$

where $d_0 = f(x_0, u_0) - A_0 x_0 - B_0 u_0$. From a mathematical point of view this corresponds to approximating the function f by its tangent plane at the point (x_0, u_0). However, the interpretation of the system (2) is conceptually different when (x_0, u_0) is not an equilibrium point. For example, it makes no sense to consider stability of the point (x_0, u_0), simply because it can only be a transient state of the system rather than an equilibrium point. One interpretation of (2) is that it approximates the (possibly transient) dynamics of the nonlinear system when the trajectory is close to (x_0, u_0). A related interpretation is that the flow of the linearized dynamics approximates the flow of the nonlinear system for states and inputs that deviate slightly from x_0 and u_0. While such an approximation is not of immediate usefulness for stability analysis, it is still very useful for stable high-performance control design, as shown in Sect. 3 and in the example in Sect. 4.

The difference between off-equilibrium linearization around a set of points and dynamic linearization along a trajectory can be seen to depend only on the granularity of the set of points in the off-equilibrium linearization. The reason for this is that the LTV system resulting from dynamic linearization depends only on the point which the trajectory passes through at a given time. Hence, off-equilibrium linearization leads to an arbitrarily close approximation of the LTV system in terms of a set of LTI systems. This of course when there exists an LTI system close to any point on the nominal trajectory of the LTV system, and the LTI systems are interpolated using a sensible interpolation mechanism Takagi-Sugeno fuzzy rules. A formal proof of this can be found in [9]. Related approximation issues are also discussed in [14, 3, 8].

In the following, consider the Takagi-Sugeno fuzzy model

$$\begin{aligned} \dot{x} &= A(\alpha)x + B(\alpha)u + d(\alpha), && (4) \\ y &= C(\alpha)x + c(\alpha), && (5) \end{aligned}$$

as an approximation to the nonlinear system (1) based on a finite number of interpolated LTI models

$$\begin{aligned} \dot{x} &= A_i x + B_i u + d_i, && (6) \\ y &= C_i x + c_i, && (7) \end{aligned}$$

with regions of validity characterized by the fuzzy sets with membership functions μ_i. It is assumed that the point of linearization is parameterized by the scheduling variable α which depends on the state x, input u and auxiliary variable θ: $\alpha = \alpha(x, u, \theta)$. The matrices are defined by

$$\begin{aligned} A(\alpha) &= \sum_{i=1}^{N} A_i w_i(\alpha), \quad & B(\alpha) &= \sum_{i=1}^{N} B_i w_i(\alpha), \quad & d(\alpha) &= \sum_{i=1}^{N} d_i w_i(\alpha), \\ C(\alpha) &= \sum_{i=1}^{N} C_i w_i(\alpha), \quad & c(\alpha) &= \sum_{i=1}^{N} c_i w_i(\alpha), \end{aligned}$$

where the weighting functions, defined by the Takagi-Sugeno inference mechanism [18], are

$$w_i(\alpha) = \frac{\mu_i(\alpha)}{\sum_{j=1}^{N} \mu_j(\alpha)}. \qquad (8)$$

3 Classical and off-equilibrium fuzzy gain scheduling

In traditional gain scheduling, e.g., [16, 15, 1, 12], a finite set (or family) of controllers is designed on the basis of classical linearizations, such as (2), around a set of equilibrium points $\mathcal{E}^\star \subset \mathcal{E}$. Compared to a single linear controller, such a fuzzy gain scheduled control design will typically extend the operating range where the control system performs satisfactorily. However, as with a single linear controller design, the resulting closed loop dynamics when operating far away from the equilibrium manifold is not taken explicitly into account in the design. Hence, the control system may have poor performance or become unstable when it is operating far away from the equilibrium manifold. At this point it is clear that off-equilibrium linearizations such as (3) will be useful, since they allow shaping of the transient dynamics away from the equilibrium manifold as well, see Fig. 1.

Off-equilibrium fuzzy gain scheduling differs from traditional gain scheduling where the scheduling variable is typically a low-dimensional vector that parameterizes the equilibrium manifold (e.g., [15, 12]) rather than the full state/input space. Consequently, a possible disadvantage with off-equilibrium linearization is that the scheduling variable may be of higher dimension than with classical linearization, and there may be additional difficulties when the full state is not measured. In this case, one might either apply an observer, or seek a sufficiently rich characterization of the operating point in terms of the input and output for the most significant transient dynamics to be captured. These problems will be further discussed in Sect. 3.4.

Off-equilibrium fuzzy gain scheduled control also differs from most approaches to Takagi-Sugeno fuzzy control, since in [19, 20, 21, 2, 23, 22] the offset term $d(\cdot)$ is systematically neglected and a single equilibrium point $x = 0$ is considered. The present framework is a systematic approach to the design of fuzzy gain scheduled control systems based on multiple linear controllers for high transient performance during regulation and tracking.

In the remainder of this section the various aspects of fuzzy gain scheduled control design, emphasising aspects related to transient performance, will be discussed. In Sects. 3.1 and 3.2 closed loop equations for the cases where the fuzzy gain scheduled controller is based on static state feedback and dynamic output feedback, respectively, will be presented. These equations will form the basis for feedforward design, considered in Sect. 3.3.

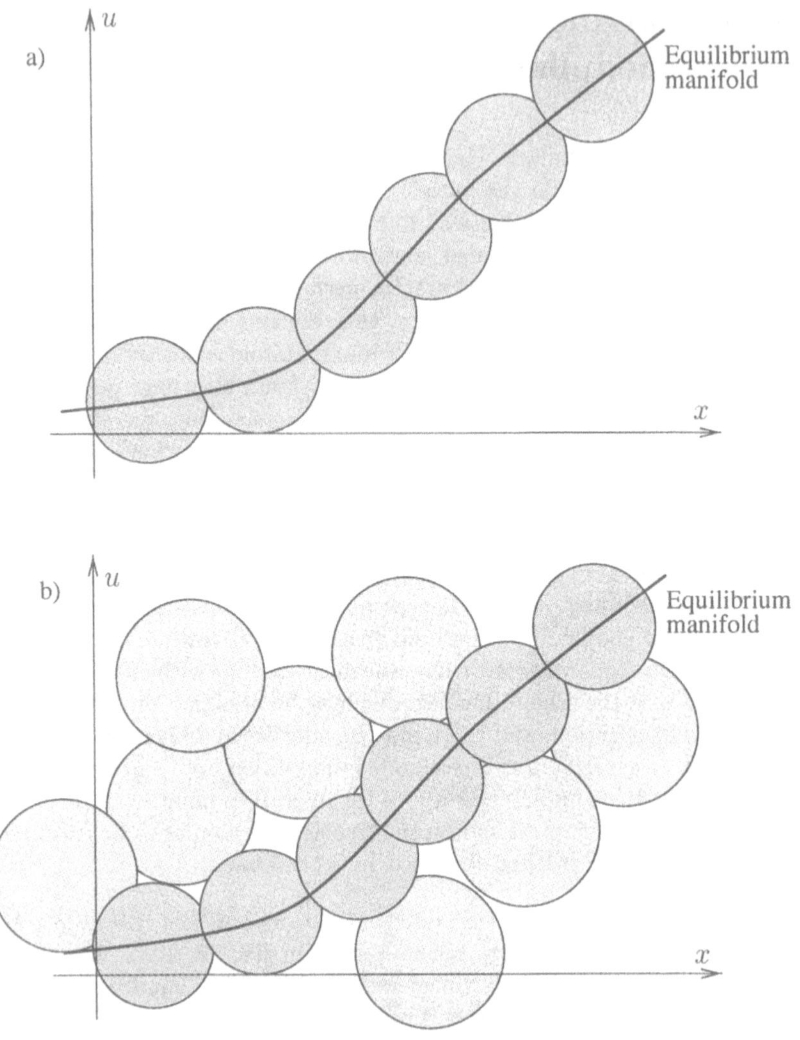

Figure 1: Classical linearization and fuzzy gain scheduling (part a) versus off-equilibrium linearization and fuzzy gain scheduling (part b).

3.1 Static state feedback

Let x^r be a state reference trajectory, and u^r a corresponding input trajectory. Consider the static state feedforward/feedback

$$u = u^r + K_c(\alpha)(x - x^r), \tag{9}$$

where $K_c(\alpha)$ is designed at a finite number of operating points α. The resulting local controllers are interpolated using the Takagi-Sugeno fuzzy inference mechanism

$$K_c(\alpha) = \sum_{i=1}^{N} K_i w_i(\alpha). \tag{10}$$

It follows immediately that the closed loop dynamics are given by

$$\dot{x} - \dot{x}^r = (A(\alpha) + B(\alpha)K_c(\alpha))(x - x^r) + \varepsilon, \tag{11}$$
$$\varepsilon = -\dot{x}^r + A(\alpha)x^r + B(\alpha)u^r + d(\alpha). \tag{12}$$

The term ε in (11) may or may not be identically zero, depending on how the trajectories x^r and u^r are generated. This problem will be discussed in Sect. 3.3, where feedforward and trajectory generation is discussed for both static state feedback and dynamic output feedback.

The controller $K_c(\alpha)$ should be designed to meet the global stability, robustness and performance requirements, both at equilibria and during transient operation. However, it is well known that due to the time-varying α, local properties (for frozen values of α) of the closed loop do not carry over to global properties [15, 17, 6]. The global properties of the nonlinear control system can be verified by extensive nonlinear simulation, certain stability criteria [20], or by explicitly taking them into the design criterion [23, 22].

Both equilibrium and off-equilibrium design can be viewed as means to reshape the flow of the open loop vector field

$$A(\alpha)x + B(\alpha)u + d(\alpha),$$

into a closed-loop vector field

$$(A(\alpha) + B(\alpha)K_c(\alpha))(x - x_r) + \varepsilon,$$

such that the vector $x - x_r$ "flows" in a way that meets the performance requirements (i.e., towards the origin, with a specific rate) and is robust to various perturbations of the vector field.

3.2 Dynamic output feedback

In contrast to the state feedback case, assume that a smooth reference output trajectory y^r is given, with corresponding state and input trajectories x^r and u^r as above. Now consider the dynamic output feedback

$$\begin{aligned} u &= u^r + K_c(\alpha)z + G_c(\alpha)(y - y^r), && (13) \\ \dot{z} &= A_c(\alpha)z + B_c(\alpha)(y - y^r), && (14) \end{aligned}$$

with internal state z. The local controller parameters $A_{c,i}$, $B_{c,i}$, $G_{c,i}$ and $K_{c,i}$ are chosen to meet the closed loop specifications locally, and then interpolated

$$A_c(\alpha) = \sum_{i=1}^{N} A_{c,i} w_i(\alpha), \qquad B_c(\alpha) = \sum_{i=1}^{N} B_{c,i} w_i(\alpha),$$

$$K_c(\alpha) = \sum_{i=1}^{N} K_{c,i} w_i(\alpha), \qquad G_c(\alpha) = \sum_{i=1}^{N} G_{c,i} w_i(\alpha).$$

The feedback (13)-(14) leads to the following closed loop dynamics

$$\begin{pmatrix} \dot{x} - \dot{x}^r \\ \dot{z} \end{pmatrix} = A_0(\alpha) \begin{pmatrix} x - x^r \\ z \end{pmatrix} + \begin{pmatrix} \varepsilon_x \\ \varepsilon_z \end{pmatrix}, \qquad (15)$$

$$A_0(\alpha) = \begin{pmatrix} A(\alpha) + B(\alpha)G_c(\alpha)C(\alpha) & B(\alpha)K_c(\alpha) \\ B_c(\alpha)C(\alpha) & A_c(\alpha) \end{pmatrix},$$

$$\begin{aligned} \varepsilon_x &= (-\dot{x}^r + A(\alpha)x^r + B(\alpha)u^r + d(\alpha)) \\ &\quad + B(\alpha)G_c(\alpha)(C(\alpha)x^r + c(\alpha) - y^r), \\ \varepsilon_z &= B_c(\alpha)(C(\alpha)x^r + c(\alpha) - y^r). \end{aligned}$$

As with static state feedback, the dynamics of the closed loop can be reshaped by local controllers at transient operating points, and the time-varying nature of α should be taken into consideration. Again, it is observed that the significance of the terms ε_x and ε_z depends on how the trajectories x^r and u^r are generated.

3.3 Trajectory generation

The controls (9) and (13)-(14) can be naturally interpreted as a 2-DOF controller with a feedforward and a feedback term. The term u^r can be viewed as a feedforward from the reference output y^r. The second term is a fuzzy gain-scheduled feedback that compensates for model uncertainties and disturbances, see Fig. 2. As discussed in Sects. 3.1 and 3.2, a major issue is how

Figure 2: The basic control structure viewed as a 2-DOF controller with trajectory generator (feedforward from the output reference trajectory) and feedback from the output or state.

to generate the trajectories x^r and u^r. This also has a significant impact on the transient behaviour of the closed loop.

Suppose a smooth output reference trajectory y^r is given (both in the state and output feedback cases). There are now several ways to generate the state and input trajectories, leading to different versions of the structure in Fig. 2.

- An **Open Loop Dynamic Trajectory Generator** could define x^r and u^r according to

$$\begin{aligned} \dot{x}^r &= A(\alpha^r)x^r + B(\alpha^r)u^r + d(\alpha^r), && (16) \\ y^r &= C(\alpha^r)x^r + c(\alpha^r), && (17) \end{aligned}$$

  where $\alpha^r = (x^r, u^r)$.

- A **Scheduled Dynamic Trajectory Generator** might define x^r and u^r according to

$$\begin{aligned} \dot{x}^r &= A(\alpha)x^r + B(\alpha)u^r + d(\alpha), && (18) \\ y^r &= C(\alpha)x^r + c(\alpha), && (19) \end{aligned}$$

  which actually introduces a feedback loop since the matrices are scheduled on α rather than α^r as in the open loop case. We observe in particular that this is the choice that gives $\varepsilon = \varepsilon_x = \varepsilon_z = 0$ in (11) and (15).

- An **Open Loop Static Trajectory Generator** may define x^r and u^r according to the steady-state equations

$$\begin{aligned} 0 &= A(\alpha^r)x^r + B(\alpha^r)u^r + d(\alpha^r), && (20) \\ y^r &= C(\alpha^r)x^r + c(\alpha^r), && (21) \end{aligned}$$

  where $\alpha^r = (x^r, u^r)$.

- A **Scheduled Static Trajectory Generator** will typically define x^r and u^r according to the steady-state equations

$$\begin{aligned} 0 &= A(\alpha)x^r + B(\alpha)u^r + d(\alpha), && (22) \\ y^r &= C(\alpha)x^r + c(\alpha), && (23) \end{aligned}$$

  which is also seen to contain a feedback.

- **No feedforward** can also be applied, i.e. $u^r = 0$, and x^r must be selected to be compatible with y^r, i.e. $y^r = g(x^r)$. This choice may require integral action to be included in the loop in order to avoid large steady-state error since $d(\cdot)$ is essentially neglected.

The various approaches to trajectory generation have different advantages and disadvantages in terms of applicability, accuracy, computational complexity and transparency. Dynamic trajectory generation will essentially invert the system, which may lead to instability when there are unstable zero-dynamics.

The generated trajectory will usually be influenced by uncertainty with respect to the current state of the system, modelling error and also changes of the nominal trajectory. In particular, there may be an error due to incorrect initialization of $x^r(0)$ in the case of dynamic trajectory generation. This may lead to significant transients that must be stabilized by the feedback compensator. Regardless of the trajectory generator strategy selected, it is highly useful that the fuzzy gain scheduled feedback compensator is designed to operate with high performance in transient operating regimes. Otherwise, the feedback loop might limit the performance of the control system. Off-equilibrium linearization and design is a useful tool to achieve this, as discussed above.

3.4 Fuzzy scheduler

The fuzzy scheduler assigns weights to each local controller depending on the current operating point.

In traditional equilibrium-based fuzzy and gain scheduled control design, the equilibrium manifold can typically be characterized by r variables, since it is defined by n equations in $n+r$ variables, assuming that the set of equations is not singular. If the number of inputs equals the number of outputs, i.e. $m = r$, the equilibrium point can be characterized by the reference output y^r.

On the other hand, with off-equilibrium based fuzzy gain scheduled control, information about the full state and input must usually be available. To achieve this, the state must be reconstructed from the output, e.g., using an observer. There will also exist special cases where neglecting a certain state in the scheduling may not have significant impact on the performance. For example, if the linearized system "parameters" $A(\cdot)$, $B(\cdot)$, $d(\cdot)$, $C(\cdot)$, and $c(\cdot)$ do not depend strongly on a certain variable, this variable need not be used for scheduling, since any linearization can be extrapolated along that particular axis without too much loss of accuracy. Also, if a certain state evolves on a much faster or much slower time-scale than the desired response of the system, it need not be important for scheduling and can be eliminated.

4 Simulation example

Consider the following unstable nonlinear dynamic system.

$$\begin{aligned} \dot{x}_1 &= x_2, \\ \dot{x}_2 &= x_1^2 + x_2^2 + u, \end{aligned} \tag{24}$$

with equilibrium manifold

$$\mathcal{E} = \left\{ (x_1^e, x_2^e, u^e) \mid x_2^e = 0 \text{ and } (x_1^e)^2 + u^e = 0 \right\}.$$

A dynamic output feedback from $y = x_1$ will be applied. It is assumed, however, that x_2 is available for the purpose of scheduling. This is reasonable, since it can be estimated by robust differentiation of y, or by the use of an observer. The input u is not required for scheduling, since the system is linear in this variable (the system matrices $A(\cdot)$ and $B(\cdot)$ do not depend on u). Notice that the dynamic output feedback will introduce additional states in the closed loop. For simplicitly, this example will mainly be concerned with the states of the system, even though the controller states are also governed by α-dependent differential equations due to the scheduling of linear controllers.

4.1 Off-equilibrium linearization

Consider the point $(x_1, x_2) = (3, 3)$, which is clearly off the equilibrium manifold, since $x_2 \neq 0$. Fig. 3b illustrates the flow of the vector field of the dynamical system

$$\begin{aligned} \dot{x}_1 &= x_2, \\ \dot{x}_2 &= 6x_1 + 6x_2 - 18, \end{aligned}$$

that results from this off-equilibrium linearization with $u = 0$. Comparing with the flow of the nonlinear system (24) in Fig. 3a, it is clear that the flow of the linearized system approximates the flow of the nonlinear system in a neighbourhood of the point of linearization, even though it is not an equilibrium.

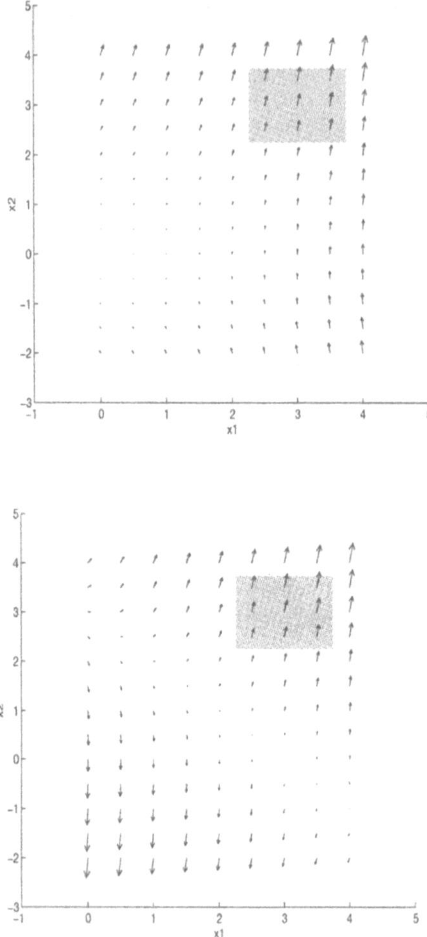

Figure 3: The left picture illustrates the flow of the nonlinear system, while the right picture illustrates the flow of the linearized system at the off-equilibrium point $(x_1, x_2) = (3.0, 3.0)$.

4.2 The effect of off-equilibrium linearization in fuzzy gain scheduled control

We consider three fuzzy gain scheduled controllers.

Controller C1 – This fuzzy gain scheduled controller contains 4 local linear dynamic output feedback controllers, designed on the basis of classical linearizations around the equilibrium points (x_1^e, x_2^e, u^e) which are in

$$\{(0.5, 0, -0.25), (1.5, 0, -2.25), (2.5, 0, -6.25), (3.5, 0, -12.25)\}.$$

The local controllers all contain integral action, and are designed such that the local closed loop poles are located at -2.5 and -5.0, and in addition there are observer poles at -10.0 and -20.0, see [6] for details on the design algorithm. The scheduling variable is $\alpha = y = x_1$, since this is sufficient to characterize the equilibrium manifold. This fuzzy gain scheduled controller is, for most practical purposes, equivalent to the gain scheduled controller used in [4, 5]. There it is also shown that this controller is superior to a single linear controller designed on the basis of linearization around a single operating point.

Controller C2 – Contains 16 local linear dynamic output feedback controllers, based on off-equilibrium linearizations around the points

$$\begin{aligned}(x_1^o, x_2^o, u^o) \in \; & \{(0.5,-1,0),(1.5,-1,0),(2.5,-1,0),(3.5,-1,0),\\ & (0.5,0,0),(1.5,0,0),(2.5,0,0),(3.5,0,0),\\ & (0.5,1,0),(1.5,1,0),(2.5,1,0),(3.5,1,0),\\ & (0.5,4,0),(1.5,4,0),(2.5,4,0),(3.5,4,0)\},\end{aligned}$$

which also covers a wide range of transient states. The design parameters are the same as for C1, but here the scheduling variable is $\alpha = (x_1, x_2)$. Notice that it is not necessary to schedule on the input because in this case $A(\cdot)$ and $B(\cdot)$ do not depend on u.

Controller C2-ff – While both C1 and C2 are purely feedback compensators, C2-ff is C2 extended with an additional open loop static trajectory generator that generates the exact equilibrium value u^r corresponding to the reference signal y^r, i.e., $u^r = -(y^r)^2$.

The fuzzy membership functions associated with the regions of validity of the linearizations and local control designs are shown in Fig. 4.

To compare the performance of these controllers a set of simulation experiments are considered:

- Experiment E1 - Step change in the reference from 0 to 1 at time 0

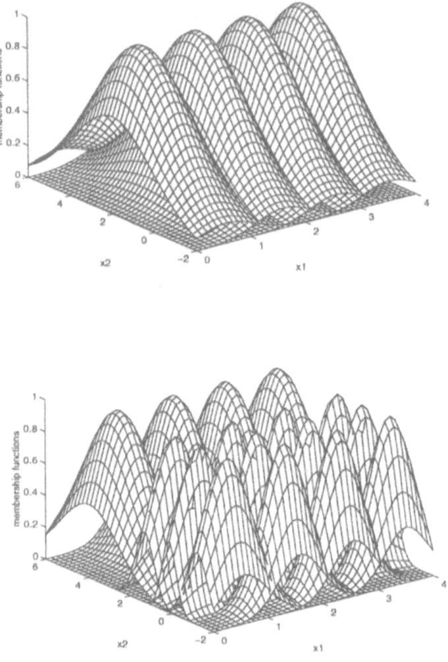

Figure 4: Membership functions for the fuzzy controllers C1 (left) and C2/C2-ff (right).

- Experiment E2 - Step change in the reference from 0 to 3 at time 0.
- Experiment E3 - Step change in the reference from 0 to 4 at time 0.

Simulation results from these three experiments using the controllers C1, C2 and C2-ff can be found in Fig. 5. From these experiments the following observations are made.

In experiment E1, the performance of C1 and C2 is roughly similar. We see from Part 3 of Fig. 5 that the system trajectory moves through transient states that are not very far away from the equilibrium manifold (the x_1-axis).

In experiment E2, the performance of C2 is clearly superior to C1. Controller C1 utilises larger control effort and there are some oscillations in both the input and output that are not present with controller C2. The problems of controller C1 in experiment E2 are caused by the fact that its transient states are significantly further away from the equilibrium manifold than in experiment E1.

In experiment E3, the closed loop becomes unstable with controller C1, while good performance and stability are still maintained with controller C2. Viewed differently, the initial state $(0, 0)$ is within the region of attraction of the desired final state $(4, 0)$ using the controller C2, while it is not with controller C1. The instability of C1 is clearly due to the inadequacy of the equilibrium based design during transient states far off from the equilibrium manifold.

In all experiments, the difference between C2 and C2-ff is minor, although the feedforward seems to make the controller meet the specifications (rise-time corresponding to closed loop poles) somewhat better.

In summary, the simulation example has clearly demonstrated the possible improvements with off-equilibrium fuzzy gain scheduled control design as opposed to fuzzy gain scheduled control design limited only to the equilibrium manifold. Improvements in both dynamic performance and region of attraction (stability) of the control system is demonstrated.

The example also demonstrates the approximately feedback linearization property of the fuzzy gain scheduled control design [6]; the response from the reference to the system output is approximately linear because the same desired closed loop design objective is applied in all operating regimes. When only classical linearizations (at the equilibrium manifold) are applied, the approximation only holds near the equilibrium manifold, which may lead to poor performance and possibly instability. Including off-equilibrium linearization and control design will extend the region of linearity to cover the full range of the design, including transient operating regimes. This brings the approxi-

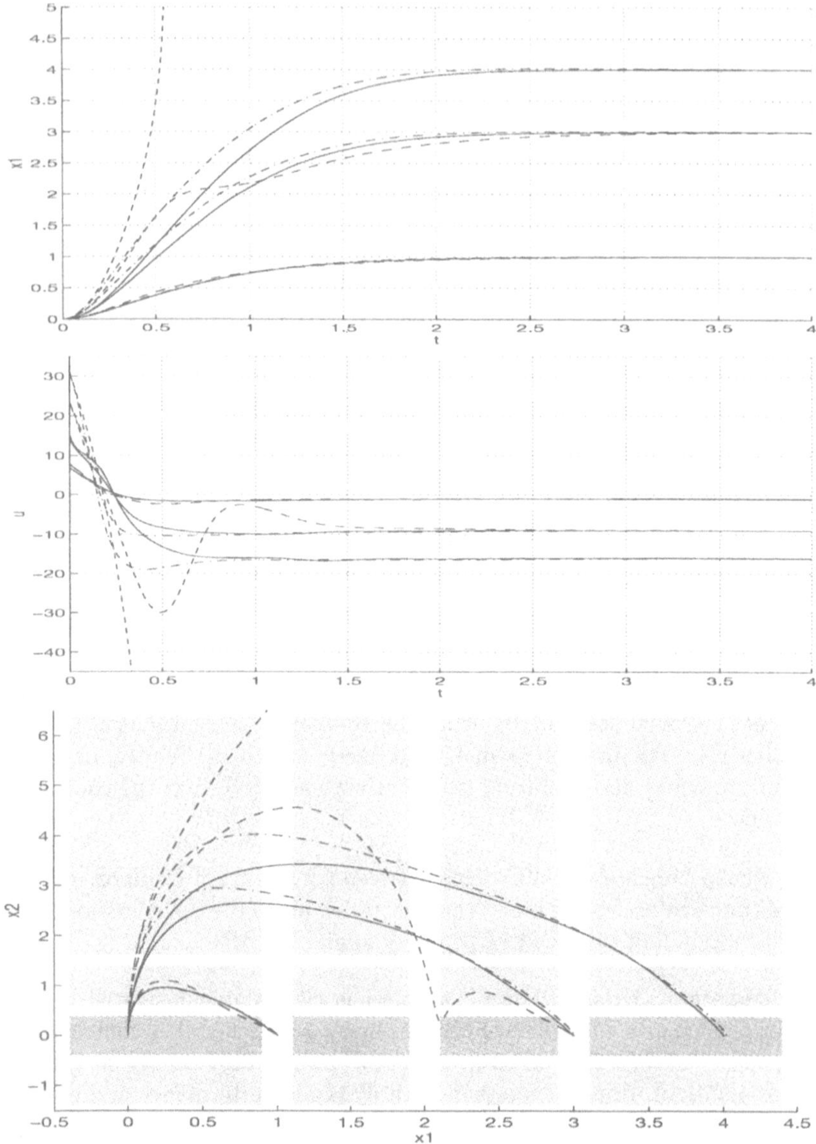

Figure 5: Simulations results with C1 (dashed lines), C2 (dashed-dotted lines) and C2-ff (solid lines). The different curves are for the experiments E1, E2 and E3, corresponding to setpoint $y^r = 1, 3, 4$. Top: The system output $y = x_1$. Middle: The control input u. Bottom: Phase plane plot (x_1, x_2). The shaded regions represent the assumed regions of validity of the local controllers, cf. Fig. 4. The darkest regions correspond to controller C1, while controllers C2 and C2-ff have local controllers in all regions. Notice that the plane (x_1, x_2) is a projection of the state-space of the closed loop, since the controller contains some states.

mate feedback linearization achieved by fuzzy gain scheduling closer to exact feedback linearization.

5 Conclusions

In this article we have studied the potential benefits of using off-equlibrium linearization in fuzzy gain scheduling control design, in addition to classical linearization. The simulation example demonstrates improvementes in performance and extended region of stability by the use of off-equilibrium linearization. Improved transient performance with this approach is also verified in a full-scale experimental vehicle speed control application, see [9].

Acknowledgment

This work was supported by the European Commission under the ESPRIT Basic Research Project 8039 NACT.

References

[1] P. Apkarian and P. Gahinet. A convex characterization of gain scheduled H_∞ controllers. *IEEE Trans. Automatic Control*, 40:853–864, 1995.

[2] S. G. Cao, N. W. Rees, and G. Feng. Stability analysis and design for a class of continuous-time fuzzy control systems. *Int. J. Control*, 64:1069–1087, 1996.

[3] D. Driankov, R. Palm, and U. Rehfuss. A Takagi-Sugeno fuzzy gain-scheduler. In *Proc. IEEE Conf. Fuzzy Systems, New Orleans*, pages 1053–1059, 1996.

[4] P. J. Gawthrop. Continuous-time local state local model networks. In *Proc. IEEE Conf. Systems, Man and Cybernetics, Vancouver, Canada*, pages 852–857, 1995.

[5] P. J. Gawthrop. Continuous-time local model networks. In R. W. Żbikowski and K. Hunt, editors, *Neural Adaptive Control Technology*, pages 41–70. World Scientific, Singapore, 1996.

[6] K. J. Hunt and T. A. Johansen. Design and analysis of gain-scheduled local controller networks. *Int. J. Control*, 66:619–651, 1997.

[7] T A. Johansen. On the optimality of the Takagi-Sugeno-Kang fuzzy inference mechanism. In *Proceedings of the 4th IEEE Conf. on Fuzzy Systems, Yokohama, Japan*, 1995.

[8] T. A. Johansen and B. A. Foss. Constructing NARMAX models using ARMAX models. *Int. J. Control*, 58:1125–1153, 1993.

[9] T. A. Johansen, K. J. Hunt, P. J. Gawthrop, and H. Fritz. Off-equilibrium linearization and design of gain scheduled control with application to vehicle speed control. Preprint, submitted for publication, 1997.

[10] I. Kaminer, A. M. Rascoal, P. P. Khargonekar, and E. E. Coleman. A velocity algorithm for the implementation of gain-scheduled controllers. *Automatica*, 31:1185–1191, 1995.

[11] H. K. Khalil. *Nonlinear Systems*. Macmillan, New York, 1992.

[12] D. A. Lawrence and W. J. Rugh. Gain scheduling dynamic linear controllers for a nonlinear plant. *Automatica*, 31:381–390, 1995.

[13] D. J. Leith and W. E. Leithead. Appropriate realization of gain-scheduled controllers with application to wind turbine regulation. *Int. J. Control*, 65:223–248, 1996.

[14] R. Palm, D. Driankov, and U. Rehfuess. Lyapunov linearization based design of Takagi-Sugeno fuzzy controllers. In *Proc. IFSA World Congress, São Paolo*, pages 513–516, 1995.

[15] W. J. Rugh. Analytical framework for gain scheduling. *IEEE Control Systems Magazine*, 11(1):79–84, 1991.

[16] J. S. Shamma and M. Athans. Analysis of gain scheduled control for nonlinear plants. *IEEE Trans. Automatic Control*, 35:898–907, 1990.

[17] J. S. Shamma and M. Athans. Gain scheduling: Potential hazards and possible remedies. *IEEE Control Systems Magazine*, 12(3):101–107, 1992.

[18] T. Takagi and M. Sugeno. Fuzzy identification of systems and its application to modeling and control. *IEEE Trans. Systems, Man, and Cybernetics*, 15:116–132, 1985.

[19] K. Tanaka and M. Sano. A robust stabilizing problem of fuzzy controller systems and its application to backing up control of a truck-trailer. *IEEE Trans. Fuzzy Systems*, 2:119–134, 1994.

[20] K. Tanaka and M. Sugeno. Stability analysis and design of fuzzy control systems. *Fuzzy Sets and Systems*, 45:135–156, 1992.

[21] H. O. Wang, K. Tanaka, and M. F. Griffin. An approach to fuzzy control of nonlinear systems: Stability and design issues. *IEEE Trans. Fuzzy Systems*, 4:14–23, 1996.

[22] J. Zhao, R. Gorez, and V. Wertz. Fuzzy control based on linear models. In R. Murray-Smith and T. A. Johansen, editors, *Multiple Model Approaches to Modelling and Control.* Taylor and Francis Ltd., London, 1997.

[23] J. Zhao, V. Wertz, and R. Gorez. Linear TS fuzzy model based robust stabilizing controller design. In *IEEE Conf. Decision and Control, New Orleans*, pages 255–260, 1995.

Design of Fuzzy Gain Schedulers

R. Palm
Siemens AG
Dept. ZT IK 4
Otto-Hahn-Ring, 81730 Munich, Germany

D. Driankov
University of Linköping
Dept. of Computer Science
581 83 Linköping, Sweden

1 Introduction

The design of a Takagi-Sugeno fuzzy controller (TSFC) is possible only if a Takagi-Sugeno fuzzy model (TSFM) of the open loop nonlinear system under control is provided. A survey of the relevant publications in this field reveals that a TSFM is either given for granted, or is identified on the basis of input-output data. Thus, it remains an open question how a TSFC can be designed in a systematic manner, given an open loop nonlinear model described in terms of differential equations. Another aspect of TSFC which has received almost no attention is related to its robust performance. The major problem here is to find the stability margins of the closed loop model with respect to unknown disturbances.

In this article we describe how a modified version of the original TSFM can be derived via Lyapunov linearization of a given , differential equations based, open loop nonlinear model. The Lyapunov linearization is performed at a finite number of appropriately selected set-points in the state space thus obtaining a set of local open loop linear models. The latter constitute the basis for an open loop linear TSFM that approximates the original open loop nonlinear model. Furthermore, we consider the control problem of stabilizing the so obtained open loop TSFM around an arbitrary set-point. In this context we show how to design a modified TSFC which guarantees the local stability of the original nonlinear model. This modified TSFC is based on local linear control laws designed by using the local open linear models. We also determine the stability margins in the case when the different local linear models are subject to unknown disturbances. Finally, we show the use of the modified TSFM and TSFC in the context of gain scheduling.

Conventional gain scheduling, applied to the control of a *slowly time varying* and *smooth* nonlinear system, is realized by means of an appropriate interpolation between a fixed set of local gains (local control laws). The local gains for predefined points in the parameter space (in the case of exogeneous scheduling variables), or predifined points in the state space (in the case of a desired trajectory), can be designed using linear control techniques. For intermediate points in the parameter space, or in the state space, the corresponding intermediate local gains have to be calculated by interpolating the already available local gains [Nichols 93]. The global control law obtained by such an interpolation is called a *gain scheduled control law.* However, if for example, the desired trajectory is not known in advance the design of the gain scheduled control law has to be carried on-line. This is not acceptable because of the large computational effort required especially in the case of multiple-input/multiple-output (MIMO) systems.

As shown in this article, one way out of this situation is to use the modified TSFC, where the space of parameters and/or the state space is divided in advance in overlapping fuzzy regions. Each fuzzy region is characterized by a different gain designed for the center of this fuzzy region so that the original nonlinear system is stabilized at this center. The global gain for any intermediate point is then obtained as a linear combination of the local gains from the centers of the fuzzy regions. In the global gain, each local gain is weighted by the degree to which an intermediate point belongs to the fuzzy region characterized by this local gain. This type of global gain is represented as a set of fuzzy rules and is called a fuzzy gain scheduled control law.

The design of the (linear) local gains for the centers of the fuzzy regions can be done by pole placement [Palm 95], or by optimal control design using LQR methods [Rehfuess 95]. However, two types of errors have to be taken into account for the proper design of the fuzzy gain scheduled control law: (i) the Lyapunov linearization of the original nonlinear system at a center of a fuzzy region results in a linearization error, (ii) the approximation of the "missing" Lyapunov linearization at an intermediate point by the weighed linear combination of the available Lyapunov linearizations at the centers of the fuzzy regions results in approximation errors. Both types of errors are cruicial for the stability and robustness analysis of the closed loop system and have to be taken into account during design.

In Sect. 2 we present the original TSFC denoted here as TSFC-1 in order to provide the basis for better understanding the differences between the TSFC-1 and its modified version TSFC-2 introduced in Sect. 3 The design of this type of a fuzzy controller depends cruicially on the identification of an open loop model in terms of a set of fuzzy rules and so far, no commonly accepted identification procedure exists. Furthermore, with the exception of the stability properties of the closed loop system incorporating a TSFC-1, very few results concerning robustness and performance have been reported and these are at an early research stage [Tanaka 93].

In Sect. 3 we present the modified version of TSFC-1, that is TSFC-2, and show its use in local stabilization of a nonlinear autonomous system described in terms of differential equations. With respect to local stabilization, a TSFC-2 is able of stabilizing a nonlinear autonomous system around **any** operating point without the need to change its gains. Since a TSFC-2 is a linear state controller all the available results from linear control theory, concerning stability, robustness, and performance characteristics, can be readily used for the design of this type of a fuzzy controller.

In Sect. 4 we describe the use of TSFC-2 in tracking a reference trajectory of a nonlinear autonomous system. With respect to tracking, the FC-2 performs gain scheduling on **any** reference state trajectory under the restriction that the reference state trajectory is slowly time varying. Here again, one can use the wealth of results in linear control theory to achieve desired robustness and performance characteristics for this type of a fuzzy controller.

In Sect. 5 we illustrate in detail the design procedure of a fuzzy gain scheduler for the tracking control of two-link robot arm and show simulation results.

2 Model based TSFC-1

The model based design of a TSFC-1 requires a fuzzy model of the nonlinear autonomous open loop system under control. Such a fuzzy model was first proposed in 1985 by Takagi and Sugeno [Takagi 85] and is the result of identification from observed input-output data. The TSFC-1 is then intended to solve the stabilization problem in the context of a given (identified) fuzzy model. Relevant publications on the fuzzy model based design and analysis of the TSFC-1 include [Tanaka 92], [Tanaka 93], [Palm 92], [Palm 95], [Rehfuess 95]. Relevant publications on the identification of the fuzzy model include [Takagi 85], [Sugeno 86], [Sugeno 88], [Sugeno 91], [Babushka 95], [Babushka 94], [Yen 95], [Herrera 95], [Bastian 95], [Park 95], [Vergara 95], [Su 95].

2.1 The open loop fuzzy model

The identified fuzzy model consists of a set of fuzzy rules where the IF-part of each fuzzy rule describes a particular fuzzy region in the state space and its corresponding THEN-part contains a linear open loop model. Thus, the $\mathbf{i}$-th fuzzy rule, describes the fuzzy model's dynamics within the fuzzy region $\mathbf{LX^i}$ specified in the rule's IF-part. Such a fuzzy rule is written as

$$\text{IF} \quad \mathbf{x} = \mathbf{LX^i} \quad \text{THEN} \quad \dot{\mathbf{x}} = \mathbf{A_i} \cdot \mathbf{x} + \mathbf{B_i} \cdot \mathbf{u}, \tag{1}$$

where

- $\mathbf{LX^i} = (LX_1^{\mathbf{i}}, \ldots, LX_n^{\mathbf{i}})^T$, where $LX_k^{\mathbf{i}}$ denotes the fuzzy value which x_k takes in the $\mathbf{i}$-th fuzzy region. Each $LX_k^{\mathbf{i}}$ is determined by a fuzzy set $\int_X LX_k^{\mathbf{i}}(x_k)/x_k$ of a standard triangular, trapezoidal, or bell-shaped type. The membership functions of any one of the previously mentioned types are nonlinear functions of x_k,
- $\dot{\mathbf{x}} = \mathbf{A_i} \cdot \mathbf{x} + \mathbf{B_i} \cdot \mathbf{u}$ is a linear autonomous model corresponding to the $\mathbf{i}$-th fuzzy region of the state space. The state vector $\dot{\mathbf{x}}$ is a $n \times 1$ linear vector-function of time and its components $x_1, x_2, \cdots, x_n$ denote time dependent state variables. The control input vector $\mathbf{u}$ is a $m \times 1$ vector function of time with components $u_1, u_2, \cdots, u_n$. The entries of the matrices $\mathbf{A_i}$ $(n \times n)$ and $\mathbf{B_i}$ $(n \times m)$ are constant.

The open loop system's dynamics corresponding to computation with a single fuzzy rule is given as

$$\dot{\mathbf{x}} = \mathbf{LX^i}(\mathbf{x}) \cdot (\mathbf{A_i} \cdot \mathbf{x} + \mathbf{B_i} \cdot \mathbf{u}). \tag{2}$$

The results computed for each individual rule via (2) are aggregated by simply taking their average, that is,

$$\mathbf{u} = \frac{\sum_{\mathbf{i}} \mathbf{LX^i}(\mathbf{x}) \cdot (\mathbf{A_i} \cdot \mathbf{x} + \mathbf{B_i} \cdot \mathbf{u})}{\sum_{\mathbf{i}} \mathbf{LX^i}(\mathbf{x})}. \tag{3}$$

The expression (3) can be simplified by normalizing the degrees of satisfaction $\mathbf{LX^i}(\mathbf{x})$ and using instead their normalized counterparts $w^{\mathbf{i}}(\mathbf{x})$ that are obtained as follows: for any $\mathbf{LX^1}(\mathbf{x}), \mathbf{LX^2}(\mathbf{x}), \ldots, \mathbf{LX^M}(\mathbf{x})$

$$w^{\mathbf{i}}(\mathbf{x}) = \frac{\mathbf{LX^i}(\mathbf{x})}{\sum_{\mathbf{i}} \mathbf{LX^i}(\mathbf{x})}. \tag{4}$$

Thus, $\sum_{\mathbf{i}} w^{\mathbf{i}}(\mathbf{x}) = 1$ and (3) can be rewritten in the more simple form,

$$\mathbf{u} = \sum_{\mathbf{i}} w^{\mathbf{i}}(\mathbf{x}) \cdot (\mathbf{A_i} \cdot \mathbf{x} + \mathbf{B_i} \cdot \mathbf{u}). \tag{5}$$

From (5) it is easily seen that the set of all fuzzy rules defines linear dynamics for all points that belong to the center of an arbitrary fuzzy region in the fuzzy state space. Take, for example, the $\mathbf{i}$-th fuzzy region. For every point that belongs to its center, i.e., $\mathbf{x} = \mathbf{x^i}$, we have that $w^{\mathbf{i}}(\mathbf{x}) = w^{\mathbf{i}}(\mathbf{x^i}) = 1$ while for each $\mathbf{j} \neq \mathbf{i}$ we have that $w^{\mathbf{j}}(\mathbf{x}) = 0$, since $\sum_{\mathbf{i}} w^{\mathbf{i}}(\mathbf{x}) = 1$. Thus we have that (5) becomes (for each $\mathbf{x} = \mathbf{x^i}$),

$$\mathbf{u} = w^{\mathbf{i}}(\mathbf{x^i}) \cdot (\mathbf{A_i} \cdot \mathbf{x^i} + \mathbf{B_i} \cdot \mathbf{u}) = \mathbf{A_i} \cdot \mathbf{x^i} + \mathbf{B_i} \cdot \mathbf{u}.$$

On the other hand, (5) defines nonlinear dynamics for all points $\mathbf{x} \neq \mathbf{x^i}$. This is so, because in this case there is no $w^{\mathbf{i}}(\mathbf{x}) = 1$, and thus each linear part $(\mathbf{A_i} \cdot \mathbf{x^i} + \mathbf{B_i} \cdot \mathbf{u})$ in (5) is multiplied by the nonlinear function $w^{\mathbf{i}}(\mathbf{x})$.

2.2 The TSFC-1

The TSFC-1 is used for the stabilization problem in the context of the fuzzy model given by (5). The authors of this subsection were unable to find any work related to the use of this type of fuzzy controller for the tracking problem. In this latter case, the linear models from the THEN-parts of the fuzzy rules defining the nonlinear open loop model have to be transformed into linear nonautonomous models (see [Slotine 91]). Thus computation with a single fuzzy rule, as well as computation with the set of all fuzzy rules, will define nonlinear nonautonomous dynamics for the open loop system.

The TSFC-1 for the stabilization problem is given as a set of fuzzy rules where each fuzzy rule is of the form

$$\text{IF} \quad \mathbf{x} = \mathbf{LX^j} \quad \text{THEN} \quad \mathbf{u} = \mathbf{K_j} \cdot \mathbf{x}, \tag{6}$$

where

- $\mathbf{LX^j} = (LX_1^{\mathbf{j}}, \ldots, LX_n^{\mathbf{j}})^T$, where $LX_k^{\mathbf{j}}$ denotes the fuzzy value which x_k takes in the **j**-th fuzzy region. Each $LX_k^{\mathbf{j}}$ is determined by a fuzzy set $\int_X LX_k^{\mathbf{j}}(x_k)/x_k$ of a standard triangular, trapezoidal, or bell-shaped type.

- $\mathbf{u} = \mathbf{K_j} \cdot \mathbf{x}$ is a linear autonomous control law corresponding to the **j**-th fuzzy region of the state space. The matrix $\mathbf{K_j}$ is of dimension $m \times n$, its elements are constant, and it is called the *gain matrix* of the **j**-th fuzzy region.

The control law's dynamics corresponding to computation with a single fuzzy rule is given as

$$\mathbf{u} = \mathbf{LX^j}(\mathbf{x}) \cdot \mathbf{K_j} \cdot \mathbf{x}. \tag{7}$$

The results computed for each individual rule via (7) are aggregated by simply taking their average, that is,

$$\dot{\mathbf{x}} = \frac{\sum\limits_{\mathbf{j}} \mathbf{LX^j}(\mathbf{x}) \cdot \mathbf{K_j} \cdot \mathbf{x}}{\sum\limits_{\mathbf{j}} \mathbf{LX^j}(\mathbf{x})}. \tag{8}$$

The expression (8) can again be simplified by normalizing the degrees of satisfaction $\mathbf{LX^j}(\mathbf{x})$ and using instead their normalized counterparts $w^{\mathbf{j}}(\mathbf{x})$. that are obtained according to (refnormalize). Since $\sum_{\mathbf{j}} w^{\mathbf{j}}(\mathbf{x}) = 1$, (8) can be rewritten in the simpler form

$$\dot{\mathbf{x}} = \sum_{\mathbf{j}} w^{\mathbf{j}}(\mathbf{x}) \cdot \mathbf{K_j} \cdot \mathbf{x}. \tag{9}$$

From (9) it is easily seen that the set of all fuzzy rules defines a control law with linear dynamics for all points that belong to the center of an arbitrary fuzzy region in the fuzzy state space. Take, for example, the **j**-th fuzzy region. For every point that belongs to its center,i.e., $\mathbf{x} = \mathbf{x^j}$, we have that $w^{\mathbf{j}}(\mathbf{x}) = w^{\mathbf{j}}(\mathbf{x^j}) = 1$ while for each $\mathbf{i} \neq \mathbf{j}$ we have that $w^{\mathbf{i}}(\mathbf{x}) = 0$, since $\sum_{\mathbf{j}} w^{\mathbf{j}}(\mathbf{x}) = 1$. Thus we have that (9) becomes (for each $\mathbf{x} = \mathbf{x^j}$),

$$\dot{\mathbf{x}} = w^{\mathbf{j}}(\mathbf{x^j}) \cdot \mathbf{K_j} \cdot \mathbf{x^j} = \mathbf{K_j} \cdot \mathbf{x^j}.$$

On the other hand, (9) defines a control law with nonlinear dynamics for all points $\mathbf{x} \neq \mathbf{x}^j$. This is so, because in this case there is no $w^j(\mathbf{x}) = 1$, and thus each linear expression $\mathbf{K_j} \cdot \mathbf{x}^j$ in (9) is multiplied by the nonlinear function $w^j(\mathbf{x})$.

2.3 The closed loop fuzzy model

So far, by using the fuzzy rules of the form

$$\text{IF} \quad \mathbf{x} = \mathbf{LX}^i \quad \text{THEN} \quad \dot{\mathbf{x}} = \mathbf{A_i} \cdot \mathbf{x} + \mathbf{B_i} \cdot \mathbf{u},$$

and

$$\text{IF} \quad \mathbf{x} = \mathbf{LX}^i \quad \text{THEN} \quad \mathbf{u} = \mathbf{K_i} \cdot \mathbf{x},$$

we have described the nonlinear autonomous open loop fuzzy model and the nonlinear TSFC-1. Furthermore, we saw that computation with the open loop fuzzy model is equivalent to

$$\dot{\mathbf{x}} = \sum_i w^i(\mathbf{x}) \cdot (\mathbf{A_i} \cdot \mathbf{x} + \mathbf{B_i} \cdot \mathbf{u}),$$

and computation with the TSFC-1 is equivalent to

$$\mathbf{u} = \sum_j w^j(\mathbf{x}) \cdot \mathbf{K_j} \cdot \mathbf{x}.$$

Thus, replacing $\mathbf{u}$ from the former equation with its equivalent term from the latter equation, we obtain the expression for the nonlinear autonomous closed loop system:

$$\dot{\mathbf{x}} = \sum_i \sum_j w^i(\mathbf{x}) \cdot w^j(\mathbf{x}) \cdot (\mathbf{A_i} + \mathbf{B_i} \cdot \mathbf{K_j}) \cdot \mathbf{x}, \tag{10}$$

Denoting, $(\mathbf{A_i} + \mathbf{B_i} \cdot \mathbf{K_j})$ by $\mathbf{A_{ij}}$, we can write

$$\dot{\mathbf{x}} = \sum_i \sum_j w^i(\mathbf{x}) \cdot w^j(\mathbf{x}) \cdot \mathbf{A_{ij}} \cdot \mathbf{x}. \tag{11}$$

From the above equation it is easily seen that the dynamics of the autonomous closed loop system is nonlinear since both $w^i(\mathbf{x})$ and $w^j(\mathbf{x})$ are nonlinear functions of the state vector $\mathbf{x}$. The dynamics of the closed loop autonomous system is again linear only for points in the fuzzy state space that belong to the center of an arbitrary fuzzy region. The problem now is to provide conditions under which the origin is an asymptotically stable equilibrium point of (11), where the domain of attraction of the origin is the whole fuzzy state space, i.e., every trajectory in the fuzzy state space converges to the origin as $t \to \infty$. The following result from [Tanaka 92] establishes these sufficient conditions as follows:

> The origin is an asymptotically stable equilibrium point of (11) with its domain of attraction being the whole fuzzy state space if there exists a common positive definite matrix $\mathbf{P}$ such that
>
> $$\mathbf{A}_{\mathbf{ij}}^T \cdot \mathbf{P} + \mathbf{P} \cdot \mathbf{A}_{\mathbf{ij}} < \mathbf{0}. \tag{12}$$

Observe here that all of $\mathbf{A}_{\mathbf{ij}}$ are Hurwitz matrices, if there exists a common positive definite matrix $\mathbf{P}$ such that $\mathbf{A}_{\mathbf{ij}}^T \cdot \mathbf{P} + \mathbf{P} \cdot \mathbf{A}_{\mathbf{ij}} < \mathbf{0}$. However, there does not always exist such a common positive definite matrix $\mathbf{P}$ even if all $\mathbf{A}_{\mathbf{ij}}$ are Hurwitz matrices (see [Vidyasagar 93] (Chap. 5, Sect. 5.4.3, Problem 5.24). In this context, a necessary condition for the existence of a common positive definite matrix $\mathbf{P}$ is stated in [Tanaka 92] as:

> If one of $\mathbf{A}_{\mathbf{ij}} \cdot \mathbf{A}_{\mathbf{kl}}$ is not a Hurwitz matrix, where $\mathbf{A}_{\mathbf{ij}}$ and $\mathbf{A}_{\mathbf{kl}}$ are Hurwitz matrices, then there does not exist a common positive definite matrix $\mathbf{P}$ such that ${\mathbf{A}_{\mathbf{ij}}}^T \cdot \mathbf{P} + \mathbf{P} \cdot \mathbf{A}_{\mathbf{ij}} < \mathbf{0}$.

2.4 The design of TSFC-1

The design problem for a TSFC-1 follows directly from the conditions under which the origin is an asymptotically stable equilibrium of (11). This condition requires the existence of a common positive matrix $\mathbf{P}$ such that for each $\mathbf{i},\mathbf{j}$, $\mathbf{A}_{\mathbf{ij}}^T \cdot \mathbf{P} + \mathbf{P} \cdot \mathbf{A}_{\mathbf{ij}} < \mathbf{0}$. In other words, it is required that $(\mathbf{A}_\mathbf{i} + \mathbf{B}_\mathbf{i} \cdot \mathbf{K}_\mathbf{j})^T \cdot \mathbf{P} + \mathbf{P} \cdot (\mathbf{A}_\mathbf{i} + \mathbf{B}_\mathbf{i} \cdot \mathbf{K}_\mathbf{j}) < \mathbf{0}$. The latter form of the required condition for asymptotic stability implies that the design problem for the TSFC-1 can be formulated as follows.

> Given $\mathbf{A}_\mathbf{i}$ and $\mathbf{B}_\mathbf{i}$ ($\mathbf{i} = 1, \ldots, \mathbf{M}$) find $\mathbf{K}_\mathbf{j}$ ($\mathbf{k} = 1, \ldots, \mathbf{M}$) such that $(\mathbf{A}_\mathbf{i} + \mathbf{B}_\mathbf{i} \cdot \mathbf{K}_\mathbf{j})^T \cdot \mathbf{P} + \mathbf{P} \cdot (\mathbf{A}_\mathbf{i} + \mathbf{B}_\mathbf{i} \cdot \mathbf{K}_\mathbf{j}) < \mathbf{0}$, where $\mathbf{P}$ is a positive definite matrix.

As already discussed , a common positive definite $\mathbf{P}$ does not exist if, $\mathbf{A}_{\mathbf{ij}} \cdot \mathbf{A}_{\mathbf{kl}}$ is not a Hurwitz matrix, though $\mathbf{A}_{\mathbf{ij}}$ and $\mathbf{A}_{\mathbf{kl}}$ are assumed to be Hurwitz matrices.

Thus the design begins with finding such $\mathbf{K_j}$, given $\mathbf{A_i}$ and $\mathbf{B_i}$, so that $(\mathbf{A_i}+\mathbf{B_i}\cdot\mathbf{K_j}) = \mathbf{A_{ij}}$ is a Hurwitz matrix. This can be achieved, for example, by constructing each $\mathbf{K_j}$ via the well known *pole assignment method* (see [Foellinger 90]).

Second, one has to check out the possible nonexistence of a common positive $\mathbf{P}$ by forming all products $\mathbf{A_{ij}}\cdot\mathbf{A_{kl}}$ and verifying that each such product is a Hurwitz matrix. If at least one $\mathbf{A_{ij}}\cdot\mathbf{A_{kl}}$ is not a Hurwitz matrix, then one has to redesign the $\mathbf{K_j}$ and $\mathbf{K_l}$ involved in $\mathbf{A_{ij}}$ and $\mathbf{A_{kl}}$ so that the product in question becomes a Hurwitz matrix. No systematic procedure for finding such $\mathbf{K_j}$'s so that all products $\mathbf{A_{ij}}\cdot\mathbf{A_{kl}}$ are Hurwitz matrices is provided in the existing literature on TSFC-1.

Observe here, that even when the nonexistence of a common positive matrix $\mathbf{P}$ such that for each $\mathbf{i},\mathbf{j}$, $\mathbf{A}_{\mathbf{ij}}^T\cdot\mathbf{P}+\mathbf{P}\cdot\mathbf{A_{ij}} < \mathbf{0}$, is ruled out, one still has to find such a common positive $\mathbf{P}$. However, since each $\mathbf{A_{ij}}$ is a Hurwitz matrix, then there exist unique positive definite $\mathbf{P_{ij}}$ such that each inequality $\mathbf{A}_{\mathbf{ij}}^T\cdot\mathbf{P_{ij}}+\mathbf{P_{ij}}\cdot\mathbf{A_{ij}} < \mathbf{0}$ holds.

Each inequality will, in general, have a different $\mathbf{P_{ij}}$ due to the different $\mathbf{A_i}$, $\mathbf{B_i}$, and $\mathbf{K_j}$ involved in the different inequalities. However, if amongst these $\mathbf{P_{ij}}$'s there is one which satisfies all inequalities, then the origin is the asymptotically stable equilibrium of (11). In this context, the trial-and-error design method proposed in [Tanaka 92] searches for such a $\mathbf{P_{ij}}$ which satisfies all inequalities, but the authors do not suggest any solution for the case when such a $\mathbf{P_{ij}}$ does not exist.

3 The model based TSFC-2

The TSFC-2 is derived from an open loop fuzzy model which in turn is derived from the Lyapunov-linearized version of a nonlinear autonomous open loop model. Both the open loop fuzzy model and the TSFC-2 have linear dynamics. Thus, the robustness and performance of the TSFC-2 can be achieved by employing the already existing techniques from linear control theory for the design of robust and optimal linear controllers. This type of fuzzy controller can be used for both the stabilization and tracking control problems. With respect to stabilization it is able of stabilizing a given nonlinear autonomous system at **any** operating point without changing the already designed gain matrices $\mathbf{K_i}$. With respect to tracking, it is able to locally track **any** trajectory again without changing the already designed gain matrices $\mathbf{K_i}$. In the latter case of tracking, it works as a gain scheduler on a given reference state trajectory, provided that trajectories are slowly time varying.

3.1 The open loop fuzzy model

Consider the nonlinear autonomous open loop system

$$\dot{\mathbf{x}} = \mathbf{f}(\mathbf{x}, \mathbf{u}), \tag{13}$$

with $(\mathbf{0}, \mathbf{0})$ being its equilibrium point. Furthermore, consider an operating point $(\mathbf{x}^{\mathbf{d}}, \mathbf{u}^{\mathbf{d}})$ around which we want to stabilize (14). In this case, (13) is transformed into the equivalent nonlinear autonomous open loop system

$$\delta\dot{\mathbf{x}} + \dot{\mathbf{x}}^{\mathbf{d}} = \mathbf{f}(\delta\mathbf{x} + \mathbf{x}^{\mathbf{d}}, \delta\mathbf{u} + \mathbf{u}^{\mathbf{d}}), \tag{14}$$

by introducing the new state vector $\delta\mathbf{x} = \mathbf{x} - \mathbf{x}^{\mathbf{d}}$ and the new input vector $\delta\mathbf{u} = \mathbf{u} - \mathbf{u}^{\mathbf{d}}$ and substituting them for $\mathbf{x}$ and $\mathbf{u}$ in (13). The equilibrium point of (14) is now $(\delta\mathbf{x} = \mathbf{0}, \delta\mathbf{u} = \mathbf{0})$.

Let us first Lyapunov-linearize the nonlinear autonomous open loop system (14) at the center $\mathbf{x}^{\mathbf{i}}$ of each fuzzy region $\mathbf{LX}^{\mathbf{i}}$ of the fuzzy state space. As a result we obtain $\mathbf{M}$ Lyapunov-linearized versions of (14) each one of the form

$$\dot{\mathbf{x}} = \mathbf{A}(\mathbf{x}^{\mathbf{i}}, \mathbf{u}^{\mathbf{i}}) \cdot (\mathbf{x} - \mathbf{x}^{\mathbf{i}}) + \mathbf{B}(\mathbf{x}^{\mathbf{i}}, \mathbf{u}^{\mathbf{i}}) \cdot (\mathbf{u} - \mathbf{u}^{\mathbf{i}}). \tag{15}$$

The fuzzy counterpart of each Lyapunov-linearized version (15) of the original nonlinear autonomous open loop system (14) is given as a fuzzy rule of the form

$$\text{IF} \quad \mathbf{x}^{\mathbf{d}} = \mathbf{LX}^{\mathbf{i}} \quad \text{THEN} \quad \dot{\mathbf{x}} = \mathbf{A}(\mathbf{x}^{\mathbf{i}}, \mathbf{u}^{\mathbf{i}}) \cdot (\mathbf{x} - \mathbf{x}^{\mathbf{d}}) + \mathbf{B}(\mathbf{x}^{\mathbf{i}}, \mathbf{u}^{\mathbf{i}}) \cdot (\mathbf{u} - \mathbf{u}^{\mathbf{d}}). \tag{16}$$

The dynamics of the open loop fuzzy model corresponding to computation with a single fuzzy rule is given as

$$\dot{\mathbf{x}} = \mathbf{LX}^{\mathbf{i}}(\mathbf{x}^{\mathbf{d}}) \cdot (\mathbf{A}(\mathbf{x}^{\mathbf{i}}, \mathbf{u}^{\mathbf{i}}) \cdot (\mathbf{x} - \mathbf{x}^{\mathbf{d}}) + \mathbf{B}(\mathbf{x}^{\mathbf{i}}, \mathbf{u}^{\mathbf{i}}) \cdot (\mathbf{u} - \mathbf{u}^{\mathbf{d}})). \tag{17}$$

Observe here that computation with the fuzzy rule

$$\text{IF} \quad \mathbf{x}^{\mathbf{d}} = \mathbf{LX}^{\mathbf{i}} \quad \text{THEN} \quad \dot{\mathbf{x}} = \mathbf{A}(\mathbf{x}^{\mathbf{i}}, \mathbf{u}^{\mathbf{i}}) \cdot (\mathbf{x} - \mathbf{x}^{\mathbf{d}}) + \mathbf{B}(\mathbf{x}^{\mathbf{i}}, \mathbf{u}^{\mathbf{i}}) \cdot (\mathbf{u} - \mathbf{u}^{\mathbf{d}}),$$

simply approximates computation with the "missing" Lyapunov-linearized version

$$\dot{\mathbf{x}} = \mathbf{A}(\mathbf{x^d}, \mathbf{u^d}) \cdot (\mathbf{x} - \mathbf{x^d}) + \mathbf{B}(\mathbf{x^d}, \mathbf{u^d}) \cdot (\mathbf{u} - \mathbf{u^d}),$$

of (14) at $(\mathbf{x^d}, \mathbf{u^d})$. Computation with the above "missing" linearized version of (14) is replaced by computation with (17) weighted with $\mathbf{LX^i}(\mathbf{x^d})$. The latter is a measure of the extent to which the "missing" Lyapunov-linearized version of (14) at $(\mathbf{x^d}, \mathbf{u^d})$ can be replaced by the given Lyapunov-linearized version of (14) at $(\mathbf{x^i}, \mathbf{u^i})$.

The results computed for each individual rule via (17) are aggregated by simply taking their weighted sum, that is,

$$\dot{\mathbf{x}} = \frac{\sum_{\mathbf{i}} \mathbf{LX^i}(\mathbf{x^d}) \cdot (\mathbf{A}(\mathbf{x^i}, \mathbf{u^i}) \cdot (\mathbf{x} - \mathbf{x^d}) + \mathbf{B}(\mathbf{x^i}, \mathbf{u^i}) \cdot (\mathbf{u} - \mathbf{u^d}))}{\sum_{\mathbf{i}} \mathbf{LX^i}(\mathbf{x^d})}. \tag{18}$$

Expression (18) can be simplified by normalizing the degrees of satisfaction $\mathbf{LX^i}(\mathbf{x^d})$ and using instead their normalized counterparts $w^{\mathbf{i}}(\mathbf{x^d})$ that are obtained according to (4). Since $\sum_{\mathbf{i}} w^{\mathbf{i}}(\mathbf{x^d}) = 1$, (18) can be rewritten in the equivalent simpler form,

$$\dot{\mathbf{x}} = \sum_{\mathbf{i}} w^{\mathbf{i}}(\mathbf{x^d}) \cdot (\mathbf{A}(\mathbf{x^i}, \mathbf{u^i}) \cdot (\mathbf{x} - \mathbf{x^d}) + \mathbf{B}(\mathbf{x^i}, \mathbf{u^i}) \cdot (\mathbf{u} - \mathbf{u^d})). \tag{19}$$

Since for any $\mathbf{x^d}$, all of $\mathbf{LX^i}(\mathbf{x^d})$ are constant and the right hand side of the above expression is a linear function, the dynamics of the fuzzy model are linear for any given $\mathbf{x^d}$. Here again, (19) is a proper substitute for the original nonlinear open loop system (13) only within a small region around $\mathbf{x^d}$.

3.2 The TSFC-2

The TSFC-2 is intended to locally asymptotically stabilize the open loop fuzzy model at any candidate operating point and thus locally asymptotically stabilize the resulting nonlinear closed loop system when this type of fuzzy controller is inserted in (14). The TSFC-2 is given as a set of fuzzy rules, each fuzzy rule being of the form

$$\text{IF} \quad \mathbf{x^d} = \mathbf{LX^j} \quad \text{THEN} \quad \mathbf{u} = \mathbf{K}(\mathbf{x^j}, \mathbf{u^j}) \cdot (\mathbf{x} - \mathbf{x^d}) + \mathbf{u^d}, \tag{20}$$

where $\mathbf{K}(\mathbf{x}^j, \mathbf{u}^j)$ is the gain matrix of the Lyapunov-linearized version of (14) for the $\mathbf{j}$-th fuzzy region. Thus the control law from the THEN-part of the above fuzzy rule is intended to locally asymptotically stabilize the open loop fuzzy model

$$\dot{\mathbf{x}} = w^j(\mathbf{x}^d) \cdot (\mathbf{A}(\mathbf{x}^j, \mathbf{u}^j) \cdot (\mathbf{x} - \mathbf{x}^d) + \mathbf{B}(\mathbf{x}^j, \mathbf{u}^j) \cdot (\mathbf{u} - \mathbf{u}^d)),$$

at $(\mathbf{x}^d, \mathbf{u}^d)$.

The control law's dynamics corresponding to computation with a single fuzzy rule is given as

$$\mathbf{u} = \mathbf{LX}^j(\mathbf{x}^d) \cdot (\mathbf{K}(\mathbf{x}^j, \mathbf{u}^j) \cdot (\mathbf{x} - \mathbf{x}^d) + \mathbf{u}^d). \tag{21}$$

Let us also point out here that computation with the above expression approximates computation with the "missing" control law

$$\mathbf{u} = \mathbf{K}(\mathbf{x}^d, \mathbf{u}^d) \cdot (\mathbf{x} - \mathbf{x}^d) + \mathbf{u}^d,$$

intended for the asymptotic stabilization of the "missing" linearized version

$$\dot{\mathbf{x}} = \mathbf{A}(\mathbf{x}^d, \mathbf{u}^d) \cdot (\mathbf{x} - \mathbf{x}^d) + \mathbf{B}(\mathbf{x}^d, \mathbf{u}^d) \cdot (\mathbf{u} - \mathbf{u}^d),$$

of (14) at $(\mathbf{x}^d, \mathbf{u}^d)$. This approximation of $\mathbf{K}(\mathbf{x}^d, \mathbf{u}^d)$ by $\mathbf{K}(\mathbf{x}^j, \mathbf{u}^j)$ is simply done by weighting the latter control law with the degree to which $\mathbf{x}^d$ satisfies the fuzzy region for which $\mathbf{K}(\mathbf{x}^j, \mathbf{u}^j)$ is intended. Thus one again can expect approximation errors reflecting the difference between the weighted $\mathbf{K}(\mathbf{x}^j, \mathbf{u}^j)$ and the "missing" $\mathbf{K}(\mathbf{x}^d, \mathbf{u}^d)$.

The results computed for each individual rule via (21) are aggregated by simply taking their weighted sum, that is,

$$\mathbf{u} = \frac{\sum_j \mathbf{LX}^j(\mathbf{x}^d) \cdot (\mathbf{K}(\mathbf{x}^j, \mathbf{u}^j) \cdot (\mathbf{x} - \mathbf{x}^d) + \mathbf{u}^d)}{\sum_j \mathbf{LX}^j(\mathbf{x}^d)}. \tag{22}$$

Thus, the above expression provides the overall result of computation with the set of fuzzy rules representing the TSFC-2. The expression (22) can again be simplified by normalizing the degrees of satisfaction $\mathbf{LX}^j(\mathbf{x}^d)$ and using instead their normalized counterparts $w^j(\mathbf{x}^d)$ obtained according to (4). Since $\sum_j w^j(\mathbf{x}^d) = 1$, (22) can be rewritten in the equivalent more simple form,

$$\mathbf{u} = \sum_{\mathbf{j}} w^{\mathbf{j}}(\mathbf{x^d}) \cdot (\mathbf{K}(\mathbf{x^j}, \mathbf{u^j}) \cdot (\mathbf{x} - \mathbf{x^d}) + \mathbf{u^d}). \tag{23}$$

3.3 The closed loop fuzzy model

So far, by using fuzzy rules of the form

$$\text{IF} \quad \mathbf{x^d} = \mathbf{LX^i} \quad \text{THEN} \quad \dot{\mathbf{x}} = \mathbf{A}(\mathbf{x^i}, \mathbf{u^i}) \cdot (\mathbf{x} - \mathbf{x^d}) + \mathbf{B}(\mathbf{x^i}, \mathbf{u^i}) \cdot (\mathbf{u} - \mathbf{u^d}),$$

and

$$\text{IF} \quad \mathbf{x^d} = \mathbf{LX^j} \quad \text{THEN} \quad \mathbf{u} = \mathbf{K}(\mathbf{x^j}, \mathbf{u^j}) \cdot (\mathbf{x} - \mathbf{x^d}) + \mathbf{u^d},$$

we have described the open loop fuzzy model and the TSFC-2. Furthermore, we saw that computation with the fuzzy model is equivalent to

$$\dot{\mathbf{x}} = \sum_{\mathbf{i}} w^{\mathbf{i}}(\mathbf{x^d}) \cdot (\mathbf{A}(\mathbf{x^i}, \mathbf{u^i}) \cdot (\mathbf{x} - \mathbf{x^d}) + \mathbf{B}(\mathbf{x^i}, \mathbf{u^i}) \cdot (\mathbf{u} - \mathbf{u^d})),$$

and computation with the TSFC-2 is equivalent to

$$\mathbf{u} = \sum_{\mathbf{j}} w^{\mathbf{j}}(\mathbf{x^d}) \cdot (\mathbf{K}(\mathbf{x^j}, \mathbf{x^j}) \cdot (\mathbf{x} - \mathbf{x^d}) + \mathbf{u^d}),$$

where both of the above expressions have linear dynamics. Thus replacing $\mathbf{u}$ from the former equation with its equivalent term from the latter one, we obtain the equation for the closed loop fuzzy model

$$\dot{\mathbf{x}} = \sum_{\mathbf{i}} \sum_{\mathbf{j}} w^{\mathbf{i}}(\mathbf{x^d}) \cdot w^{\mathbf{j}}(\mathbf{x^d}) \cdot (\mathbf{A}(\mathbf{x^i}, \mathbf{u^i}) + \mathbf{B}(\mathbf{x^i}, \mathbf{u^i}) \cdot \mathbf{K}(\mathbf{x^j}, \mathbf{u^j})) \cdot (\mathbf{x} - \mathbf{x^d}). \tag{24}$$

Now let us derive the conditions under which (24) is asymptotically stable. In order to do this let us first present the following result about robust stability reported in [Zhou 87]. Consider a linear system with linear perturbations

$$\dot{\mathbf{x}} = \mathbf{A} \cdot \mathbf{x} + \sum_{i=1}^{M} k_i \delta \mathbf{A}_i \cdot \mathbf{x}, \tag{25}$$

where $\mathbf{A}$ is a Hurwitz matrix, $\delta\mathbf{A}_i$ are constant matrices of the same dimension as $\mathbf{A}$, and k_i are uncertain parameters with values in an arbitrary interval around zero. Let $\mathbf{P}$ be the unique solution of the Lyapunov equation

$$\mathbf{A}^T \cdot \mathbf{P} + \mathbf{P} \cdot \mathbf{A} = -2 \cdot \mathbf{I}, \tag{26}$$

and let us define the matrices $\mathbf{P}_i$ as

$$\mathbf{P}_i = \frac{\delta\mathbf{A}_i^T \cdot \mathbf{P} + \mathbf{P} \cdot \delta\mathbf{A}_i}{2}. \tag{27}$$

Then the linear system (25) is asymptotically stable if

$$\sum_{i=1}^{M} |k_i| \cdot \sigma_{max}(\mathbf{P}_i) < 1, \tag{28}$$

where $\sigma_{max}(\bullet)$ denotes the largest *singular value* of a matrix. See [Press 90] (Chap. 2, Sect. 2.9) about computation of singular values.

Now consider (24) and let for each $\mathbf{i} \neq \mathbf{j}$

$$\begin{aligned} \mathbf{A}(\mathbf{x^i}, \mathbf{u^i}) + \mathbf{B}(\mathbf{x^i}, \mathbf{u^i}) \cdot \mathbf{K}(\mathbf{x^i}, \mathbf{u^i}) &= \mathbf{A}(\mathbf{x^j}, \mathbf{u^j}) + \mathbf{B}(\mathbf{x^j}, \mathbf{u^j}) \cdot \mathbf{K}(\mathbf{x^j}, \mathbf{u^j}) \\ &= \mathbf{A}, \end{aligned} \tag{29}$$

be Hurwitz matrices. Furthermore, let us denote $(\mathbf{A}(\mathbf{x^i}, \mathbf{u^i}) + \mathbf{B}(\mathbf{x^i}, \mathbf{u^i}) \cdot \mathbf{K}(\mathbf{x^j}, \mathbf{u^j})$ by $\mathbf{A_{ij}}$ and define $\delta\mathbf{A_{ij}}$ as

$$\delta\mathbf{A_{ij}} + \mathbf{A} = \mathbf{A_{ij}}. \tag{30}$$

Now (24) can be rewritten as

$$\dot{\mathbf{x}} = \mathbf{A} \cdot (\mathbf{x} - \mathbf{x^d}) + \sum_{\mathbf{i}} \sum_{\mathbf{j}} w^{\mathbf{i}}(\mathbf{x^d}) \cdot w^{\mathbf{j}}(\mathbf{x^d}) \cdot \delta\mathbf{A_{ij}} \cdot (\mathbf{x} - \mathbf{x^d}). \tag{31}$$

It is easily seen that the above equation is of the same form as (25) where the constant k_i corresponds to the constant product $w^{\mathbf{i}}(\mathbf{x^d}) \cdot w^{\mathbf{j}}(\mathbf{x^d})$ and $\delta\mathbf{A}_i$ corresponds to $\delta\mathbf{A_{ij}}$. However, observe here that at least one product $w^{\mathbf{i}}(\mathbf{x^d}) \cdot w^{\mathbf{j}}(\mathbf{x^d})$ is greater than zero and less or equal to one since at least one $w^{\mathbf{i}}(\mathbf{x^d})$ is greater than zero and less or equal to one and $\sum_{\mathbf{i}} \sum_{\mathbf{j}} w^{\mathbf{i}}(\mathbf{x^d}) \cdot w^{\mathbf{j}}(\mathbf{x^d}) = 1$. With these observations in mind let us now reformulate the result about the robust asymptotic stability of the linear system (25) in terms of the linear system (31).

Consider the linear system with linear perturbations

$$\dot{\mathbf{x}} = \mathbf{A} \cdot (\mathbf{x} - \mathbf{x^d}) + \sum_{\mathbf{i}} \sum_{\mathbf{j}} w^{\mathbf{i}}(\mathbf{x^d}) \cdot w^{\mathbf{j}}(\mathbf{x^d}) \cdot \delta \mathbf{A_{ij}} \cdot (\mathbf{x} - \mathbf{x^d}), \tag{32}$$

where $\mathbf{A}$ is a Hurwitz matrix, $\delta\mathbf{A_{ij}}$ are constant matrices of the same dimension as $\mathbf{A}$, and $w^{\mathbf{i}}(\mathbf{x^d}) \cdot w^{\mathbf{j}}(\mathbf{x^d})$ are uncertain parameters with values in an arbitrary interval around zero. Let $\mathbf{P}$ be the unique solution of the Lyapunov equation

$$\mathbf{A}^T \cdot \mathbf{P} + \mathbf{P} \cdot \mathbf{A} = -2 \cdot \mathbf{I}, \tag{33}$$

and let us define the matrices $\mathbf{P_{ij}}$ as

$$\mathbf{P_{ij}} = \frac{\delta\mathbf{A_{ij}}^T \cdot \mathbf{P} + \mathbf{P} \cdot \delta\mathbf{A_{ij}}}{2}. \tag{34}$$

Then the linear system (31) (equivalent to (24)) is asymptotically stable if

$$\sum_i \sum_j w^{\mathbf{i}}(\mathbf{x^d}) \cdot w^{\mathbf{j}}(\mathbf{x^d}) \cdot \sigma_{max}(\mathbf{P_{ij}}) < 1. \tag{35}$$

Since in our case the products $w^{\mathbf{i}}(\mathbf{x^d}) \cdot w^{\mathbf{j}}(\mathbf{x^d})$ are known, at least one such product is greater than zero, and $\sum_i \sum_j w^{\mathbf{i}}(\mathbf{x^d}) \cdot w^{\mathbf{j}}(\mathbf{x^d}) = 1$, it is obvious that the above condition holds if the stronger condition

$$\sigma_{max}(\mathbf{P_{ij}}) < 1, \quad \text{for each} \quad \mathbf{i}, \mathbf{j} = \mathbf{1}, \mathbf{2}, \ldots, \mathbf{M}. \tag{36}$$

also holds. It is easilly seen that the weaker condition (35) is easy to use when the operating point $\mathbf{x^d}$ is known. In this case all of the $w^{\mathbf{i}}(\mathbf{x^d})$'s are available as well as the matrices $\mathbf{P_{ij}}$. However, when the operating point $\mathbf{x^d}$ is not known, then the only known component of (35) is $\sigma_{max}(\mathbf{P_{ij}})$ since the matrices $\mathbf{P_{ij}}$ are available even without $\mathbf{x^d}$ being known.

Observe here that both conditions guarantee the asymptotic stability of (24) only if (29) and (30) hold. Based on this observation we will present the design of a TSFC-2 in the next subsection.

3.4 The design of TSFC-2

As already described in the previous subsection the original nonlinear autonomous system (13) can be locally asymptotically stabilized at an arbitrary set point $\mathbf{x^d}$ if we can design such gain matrices $\mathbf{K}(\mathbf{x^i}, \mathbf{u^i})$ which can locally asymptotically stabilize the closed loop fuzzy system (24) at $\mathbf{x^d}$. Furthermore, in order to derive condition (36) which guaranteed that $\mathbf{x^d}$ is the locally asymptotically stable equilibrium point of the above system, we transformed this system in the form (31) where $\mathbf{A}$ was a Hurwitz matrix defined by (29) and $\delta\mathbf{A_{ij}}$ were defined by (30). In other words, according to (29), we have to make each one of $\mathbf{A}(\mathbf{x^i},\mathbf{u^i}) + \mathbf{B}(\mathbf{x^i},\mathbf{u^i}) \cdot \mathbf{K}(\mathbf{x^i},\mathbf{u^i})$ a Hurwitz matrix. Both $\mathbf{A}(\mathbf{x^i},\mathbf{u^i})$ and $\mathbf{B}(\mathbf{x^i},\mathbf{u^i})$ are known since they were obtained via the Lyapunov-linearization of (14) at $(\mathbf{x^i},\mathbf{u^i})$. Then one can determine the unknown matrices $\mathbf{K}(\mathbf{x^i},\mathbf{u^i})$ by the pole assignment method so that each one of $\mathbf{A}(\mathbf{x^i},\mathbf{u^i}) + \mathbf{B}(\mathbf{x^i},\mathbf{u^i}) \cdot \mathbf{K}(\mathbf{x^i},\mathbf{u^i})$ becomes a Hurwitz matrix. Furthermore, by using the same set of desired poles to determine the unknown $\mathbf{K}(\mathbf{x^i},\mathbf{u^i})$ one obtains exactly the required equalities from (29). Observe here, that using the same set of desired poles we still obtain different $\mathbf{K}(\mathbf{x^i},\mathbf{u^i})$ because for each $\mathbf{i} \neq \mathbf{j}$ we have that $\mathbf{A}(\mathbf{x^i},\mathbf{u^i}) \neq \mathbf{A}(\mathbf{x^j},\mathbf{u^j})$ and $\mathbf{B}(\mathbf{x^i},\mathbf{u^i}) \neq \mathbf{B}(\mathbf{x^j},\mathbf{u^j})$. However, since the set of desired poles is the same we have that for each $\mathbf{i} \neq \mathbf{j}$, $\mathbf{A}(\mathbf{x^i},\mathbf{u^i}) + \mathbf{B}(\mathbf{x^i},\mathbf{u^i}) \cdot \mathbf{K}(\mathbf{x^i},\mathbf{u^i}) = \mathbf{A}(\mathbf{x^j},\mathbf{u^j}) + \mathbf{B}(\mathbf{x^j},\mathbf{u^j}) \cdot \mathbf{K}(\mathbf{x^j},\mathbf{u^j})$. Thus by using the pole assignment method on the same set of desired poles we can design the gain matrices $\mathbf{K}(\mathbf{x^i},\mathbf{u^i})$ $(\mathbf{i} = \mathbf{1}, \mathbf{2}, \ldots, \mathbf{M})$ such that (29) holds.

Next, in order to obtain the matrices $\mathbf{P_{ij}}$ whose largest singular values have to obey condition (36) for robust asymptotic stability, we have to construct the matrices $\delta\mathbf{A_{ij}}$. Using (30) these can be determined as

$$\delta\mathbf{A_{ij}} = \mathbf{B}(\mathbf{x^i},\mathbf{u^i}) \cdot (\mathbf{K}(\mathbf{x^j},\mathbf{u^j}) - \mathbf{K}(\mathbf{x^i},\mathbf{u^i})).$$

where $\mathbf{B}(\mathbf{x^i},\mathbf{u^i})$ and $\mathbf{K}(\mathbf{x^i},\mathbf{u^i})$ are already known.

Finally, we have to determine the matrices $\mathbf{P_{ij}}$ from condition (36). These were defined in (34) where $\mathbf{P}$ was the unique solution of the Lyapunov equation (33). Thus, the determination of the matrices $\mathbf{P_{ij}}$ requires solving the Lyapunov equation (33) for $\mathbf{P}$. This equation has a unique solution if and only if $\mathbf{A}$ is a Hurwitz matrix. In our case, the matrix $\mathbf{A}$ is such that, for each $\mathbf{i}$

$$\mathbf{A} = \mathbf{A}(\mathbf{x^i},\mathbf{u^i}) + \mathbf{B}(\mathbf{x^i},\mathbf{u^i}) \cdot \mathbf{K}(\mathbf{x^i},\mathbf{u^i}) = \mathbf{A_{ii}}, \tag{37}$$

and is also a Hurwitz matrix because each one of $\mathbf{A_{ii}}$ was already designed as a Hurwitz matrix. Thus the above Lyapunov equation does have a unique solution $\mathbf{P}$. Furthermore, since the different $\delta\mathbf{A_{ij}}$ are known, the matrices $\mathbf{P_{ij}}$ can be determined, according to (34) as

$$\mathbf{P_{ij}} = \frac{(\mathbf{B}(\mathbf{x^i},\mathbf{u^i}) \cdot (\mathbf{K}(\mathbf{x^j},\mathbf{u^j}) - \mathbf{K}(\mathbf{x^i},\mathbf{u^i})))^T \cdot \mathbf{P}}{2}$$

$$+ \quad \frac{\mathbf{P} \cdot \mathbf{B}(\mathbf{x^i}, \mathbf{u^i}) \cdot (\mathbf{K}(\mathbf{x^j}, \mathbf{u^j}) - \mathbf{K}(\mathbf{x^i}, \mathbf{u^i}))}{2}.$$

Now, in order to verify whether the gain matrices $\mathbf{K}(\mathbf{x^i}, \mathbf{u^i})$, designed via the pole assignment method, satisfy condition (36) for local asymptotic stability, we have to compute the set of singular values for each $\mathbf{P_{ij}}$, take the largest singular value $\sigma_{max}(\mathbf{P_{ij}})$ for each $\mathbf{P_{ij}}$, and check whether the inequalities (36) are satisfied. If each of the inequalities from (36) is satisfied, then we have that the gain matrices $\mathbf{K}(\mathbf{x^i}, \mathbf{u^i})$, designed via the pole assignment method, are such that the fuzzy closed loop system (24) has the given operating point $\mathbf{x^d}$ as its locally asymptotically stable equilibrium point. This in turn, guarantees that $\mathbf{x^d}$ is also the locally asymptotically stable equilibrium point of the original nonlinear autonomous system (13).

It has to be stressed here, that the design method for the TSFC-2, as presented in this subsection, does not guarantee that the gain matrices $\mathbf{K}(\mathbf{x^i}, \mathbf{u^i})$, designed via the pole assignment method, are such that (36) is automatically satisfied once these gain matrices are determined. This in turn implies that $\mathbf{K}(\mathbf{x^i}, \mathbf{u^i})$ may have to be redesigned over and over again until such gain matrices are found that satisfy the above inequalities.

However, once the proper (in the context of (36) gain matrices are found they can be used to locally asymptotically stabilize the original nonlinear autonomous system (13) for **any** operating point $(\mathbf{x^d}, \mathbf{u^d})$. If this operating point is one of $(\mathbf{x^i}, \mathbf{u^i})$, i.e. it belongs to the center of a fuzzy region, this implies that only the corresponding, already available, gain matrix $\mathbf{K}(\mathbf{x^i}, \mathbf{u^i})$ will be used for stabilization since in this case $w^{\mathbf{i}}(\mathbf{x^d}) = 1$ while $\forall \mathbf{j} \neq \mathbf{i}$ $w^{\mathbf{j}}(\mathbf{x^d}) = 0$. If $(\mathbf{x^d}, \mathbf{u^d})$ is not amongst the $(\mathbf{x^i}, \mathbf{u^i})$ then it is an intermediate operating point which satisfies a number of fuzzy regions, each to a different degree of satisfaction. In this case a weighted combination of the already available gain matrices for these fuzzy regions will be used to locally asymptotically stabilize (13). Thus for any operating point in the fuzzy state space there always exists an already available gain matrix, or a weighted combination of already available gain matrices.

3.5 Approximation errors

The so called *approximation errors* will now be considered in detail.

Let $(\mathbf{x^d}, \mathbf{u^d})$ be an arbitrary intermediate operating point, that is, it is not amongst the already available operating points $(\mathbf{x^i}, \mathbf{u^i})$, $(\mathbf{i} = \mathbf{1}, \mathbf{2}, \ldots, \mathbf{M})$. As we saw in Sect. 4.2.4, the fuzzy closed loop linear system for such an arbitrary intermediate operating point is given by

$$\dot{\mathbf{x}} = \sum_{\mathbf{i}}\sum_{\mathbf{j}} w^{\mathbf{i}}(\mathbf{x^d})\cdot w^{\mathbf{j}}(\mathbf{x^d})\cdot(\mathbf{A}(\mathbf{x^i},\mathbf{u^i})+\mathbf{B}(\mathbf{x^i},\mathbf{u^i})\cdot\mathbf{K}(\mathbf{x^j},\mathbf{u^j}))\cdot(\mathbf{x}-\mathbf{x^d}), \quad (38)$$

If, instead of using (38), we Lyapunov linearize the original nonlinear autonomous system (13) at $(\mathbf{x^d}, \mathbf{u^d})$ and then design an appropriate gain matrix $\mathbf{K}(\mathbf{x^d}, \mathbf{u^d})$ we will obtain the following closed loop system

$$\dot{\mathbf{x}} = (\mathbf{A}(\mathbf{x^d},\mathbf{u^d}) + \mathbf{B}(\mathbf{x^d},\mathbf{u^d})\cdot\mathbf{K}(\mathbf{x^d},\mathbf{u^d}))\cdot(\mathbf{x}-\mathbf{x^d}). \quad (39)$$

The difference between the matrix

$$\sum_{\mathbf{i}}\sum_{\mathbf{j}} w^{\mathbf{i}}(\mathbf{x^d})\cdot w^{\mathbf{j}}(\mathbf{x^d})\cdot(\mathbf{A}(\mathbf{x^i},\mathbf{u^i})+\mathbf{B}(\mathbf{x^i},\mathbf{u^i})\cdot\mathbf{K}(\mathbf{x^j},\mathbf{u^j})),$$

from (38) and the matrix

$$\mathbf{A}(\mathbf{x^d},\mathbf{u^d}) + \mathbf{B}(\mathbf{x^d},\mathbf{u^d})\cdot\mathbf{K}(\mathbf{x^d},\mathbf{u^d}),$$

from (39), is constant for any $\mathbf{x}$ since both (38) and (39) describe linear behaviors. This difference constitutes the approximation error matrix due to the use of (38) instead of (39). We treat this approximation error as an unknown disturbance matrix $\mathbf{E}$ in the context of the result about robust stability, reported in [Zhou 87], and already used in the design of the TSFC-2 in subsection 2.7.4. In order to utilize this result for the particular case of an approximation error we proceed as follows.

Suppose that the condition (36) which guaranteed the stability of (38) holds, and thus the latter linear closed loop system has $\mathbf{x^d}$ as its local asymptotically stable equilibrium point. Now we have to establish the condition under which (38) remains asymptotically stable at $\mathbf{x^d}$ despite of an approximation error in terms of an unknown perturbation matrix $\mathbf{E}$. For the sake of simplicity, let us denote the right hand side of (38) by $\mathbf{A}'$. Thus (38) can be rewritten in the equivalent form

$$\dot{\mathbf{x}} = \mathbf{A}'\cdot(\mathbf{x}-\mathbf{x^d}), \quad (40)$$

where

$$\mathbf{A}' = \sum_{\mathbf{i}}\sum_{\mathbf{j}} w^{\mathbf{i}}(\mathbf{x^d})\cdot w^{\mathbf{j}}(\mathbf{x^d})\cdot(\mathbf{A}(\mathbf{x^i},\mathbf{u^i})+\mathbf{B}(\mathbf{x^i},\mathbf{u^i})\cdot\mathbf{K}(\mathbf{x^j},\mathbf{u^j})).$$

In other words, the problem of studying the stability of (38) in the presence of an unknown perturbation matrix $\mathbf{E}$, can be reduced to studying the stability of the linear system

$$\dot{\mathbf{x}} = (\mathbf{A}' + \mathbf{E}) \cdot (\mathbf{x} - \mathbf{x}^{\mathbf{d}}), \tag{41}$$

where the unknown perturbation matrix $\mathbf{E}$ with unknown elements e_{ij} is of the same dimension as $\mathbf{A}'$, i.e., $(n \times n)$ and is defined as

$$\mathbf{E} = \sum_{m=1}^{n\times n} k_m \cdot \mathbf{E}^m, \tag{42}$$

where k_m are unknown parameters, and the elements e_{ij}^m, $(i, j = 1, 2, ..., n)$ of each E^m are obtained as follows. Let us number the elements a_{ij} of $\mathbf{A}'$ from 1 to $n \times n$. Thus we can rewrite each element of $\mathbf{A}'$ as a_{ij}^m where $m = 1, 2, \ldots, n \times n$. Then the elements e_{ij}^m of each $\mathbf{E}^m$ are defined as

if $a_{ij}^m \neq 0$ then $e_{ij}^m = 1$, and $e_{pq}^m = 0$, $\forall p, q \neq i, j$, otherwise $e_{ij}^m = 0, \forall i, j = 1, 2, \ldots, n$.

It is easily seen that if $a_{ij}^m \neq 0$, then $\mathbf{E}^m$ has only one element equal to one, namely e_{ij}^m, and the rest of its elements are equal to zero.

Thus, according to (42), for each i, j we have that

$$e_{ij} = \sum_{m=1}^{n\times n} k_m \cdot e_{ij}^m = k_m,$$

if $a_{ij}^m \neq 0$, otherwise we have that $e_{ij} = 0$. Hence, for each i, j, the elements $a_{ij}^m + e_{ij}$ of $\mathbf{A}' + \mathbf{E}$ are equal to zero whenever $a_{ij}^m = 0$ and equal to $a_{ij}^m + k_m$ whenever $a_{ij}^m \neq 0$.

Now, in the context of (28) we have that

- $\mathbf{A}'$ is stable, since the condition (36) was assumed to hold, and thus is a Hurwitz matrix,
- $\mathbf{E}^{ij}$ are known perturbation matrices (by construction),
- k_{ij} are unknown parameters in an interval around zero.

In this case, the system (41) is stable if the condition (28) is fulfilled. In our case this condition means that the following inequality should hold for each i, j

$$\sum_{i=1}^{n}\sum_{j=1}^{n} |k_{ij}| \cdot \sigma_{max}(\mathbf{F}_{ij}) < 1, \tag{43}$$

where

$$\mathbf{F}_{ij} = \frac{\mathbf{E}^{ijT} \cdot \mathbf{P} + \mathbf{P} \cdot \mathbf{E}^{ij}}{2}, \tag{44}$$

where $\mathbf{P}$ is the unique solution of the Lyapunov equation

$$\mathbf{A}'^{T} \cdot \mathbf{P} + \mathbf{P} \cdot \mathbf{A}' = -2 \cdot \mathbf{I}. \tag{45}$$

Thus, as long as the approximation error, in terms of the perturbation matrix $\mathbf{E}$, is such that the above inequality holds, the system (41) is stable. In other words, despite of the approximation error the matrix $\mathbf{A}'$ remains a Hurwitz matrix. Since $\mathbf{A}'$ remains a Hurwitz matrix, then (40) also remains stable. However, (40) is equivalent to (38). Thus, whenever the approximation error in terms of the perturbation matrix $\mathbf{E}$ is such that (43) is fulfilled, the system (38) remains stable.

4 Gain scheduling with TSFC-2

The tracking control problem for a nonlinear autonomous system can be transformed into a stabilization control problem for a nonautonomous system. This transformation is done as follows.

Consider a nonlinear autonomous system of the form

$$\dot{\mathbf{x}} = \mathbf{f}(\mathbf{x}, \mathbf{u}), \tag{46}$$

and let $\mathbf{x}^{\mathbf{d}}(t)$ be a known state reference trajectory (a solution of (46)) starting at initial time $t_0 = 0$ and initial condition (or initial state) $\mathbf{x}(0)$) and corresponding to a given reference input $\mathbf{u}^{\mathbf{d}}(t)$. The tracking control problem for (46) is then transformed into the stabilization problem for the nonlinear nonautonomous system

$$\delta\dot{\mathbf{x}} = \mathbf{f}(\delta\mathbf{x} + \mathbf{x}^{\mathbf{d}}, \delta\mathbf{u} + \mathbf{u}^{\mathbf{d}}, t) - \mathbf{f}(\mathbf{x}^{\mathbf{d}}, \mathbf{u}^{\mathbf{d}}, t), \tag{47}$$

where $\delta\mathbf{x} = \mathbf{x}(t) - \mathbf{x^d}(t)$, $\delta\mathbf{u} = \mathbf{u}(t) - \mathbf{u^d}(t)$, and $\delta\dot{\mathbf{x}} = \dot{\mathbf{x}}(t) - \dot{\mathbf{x}}^{\mathbf{d}}(t)$. Observe here, that (47) is now a nonlinear nonautonomous open loop system the stability of which is to be studied at $(\delta\mathbf{x} = \mathbf{0}, \delta\mathbf{u} = \mathbf{0})$ rather than along the state reference trajectory $\mathbf{x^d}(t)$. The Lyapunov-linearized version of (47) around $(\mathbf{0}, \mathbf{0})$ results in the linear nonautonomous system

$$\delta\dot{\mathbf{x}} = \mathbf{A}(\mathbf{x^d}, \mathbf{u^d}, t) \cdot \delta\mathbf{x} + \mathbf{B}(\mathbf{x^d}, \mathbf{u^d}, t) \cdot \delta\mathbf{u}, \tag{48}$$

which is equivalent to

$$\dot{\mathbf{x}} = \mathbf{A}(\mathbf{x^d}, \mathbf{u^d}, t) \cdot (\mathbf{x}(t) - \mathbf{x^d}(t)) + \mathbf{B}(\mathbf{x^d}, \mathbf{u^d}, t) \cdot (\mathbf{u}(t) - \mathbf{u^d}(t)), \tag{49}$$

where the equilibrium point of the latter system is now the trajectory $\mathbf{x^d}(t)$. Because of the equivalence between (48) and (49), the local uniform exponential stability of (49) at $\mathbf{x^d}(t)$ guarantees under certain conditions (see Sect. 2.4.3) that $\mathbf{x^d}(t)$ is also the locally uniformly exponentially stable trajectory of (47) or (46) Thus the control problem is the design of such a nonautonomous control law which, when inserted in (49), will locally uniformly exponentially stabilize the resulting linear nonautonomous closed loop system at the state reference trajectory $\mathbf{x^d}(t)$. Gain scheduling is one approach to the design of the above discussed control law. In the next subsection we will summarize the essential steps of the gain scheduling design method as described in [Rugh 91], [Shamma 88], [Nichols 93].

4.1 The gain scheduling design method

The gain scheduling design is done as follows (see [Rugh 91], [Nichols 93], [Shamma 88]). Consider first the known reference trajectory $\mathbf{x^d}(t)$ used in the derivation of (47). If this trajectory is evaluated at a particular *frozen time* τ then the value $\mathbf{x^d}(\tau)$ is obtained and is called the *frozen time value* of $\mathbf{x^d}(t)$ for $t = \tau$. A set $\{\mathbf{x^d}(\tau)\}$ of such frozen values obtained for equidistant frozen times $\tau = \tau_1, \tau_2, \ldots, \tau_N$, is chosen in advance.

Second, the linear nonautonomous open loop system (49) is obtained via Lyapunov-linearization of the nonlinear nonautonomous open loop system (48) at $\mathbf{x^d}(t)$ and its corresponding $\mathbf{u^d}(t)$. Then, N frozen time linear autonomous open loop systems of the form (49) are obtained by replacing $\mathbf{x^d}(t)$ and $\mathbf{u^d}(t)$ from (49) by pairs of their frozen time values

$$(\mathbf{x^d}(\tau_1), \mathbf{u^d}(\tau_1)), (\mathbf{x^d}(\tau_2), \mathbf{u^d}(\tau_2)), \ldots, (\mathbf{x^d}(\tau_N), \mathbf{u^d}(\tau_N)).$$

That is,

$$\dot{\mathbf{x}} = \mathbf{A}(\mathbf{x}^{\mathbf{d}}(\tau), \mathbf{u}^{\mathbf{d}}(\tau)) \cdot (\mathbf{x} - \mathbf{x}^{\mathbf{d}}(\tau)) + \mathbf{B}(\mathbf{x}^{\mathbf{d}}(\tau), \mathbf{u}^{\mathbf{d}}(\tau)) \cdot (\mathbf{u} - \mathbf{u}^{\mathbf{d}}(\tau)), \quad (50)$$

where $\tau = \tau_1, \tau_2, \ldots, \tau_N$.

Third, N linear autonomous control laws are designed, one for each of the frozen time linear autonomous open loop systems. These control laws are of the form (for $\tau = \tau_1, \tau_2, \ldots, \tau_N$)

$$\mathbf{u} = \mathbf{K}(\mathbf{x}^{\mathbf{d}}(\tau), \mathbf{u}^{\mathbf{d}}(\tau)) \cdot (\mathbf{x} - \mathbf{x}^{\mathbf{d}}(\tau)) + \mathbf{u}^{\mathbf{d}}(\tau). \quad (51)$$

Thus one obtains **N** frozen time closed loop systems

$$\dot{\mathbf{x}} = (\mathbf{A}(\mathbf{x}^{\mathbf{d}}(\tau), \mathbf{u}^{\mathbf{d}}(\tau)) + \mathbf{B}(\mathbf{x}^{\mathbf{d}}(\tau), \mathbf{u}^{\mathbf{d}}(\tau)) \cdot \mathbf{K}(\mathbf{x}^{\mathbf{d}}(\tau), \mathbf{u}^{\mathbf{d}}(\tau))) \cdot (\mathbf{x} - \mathbf{x}^{\mathbf{d}}(\tau)). \quad (52)$$

The control law (51) is designed in such a manner so that each $\mathbf{x}^{\mathbf{d}}(\tau_i)$ is a local uniformly exponentially stable equilibrium point of its corresponding (52).

Fourth, the already available autonomous control laws from (51) are interpolated (or gain scheduled). The so obtained autonomous gain scheduled control law should be such that each intermediate operating point $\mathbf{x}^{\mathbf{d}}(\tau)$, where $\tau_i < \tau < \tau_{i+1}$, is a local uniformly exponentially stable equilibrium point of the frozen time Lyapunov-linearized version of (46) at $\mathbf{x}^{\mathbf{d}}(\tau)$ and its corresponding $\mathbf{u}^{\mathbf{d}}(\tau)$. Observe here that the autonomous gain scheduled control law for an intermediate operating point $\mathbf{x}^{\mathbf{d}}(\tau)$ is not derived on the basis of the *interpolated* version of (50) for the intermediate point $\mathbf{x}^{\mathbf{d}}(\tau)$ and its corresponding $\mathbf{u}^{\mathbf{d}}(\tau)$. This interpolated version of (50) can, in principle, be constructed from the frozen time linear autonomous open loop systems of the form (50) for $\tau = \{\tau_i, \tau_{i+1}\}$. Instead, the autonomous gain scheduled control law is obtained directly by interpolating the already available control laws from (52) for $\tau = \{\tau_i, \tau_{i+1}\}$.

However, even if the gain scheduled control law is able to locally uniformly exponentially stabilize each frozen time autonomous closed loop system from (52) and each such system obeys certain robustness and performance requirements, these properties need not carry over to the over to the linear nonautonomous closed loop system. That is, the system (49) in which the gain scheduled control law is inserted. However, under the condition that the linear nonautonomous open loop system (49) is slowly time varying then all of the above mentioned properties of the time frozen linear autonomous closed loop systems are inherited by the the linear nonautonomous closed loop system. This implies in turn that the same local properties carry over to the nonlinear autonomous closed loop system obtained by inserting the gain scheduled control law in (47) or (46). Observe here that in the case of gain scheduling on a state reference trajectory the condition requiring that (49) be slowly time varying is equivalent to the requirement that the state reference trajectory $\mathbf{x}^{\mathbf{d}}(t)$ should vary slowly [Rugh 91] (Chap. 4, 4.2.2, Theorem 4.2–1).

It has to be stressed here, that no general and formally motivated method for the interpolation of the already available autonomous control laws from (51) into a gain scheduled control law is available. The interpolation is done in an ad hoc manner by the use of heuristic guidelines [Nichols 93] and additional information about the plant [Rugh 91]. One consequence is that at intermediate operating points the poles locations of the linear closed loop system obtained by inserting the gain scheduled control law in the frozen time Lyapunov-linearized version of (46) are, in general, incorrect [Rugh 91].

In this context, the fuzzy gain scheduler described in the next subsection, is immune to incorrect poles since the gain scheduled control law is derived on the basis of the fuzzily interpolated Lyapunov-linearized version of (46) at a given frozen time value of $\mathbf{x}^{\mathbf{d}}(t)$. This fuzzily interpolated Lyapunov-linearized version of (46) is constructed from the frozen time linear autonomous open loop systems of the form (50). Hence, the gain scheduled control law is designed with a particular frozen time open loop system in mind and thus, the poles of the resulting frozen time closed loop system do not migrate from the desired pole locations used for the design of the gain scheduled control law. Thus, the design of the gain scheduled control law of a fuzzy gain scheduler is formally motivated.

Yet another advantage of the fuzzy gain scheduler is that the method for determining the weights used in the construction of the fuzzily interpolated Lyapunov-linearized version of (46) from the frozen time linear autonomous open loop systems (50) is general and computationally efficient. These weights are simply obtained as the degrees of satisfaction of the different fuzzy regions from the fuzzy state space by a given intermediate frozen time value of $\mathbf{x}^{\mathbf{d}}(t)$.

4.2 The design of a fuzzy gain scheduler

Suppose first, that given an arbitrary state reference trajectory $\mathbf{x}^{\mathbf{d}}(t)$ together with its corresponding reference input $\mathbf{u}^{\mathbf{d}}(t)$, the nonlinear autonomous open loop system (46) is transformed into the linear nonautonomous open loop system (49). Furthermore, consider a set $\{\mathbf{x}^{\mathbf{d}}(\tau)\}$ of frozen time values of $\mathbf{x}^{\mathbf{d}}(t)$ obtained for equidistant frozen times $\tau_1, \tau_2, \ldots, \tau_N$ and the corresponding set $\{\mathbf{u}^{\mathbf{d}}(\tau)\}$ of frozen time values of $\mathbf{u}^{\mathbf{d}}(t)$. The set $\{\mathbf{x}^{\mathbf{d}}(\tau)\}$ of frozen time values is then used in the derivation of the $\mathbf{N}$ frozen time autonomous open loop systems from (50) which were of the form

$$\dot{\mathbf{x}} = \mathbf{A}(\mathbf{x^d}(\tau), \mathbf{u^d}(\tau)) \cdot (\mathbf{x} - \mathbf{x^d}(\tau)) + \mathbf{B}(\mathbf{x^d}(\tau), \mathbf{u^d}(\tau)) \cdot (\mathbf{u} - \mathbf{u^d}(\tau)).$$

Now recall here that the linear autonomous open loop fuzzy model (19) used in the design of the TSFC-2 (see Sect. 2.2.2) was of the form

$$\dot{\mathbf{x}} = \sum_{\mathbf{i}} w^{\mathbf{i}}(\mathbf{x^d}) \cdot (\mathbf{A}(\mathbf{x^i}, \mathbf{u^i}) \cdot (\mathbf{x} - \mathbf{x^d}) + \mathbf{B}(\mathbf{x^i}, \mathbf{u^i}) \cdot (\mathbf{u} - \mathbf{u^d})), \tag{53}$$

and was obtained by Lyapunov-linearization of the autonomous counterpart (14) of (47) at the points $(\mathbf{x^i}, \mathbf{u^i})$ where $\mathbf{x^i}$ were the centers of the fuzzy regions from the fuzzy state space and $\mathbf{u^i}$ were the input values corresponding to these centers. In the case of the nonlinear autonomous open loop system (46), the above expression can be derived by considering each pair $(\mathbf{x^i}, \mathbf{u^i})$ as a pair of frozen time values associated with the centers of the fuzzy regions from the fuzzy state space of (46) and different from $(\mathbf{x^d}(\tau), \mathbf{u^d}(\tau))$. Then, by replacing $\mathbf{x^d}(t)$ and $\mathbf{u^d}(t)$ from (49) with the pairs $(\mathbf{x^i}, \mathbf{u^i})$ we obtain $\mathbf{M}$ linear autonomous open loop systems

$$\dot{\mathbf{x}} = \mathbf{A}(\mathbf{x^i}, \mathbf{u^i}) \cdot (\mathbf{x} - \mathbf{x^i}) + \mathbf{B}(\mathbf{x^i}, \mathbf{u^i}) \cdot (\mathbf{u} - \mathbf{u^i}). \tag{54}$$

Observe here that (50) and (54) are equivalent for $\mathbf{x^d}(\tau) = \mathbf{x^i}$ and $\mathbf{u^d}(\tau) = \mathbf{u^i}$. However, for $\mathbf{x^d}(\tau) \neq \mathbf{x^i}$ and $\mathbf{u^d}(\tau) \neq \mathbf{u^i}$ each one of (50) can be approximated using the already available (54). This is so because the linear autonomous fuzzy open loop expression (53) approximates the behavior of a "missing"(autonomous) Lyapunov-linearized system at any intermediate operating point $(\mathbf{x^d}, \mathbf{u^d})$. Thus, (53) can be used to approximate the behavior of the "missing" (from (54)) frozen time Lyapunov-linearized versions (50) for any intermediate operating points $(\mathbf{x^d}, \mathbf{u^d}) = (\mathbf{x^d}(\tau), \mathbf{u^d}(\tau))$. Hence, the expression

$$\dot{\mathbf{x}} = \sum_{\mathbf{i}} w^{\mathbf{i}}(\mathbf{x^d}(\tau)) \cdot (\mathbf{A}(\mathbf{x^i}, \mathbf{u^i}) \cdot (\mathbf{x} - \mathbf{x^d}(\tau)) + \mathbf{B}(\mathbf{x^i}, \mathbf{u^i}) \cdot (\mathbf{u} - \mathbf{u^d}(\tau))). \tag{55}$$

approximates the behavior of each one of (50) by a weighted combination of (54).

Recall also that the set of all rules of the TSFC-2 defines a control law of the form

$$\mathbf{u} = \sum_{\mathbf{j}} w^{\mathbf{j}}(\mathbf{x^d}) \cdot (\mathbf{K}(\mathbf{x^j}, \mathbf{u^j}) \cdot (\mathbf{x} - \mathbf{x^d}) + \mathbf{u^d}), \tag{56}$$

which can be designed in such a way (see Sect. 2.2.5–2.2.6) so that it can locally uniformly exponentially stabilize (53) at an arbitrary operating point $\mathbf{x^d}$ and its corresponding $\mathbf{u^d}$, despite approximation errors of known magnitude. That is, even around operating points $\mathbf{x^d} = \mathbf{x^d}(\tau)$. Thus, the control law of the form

$$\mathbf{u} = \sum_{\mathbf{j}} w^{\mathbf{j}}(\mathbf{x^d}(\tau)) \cdot (\mathbf{K}(\mathbf{x^j}, \mathbf{u^j}) \cdot (\mathbf{x} - \mathbf{x^d}(\tau)) + \mathbf{u^d}(\tau)), \tag{57}$$

will be able to locally uniformly exponentially stabilize (55) at an arbitrary operating point $\mathbf{x^d}(\tau)$ as well since the latter system is obtained from (53) only by substituting the operating point $\mathbf{x^d}$ with the different operating point $\mathbf{x^d}(\tau)$ and $\mathbf{u^d}$ with $\mathbf{u^d}(\tau)$.

Recall also that the closed loop system obtained by inserting (56) in (53) was of the form

$$\dot{\mathbf{x}} = \sum_{\mathbf{i}} \sum_{\mathbf{j}} w^{\mathbf{i}}(\mathbf{x^d}) \cdot w^{\mathbf{j}}(\mathbf{x^d}) \cdot (\mathbf{A}(\mathbf{x^i}, \mathbf{u^i}) + \mathbf{B}(\mathbf{x^i}, \mathbf{u^i}) \cdot \mathbf{K}(\mathbf{x^j}, \mathbf{u^j})) \cdot (\mathbf{x} - \mathbf{x^d}), \tag{58}$$

and as long as the approximation errors for (58), (see Sect. 2.2.5–2.2.6), are in certain range the above closed loop system has $\mathbf{x^d}$ as its local uniformly exponentially stable equilibrium point. Now inserting (57) in (55)) we obtain the fuzzily approximated version of (52)

$$\dot{\mathbf{x}} = \sum_{\mathbf{i}} \sum_{\mathbf{j}} w^{\mathbf{i}}(\mathbf{x^d}(\tau)) \cdot w^{\mathbf{j}}(\mathbf{x^d}(\tau)) \cdot (\mathbf{A}(\mathbf{x^i}, \mathbf{u^i}) + \mathbf{B}(\mathbf{x^i}, \mathbf{u^i}) \cdot \mathbf{K}(\mathbf{x^j}, \mathbf{u^j})) \cdot (\mathbf{x} - \mathbf{x^d}(\tau)). \tag{59}$$

It is easily seen that the latter closed loop system is obtained from (58) only by substituting the operating point $\mathbf{x^d}$ with the different operating point $\mathbf{x^d}(\tau)$ and $\mathbf{u^d}$ with $\mathbf{u^d}(\tau)$. Thus, (59) will have $\mathbf{x^d}(\tau)$ as its local uniformly exponentially stable equilibrium point provided that the approximation errors are in a certain range.

Thus we have seen that a Takagi-Sugeno FC-2 such that, computation with all of its rules can be represented by (57), is able to locally uniformly exponentially stabilize the frozen time closed loop system (52) at all points $\mathbf{x^d}(\tau)$. Furthermore, under the assumption that the linear nonatonomous system (49) is slowly time varying, such a TSFC-2 can also locally uniformly exponentially stabilize the latter system, when inserted in it, since it can locally uniformly exponentially stabilize all of its frozen time counterparts (52).

It is easily seen that a Takagi-Sugeno FC-2 such that, computation with all of its rules results can be represente= d by (57), has rules of the form

$$\text{IF} \quad \mathbf{x^d}(\tau) = \mathbf{LX^j} \quad \text{THEN} \quad \mathbf{u} = \mathbf{K}(\mathbf{x^j}, \mathbf{u^j}) \cdot (\mathbf{x} - \mathbf{x^d}(\tau)) + \mathbf{u^d}(\tau).$$

Obviouslly, the construction of the gain matrices of a TSFC-2 with rules of the above type requires an open loop system the fuzzy rule based representation of which is of the form

$$\begin{aligned} &\text{IF} \quad \mathbf{x^d}(\tau) = \mathbf{LX^i} \\ &\text{THEN} \quad \dot{\mathbf{x}} = \mathbf{A}(\mathbf{x^i}, \mathbf{u^i}) \cdot = (\mathbf{x} - \mathbf{x^d}(\tau)) + \mathbf{B}(\mathbf{x^i}, \mathbf{u^i}) \cdot (\mathbf{u} - \mathbf{u^d}(\tau)), \end{aligned}$$

since computation with all the above rules results exactly in (55).

Both of the above types of fuzzy rules can be designed in exactly the same manner as in the case of the stabilization control problem from the previous subsection. This is so, since the only difference between the use of the TSFC-2 for stabilization around an operating point and its use as a gain scheduler for= tracking a state reference trajectory is purely syntactic, i.e., the use of $\mathbf{x^d}$ instead of $\mathbf{x^d}(\tau)$ and vice versa.

5 A robot arm example

5.1 Basic equations

Let the basic equation describing the motion of a robot arm be

$$\mathbf{M}(\mathbf{q}) \cdot \ddot{\mathbf{q}} + \mathbf{N}(\mathbf{q}, \dot{\mathbf{q}}) = \mathbf{u}, \tag{60}$$

where

$\mathbf{q}$ is a $(k \times 1)$ position vector,

$\dot{\mathbf{q}}$ is a $(k \times 1)$ velocity vector,

$\ddot{\mathbf{q}}$ is a $(k \times 1)$ accelleleration vector,

$\mathbf{M}(\mathbf{q})$ is a $(k \times k)$ matrix of inertia (invertable),

$\mathbf{N}(\mathbf{q}, \dot{\mathbf{q}})$ is a $(k \times 1)$ vector of damping, centrifugal, coriolis, gravitational forces,

$\mathbf{u}$ is a $(k \times 1)$ vector of generalized forces and torques.

Furthermore, let the open loop system (60) have k degrees of freedom (d.o.f.). Normally, $k = 6$, so that the number of d.o.f. in the cartesian space $\mathbf{R}^{\mathbf{k}}_{\mathbf{x}}$ is equal to the number of d.o.f. in the joint space $\mathbf{R}^{\mathbf{k}}_{\mathbf{q}}$. The control problem is to follow a given trajectory $\mathbf{q}^{\mathbf{d}}(t)$ and to produce a torque vector $\mathbf{u}$ such that the tracking error approaches 0 as $t \to \infty$.

In order to apply the TSFC-2 approach we rewrite the equations of motion (60) into the following set of state equations

$$\mathbf{M}(\mathbf{x}) \cdot \dot{\mathbf{x}} + \mathbf{N}(\mathbf{x}) = \mathbf{u}(\mathbf{x}), \tag{61}$$

with

$\mathbf{x} = (x_1, \ldots, x_n)^T = (q_1, \dot{q}_1, \ldots, q_k, \dot{q}_k)^T$ is the $(n \times 1)$ state vector $(n = 2k)$,

$\mathbf{u}(\mathbf{x})$ is the $(n \times 1)$ vector of generalized forces and torques,

$\mathbf{M}(\mathbf{x})$ is the $(n \times n)$ matrix of inertia (invertable),

$\mathbf{N}(\mathbf{x})$ is the $(n \times 1)$ vector of damping, centrifugal, coriolis, gravitational forces.

The Lyapunov linearization of (61) around a desired operating point $\mathbf{x}^{\mathbf{d}}$ or a slowly varying trajectory $\mathbf{x}^{\mathbf{d}}(\tau)$ results in

$$\mathbf{M}(\mathbf{x}^{\mathbf{d}} + \delta\mathbf{x}) \cdot (\dot{\mathbf{x}} + \delta\dot{\mathbf{x}}) = -\mathbf{N}(\mathbf{x}^{\mathbf{d}} + \delta\mathbf{x}) + \mathbf{u}^{\mathbf{d}} + \delta\mathbf{u}, \tag{62}$$

where

$\mathbf{x} = \mathbf{x}^{\mathbf{d}} + \delta\mathbf{x}$,

$\mathbf{u} = \mathbf{u}^{\mathbf{d}} + \delta\mathbf{u}$,

$\mathbf{x}^{\mathbf{d}}$ is the desired state vector,

$\mathbf{u}^{\mathbf{d}}$ is the control vector corresponding to $\mathbf{x}^{\mathbf{d}}$,

$\delta\mathbf{x}$ is the error,

$\delta\mathbf{u}$ is the control input error.

In what follows we neglect higher order terms of $\delta\mathbf{x}$ and $\delta\mathbf{u}$ and assume further that

$$\mathbf{M}(\mathbf{x^d} + \delta\mathbf{x}) \cdot \dot{\mathbf{x}} \approx \mathbf{M}(\mathbf{x^d}) \cdot \dot{\mathbf{x}}.$$

This simplification cane be done for slow changes of $\mathbf{x^d}$ and, with that, of $\mathbf{M}(\mathbf{x^d})$. Then we obtain from (62)

$$\mathbf{M}(\mathbf{x^d}) \cdot (\dot{\mathbf{x}}^\mathbf{d} + \delta\dot{\mathbf{x}}) = -\mathbf{N}(\mathbf{x^d}) + \mathbf{A}(\mathbf{x^d}) \cdot \delta\mathbf{x} + \mathbf{u^d} + \delta\mathbf{u}, \tag{63}$$

where

$\mathbf{A}(\mathbf{x_d}) = - \left.\frac{\partial\mathbf{N}(\mathbf{x})}{\partial\mathbf{x}}\right|_{\mathbf{x}=\mathbf{x^d}}$ is the Jacobian.

From (61) we directly obtain the specific solution

$$\mathbf{M}(\mathbf{x^d}) \cdot \dot{\mathbf{x}}^\mathbf{d} = -\mathbf{N}(\mathbf{x^d}) + \mathbf{u^d}, \tag{64}$$

so that (63) transforms into

$$\mathbf{M}(\mathbf{x^d}) \cdot \delta\dot{\mathbf{x}} = \mathbf{A}(\mathbf{x^d}) \cdot \delta\mathbf{x} + \delta\mathbf{u}, \tag{65}$$

or equivalently into

$$\delta\dot{\mathbf{x}} = \mathbf{M}^{-1}(\mathbf{x^d}) \cdot \mathbf{A}(\mathbf{x^d}) \cdot \delta\mathbf{x} + \mathbf{M}^{-1}(\mathbf{x^d}) \cdot \delta\mathbf{u}. \tag{66}$$

This is a linear state equation with an equilibrium point at $\delta\mathbf{x} = \mathbf{0}$ for $\delta\mathbf{u} = 0$. In (66) we introduce the linear control law

$$\delta\mathbf{u} = \mathbf{K}(\mathbf{x^d}) \cdot \delta\mathbf{x}, \tag{67}$$

and thus we obtain

$$\delta\dot{\mathbf{x}} = \mathbf{M}^{-1}(\mathbf{x^d}) \cdot (\mathbf{A}(\mathbf{x^d}) + \mathbf{K}(\mathbf{x^d})) \cdot \delta\mathbf{x}. \tag{68}$$

The control law (67) is now reformulated into

$$\mathbf{u} = \mathbf{u^d} + \mathbf{K}(\mathbf{x^d}) \cdot \delta\mathbf{x}, \tag{69}$$

where the desired control input $\mathbf{u^d}$ is chosen to be

$$\mathbf{u^d} = \mathbf{M}(\mathbf{x^d}) \cdot \dot{\mathbf{x}}^\mathbf{d} + \mathbf{N}(\mathbf{x^d}). \tag{70}$$

Since the actual values of $\mathbf{M}(\mathbf{x^d})$ and $\mathbf{N}(\mathbf{x^d})$ can only be estimated one has to take into account the estimate $\hat{\mathbf{u}}^\mathbf{d}$ instead of $\mathbf{u^d}$, that is

$$\hat{\mathbf{u}}^\mathbf{d} = \hat{\mathbf{M}}(\mathbf{x^d}) \cdot \dot{\mathbf{x}}^\mathbf{d} + \hat{\mathbf{N}}(\mathbf{x^d}). \tag{71}$$

With that, the system equation (68) transforms into

$$\begin{aligned}\delta\dot{\mathbf{x}} &= \mathbf{M}^{-1}(\mathbf{x}^{\mathbf{d}})\cdot(\mathbf{A}(\mathbf{x}^{\mathbf{d}})+\mathbf{K}(\mathbf{x}^{\mathbf{d}}))\cdot\delta\mathbf{x}+\mathbf{M}^{-1}(\mathbf{x}^{\mathbf{d}})\cdot\boldsymbol{\Delta}\mathbf{u}^{\mathbf{d}}, \\ \boldsymbol{\Delta}\mathbf{u}^{\mathbf{d}} &= \hat{\mathbf{u}}^{\mathbf{d}}-\mathbf{u}^{\mathbf{d}} = (\hat{\mathbf{M}}(\mathbf{x}^{\mathbf{d}})-\mathbf{M}(\mathbf{x}^{\mathbf{d}}))\cdot\dot{\mathbf{x}}^{\mathbf{d}}+(\hat{\mathbf{N}}(\mathbf{x}^{\mathbf{d}})-\mathbf{N}(\mathbf{x}^{\mathbf{d}})).\end{aligned} \tag{72}$$

Observe here that $\mathbf{M}(\mathbf{x}^{\mathbf{d}})$ and $\mathbf{N}(\mathbf{x}^{\mathbf{d}})$ should be estimated as accurate as possible in order not to invalidate the linearization results.

Every possible control law (67) in the bounded state space is now approximated by a Takagi-Sugeno FLC-2 being represented by $\mathbf{M}$ individual control laws

$$\delta\mathbf{u} = \mathbf{K}(\mathbf{x}^{\mathbf{j}})\cdot\delta\mathbf{x} \quad \mathbf{j}=1,\ldots,\mathbf{M}. \tag{73}$$

Considering (69) and (71), this control law can be reformulated as

$$\begin{aligned}\mathbf{u} &= \hat{\mathbf{u}}^{\mathbf{d}}(\mathbf{x}^{\mathbf{j}})+\mathbf{K}(\mathbf{x}^{\mathbf{j}})\cdot\delta\mathbf{x} \quad \mathbf{j}=1,\ldots,\mathbf{M}, \\ \hat{\mathbf{u}}^{\mathbf{d}}(\mathbf{x}^{\mathbf{j}}) &= \hat{\mathbf{M}}(\mathbf{x}^{\mathbf{j}})\cdot\dot{\mathbf{x}}^{\mathbf{d}}+\mathbf{N}(\mathbf{x}^{\mathbf{j}}).\end{aligned} \tag{74}$$

The corresponding fuzzy controller rules are

$$R_{\mathrm{C}}^{\mathbf{j}}: \quad \text{IF} \quad \mathbf{x}^{\mathbf{d}} = \mathbf{LX}^{\mathbf{j}} \quad \text{THEN} \quad \mathbf{u} = \mathbf{K}(\mathbf{x}^{\mathbf{j}})\cdot(\mathbf{x}-\mathbf{x}^{\mathbf{d}})+\hat{\mathbf{M}}(\mathbf{x}^{\mathbf{i}})\cdot\dot{\mathbf{x}}^{\mathbf{d}}+\mathbf{N}(\mathbf{x}^{\mathbf{i}}).$$

The fuzzy system rules which approximate (66) are

$$R_{\mathrm{S}}^{\mathbf{i}}: \quad \text{IF} \quad \mathbf{x}^{\mathbf{d}} = \mathbf{LX}^{\mathbf{i}} \quad \text{THEN} \quad \delta\dot{\mathbf{x}} = \mathbf{M}^{-1}(\mathbf{x}^{\mathbf{i}})\cdot\mathbf{A}(\mathbf{x}^{\mathbf{i}})\cdot\delta\mathbf{x}+\mathbf{M}^{-1}(\mathbf{x}^{\mathbf{i}})\cdot\delta\mathbf{u}$$

Following the results from Sect. 4.2.4 we obtain the approximated version of (72), that is

$$\begin{aligned}\delta\dot{\mathbf{x}} = \sum_{\mathbf{i}}\sum_{\mathbf{j}} & w^{\mathbf{i}}(\mathbf{x}^{\mathbf{d}})\cdot w^{\mathbf{j}}(\mathbf{x}^{\mathbf{d}})\cdot\mathbf{M}^{-1}(\mathbf{x}^{\mathbf{i}})\cdot[(\mathbf{A}(\mathbf{x}^{\mathbf{i}})+\mathbf{K}(\mathbf{x}^{\mathbf{j}}))\cdot\delta\mathbf{x} \\ & +(\hat{\mathbf{M}}(\mathbf{x}^{\mathbf{j}})-\mathbf{M}(\mathbf{x}^{\mathbf{i}}))\cdot\dot{\mathbf{x}}^{\mathbf{d}}+(\hat{\mathbf{N}}(\mathbf{x}^{\mathbf{j}})-\mathbf{N}(\mathbf{x}^{\mathbf{i}}))],\end{aligned} \tag{75}$$

In the following we assume the modeling error $\boldsymbol{\Delta}$, that is,

$$\sum_{\mathbf{i}}\sum_{\mathbf{j}} w^{\mathbf{i}}(\mathbf{x}^{\mathbf{d}})\cdot w^{\mathbf{j}}(\mathbf{x}^{\mathbf{d}})\cdot\mathbf{M}^{-1}(\mathbf{x}^{\mathbf{i}})\cdot[(\hat{\mathbf{M}}(\mathbf{x}^{\mathbf{j}})-\mathbf{M}(\mathbf{x}^{\mathbf{i}}))\cdot\dot{\mathbf{x}}^{\mathbf{d}}+(\hat{\mathbf{N}}(\mathbf{x}^{\mathbf{j}})-\mathbf{N}(\mathbf{x}^{\mathbf{i}}))],$$

to be small enough to keep the linearization intact.

With regard to the design principles discussed in Sect. 4.2.4 we choose the control laws (74) in such a way so that the system (66) is robustly stable regarding approximation errors. In this connection it is important to divide the state space in an appropriate number of fuzzy regions so that on the one hand the fuzzy approximation is fine enough and, on the other hand, the set of fuzzy controller rules is restricted to a reasonable size.

5.2 The two-link robot arm

In this subsection we concentrate on the equation of motion of the two-link robot arm, the linearization of the nonlinear system, and on the computation of the fuzzy gain scheduler.

The equation of motion

Consider the following two-link robot arm its masses being concentrated at the ends of each link and the motor inertias being neglected (see Fig. 1):

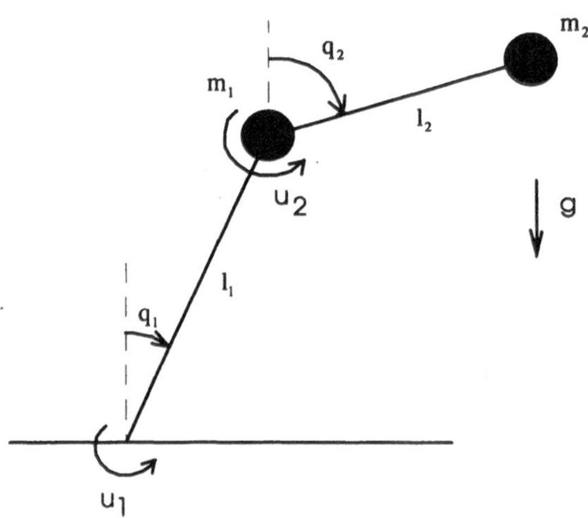

Figure 1: Scheme of a two-link robot arm.

According to (60) we have the equation of motion

$$\mathbf{M}(\mathbf{q}) \cdot \ddot{\mathbf{q}} = -\mathbf{N}(\mathbf{q}, \dot{\mathbf{q}}) + \mathbf{u}. \tag{76}$$

In this equation

$$\begin{aligned}\mathbf{M}(\mathbf{q}) &= \begin{pmatrix} m_{11} & m_{12} \\ m_{21} & m_{22} \end{pmatrix} \\ &= \begin{pmatrix} (m_1+m_2)\cdot l_1^2 & m_2\cdot l_1\cdot l_2\cdot\cos(q_1-q_2) \\ m_2\cdot l_1\cdot l_2\cdot\cos(q_1-q_2) & m_2 l_2^2 \end{pmatrix}, \end{aligned} \tag{77}$$

and

$$\mathbf{N}(\mathbf{q},\dot{\mathbf{q}}) = \begin{pmatrix} n_1 \\ n_2 \end{pmatrix}, \tag{78}$$

where the matrix on the right hand side of (78) is equal to

$$\begin{pmatrix} \dot{q}_2^2\cdot m_2\cdot l_1\cdot l_2\cdot\sin(q_1-q_2)-(m_1+m_2)\cdot g\cdot l_1\cdot\sin q_1+K_{q1}\cdot\dot{q}_1 \\ -\dot{q}_1^2\cdot m_2\cdot l_1\cdot l_2\cdot\sin(q_1-q_2)-m_2\cdot g\cdot l_2\cdot\sin q_2+K_{q2}\cdot\dot{q}_2 \end{pmatrix},$$

with K_{q1} and K_{q2} being the damping coefficients for the coordinates q_1 and q_2, respectively.

The inverse of $\mathbf{M}(\mathbf{q})$ in the control law is given as

$$\mathbf{M}^{-1}(\mathbf{q}) = \begin{pmatrix} m_{11}^{-1} & m_{12}^{-1} \\ m_{21}^{-1} & m_{22}^{-1} \end{pmatrix}, \tag{79}$$

where the matrix on the right hand side of (79) is equal to

$$\begin{pmatrix} m_2 l_2^2/D & -m_2\cdot l_1\cdot l_2\cdot\cos(q_1-q_2)/D \\ -m_2\cdot l_1\cdot l_2\cdot\cos(q_1-q_2)/D & (m_1+m_2)\cdot l_1^2/D \end{pmatrix},$$

with

$$D = m_2\cdot l_1^2\cdot l_2^2\cdot(m_1+m_2\cdot\sin^2(q_1-q_2)). \tag{80}$$

According to (61) we have the equations of motion in state space representation

$$\mathbf{M}(\mathbf{x})\cdot\dot{\mathbf{x}}+\mathbf{N}(\mathbf{x}) = \mathbf{u}(\mathbf{x}), \tag{81}$$

where

$\mathbf{x} = (x_1, x_2, x_3, x_4)^T = (q_1, \dot{q}_1, q_2, \dot{q}_2)^T$ is the (4×1) state vector,

$\mathbf{u}(\mathbf{x})$ is the (4×1) vector of generalized forces and torques,

$\mathbf{M}(\mathbf{x})$ is the (4×4) matrix of inertia (invertable),

$\mathbf{N}(\mathbf{x})$ is the (4×1) vector of damping, centrifugal, coriolis, and gravitational forces.

The next step is to linearize (81) which results in (66)

$$\delta\dot{\mathbf{x}} = \mathbf{M}^{-1}(\mathbf{x^d}) \cdot \mathbf{A}(\mathbf{x^d}) \cdot \delta\mathbf{x} + \mathbf{M}^{-1}(\mathbf{x^d}) \cdot \delta\mathbf{u}. \tag{82}$$

The inverse of $\mathbf{M}(\mathbf{x^d})$ is

$$\mathbf{M}^{-1}(\mathbf{x^d}) = \begin{pmatrix} 1 & 0 & 0 & 0 \\ 0 & m_{11}^{-1} & 0 & m_{12}^{-1} \\ 0 & 0 & 1 & 0 \\ 0 & m_{21}^{-1} & 0 & m_{22}^{-1} \end{pmatrix}, \tag{83}$$

where the elements m_{ij}^{-1} are functions of the desired state vector

$$\mathbf{x^d} = (x_1^d, x_2^d, x_3^d, x_4^d)^T = (q_1^d, \dot{q}_1^d, q_2^d, \dot{q}_2^d)^T.$$

The matrix $\mathbf{A}(\mathbf{x^d})$ of (82) is derived as

$$\mathbf{A}(\mathbf{x^d}) = \begin{pmatrix} 0 & 1 & 0 & 0 \\ A_{21} & A_{22} & A_{23} & A_{24} \\ 0 & 0 & 0 & 1 \\ A_{41} & A_{42} & A_{43} & A_{44} \end{pmatrix}_{\mathbf{x}=\mathbf{x^d}}, \tag{84}$$

where

$$\begin{aligned}
A_{21} &= -\frac{\partial n_1}{\partial x_1} = -[x_4^2 m_2 l_1 l_2 \cos(x_1 - x_3) - (m_1 + m_2) g l_1 \cos x_1], \\
A_{22} &= -\frac{\partial n_1}{\partial x_2} = -K_{x_1}, \\
A_{23} &= -\frac{\partial n_1}{\partial x_3} = -[-x_4^2 m_2 l_1 l_2 \cos(x_1 - x_3)], \\
A_{24} &= -\frac{\partial n_1}{\partial x_4} = -[2 x_4 m_2 l_1 l_2 \sin(x_1 - x_3)], \\
A_{41} &= -\frac{\partial n_2}{\partial x_1} = -[-x_2^2 m_2 l_1 l_2 \cos(x_1 - x_3)], \\
A_{42} &= -\frac{\partial n_2}{\partial x_2} = -[-2 x_2 m_2 l_1 l_2 \sin(x_1 - x_3)], \\
A_{43} &= -\frac{\partial n_2}{\partial x_3} = -[x_2^2 m_2 l_1 l_2 \cos(x_1 - x_3) - m_2 g l_2 \cos x_3], \\
A_{44} &= -\frac{\partial n_2}{\partial x_4} = -K_{x_3}.
\end{aligned} \tag{85}$$

Recall that the control law corresponding to (82) was

$$\delta \mathbf{u} = \mathbf{K}(\mathbf{x^d}) \cdot \delta \mathbf{x}, \tag{86}$$

(see (67)) where $\mathbf{K}(\mathbf{x^d})$ has to be designed such that

$$\tilde{\mathbf{A}} = \mathbf{M^{-1}}(\mathbf{x^d}) \cdot (\mathbf{A}(\mathbf{x^d}) + \mathbf{K}(\mathbf{x^d})),$$

has negative eigenvalues. Moreover, according to (73) one has to design **M** different matrices $\mathbf{K}(\mathbf{x^j})$. In order to do so we choose the eigenvalues for each center of a region to be the same

$$\mathbf{p} = (-100, -5, -250, -3)^T.$$

Furthermore, we partition $q_1, \dot{q}_1, q_2, \dot{q}_2$ as shown in table 5.2. The corresponding membership functions are shown in Fig. 2.

Table 1: Partitioning of the fuzzy state space

variable	linguistic term	centers of regions
q_1	LOW, MED, HIGH	(-1.047, 0, 1.047)
$\dot{q}_1$	N, P	(-1.57, 1.57)
q_2	LOW, MED, HIGH	(0.524, 1.57, 2.617)
$\dot{q}_2$	N, P	(-1.57, 1.57)

This leads to $3 \times 2 \times 3 \times 2 = 36$ different fuzzy regions with the same number of control matrices $\mathbf{K}(\mathbf{x^j})$. However, it must be emphasized that the number of rules is reduced to at least 9 different rules because of specific symmetries of matrix **A** regarding x_2 and x_4. Matrix $\mathbf{K}(\mathbf{x^j})$ has the following structure

$$\mathbf{K}(\mathbf{x^j}) = \begin{pmatrix} 0 & 0 & 0 & 0 \\ k_{21} & k_{22} & k_{23} & k_{24} \\ 0 & 0 & 0 & 0 \\ k_{41} & k_{42} & k_{43} & k_{44} \end{pmatrix}^j, \tag{87}$$

so that

$$\delta \mathbf{u}(\mathbf{x^j}) = \begin{pmatrix} 0 \\ k_{21} \cdot \delta q_1 + k_{22} \cdot \delta \dot{q}_1 + k_{23} \cdot \delta q_2 + k_{24} \cdot \delta \dot{q}_2 \\ 0 \\ k_{41} \cdot \delta q_1 + k_{42} \cdot \delta \dot{q}_1 + k_{43} \cdot \delta q_2 + k_{44} \cdot \delta \dot{q}_2 \end{pmatrix}^j. \tag{88}$$

The mechanical parameters of the robot are

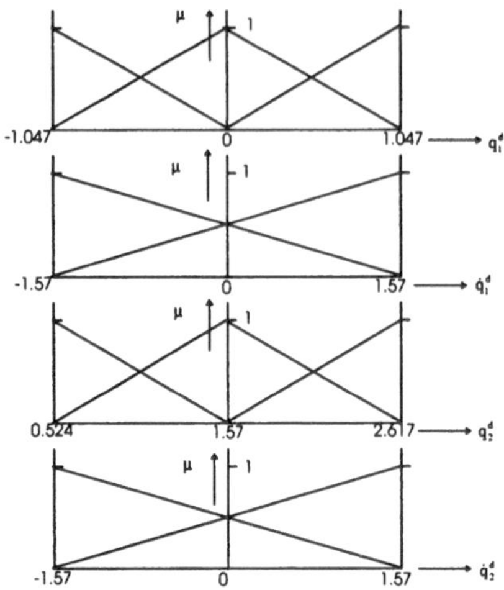

Figure 2: Membership functions for q_1, $\dot{q}_1$, q_2, and $\dot{q}_2$

$$m_1 = 1.5kg,$$

$$m_2 = 1kg,$$

$$l_1 = 0.2m,$$

$$l_2 = 0.2m,$$

$$K_{q1} = 10\frac{kgm^2}{s},$$

$$K_{q2} = 10\frac{kgm^2}{s}.$$

The pole placement routine leads to 36 control rules shown in Tables 2, 3, 4, and 5.

In the following we present some simulations showing the behavior of the system under control with different parameter fluctuations. The desired robot trajectory is

$$(q_1^d, \dot{q}_1^d, q_2^d, \dot{q}_2^d) = (0.5-0.5\cdot\sin 3t, -0.5\cdot 3\cdot\cos 3t, 1.57+0.5\cdot\sin 3t, 0.5\cdot 3\cdot\cos 3t).$$

Figure 3 shows the simulation result for perfect compensation at the operating points and a payload $m_2 = 1kg$. The resulting errors are in the range of $(-0.02, 0.02)$.

Figure 4 shows a simulation for partial compensation ($\hat{u}_i^d = 0.9 \cdot u_i^d$, $i = 1, 2$) and a payload $m_2 = 1$ kg.

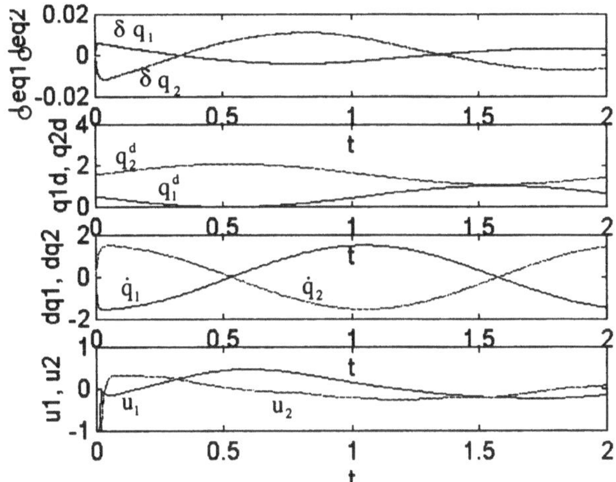

Figure 3: Simulation results for $\omega = 3\ \mathrm{s}^{-1}$, $m_2 = 1$ kg and perfect compensation ($\hat{u}_i^d = 1.0 \cdot u_i^d$, $i = 1, 2$)

Figure 5 refers to an example where u_i^d are over-compensated at the operating points ($\hat{u}_i^d = 1.1 \cdot u_i^d$, $i = 1, 2$). In this case the system becomes unstable.

When slowing down the frequency with which the robot links move from $\omega = 3\ \mathrm{s}^{-1}$ to $\omega = 3\ \mathrm{s}^{-1}$ we reach again a stable behavior. The new desired robot trajectory is then

$$(q_1^d, \dot{q}_1^d, q_2^d, \dot{q}_2^d) = (0.5 - 0.5 \cdot \sin t, -0.5 \cdot 1 \cdot \cos t, 1.57 + 0.5 \cdot \sin t, 0.5 \cdot 1 \cdot \cos t)$$

The corresponding simulation results are shown in Fig. 6

Figures 7 and 8 refer to examples with different payloads, but with perfect compensation. The errors δq_1 and δq_2 are in the range of $(-0.05, 0.05)$ for $m_2 = 2.5 kg$ and $(-0.01, 0.01)$ for $m_2 = 0.2$ kg.

For what concerns stability analysis, it should be pointed out that the desired trajectory and, with that, system matrix $\tilde{\mathbf{A}}$ is slowly time varying. The largest time constants of the robot links controlled are $\tau_{q_1} = 0.2$ s sinusoidal signal is $\tau_{signal} = 2\pi/\omega = 2.09$ s which is $6 - 10$ times longer than the time constants of the links. Therefore, one can assume that the desired exogeneous trajectory is slowly time varying which is the basic condition for using the frozen time stability analysis.

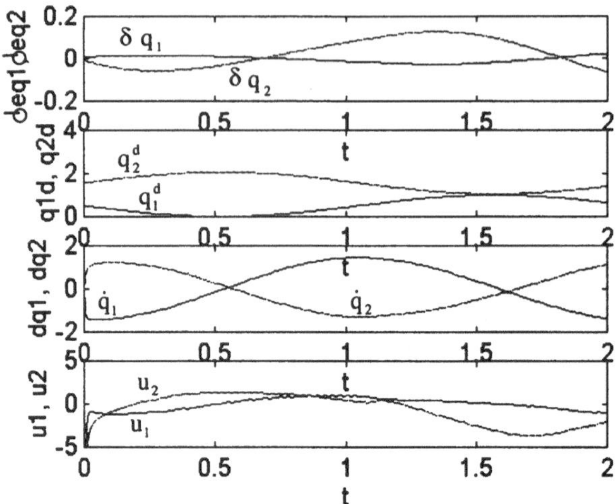

Figure 4: Simulation results for $\omega = 3\ \mathrm{s}^{-1}$, $m_2 = 1$ kg and partial compensation ($\hat{u}_i^d = 0.9 \cdot u_i^d$, $i = 1, 2$)

6 Conclusions

In fuzzy gain scheduling the space of exogenous parameters and/or the state space is divided in advance into overlapping fuzzy regions. Each fuzzy region is characterized by a different gain designed for the center of this fuzzy region so that the the original nonlinear system is stable at this center. The global control law for any intermediate point between the centers of the fuzzy regions is then obtained as a linear combination of the local control laws at these centers weighted by the degrees of satisfaction of the fuzzy regions to which the centers belong. This type of global control law is represented as a set of fuzzy rules and is called a fuzzy gain scheduled control law.

In this article we used some results about robust stability for linearly perturbed systems and the Lyapunov second method to determine the range of approximation errors within which the fuzzy gain scheduled closed loop system is stable and robust. However this method requires the perturbation matrices for all possible combinations of centers of the fuzzy regions. This number of combinations may be extremely large for MIMO systems. Furthermore, the method requires identical pole placement at the centers of all fuzzy regions due to which the performance of the different local control laws cannot be optimized. Last but not least, uncertainties in the local control laws and the gain scheduled control law are not taken into account.

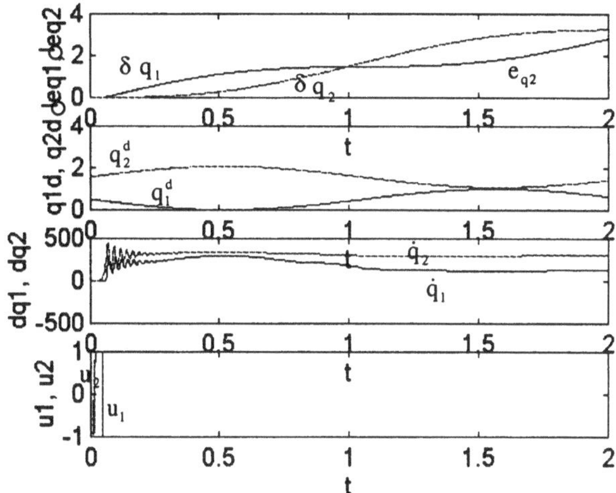

Figure 5: Simulation results for $\omega = 3\ \mathrm{s}^{-1}$, $m_2 = 1$ kg and partial compensation ($\hat{u}_i^d = 1.1 \cdot u_i^d$, $i = 1, 2$)

References

[Babushka 95] Babuska R., and Verbruggen H.B., "A new Identification Method for Linguistic Fuzzy Models", *Proceedings FUZZ-IEEE/IFES'95*, Yokohama, Japan, March 20–24, 1995, pp. 905-912.

[Babushka 94] Babushka R., and Verbruggen H.B. "Comparing Different Methods for Premise Identification in Sugeno-Takagi Models", *Proceedings EUFIT'94*, Aachen, Germany, September 20–23, 1994, pp. 1188–1192.

[Bastian 95] Bastian A."Towards a Fuzzy System Identification Theory", *Proceedings IFSA'95*, Sao Paolo, Brasil, July 21–28, 1995, pp. 69–72.

[Foellinger 90] Foellinger O."Regelungstechnik", Huethig Buch Verlag GMBH, Heildelberg, 1990.

[Herrera 95] Herrera F, Lozano M., and Verdegay J.L., "Design of Control Rules Base Based on Genetic Algorithms", *Proceedings IFSA'95*, Sao Paolo, Brasil, July 21–28, 1995, pp. 265–268.

[Hirota K 1993] Hirota K. "Industrial Application of Fuzzy Technology" Springer-Verlag, 1993

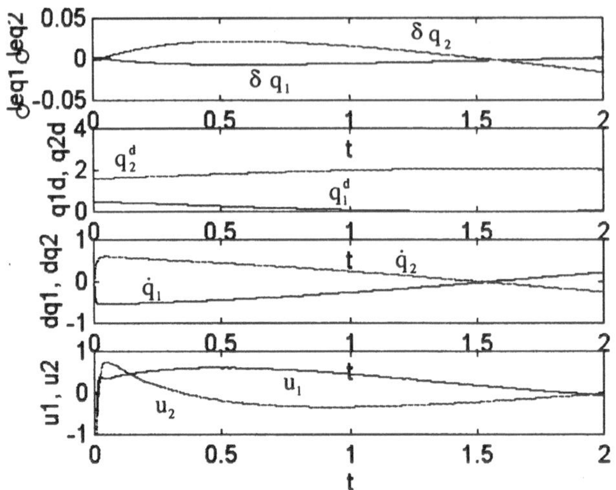

Figure 6: Simulation results for $\omega = 1\ \mathrm{s}^{-1}$, $m_2 = 1$ kg and partial compensation ($\hat{u}_i^d = 1.1 \cdot u_i^d$, $i = 1, 2$)

[Nichols 93] Nichols R.A., ReichertR.T. and Rugh W.J. "Gain Scheduling for H-Infinity Controllers: A Flight Control Example". *IEEE Trans. on Control Systems Technology* Vol. 1, No.2, June 1993 pp.69-79

[Palm 92] Palm R. "Sliding Mode Fuzzy Control", *IEEE International Conference on Fuzzy Systems 1992, Fuzz-IEEE'92 - Proceedings* San Diego March 8-12, pp.519-526

[Palm 95] Palm, R., Driankov, D, and Rehfuess, U., "Lyapunov Linearization Based Design of Takagi-Sugeno Controllers ", *Proceedings FUZZ-IEEE'95*, Sao Paolo, Brasil, July 21–28, 1995, pp. 513–516.

[Park 95] Park, M., Seunghwan, Ji, and Mignon, P., "A New Approach to the Identification of a Fuzzy Model", *Proceedings FUZZ-IEEE/IFES'95*, Yokohama, Japan, March 20–24, 1995, pp. 2159–2164.

[Press 90] Press, W.H., et al. " Numerical recipes in C", Cambridge University Press, Cambridge, 1990.

[Rehfuess 95] Rehfuess, U., and Palm, R. " Design of Takagi-Sugeno Controllers Based on Linear Quadratic Control", *Proceedings First International Symposium on Fuzzy Logic*, Zurich, Switzerland, May 26–27, 1995, pp. C10–C15.

[Rugh 91] Rugh, W.J. "Analytical Framework for Gain Scheduling", *IEEE Control Systems Magazine*, **11**(1)(1991)79–84.

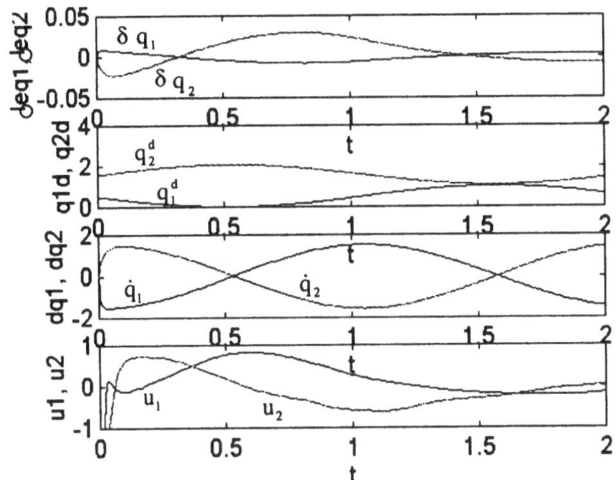

Figure 7: Simulation results for $\omega = 3\frac{1}{s}$, $m_2 = 2.5$ kg and perfect compensation ($\hat{u}_i^d = 1.0 \cdot u_i^d$, $i = 1, 2$)

[Shamma 88] Shamma J.S. "Analysis and Design of Gain Scheduled Control Systems" PhD Thesis No. LIDS-TH-1770 Lab. for Information and Decision Sciences MIT Cambridge MA 02139 1988.

[Slotine 91] Slotine J-J. E., W. Li. "Applied Nonlinear Control". *Prentice Hall, New Jersey* 1991

[Smith 90] Smith S.M. and Comer D.J. "Self-tuning of a Fuzzy Logic Controller Using a Cell State Space Algorithm". *Proceedings of the IEEE International Conference on Systems, Man and Cybernetics* 1990, pp. 445-450.

[Su 95] Su, Mu-Cun et al, "Rule Extraction Using a Novel Class of Fuzzy Degraded Hyperellipsoidal Composite Neural Networks", *Proceedings FUZZ-IEEE/IFES'95*, Yokohama, Japan, March 20–24, 1995, pp. 233–238.

[Sugeno 86] Sugeno M., and Kang G.T., "Fuzzy modelling and Control of Multilayer Incinerator", *Fuzzy sets and Systems*, **18**(1986),329–346.

[Sugeno 88] Sugeno M., and Kang G.T. "Structure Identification of a Fuzzy model", *Fuzzy Sets and Systems*,**28**(1988)15–33.

[Sugeno 91] Sugeno M., and Tanaka K. "Successive Identification of Fuzzy Model and its Application to Prediction of Complex System", *Fuzzy Sets and Systems*, **42**(1991)315–344.

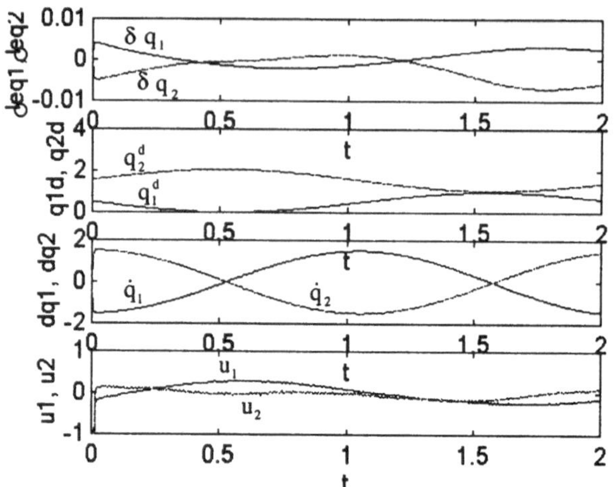

Figure 8: Simulation results for $\omega = 3\ \mathrm{s}^{-1}$, $m_2 = 0.2$ kg and perfect compensation ($\hat{u}_i^d = 1.0 \cdot u_i^d$, $i = 1, 2$)

[Sugeno M 1995] Sugeno M. "Complicated Systems Control by Language" (in Intelligent Information Processing by Fuzzy Thinking: Reports of LIFE),pp 465/483, Computer Age Co., 1995/4 (in Japanese)

[Takagi 85] Takagi T., and Sugeno M. " Fuzzy Identification of Systems and Its Applications to Modelling and Control". *IEEE Trans. on Syst., Man, and Cyb.* Vol. SMC-15. No.1 January/February 1985, pp.116-132

[Tanaka 93] Tanaka, K., and Sano, M., "Concept of Stability Margin for Fuzzy systems and Design of Robust Fuzzy controllers", *Proceedings FUZZ-IEEE'93*, San Francisco, CA, March 28–April 1, 1993, pp. 29–34.

[Tanaka 92] Tanaka K., and Sugeno M. "Stability Analysis and Design of Fuzzy Control Systems". *Fuzzy Sets and Systems* 45 (1992) North-Holland, pp. 135-156

[Vergara 95] Vergara V., and Moraga C., "Optimal Fuzzy Identification Models", *Proceedings IFSA '95*, Sao Paolo, Brasil, July 21–28, 1995, pp. 109–112.

[Vidyasagar 93] Vidyasagar M., *Nonlinear Systems Analysis*, Prentice Hall, Inc., Englewood Cliffs, New Jersey 07632, 1993.

[Yen 95] Yen J., and Gillespie W., "Integrating Global and Local Evaluations for Fuzzy Model Identification using Genetic Algorithms", *Proceedings IFSA '95*, Sao Paolo, Brasil, July 21–28, 1995, pp. 121–124.

[Zhou 87] Zhou K., and Khargonekar P., "Stability Robustness for Linear State-Space Models with Structured Uncertainty", *IEEE Transactions on Automatic Control*, **AC-32**(7)(1987)621–623.

Table 2: Rules 1-10 for the two link robot arm

Rule1
IF (q_1^d IS LOW) AND ($\dot{q}_1^d$ IS N) AND (q_2^d IS LOW) AND ($\dot{q}_2^d$ IS N) THEN
u_1 = -123.6033 * δq_1 -15.4631 * $\delta \dot{q}_1$ -15.4399 * δq_2 -0.8376 * $\delta \dot{q}_2$
u_2 = -3.2263 * δq_1 -0.0358 * $\delta \dot{q}_1$ -14.4916 * δq_2 + 5.8652 * $\delta \dot{q}_2$

Rule2
IF (q_1^d IS LOW) AND ($\dot{q}_1^d$ IS N) AND (q_2^d IS LOW) AND ($\dot{q}_2^d$ IS P) THEN
u_1 = -123.6033 * δq_1 -15.4631 * $\delta \dot{q}_1$ -15.4399 * δq_2 -0.8376 * $\delta \dot{q}_2$
u_2 = -3.2263 * δq_1 -0.0358 * $\delta \dot{q}_1$ -14.4916 * δq_2 + 5.8652 * $\delta \dot{q}_2$

Rule3
IF (q_1^d IS LOW) AND ($\dot{q}_1^d$ IS N) AND (q_2^d IS MED) AND ($\dot{q}_2^d$ IS N) THEN
u_1 = -120.8947 * δq_1 -15.2611 * $\delta \dot{q}_1$ -4.2759 * δq_2 + 2.8419 * $\delta \dot{q}_2$
u_2 = 38.8270 * δq_1 + 8.6862 * $\delta \dot{q}_1$ -7.5294 * δq_2 + 6.0870 * $\delta \dot{q}_2$

Rule4
IF (q_1^d IS LOW) AND ($\dot{q}_1^d$ IS N) AND (q_2^d IS MED) AND ($\dot{q}_2^d$ IS P) THEN
u_1 = -120.8947 * δq_1 -15.2611 * $\delta \dot{q}_1$ -4.2759 * δq_2 + 2.8419 * $\delta \dot{q}_2$
u_2 = 38.8270 * δq_1 + 8.6862 * $\delta \dot{q}_1$ -7.5294 * δq_2 + 6.0870 * $\delta \dot{q}_2$

Rule5
IF (q_1^d IS LOW) AND ($\dot{q}_1^d$ IS N) AND (q_2^d IS HIGH) AND ($\dot{q}_2^d$ IS N) THEN
u_1 = -120.8947 * δq_1 -15.2611 * $\delta \dot{q}_1$ -4.2759 * δq_2 + 3.0393 * $\delta \dot{q}_2$
u_2 = 38.8270 * δq_1 + 8.4888 * $\delta \dot{q}_1$ -5.8302 * δq_2 + 6.0870 * $\delta \dot{q}_2$

Rule6
IF (q_1^d IS LOW) AND ($\dot{q}_1^d$ IS N) AND (q_2^d IS HIGH) AND ($\dot{q}_2^d$ IS P) THEN
u_1 = -120.8947 * δq_1 -15.2611 * $\delta \dot{q}_1$ -4.2759 * δq_2 + 3.0393 * $\delta \dot{q}_2$
u_2 = 38.8270 * δq_1 + 8.4888 * $\delta \dot{q}_1$ -5.8302 * δq_2 + 6.0870 * $\delta \dot{q}_2$

Rule7
IF (q_1^d IS LOW) AND ($\dot{q}_1^d$ IS P) AND (q_2^d IS LOW) AND ($\dot{q}_2^d$ IS N) THEN
u_1 = -123.6033 * δq_1 -15.4631 * $\delta \dot{q}_1$ -15.4399 * δq_2 -0.8376 * $\delta \dot{q}_2$
u_2 = -3.2263 * δq_1 -0.0358 * $\delta \dot{q}_1$ -14.4916 * δq_2 + 5.8652 * $\delta \dot{q}_2$

Rule8
IF (q_1^d IS LOW) AND ($\dot{q}_1^d$ IS P) AND (q_2^d IS LOW) AND ($\dot{q}_2^d$ IS P) THEN
u_1 = -123.6033 * δq_1 -15.4631 * $\delta \dot{q}_1$ -15.4399 * δq_2 -0.8376 * $\delta \dot{q}_2$
u_2 = -3.2263 * δq_1 -0.0358 * $\delta \dot{q}_1$ -14.4916 * δq_2 + 5.8652 * $\delta \dot{q}_2$

Rule9
IF (q_1^d IS LOW) AND ($\dot{q}_1^d$ IS P) AND (q_2^d IS MED) AND ($\dot{q}_2^d$ IS N) THEN
u_1 = -120.8947 * δq_1 -15.2611 * $\delta \dot{q}_1$ -4.2759 * δq_2 + 2.8419 * $\delta \dot{q}_2$
u_2 = 38.8270 * δq_1 + 8.6862 * $\delta \dot{q}_1$ -7.5294 * δq_2 + 6.0870 * $\delta \dot{q}_2$

Rule10
IF (q_1^d IS LOW) AND ($\dot{q}_1^d$ IS P) AND (q_2^d IS MED) AND ($\dot{q}_2^d$ IS P) THEN
u_1 = -120.8947 * δq_1 -15.2611 * $\delta \dot{q}_1$ -4.2759 * δq_2 + 2.8419 * $\delta \dot{q}_2$
u_2 = 38.8270 * δq_1 + 8.6862 * $\delta \dot{q}_1$ -7.5294 * δq_2 + 6.0870 * $\delta \dot{q}_2$

Table 3: Rules 11-20 for the two link robot arm

Rule11
IF (q_1^d IS LOW) AND ($\dot{q}_1^d$ IS P) AND (q_2^d IS HIGH) AND ($\dot{q}_2^d$ IS N) THEN
$u_1 = -120.8947 * \delta q_1 -15.2611 * \delta\dot{q}_1 -4.2759 * \delta q_2 + 3.0393 * \delta\dot{q}_2$
$u_2 = 38.8270 * \delta q_1 + 8.4888 * \delta\dot{q}_1 -5.8302 * \delta q_2 + 6.0870 * \delta\dot{q}_2$

Rule12
IF (q_1^d IS LOW) AND ($\dot{q}_1^d$ IS P) AND (q_2^d IS HIGH) AND ($\dot{q}_2^d$ IS P) THEN
$u_1 = -120.8947 * \delta q_1 -15.2611 * \delta\dot{q}_1 -4.2759 * \delta q_2 + 3.0393 * \delta\dot{q}_2$
$u_2 = 38.8270 * \delta q_1 + 8.4888 * \delta\dot{q}_1 -5.8302 * \delta q_2 + 6.0870 * \delta\dot{q}_2$

Rule13
IF (q_1^d IS MED) AND ($\dot{q}_1^d$ IS N) AND (q_2^d IS LOW) AND ($\dot{q}_2^d$ IS N) THEN
$u_1 = -128.7644 * \delta q_1 -15.6650 * \delta\dot{q}_1 -26.6040 * \delta q_2 -4.3197 * \delta\dot{q}_2$
$u_2 = -45.2796 * \delta q_1 -8.9551 * \delta\dot{q}_1 -19.7547 * \delta q_2 + 5.6435 * \delta\dot{q}_2$

Rule14
IF (q_1^d IS MED) AND ($\dot{q}_1^d$ IS N) AND (q_2^d IS LOW) AND ($\dot{q}_2^d$ IS P) THEN
$u_1 = -128.7644 * \delta q_1 -15.6650 * \delta\dot{q}_1 -26.6040 * \delta q_2 -4.3197 * \delta\dot{q}_2$
$u_2 = -45.2796 * \delta q_1 -8.9551 * \delta\dot{q}_1 -19.7547 * \delta q_2 + 5.6435 * \delta\dot{q}_2$

Rule15
IF (q_1^d IS MED) AND ($\dot{q}_1^d$ IS N) AND (q_2^d IS MED) AND ($\dot{q}_2^d$ IS N) THEN
$u_1 = -126.0558 * \delta q_1 -15.4631 * \delta\dot{q}_1 -15.4399 * \delta q_2 -0.8376 * \delta\dot{q}_2$
$u_2 = -3.2263 * \delta q_1 -0.0358 * \delta\dot{q}_1 -12.7924 * \delta q_2 + 5.8652 * \delta\dot{q}_2$

Rule16
IF (q_1^d IS MED) AND ($\dot{q}_1^d$ IS N) AND (q_2^d IS MED) AND ($\dot{q}_2^d$ IS P) THEN
$u_1 = -126.0558 * \delta q_1 -15.4631 * \delta\dot{q}_1 -15.4399 * \delta q_2 -0.8376 * \delta\dot{q}_2$
$u_2 = -3.2263 * \delta q_1 -0.0358 * \delta\dot{q}_1 -12.7924 * \delta q_2 + 5.8652 * \delta\dot{q}_2$

Rule17
IF (q_1^d IS MED) AND ($\dot{q}_1^d$ IS N) AND (q_2^d IS HIGH) AND ($\dot{q}_2^d$ IS N) THEN
$u_1 = -123.3472 * \delta q_1 -15.2611 * \delta\dot{q}_1 -4.2759 * \delta q_2 + 2.8419 * \delta\dot{q}_2$
$u_2 = 38.8270 * \delta q_1 + 8.6862 * \delta\dot{q}_1 -5.8302 * \delta q_2 + 6.0870 * \delta\dot{q}_2$

Rule18
IF (q_1^d IS MED) AND ($\dot{q}_1^d$ IS N) AND (q_2^d IS HIGH) AND ($\dot{q}_2^d$ IS P) THEN
$u_1 = -123.3472 * \delta q_1 -15.2611 * \delta\dot{q}_1 -4.2759 * \delta q_2 + 2.8419 * \delta\dot{q}_2$
$u_2 = 38.8270 * \delta q_1 + 8.6862 * \delta\dot{q}_1 -5.8302 * \delta q_2 + 6.0870 * \delta\dot{q}_2$

Rule19
IF (q_1^d IS MED) AND ($\dot{q}_1^d$ IS P) AND (q_2^d IS LOW) AND ($\dot{q}_2^d$ IS N) THEN
$u_1 = -128.7644 * \delta q_1 -15.6650 * \delta\dot{q}_1 -26.6040 * \delta q_2 -4.3197 * \delta\dot{q}_2$
$u_2 = -45.2796 * \delta q_1 -8.9551 * \delta\dot{q}_1 -19.7547 * \delta q_2 + 5.6435 * \delta\dot{q}_2$

Rule20
IF (q_1^d IS MED) AND ($\dot{q}_1^d$ IS P) AND (q_2^d IS LOW) AND ($\dot{q}_2^d$ IS P) THEN
$u_1 = -128.7644 * \delta q_1 -15.6650 * \delta\dot{q}_1 -26.6040 * \delta q_2 -4.3197 * \delta\dot{q}_2$
$u_2 = -45.2796 * \delta q_1 -8.9551 * \delta\dot{q}_1 -19.7547 * \delta q_2 + 5.6435 * \delta\dot{q}_2$

Table 4: Rules 21-30 for the two link robot arm

Rule21
IF (q_1^d IS MED) AND ($\dot{q}_1^d$ IS P) AND (q_2^d IS MED) AND ($\dot{q}_2^d$ IS N) THEN
u_1 = -126.0558 * δq_1 -15.4631 * $\delta \dot{q}_1$ -15.4399 * δq_2 -0.8376 * $\delta \dot{q}_2$
u_2 = -3.2263 * δq_1 -0.0358 * $\delta \dot{q}_1$ -12.7924 * δq_2 + 5.8652 * $\delta \dot{q}_2$

Rule22
IF (q_1^d IS MED) AND ($\dot{q}_1^d$ IS P) AND (q_2^d IS MED) AND ($\dot{q}_2^d$ IS P) THEN
u_1 = -126.0558 * δq_1 -15.4631 * $\delta \dot{q}_1$ -15.4399 * δq_2 -0.8376 * $\delta \dot{q}_2$
u_2 = -3.2263 * δq_1 -0.0358 * $\delta \dot{q}_1$ -12.7924 * δq_2 + 5.8652 * $\delta \dot{q}_2$

Rule23
IF (q_1^d IS MED) AND ($\dot{q}_1^d$ IS P) AND (q_2^d IS HIGH) AND ($\dot{q}_2^d$ IS N) THEN
u_1 = -123.3472 * δq_1 -15.2611 * $\delta \dot{q}_1$ -4.2759 * δq_2 + 2.8419 * $\delta \dot{q}_2$
u_2 = 38.8270 * δq_1 + 8.6862 * $\delta \dot{q}_1$ -5.8302 * δq_2 + 6.0870 * $\delta \dot{q}_2$

Rule24
IF (q_1 IS MED) AND ($\dot{q}_1^d$ IS P) AND (q_2^d IS HIGH) AND ($\dot{q}_2^d$ IS P) THEN
u_1 = -123.3472 * δq_1 -15.2611 * $\delta \dot{q}_1$ -4.2759 * δq_2 + 2.8419 * $\delta \dot{q}_2$
u_2 = 38.8270 * δq_1 + 8.6862 * $\delta \dot{q}_1$ -5.8302 * δq_2 + 6.0870 * $\delta \dot{q}_2$

Rule25
IF (q_1^d IS HIGH) AND ($\dot{q}_1^d$ IS N) AND (q_2^d IS LOW) AND ($\dot{q}_2^d$ IS N) THEN
u_1 = -126.3119 * δq_1 -15.6650 * $\delta \dot{q}_1$ -26.6040 * δq_2 -4.1223 * $\delta \dot{q}_2$
u_2 = -45.2796 * δq_1 -9.1525 * $\delta \dot{q}_1$ -19.7547 * δq_2 + 5.6435 * $\delta \dot{q}_2$

Rule26
IF (q_1^d IS HIGH) AND ($\dot{q}_1^d$ IS N) AND (q_2^d IS LOW) AND ($\dot{q}_2^d$ IS P) THEN
u_1 = -126.3119 * δq_1 -15.6650 * $\delta \dot{q}_1$ -26.6040 * δq_2 -4.1223 * $\delta \dot{q}_2$
u_2 = -45.2796 * δq_1 -9.1525 * $\delta \dot{q}_1$ -19.7547 * δq_2 + 5.6435 * $\delta \dot{q}_2$

Rule27
IF (q_1^d IS HIGH) AND ($\dot{q}_1^d$ IS N) AND (q_2^d IS MED) AND ($\dot{q}_2^d$ IS N) THEN
u_1 = -126.3119 * δq_1 -15.6650 * $\delta \dot{q}_1$ -26.6040 * δq_2 -4.3197 * $\delta \dot{q}_2$
u_2 = -45.2796 * δq_1 -8.9551 * $\delta \dot{q}_1$ -18.0555 * δq_2 + 5.6435 * $\delta \dot{q}_2$

Rule28
IF (q_1^d IS HIGH) AND ($\dot{q}_1^d$ IS N) AND (q_2^d IS MED) AND ($\dot{q}_2^d$ IS P) THEN
u_1 = -126.3119 * δq_1 -15.6650 * $\delta \dot{q}_1$ -26.6040 * δq_2 -4.3197 * $\delta \dot{q}_2$
u_2 = -45.2796 * δq_1 -8.9551 * $\delta \dot{q}_1$ -18.0555 * δq_2 + 5.6435 * $\delta \dot{q}_2$

Rule29
IF (q_1^d IS HIGH) AND ($\dot{q}_1^d$ IS N) AND (q_2^d IS HIGH) AND ($\dot{q}_2^d$ IS N) THEN
u_1 = -123.6033 * δq_1 -15.4631 * $\delta \dot{q}_1$ -15.4399 * δq_2 -0.8376 * $\delta \dot{q}_2$
u_2 = -3.2263 * δq_1 -0.0358 * $\delta \dot{q}_1$ -11.0933 * δq_2 + 5.8652 * $\delta \dot{q}_2$

Rule30
IF (q_1^d IS HIGH) AND ($\dot{q}_1^d$ IS N) AND (q_2^d IS HIGH) AND ($\dot{q}_2^d$ IS P) THEN
u_1 = -123.6033 * δq_1 -15.4631 * $\delta \dot{q}_1$ -15.4399 * δq_2 -0.8376 * $\delta \dot{q}_2$
u_2 = -3.2263 * δq_1 -0.0358 * $\delta \dot{q}_1$ -11.0933 * δq_2 + 5.8652 * $\delta \dot{q}_2$

Table 5: Rules 31-36 for the two link robot arm

Rule31

IF (q_1^d IS HIGH) AND ($\dot{q}_1^d$ IS P) AND (q_2^d IS LOW) AND ($\dot{q}_2^d$ IS N) THEN

u_1 = -126.3119 * δq_1 -15.6650 * $\delta \dot{q}_1$ -26.6040 * δq_2 -4.1223 * $\delta \dot{q}_2$

u_2 = -45.2796 * δq_1 -9.1525 * $\delta \dot{q}_1$ -19.7547 * δq_2 + 5.6435 * $\delta \dot{q}_2$

Rule32

IF (q_1^d IS HIGH) AND ($\dot{q}_1^d$ IS P) AND (q_2^d IS LOW) AND ($\dot{q}_2^d$ IS P) THEN

u_1 = -126.3119 * δq_1 -15.6650 * $\delta \dot{q}_1$ -26.6040 * δq_2 -4.1223 * $\delta \dot{q}_2$

u_2 = -45.2796 * δq_1 -9.1525 * $\delta \dot{q}_1$ -19.7547 * δq_2 + 5.6435 * $\delta \dot{q}_2$

Rule33

IF (q_1^d IS HIGH) AND ($\dot{q}_1^d$ IS P) AND (q_2^d IS MED) AND ($\dot{q}_2^d$ IS N) THEN

u_1 = -126.3119 * δq_1 -15.6650 * $\delta \dot{q}_1$ -26.6040 * δq_2 -4.3197 * $\delta \dot{q}_2$

u_2 = -45.2796 * δq_1 -8.9551 * $\delta \dot{q}_1$ -18.0555 * δq_2 + 5.6435 * $\delta \dot{q}_2$

Rule34

IF (q_1^d IS HIGH) AND ($\dot{q}_1^d$ IS P) AND (q_2^d IS MED) AND ($\dot{q}_2^d$ IS P) THEN

u_1 = -126.3119 * δq_1 -15.6650 * $\delta \dot{q}_1$ -26.6040 * δq_2 -4.3197 * $\delta \dot{q}_2$

u_2 = -45.2796 * δq_1 -8.9551 * $\delta \dot{q}_1$ -18.0555 * δq_2 + 5.6435 * $\delta \dot{q}_2$

Rule35

IF (q_1^d IS HIGH) AND ($\dot{q}_1^d$ IS P) AND (q_2^d IS HIGH) AND ($\dot{q}_2^d$ IS N) THEN

u_1 = -123.6033 * δq_1 -15.4631 * $\delta \dot{q}_1$ -15.4399 * δq_2 -0.8376 * $\delta \dot{q}_2$

u_2 = -3.2263 * δq_1 -0.0358 * $\delta \dot{q}_1$ -11.0933 * δq_2 + 5.8652 * $\delta \dot{q}_2$

Rule36

IF (q_1^d IS HIGH) AND ($\dot{q}_1^d$ IS P) AND (q_2^d IS HIGH) AND ($\dot{q}_2^d$ IS P) THEN

u_1 = -123.6033 * δq_1 -15.4631 * $\delta \dot{q}_1$ -15.4399 * δq_2 -0.8376 * $\delta \dot{q}_2$

u_2 = -3.2263 * δq_1 -0.0358 * $\delta \dot{q}_1$ -11.0933 * δq_2 + 5.8652 * $\delta \dot{q}_2$